# Thin Films and Nanostructures

*Electronic Excitations in Organic Based Nanostructures*

V.M. Agranovich
and
G.F. Bassani

VOLUME 31

2003

ELSEVIER
ACADEMIC
PRESS

Amsterdam   Boston   Heidelberg   London   New York   Oxford   Paris
San Diego   San Francisco   Singapore   Sydney   Tokyo

| ELSEVIER B.V. | ELSEVIER Inc. | ELSEVIER Ltd | ELSEVIER Ltd |
|---|---|---|---|
| Sara Burgerhartstraat 25 | 525 B Street, Suite 1900 | The Boulevard, Langford Lane | 84 Theobalds Road |
| P.O. Box 211, 1000 AE | San Diego, CA 92101-4495 | Kidlington, Oxford OX5 1GB | London WC1X 8RR |
| Amsterdam, The Netherlands | USA | UK | UK |

© 2003 Elsevier Inc. All rights reserved

This work is protected under copyright by Elsevier, and the following terms and conditions apply to its use:

Photocopying
Single photocopies of single chapters may be made for personal use as allowed by national copyright laws. Permission of the Publisher and payment of a fee is required for all other photocopying, including multiple or systematic copying, copying for advertising or promotional purposes, resale, and all forms of document delivery. Special rates are available for educational institutions that wish to make photocopies for non-profit educational classroom use.

Permissions may be sought directly from Elsevier's Science & Technology Rights Department in Oxford, UK: phone: (+44) 1865 843830, fax: (+44) 1865 853333, e-mail: permissions@elsevier.com. You may also complete your request on-line via the Elsevier homepage (http://www.elsevier.com), by selecting 'Customer Support' and then 'Obtaining Permissions'.

In the USA, users may clear permissions and make payments through the Copyright Clearance Center, Inc., 222 Rosewood Drive, Danvers, MA 01923, USA; phone: (+1) (978) 7508400, fax: (+1) (978) 7504744, and in the UK through the Copyright Licensing Agency Rapid Clearance Service (CLARCS), 90 Tottenham Court Road, London W1P 0LP, UK; phone: (+44) 207 631 5555; fax: (+44) 207 631 5500. Other countries may have a local reprographic rights agency for payments.

Derivative Works
Tables of contents may be reproduced for internal circulation, but permission of Elsevier is required for external resale or distribution of such material.
Permission of the Publisher is required for all other derivative works, including compilations and translations.

Electronic Storage or Usage
Permission of the Publisher is required to store or use electronically any material contained in this work, including any chapter or part of a chapter.

Except as outlined above, no part of this work may be reproduced, stored in a retrieval system or transmitted in any form or by any means, electronic, mechanical, photocopying, recording or otherwise, without prior written permission of the Publisher.
Address permissions requests to: Elsevier's Science & Technology Rights Department, at the phone, fax and e-mail addresses noted above.

Notice
No responsibility is assumed by the Publisher for any injury and/or damage to persons or property as a matter of products liability, negligence or otherwise, or from any use or operation of any methods, products, instructions or ideas contained in the material herein. Because of rapid advances in the medical sciences, in particular, independent verification of diagnoses and drug dosages should be made.

First edition 2003

ISBN: 0-12-533031-6
ISSN: 1079-4050 (Series)

∞ The paper used in this publication meets the requirements of ANSI/NISO Z39.48-1992 (Permanence of Paper).
Printed in Great Britain.

**Serial Editors**

VLADIMIR AGRANOVICH
*Institute of Spectroscopy*
*Russian Academy of Sciences*
*Moscow, Russia*

DEBORAH TAYLOR
*Motorola*
*Austin, Texas*

**Honorary Editors**

MAURICE H. FRANCOMBE
*Department of Physics*
*and Astronomy*
*Georgia State University*
*Atlanta, Georgia*

STEPHEN M. ROSSNAGEL
*IBM Corporation,*
*T. J. Watson Research Center*
*Yorktown Heights, New York*

ABRAHAM ULMAN
*Alstadt-Lord-Mark Professor*
*Department of Chemistry*
*Polymer Research Institute*
*Polytechnic University*
*Brooklyn, New York*

**Editorial Board**

DAVID L. ALLARA
Pennsylvania State University

ALLEN J. BARD
University of Texas, Austin

FRANCO BASSANI
Scuola Normale Superiore, Pisa

MASAMICHI FUJIHIRA
Tokyo Institute of Technology

GEORGE GAINS
Rensselaer Polytechnic Institute

PHILLIP HODGE
University of Manchester

JACOB N. ISRAELACHIVILI
University of California
Santa Barbara

MICHAEL L. KLEIN
University of Pennsylvania

HANS KUHN
MPI Gottingen

JEROME B. LANDO
Case Western Reserve University

HELMUT MOHWALD
University of Mainz

NICOLAI PLATE
Russian Academy of Sciences

HELMUT RINGSDORF
University of Mainz

GIACINTO SCOLES
Princeton University

JEROME D. SWALEN
International Business
Machines Corporation

MATTHEW V. TIRRELL
University of Minnesota,
Minneapolis

CLAUDE WEISBUCH
Ecole Politechnique, Paris

GEORGE M. WHITESIDES
Harvard University

ANVAR ZAKHIDOV
University of Texas at Dallas

*Recent volumes in this serial appear at the end of this volume*

# Thin Films and Nanostructures

*Electronic Excitations in Organic Based Nanostructures*

*Volume 31*

# Contents

Contributors . . . . . . . . . . . . . . . . . . . . . . . . . . . . . . . . . . ix
Preface . . . . . . . . . . . . . . . . . . . . . . . . . . . . . . . . . . . . . . xi

## Chapter 1. Frenkel and Charge-Transfer Excitons in Organic Solids
*Jasper Knoester and Vladimir M. Agranovich*

1. Introduction . . . . . . . . . . . . . . . . . . . . . . . . . . . . . . . . . 1
2. Microscopic Theory of Frenkel Excitons . . . . . . . . . . . . . . . . . . 4
3. Dielectric Theory of Frenkel Excitons . . . . . . . . . . . . . . . . . . . 34
4. Diffusion of Frenkel Excitons . . . . . . . . . . . . . . . . . . . . . . . . 45
5. Self-Trapping of Excitons: Spectra and Transport . . . . . . . . . . . . . 58
6. Charge-Transfer Excitons in Organic Solids . . . . . . . . . . . . . . . . 63
7. Molecular Aggregates: Low-Dimensional Exciton Systems . . . . . . . . 67
8. Excitons in Biological Systems . . . . . . . . . . . . . . . . . . . . . . . 87
9. Concluding Remarks . . . . . . . . . . . . . . . . . . . . . . . . . . . . . 90
   Acknowledgement . . . . . . . . . . . . . . . . . . . . . . . . . . . . . . 90
   References . . . . . . . . . . . . . . . . . . . . . . . . . . . . . . . . . . 90

## Chapter 2. Wannier–Mott Excitons in Semiconductors
*G.C. La Rocca*

1. Introduction . . . . . . . . . . . . . . . . . . . . . . . . . . . . . . . . . 97
2. Wannier–Mott Excitons in Bulk Semiconductors . . . . . . . . . . . . . 98
3. Quantum Confined Wannier–Mott Excitons . . . . . . . . . . . . . . . . 106
4. Quantum Well Exciton Optical Nonlinearities . . . . . . . . . . . . . . . 115
   References . . . . . . . . . . . . . . . . . . . . . . . . . . . . . . . . . . 126

## Chapter 3. Polaritons
*Franco Bassani*

1. Introduction and General Concepts . . . . . . . . . . . . . . . . . . . . 129
2. Classical Theory of Polaritons . . . . . . . . . . . . . . . . . . . . . . . 132
3. Quantum Theory of Polaritons . . . . . . . . . . . . . . . . . . . . . . . 135
4. Real Space Density Matrix Approach . . . . . . . . . . . . . . . . . . . 137
5. Experiments on Polaritons . . . . . . . . . . . . . . . . . . . . . . . . . 140
6. Surface Polaritons . . . . . . . . . . . . . . . . . . . . . . . . . . . . . . 141

| 7. Quantum Well Polaritons | 147 |
|---|---|
| 8. Quantum Wire Polaritons | 161 |
| 9. Exciton-Polaritons in Microcavities | 166 |
| Acknowledgements | 180 |
| References | 180 |

## Chapter 4. Optics and Nonlinearities of Excitons in Organic Multilayered Nanostructures and Superlattices

*V.M. Agranovich and A.M. Kamchatnov*

| 1. Introduction | 185 |
|---|---|
| 2. Dielectric Constant Tensor of Long Period Organic Superlattices with Isotropic Layers | 186 |
| 3. Dielectric Constant Tensor of Long Period Organic Superlattices with Anisotropic Layers | 190 |
| 4. Optical Nonlinearities in Organic Multilayers | 194 |
| 5. Dielectric Tensor for Short Period Organic Superlattices | 197 |
| 6. Gas-Condensed Matter Shift and Possibility to Govern Spectra of Frenkel Excitons | 201 |
| 7. Fermi Resonance Interface Modes in Organic Superlattices | 207 |
| References | 219 |

## Chapter 5. Mixing of Frenkel and Charge-Transfer Excitons and Their Quantum Confinement in Thin Films

*Michael Hoffmann*

| 1. Introduction | 221 |
|---|---|
| 2. Electronic Frenkel and Charge-Transfer Excitons in Rigid One-Dimensional Crystals | 226 |
| 3. Strong Coupling of the Electronic Excitations with Internal Phonon Modes | 245 |
| 4. Applications and Consequences for Quantum Confinement | 276 |
| 5. Conclusion | 288 |
| Acknowledgements | 289 |
| References | 290 |

## Chapter 6. Two-Dimensional Charge-Transfer Excitons at a Donor–Acceptor Interface

*V.M. Agranovich and G.C. La Rocca*

| 1. Phase Transition from Dielectric to Conducting State (Cold Photoconductivity) | 293 |
|---|---|
| 2. Cumulative Photovoltage in Asymmetrical Donor–Acceptor Organic Superlattices | 304 |
| 3. Nonlinear Optical Response of Charge-Transfer Excitons at Donor–Acceptor Interface | 311 |
| References | 315 |

## Chapter 7. Hybridization of Frenkel and Wannier–Mott Excitons in Organic-Inorganic Heterostructures. Strong Coupling Regime

*V.M. Agranovich, V.I. Yudson and P. Reineker*

| | |
|---|---|
| 1. Introduction | 317 |
| 2. Hybrid 2D Frenkel Wannier–Mott Excitons at the Interface of Organic and Inorganic Quantum Wells | 321 |
| 3. Nonlinear Optics of 2D Hybrid Frenkel–Wannier–Mott Excitons | 331 |
| 4. Hybrid Excitons in Parallel Organic and Inorganic Semiconductor Quantum Wires | 339 |
| 5. On the Hybridization of "Zero-Dimensional" Frenkel and Wannier–Mott Excitons | 344 |
| 6. Hybridization of Excitons in Microcavity Configurations | 345 |
| References | 352 |

## Chapter 8. Strong Optical Coupling in Organic Semiconductor Microcavities

*David G. Lidzey*

| | |
|---|---|
| 1. Introduction | 355 |
| 2. Organic Semiconductor Microcavities | 356 |
| 3. Organic Semiconductors for Strong Optical Coupling | 367 |
| 4. Optical Measurement Techniques | 373 |
| 5. Dispersion of Cavity Polaritons | 374 |
| 6. Cavity Emission Following Non-Resonant Laser Excitation | 381 |
| 7. Photon Emission Following Resonant Excitation | 389 |
| 8. Photon-Mediated Hybridisation between Frenkel Excitons | 393 |
| 9. Future Prospects | 397 |
| References | 400 |

## Chapter 9. Electronic Energy Transfer in a Planar Microcavity

*D.M. Basko*

| | |
|---|---|
| 1. Energy Transfer: an Introductory Discussion | 404 |
| 2. Semiclassical Description of the Transfer | 406 |
| 3. Modeling a Specific Structure | 413 |
| 4. Weak Absorption Regime | 422 |
| 5. Numerical Results | 426 |
| 6. Discussion | 439 |
| 7. Concluding Remarks | 444 |
| Acknowledgements | 445 |
| References | 445 |

## Chapter 10. Energy Transfer from a Semiconductor Nanostructure to an Organic Material and a New Concept for Light-Emitting Devices

*D.M. Basko*

| | | |
|---|---|---|
| 1. | Introduction | 447 |
| 2. | The General Calculation Scheme for the Energy Transfer | 449 |
| 3. | Transfer in the Planar Geometry: Quantum Wells | 456 |
| 4. | Transfer in the Spherical Geometry: Quantum Dots | 469 |
| 5. | Application to Light-Emitting Devices | 480 |
| 6. | Summary | 483 |
| | References | 485 |

Index . . . . . . . . . . . . . . . . . . . . . . . . . . . . . . . . 487

## Contributors

Chapter 1. *Frenkel and Charge-Transfer Excitons in Organic Solids:* Jasper Knoester, Institute for Theoretical Physics and Materials Science Center, University of Groningen, Nijenborgh 4, 9747 AG Groningen, The Netherlands (e-mail: knoester@phys.rug.nl), and Vladimir M. Agranovich, Institute of Spectroscopy, Russian Academy of Sciences, Troitsk, Moscow Obl., 142100, Russia (e-mail: agran@isan.troitsk.ru)

Chapter 2. *Wannier–Mott Excitons in Semiconductors:* G.C. La Rocca, Scuola Normale Superiore and I.N.F.M., Piazza dei Cavalieri 7, 56126 Pisa, Italy (e-mail: larocca@sns.it)

Chapter 3. *Polaritons:* Franco Bassani, Scuola Normale Superiore and I.N.F.M., Piazza dei Cavalieri 7, 56126 Pisa, Italy (e-mail: bassani@sns.it)

Chapter 4. *Optics and Nonlinearities of Excitons in Organic Multilayered Nanostructures and Superlattices:* V.M. Agranovich and A.M. Kamchatnov, Institute of Spectroscopy, Russian Academy of Sciences, Troitsk, Moscow Reg., 142190, Russia (e-mails: agran@isan.troitsk.ru, kamch@isan.troitsk.ru)

Chapter 5. *Mixing of Frenkel and Charge-Transfer Excitons and Their Quantum Confinement in Thin Films:* Michael Hoffmann, Institut für Angewandte Photophysik, Technische Universität Dresden, 01062 Dresden, Germany (e-mail: mi-hoff@iapp.de)

Chapter 6. *Two-Dimensional Charge-Transfer Excitons at a Donor–Acceptor Interface:* V.M. Agranovich, Institute of Spectroscopy, Russian Academy of Sciences, Troitsk, Moscow Reg., 142190, Russia (e-mail: agran@isan.troitsk.ru), and G.C. La Rocca, Scuola Normale Superiore and I.N.F.M., Piazza dei Cavalieri 7, 56126 Pisa, Italy (e-mail: larocca@sns.it)

Chapter 7. *Hybridization of Frenkel and Wannier–Mott Excitons in Organic-Inorganic Heterostructures. Strong Coupling Regime:* V.M. Agranovich, V.I. Yudson, Institute for Spectroscopy, Russian Academy of Sciences, Troitsk, Moscow obl., 142190 Russia (e-mails: agran@isan.troitsk.ru, yudson@isan.troitsk.ru), and P. Reineker, Department of Theoretical Physics, University of Ulm, 69089 Ulm, Germany (e-mail: peter.reineker@physik.uni-ulm.de)

Chapter 8. *Strong Optical Coupling in Organic Semiconductor Microcavities:* David G. Lidzey, Department of Physics and Astronomy, University of Sheffield, Hicks Building, Hounsfield, Sheffield, S37RH, UK (e-mail: d.g.lidzey@sheffield.ac.uk)

Chapter 9. *Electronic Energy Transfer in a Planar Microcavity:* D.M. Basko, ICTP Trieste, International Centre for Theoretic Physics, Strada Costiera 11, Trieste, I-34000, Italy (e-mail: basko@ictp.trieste.it)

Chapter 10. *Energy Transfer from a Semiconductor Nanostructure to an Organic Material and a New Concept for Light-Emitting Devices:* D.M. Basko, ICTP Trieste, International Centre for Theoretic Physics, Strada Costiera 11, Trieste, I-34000, Italy (e-mail: basko@ictp.trieste.it)

# Preface

A lot of efforts have been devoted in the last five-ten years to the growth of organic crystalline layered structures (including organic quantum wells and superlattices). Improvement in the technique of molecular beam deposition has led to a variety of good quality organic thin films, multilayered structures and heterostructures based on molecular solids, as well as combinations of organic and inorganic semiconductors. The possibility of growing tailor-made systems incorporating different organic crystalline materials with even more flexibility than for multiple quantum wells based on inorganic semiconductors alone, opens a promising field of research from the point of view of fundamental as well as applied physics. The advent of such a new class of organic crystalline materials prompted scientists to investigate their nonlinear and electro-optical properties, in order to understand their possible advantages in comparison with the usual organic or inorganic materials.

This volume, written by recognized experts in the field, is the first book devoted to a systematic discussion of the properties of organic crystalline multilayers and organic based nano- and multilayer heterostructures. It has been demonstrated that for such structures we can expect many new interesting optical effects and phenomena which can be important also for applications. Some of these predictions have already prompted experimental investigations (see, for example, Chapter 8, devoted to the observation of a strong Frenkel exciton-cavity photon coupling and giant Rabi splitting for Frenkel excitons in microcavities) and, no doubt, in the nearest future the experimental study of effects described in the book will continue.

We hope that this volume will come into use not only among physicists but also chemists and biologists. To help the nonspecialist reader we have included in the book three chapters (Chapters 1, 2, 3) which contain a tutorial introduction to the physics of electronic excitations in organic and inorganic solids.

All the chapters in the book are self-contained, and each concerns a specific aspect of the up-dated developments in the field. The list of contents will provide the reader with a detailed account of all the material included in the present book, in the following we limit ourselves to a few general remarks.

The optical and electro-optical effects are discussed in two classes of materials: (a) multilayer organic structures, (b) organic based heterostructures in the region of excitonic resonances.

Multilayer structures are systems with "condensed" interfaces, and under this conditions the specific surface and quasi two-dimensional effects at interfaces may play an important role in determining the sample properties. This contribution is particularly important in the cases where new excited states arise along the interfaces. In the study of nonlinear dynamics of interfaces we show in Chapter 4 that intermolecular anharmonicity across the interface produces new states: "the Fermi Resonance Interface modes", and leads to bistability and multistability in the energy transmission through the interface.

In Chapter 5 we describe the electronic excitations in quasi one-dimensional organic crystals with strong orbital overlap between neighboring molecules. In such crystals, the energy difference between the lowest Frenkel exciton and the nearest-neighbor charge-transfer excitons becomes small and their strong mixing determines the nature of the lowest energy states. The theory is used for understanding the optical properties of crystalline thin films of PTCDA(3,4,9,10-perylenetetra-carboxylic dianhydride) and MePTCDI(N-N''-dimethylperylene-3,4,9,10-dicarbiximide). We discuss also the new surface exciton states which arise due to mixing of Frenkel and charge-transfer excitons. These states can be important for the investigation of quantum confinement of exciton states in very thin layers of quasi-one-dimensional organic crystals. Since the synthetic tailoring of new organic compounds has endless possibilities because the growth of organic multilayer structures is not limited by lattice matching restrictions, such systems are expected to demonstrate a variety of potentially useful properties.

In the case of donor-acceptor multilayer structures described in Chapter 6, the peculiar resonant optical nonlinearities and the photo-voltaic effect associated with the interface charge-transfer excitons characteristic of such systems at moderate excitation densities are discussed. We also consider, at higher excitation densities, the ionization instability leading to photoconductivity even at low temperature (cold photoconductivity).

We consider the properties of electronic excitations in hetero-nano-structures based on combinations of organic materials and inorganic semiconductors, having respectively Frenkel excitons and Wannier–Mott excitons with nearly equal energies. We show that in this case the resonant coupling between them in quantum wells (or wires or dots) may lead to novel striking effects, such as a splitting of the excitonic spectrum and an enhancement of the resonant optical nonlinearities. We describe the properties of hybrid Frenkel–Wannier–Mott excitons, which appear when the energy splitting of the excitonic spectrum is large compared to the width of the exciton resonances (Chapter 7, the strong resonant coupling). Such peculiar excitations share at the same time the properties of the Wannier excitons (e.g., the large radius) and of the Frenkel excitons (e.g., the large oscillator strength). For this reason hybrid excitons are expected to have resonant optical nonlinearities significantly enhanced with respect to traditional inorganic or organic systems.

The structures mentioned above depend on the technologically challenging problem of growing high quality organic-inorganic heterojunctions only a few nanometers apart. A simpler way of realizing a hybrid exciton system is to couple Frenkel and Wannier excitons through a microcavity (MC) electromagnetic field. Strong exciton-radiation interactions are observed in microcavities, because the hybridization is not due to the Coulomb short-range interaction, but to strong long-range interaction stemming from cavity photons. For cavity embedded quantum wells, the fabrication would be much easier as their separation can be of the order of an optical wavelength. This situation is qualitatively equivalent to that of two coupled microcavities for which the growth conditions could be separately optimized for the organic and inorganic well. Planar microcavity structures provide a versatile mean to control the optical properties of semiconductors and to enhance the performance of opto-electronic devices. In Chapter 8, following the main results of the theory of polaritons in planar microcavity given in the introductory Chapter 3, and in Chapter 7, we describe experiments which demonstrate the giant Rabi splitting of Frenkel excitons. This giant Rabi splitting (which can be of the order of 100 meV) opens a new channels of cavity polariton relaxation and strongly affects absorption, transmission and photoluminescence. The results of these experiments represent a further step towards new exciton controlled devices.

In Chapter 9 the case of weak resonant coupling between Frenkel excitons in organics and Wannier–Mott excitons in inorganic semiconductors is considered. For this case the Foerster mechanism of energy transfer from an inorganic quantum well to an organic overlayer is of great interest. Such an effect may be especially useful for applications: the electrical pumping of excitons in the semiconductor quantum well can be used to efficiently turn on the organic material luminescence. Using this effect we propose a new concept for light emitting devices.

The strategy of combining organic and inorganic materials in the same nanostructures, as shown by the above examples, may lead to many novel devices which take advantage of the good properties of both classes of materials, overcoming the basic limitations of each individual class.

In the concluding Chapter 10 the electronic energy transfer in a microcavity is discussed. It is well known that the retarded interaction, along with the Coulomb interaction, contributes to the energy transfer between donors and acceptors. In a planar microcavity, whose thickness is of the order of the light wavelength, we can expect the retarded interaction to be enhanced due to the cavity photon contribution. The cavity enhancement of the energy transfer is expected to stimulate experimental investigations of energy transfer in microcavity. In this chapter a microcavity whose optical properties may be modified by the presence of absorbing acceptors (the donors are assumed not to affect the optical properties strongly) is considered. It is shown that different situations may be realized depending on the acceptor absorption: in the case of strong and broad absorption the cavity mode is practically destroyed; in the regime of weak absorption, the cavity mode is still well defined and just acquires some additional broadening. We also consider the strong coupling regime, when the acceptor absorption has the shape of a strong and narrow peak and two polariton branches appear due to coherent mixing of the acceptor excitations and the cavity mode. For all these cases the retarded interaction is responsible for the energy transfer. The role of different dissipative processes which may compete with the energy transfer, is also analyzed in detail. The comparison of the theory with the results of recent experiments is given.

In organising the material presented in this book we greatly enjoyed the collaboration with the authors of the various chapters and with other experts in this field. To all of them we wish to express our thanks.

VLADIMIR AGRANOVICH<br>
FRANCO BASSANI

Chapter 1

# Frenkel and Charge-Transfer Excitons in Organic Solids

JASPER KNOESTER

*Institute for Theoretical Physics and Materials Science Center, University of Groningen, Nijenborgh 4, 9747 AG Groningen, The Netherlands*

VLADIMIR M. AGRANOVICH

*Institute of Spectroscopy, Russian Academy of Sciences, Troitsk, Moscow Obl., 142100, Russia*

1. Introduction . . . . . . . . . . . . . . . . . . . . . . . . . . . . . . . . . . . . . 1
2. Microscopic Theory of Frenkel Excitons . . . . . . . . . . . . . . . . . . . . . 4
3. Dielectric Theory of Frenkel Excitons . . . . . . . . . . . . . . . . . . . . . . 34
4. Diffusion of Frenkel Excitons . . . . . . . . . . . . . . . . . . . . . . . . . . 45
5. Self-Trapping of Excitons: Spectra and Transport . . . . . . . . . . . . . . . . 58
6. Charge-Transfer Excitons in Organic Solids . . . . . . . . . . . . . . . . . . . 63
7. Molecular Aggregates: Low-Dimensional Exciton Systems . . . . . . . . . . . 67
8. Excitons in Biological Systems . . . . . . . . . . . . . . . . . . . . . . . . . 87
9. Concluding Remarks . . . . . . . . . . . . . . . . . . . . . . . . . . . . . . 90
   Acknowledgement . . . . . . . . . . . . . . . . . . . . . . . . . . . . . . . 90
   References . . . . . . . . . . . . . . . . . . . . . . . . . . . . . . . . . . . 90

## 1. Introduction

Excitons are electronic excitations of dielectric solids and clusters that play a crucial rule in the optical response of these materials. They represent bound electron–hole pairs, that may be generated by absorption of light or by relaxation of free electrons and holes after optical or electrical pumping. Among the optical properties of exciton systems that have aroused much interest, are cooperative spontaneous emission [1,2] (including application in light-emitting diodes [3] and lasers [4]), strong optical nonlinearities [5–7], and optical bistability [8]. Also the possibility to optically create Bose–Einstein condensates of (pairs of) excitons has attracted much attention [9,10].

Being electron–hole pairs, excitons carry no charge, which means that they do not contribute to electrical conduction. They do carry excitation energy, however, and their mobility therefore is responsible for energy transport processes. This transport not only is of much interest in solids, like semiconductors and insulating organic crystals, but also in mesoscopic and nanoscopic systems, such as the molecular cyanine aggregates that are responsible for light sensitization of silver-halide crystals in color photography [11,12] and molecular light-harvesting complexes in the photosynthetic systems of bacteria and higher plants [13].

Excitons are usually distinguished in two classes: Frenkel excitons and Wannier–Mott excitons [14]. The distinction lies in the typical separation between electron and hole. For Frenkel excitons, this separation is essentially zero (electron and hole occur on the same molecule or atom) and their binding energy is large ($\sim 1$ eV). For the Wannier–Mott exciton the electron–hole separation is much larger than a single molecule or atom and the binding energy is small ($\sim 1$ meV). Physically, the distinction originates from the competition between two energy scales: the electron–hole coupling and the rates for electron and hole hopping between different molecules or atoms [15]. The hopping rates allow the electrons and holes to move individually between different molecules in the solid or aggregate. This charge hopping arises from the overlap between the electron and hole orbitals on neighboring molecules. If the Coulomb coupling between electrons and holes can be neglected, this leads to freely moving electrons and holes, that may conduct electricity. The Coulomb interaction between electron and hole competes with their possibility to move independently through the system and if the interaction is strong enough it gives rise to bound electron–hole states, which appear as discrete levels below the continuum of "ionized" states in which electron and hole are essentially free. In fact, if the Coulomb attraction by far exceeds the hopping rates, the binding between electron and hole becomes too strong for them to separate and they always occupy the same molecule or atom. This is the Frenkel exciton limit, in which the exciton essentially is a molecular (or atomic) excitation, with an electron in the lowest unoccupied molecular orbital (LUMO) and a hole in the highest occupied molecular orbital (HOMO), that propagates through the crystal driven by electrostatic interactions.

From the above, it is clear that systems in which the charge overlap between neighboring molecules or atoms is small, typically carry Frenkel excitons. Examples are Van der Waals solids (molecular crystals and their self-assembled mesoscopic analogues, molecular aggregates), but also alkali-halide crystals and noble-gas crystals, such as argon. It should be noted that in general only the lowest molecular singlet excited state gives rise to real Frenkel excitons; higher excited states in general have sufficient charge overlap to allow for a finite electron–hole separation. Although the electron and hole in a Frenkel exciton reside on the same molecule, they do have the ability to move through the crystal as a pair. This center-of-mass motion of the pair arises from electrostatic interactions between the electrons on different molecules. Generally the strongest contribution to this propagation derives from the interactions between the transition dipoles of the individual molecular transitions. This gives rise to a long-range ($1/r^3$) excitation transfer interaction between molecules, which in turn is responsible for the formation of an exciton band. We stress that this interaction does not involve charge overlap.

In systems with a stronger charge overlap between neighboring molecules or atoms, the relative motion of electron and hole introduces extra degrees of freedom. For Wannier–Mott excitons, the electron–hole separation is of the order of hundreds to thousands of Ångstroms, so that one may regard the underlying system essentially as a continuum characterized by a dielectric constant. This gives rise to a series of hydrogen-like states for the relative motion of electron and hole. The smallness of the binding energy and the appreciable size of the radius of these orbitals as opposed to the real hydrogen orbitals, originates from the small effective mass of the electron and hole in semiconductors and the typically large value of the dielectric constant.

While the names Frenkel and Wannier–Mott excitons refer to two extreme cases with regards to the length scale of relative electron–hole motion (the internal exciton structure), this scale in practice does allow for all the intermediate cases. The intermediate case that has acquired a special status is the charge-transfer exciton (CTE), which in its lowest-energy variation has an electron–hole separation of one molecule. Thus, the hole is located at one molecule and the electron on the neighboring one. This is often referred to as a "donor–acceptor (D-A) complex". CTEs may occur in systems with an alternating structure of two types of molecules [16], but also in crystals with just one type of molecule (such as anthracene, naphthalene, and many others). In the latter case, any molecule in the crystal can play the role of donor or acceptor. CTEs currently are considered important intermediate states in the photo-conductivity of organic crystals [17]. In this process, they are essential in the creation of free carriers from photo-generated Frenkel excitons.

In organic crystals and aggregates, we typically deal with Frenkel and charge-transfer excitons as the important optically accessible excitations. In some cases, for example, in quasi-one-dimensional organic solids with very small intermolecular distances within the stacks, the mixing of these two types of excitations is important (see the chapter by M. Hoffmann in this book). It is worth mentioning here that the low-temperature spectroscopy of organic molecular crystals grew out of classical experiments carried out in the late 1920s by Pringsheim and Kronenberger [18], and Obreimov and de Haas [19,20]. At that time only the Bloch band scheme for electronic states in crystals was known. This concept predicted very broad absorption bands, in contradiction to the narrow lines observed in the cited experiments. It is known that I.V. Obreimov attracted the attention of Ya.I. Frenkel to this problem and in the first 1931 paper of Frenkel [21], where the concept of excitons in a molecular crystal was formulated for the first time, the reader may find the corresponding acknowledgment. The name "exciton" was introduced by Frenkel in 1936 [22]. After these first steps, a growing list of important experimental observations have been reported by Obreimov and Prikhotko [23] and many others. A classical review of experiments on excitons in organic crystals is the book by Broude, Rashba, and Sheka [24]. More recently, Frenkel excitons and their associated optical and energy transport properties have attracted much attention in mesoscopic and nanoscopic organic systems, such as (bio-)molecular aggregates [13,25].

In this chapter, we give an introduction into the properties of electronic excitations in organic solids, with an emphasis on Frenkel excitons. We explain the basic theory and discuss its implication for and relation to experiments. A similar introduction into the properties of Wannier–Mott excitons, which are typical for inorganic semiconductors, will be given in the chapter by G.C. La Rocca in this book. Both chapters together serve as basis for the following chapters, which are concerned with the photo-physics of organic multi-layers and hetero- and nanostructures involving organic components.

The outline of this chapter is as follows. In Section 2, we describe in some detail the microscopic theory of excitons in molecular systems. We start by considering the tutorial example of the molecular dimer, then introduce a second-quantized notation, and make the step to multi-molecular systems. After the basic properties of the corresponding Hamiltonian have been studied, we specialize to crystal structures, for which we deal with the topics of Davydov splitting, microscopic calculation of the dielectric function, exciton-polaritons, nonlinear optics, and exciton–phonon interaction. In Section 3, we describe the dielectric theory of Frenkel excitons and demonstrate that this microscopic classical theory of excitons in many respects is very powerful and easier to deal with than a fully quantum mechanical microscopic theory. The diffusive motion of Frenkel excitons in molecular crystals is dealt with in Section 4, with special distinction between coherent and incoherent excitons. Related to the topic of exciton motion is exciton self-trapping, which occurs in the case of strong exciton–phonon interaction. This phenomenon is discussed in Section 5. In Section 6, we briefly address charge-transfer excitons. Most of the material in Sections 2–6 deals explicitly with bulk crystals. In Section 7 we consider special properties of excitons in lower-dimensional structure, with particular attention to the recently much studied examples of one-dimensional aggregates of organic molecules. Here the role of disorder will get special attention. The most recent applications of Frenkel exciton theory to bio-molecular aggregates, such as those that occur in photosynthetic antenna complexes, will be the topic of Section 8. Finally, we conclude in Section 9.

## 2. Microscopic Theory of Frenkel Excitons

### 2.1. SINGLE MOLECULE: EIGENSTATES AND TWO-LEVEL APPROXIMATION

As Frenkel excitons are essentially molecular excitations, the best way to introduce them is by starting out from a single molecule in the gas phase. The molecule is described by a Hamiltonian $\widehat{H}_{\text{mol}}$, which contains the kinetic energy of all electrons and nuclei and all Coulomb couplings between these charged particles. We will assume that the (adiabatic) eigenstates are known. In particular, we will assume that the ground state is nondegenerate; it is denoted $|g\rangle$ and has energy $\hbar\omega_g$. The excited states will be denoted $|f\rangle$, with energy $\hbar\omega_f$. We will be particularly interested in the vibrationally relaxed electronic states, although the exciton theory which we are about to describe is also valid for vibronic states (i.e., an electronic excitation which is strongly coupled to a local vibrational mode). The coupling of the electronic states to phonons in general will be discussed at a later stage (Section 2.9).

The linear optical (electromagnetic) response of the molecule has contributions from all possible transitions between the ground state and the excited states. The strength of each contribution is determined by the oscillator strength (basically the squared transition dipole element) of the transition and the detuning $(\omega_f - \omega_g) - \omega$, where $\omega$ is the frequency of the exciting light pulse. If one of these transitions, say the one from $|g\rangle$ to the particular state $|e\rangle$, dominates the others, it is useful to restrict all considerations just to the pair of states $|g\rangle$ and $|e\rangle$. We will make this reduction to a "two-level molecule" in most of the microscopic theory in this chapter, as this suffices to explain the essential properties of Frenkel excitons. The effects of going beyond this approximation will be addressed in Sections 2.5.4 and 3.4.

On the subspace of the two selected levels, the Hamiltonian may now be written:

$$\widehat{H}_{\text{mol}} = \hbar\omega_g |g\rangle\langle g| + \hbar\omega_e |e\rangle\langle e| = \hbar\omega_g + \hbar\omega_0 |e\rangle\langle e|, \quad (1)$$

where $\omega_0 = \omega_e - \omega_g$, the transition frequency. Within the same subspace, we may write the dipole operator $\hat{\mu}$ of the molecule as

$$\hat{\mu} = \mu(|e\rangle\langle g| + |g\rangle\langle e|) + \mu_g |g\rangle\langle g| + \mu_e |e\rangle\langle e|. \quad (2)$$

Here, $\mu$ is the transition dipole between both states, i.e., the matrix element $\langle e|\hat{\mu}|g\rangle$. Throughout this chapter, we will assume that the molecular wave functions are real, so that also this matrix element is real. For nondegenerate excited states (in particular for singlet states), this choice may always be made. Furthermore, $\mu_g$ and $\mu_e$ are the permanent dipoles in the ground and excited states, respectively. Note that for molecules with inversion symmetry, such permanent dipoles vanish.

The optical absorption spectrum of this two-level molecule, which may be calculated using Fermi's golden rule, is a delta peak at $\omega = \omega_0$ with a total area proportional to $\mu^2$, the so-called (oscillator) strength of the transition. In practice, this delta-peak will be broadened for two reasons [26]. First, the optical transition is homogeneously broadened due to the finite lifetime of the excited level and due to dephasing processes resulting from interactions with a heat bath. Second, one usually measures on an inhomogeneous ensemble of molecules, i.e., an ensemble in which the value of $\omega_0$ varies from molecule to molecule. In the gas phase, such inhomogeneity may be due to different Doppler shifts of the absorption lines, arising from the fact that the velocities of the molecules differ from each other. In condensed phases (solutions or disordered solids) inhomogeneity derives from the fact that each molecule has a (slightly) different environment, leading to random shifts of the molecular transition frequencies. The probability distribution of molecular frequencies is referred to as the inhomogeneous distribution and is often taken to be Gaussian. The resulting total absorption lineshape is the convolution of the homogeneous lineshape and the inhomogeneous distribution.

## 2.2. Dimer of Two-Level Molecules

We now turn to a dimer of two identical molecules. The molecules are close enough to allow the electrons and nuclei to interact with each other, but at the same time, we will assume that the separation is large enough to neglect overlap of the electronic orbitals of interest. As mentioned already in the Introduction, in molecular crystals this generally holds to a good approximation for the ground state and the lowest (few) excited state(s). Excitons in crystals with strong orbital overlap will be considered in the chapter by M. Hoffman in this book. The Hamiltonian of the dimer now reads:

$$\widehat{H}_{\text{dim}} = \widehat{H}_{\text{mol},1} + \widehat{H}_{\text{mol},2} + \widehat{V}_{1,2}, \tag{3}$$

where $\widehat{H}_{\text{mol},n}$ denotes the Hamiltonian for molecule $n$ (compare previous section), containing the kinetic energy of electrons and nuclei within molecule $n$ ($= 1, 2$) and all Coulomb interactions between these electrons and nuclei. The operator $\widehat{V}_{1,2}$ contains all Coulomb interactions between pairs of charged particles of which one belongs to molecule 1 and the other to molecule 2. Note that this distinction of terms can only be made if it is clear to which molecule a given electron belongs, which means that charge transfer effects (orbital overlap) should be negligible.

We will restrict ourselves again to the two-level picture of the molecules. For the dimer, this yields a basis of four states: $|g_1 g_2\rangle$ in which both molecules reside in their ground state, $|e_1 g_2\rangle$, in which molecule 1 is in the excited state $|e\rangle$ and molecule 2 is in its ground state; $|g_1 e_2\rangle$ (vice versa), and finally $|e_1 e_2\rangle$, in which both molecules are in their respective excited states. On this basis, the Hamiltonian reads:

$$\widehat{H}_{\text{dim}} = \sum_{u,v} H_{u,v} |u\rangle\langle v|, \tag{4}$$

with $H_{u,v} = \langle u|\widehat{H}_{\text{dim}}|v\rangle$ and $|u\rangle$ and $|v\rangle$ running over the four basis states. Of the 16 matrix elements only 10 are independent. We will not calculate all of these matrix elements explicitly, but limit ourselves to those that conserve the number of excitation quanta within the dimer. Thus, we will neglect at this moment the coupling between $|g_1g_2\rangle$ and the singly and doubly excited states, etc. The reason is that such states differ by an energy of the order of a molecular transition energy (typically 10,000–30,000 cm$^{-1}$), while for molecular systems the coupling matrix elements are typically of the order of 100–1000 cm$^{-1}$ for singlet excitations and even smaller for triplet ones. Thus, the mixing between states with a different number of excitation quanta will be very small. We will come back to this in the context of the Heitler–London approximation (Section 2.4.4) and nonlinear optical response (Section 2.8).

The Hamiltonian of the dimer may now be written:

$$\widehat{H}_{\text{dim}} = (2\hbar\omega_g + V_{gg,gg})$$
$$+ (\hbar\omega_0 + D_{1,2})|e_1g_2\rangle\langle e_1g_2| + (\hbar\omega_0 + D_{2,1})|g_1e_2\rangle\langle g_1e_2|$$
$$+ J_{1,2}(|e_1g_2\rangle\langle g_1e_2| + |g_1e_2\rangle\langle e_1g_2|)$$
$$+ (2\hbar\omega_0 + D')|e_1e_2\rangle\langle e_1e_2|. \tag{5}$$

Here, the first line represents the energy of the dimer's ground state, which in the current approximation is simply the state $|g_1g_2\rangle$. Its energy consists of twice the molecular ground state energy plus the electromagnetic interactions between the molecules in each of their ground states: $V_{gg,gg} = \langle g_1g_2|\widehat{V}_{1,2}|g_1g_2\rangle$. To lowest order in the inverse intermolecular distance, this is the interaction between the permanent ground state dipoles of each of the molecules. For centrosymmetric molecules this matrix element is determined by quadrupolar and higher-order interactions.

The second and third lines in Eq. (5) describe the Hamiltonian in the subspace of singly excited states (the so-called one-exciton space). The second line contains the single-molecule terms, whose energies above the ground state are shifted away from $\hbar\omega_0$. For molecule 1, this shift is $D_{1,2} = \langle e_1g_2|\widehat{V}_{1,2}|e_1g_2\rangle - \langle g_1g_2|\widehat{V}_{1,2}|g_1g_2\rangle$, which is the difference in the Coulomb interaction between the ground state molecule 2 and the excited and unexcited molecule 1, respectively. $D_{2,1}$ has a similar meaning, with the role of molecules 1 and 2 interchanged. In molecular crystals, these shifts are referred to as the gas-condensed matter shifts. For centrosymmetric molecules, the lowest-order contributions to this shift are of quadrupolar nature. The third line is the crucial one for Frenkel exciton systems: it describes the transfer of the excitation from one molecule to the other due to the Coulomb interactions (Figure 1). The corresponding matrix element is defined through $J_{1,2} = \langle e_1g_2|\widehat{V}_{1,2}|g_1e_2\rangle = \langle g_1e_2|\widehat{V}_{1,2}|e_1g_2\rangle$, where the equality holds, because we assume the wave functions to be real. This term is also often referred to as the "resonant" interaction. For dipole allowed transitions it is dominated by the interaction between the molecular transition dipoles,

$$J_{1,2}^{\text{dip}} = ((\boldsymbol{\mu}_1 \cdot \boldsymbol{\mu}_2)|\mathbf{r}_{12}|^2 - 3(\boldsymbol{\mu}_1 \cdot \mathbf{r}_{12})(\boldsymbol{\mu}_2 \cdot \mathbf{r}_{12}))/|\mathbf{r}_{12}|^5 \tag{6}$$

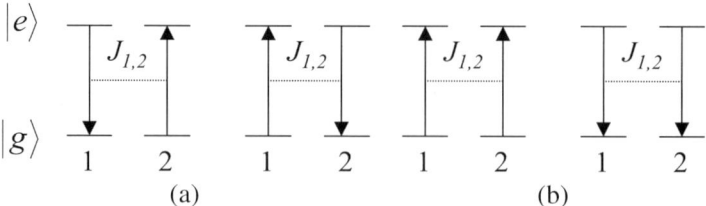

Fig. 1. Resonant (excitation transfer) interactions (a) and off-resonant interactions (b) within a dimer of two-level molecules. $|g\rangle$ is the molecular ground state, $|e\rangle$ the excited state.

where $\mathbf{r}_{12}$ is the position vector between molecules 1 and 2, and $\boldsymbol{\mu}_n$ denotes the transition dipole of molecule $n$.

Finally, the fourth line of Eq. (5) describes the two-exciton subspace, which for a dimer only contains one basis state. Naturally, its energy above the overall ground state energy is twice the molecular excitation energy plus a contribution due to the difference between the Coulomb interactions in the doubly excited state and the dimer's ground state: $D' = \langle e_1 e_2 | \widehat{V}_{1,2} | e_1 e_2 \rangle - \langle g_1 g_2 | \widehat{V}_{1,2} | g_1 g_2 \rangle$. Again, for centrosymmetric molecules, this shift depends on quadrupolar and higher-order interactions.

To end this section, we consider the one-exciton subspace in more detail. These singly excited states are the ones that are traditionally understood to be the Frenkel excitons. To get the actual exciton states for the dimer, we have to diagonalize the second and third lines of Eq. (5). Let us assume that the molecules are positioned in such a symmetric way that $D_{1,2} = D_{2,1} = D$. Then the one-exciton eigenstates are the symmetric and antisymmetric combinations of the molecular excited states:

$$|\pm\rangle = \bigl(|e_1 g_2\rangle \pm |g_1 e_2\rangle\bigr)/\sqrt{2}, \tag{7}$$

with energies (relative to the overall ground state)

$$E_\pm = \hbar\omega_0 + D \pm J_{1,2}, \tag{8}$$

i.e., the originally degenerate pair of molecular excited states mixes and splits into a pair of states delocalized over the dimer with an energy splitting $2|J_{1,2}|$. The transition dipoles are given by

$$\boldsymbol{\mu}_\pm = (\boldsymbol{\mu}_1 \pm \boldsymbol{\mu}_2)/\sqrt{2}. \tag{9}$$

As both molecules are identical, their transition dipoles may only differ in orientation, from which it is easily derived that $\boldsymbol{\mu}_+$ and $\boldsymbol{\mu}_-$ are oriented perpendicular to each other. We thus find that the coupling $\widehat{V}$ of both molecules in the dimer gives

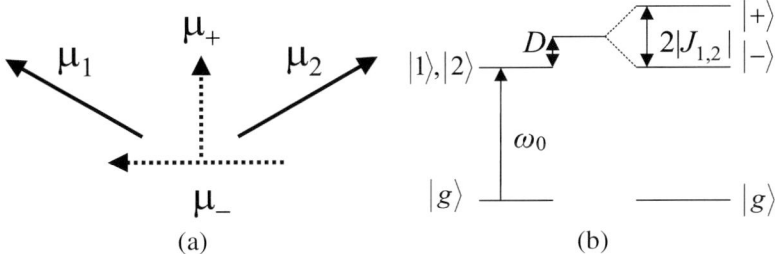

Fig. 2. (a) Transition dipoles of the individual molecules 1 and 2 in a dimer and of the dimer eigenstates $|+\rangle$ and $|-\rangle$. (b) One-exciton level diagram of a homogeneous dimer before (left) and after (right) taking the intermolecular interactions into account.

rise to an absorption spectrum in which the molecular peak at $\hbar\omega_0$ is split into two perpendicularly polarized absorption bands, with strength $|\boldsymbol{\mu}_+|^2$ and $|\boldsymbol{\mu}_-|^2$, centered around $\hbar\omega_0 + D$ and separated by $2|J_{1,2}|$ (Figure 2). This splitting is the dimer analog of the Davydov splitting in a bulk crystal which we will encounter in Section 2.5.3.

### 2.3. SECOND QUANTIZATION

So far, we have used state vectors, like $|g_1 g_2\rangle$, $|e_1 g_2\rangle$, etc., to describe the states and the interaction operators, like Eq. (5). For a system consisting of many molecules and with possibly many excitations, such a notation is not very convenient. Instead, it is easier to work with creation and annihilation operators (second quantization). At a fundamental level, the operators of interest to describe the electronic states, are the electron creation and annihilation operators, which add or take away an electron from a certain electronic orbital (single-electron state) in a particular molecule. As electrons are fermions, these operators obey Fermi anticommutation relations. Since in most of this chapter, we will be dealing with two-level molecules and we will neglect the possibility of charge transfer between molecules, we can work with a simpler set of operators, however. Each molecule can be in only two states: the ground state $|g\rangle$ and the excited state $|e\rangle$. Within a single-electron picture (Hartree–Fock approximation) the excited state is formed by annihilating an electron from a particular one of the electron orbitals that is occupied in the ground state and creating an electron in a particular unoccupied orbital. Thus, the excitation of a molecule is a product of an annihilation and a creation of an electron [27,28] (of course, many different occupied and unoccupied orbitals could be combined like this, but remember that we assumed a situation where we can limit ourselves to one dominant molecular transition). Neglecting intermolecular charge transfer, implies that we do not need to account for excitations in which after annihilation an electron is created in an excited orbital of

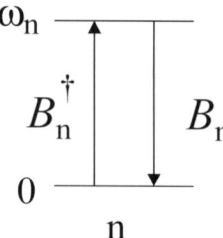

Fig. 3. Action of Pauli creation ($B_n^\dagger$) and annihilation ($B_n$) operators in a two-level molecule.

a different molecule. As mentioned in the Introduction, this type of excitation, in which the ground state hole and the excited state electron are always on the same molecule, is called the *Frenkel exciton* [21]. From the point of view of book keeping, it is a simple excitation, as we only need one coordinate to describe it (position of electron and hole are the same). In the case of possible charge transfer (like in charge transfer excitons or in Wannier excitons), the position of electron and hole should be considered as two separate degrees of freedom, complicating the description, as the exciton acquires internal degrees of freedom through the relative motion of electron and hole. In the Introduction, we have already described the physics underlying the distinction between Frenkel and Wannier–Mott excitons.

The creation and annihilation of an excitation on a two-level molecule are described by so-called Pauli creation and annihilation operators $B_n^\dagger$ and $B_n$, respectively (see Figure 3) [27,28]. Here, $n$ labels the molecule (for instance, $n = 1$ or $n = 2$ if we deal with a dimer). The operator $B_n^\dagger$ changes the state of molecule $n$ from the ground state to its excited state, while $B_n$ does the reverse. The name Pauli operator comes from the algebra of spin 1/2 operators; our two-level molecule (or in fact any two-level system) is formally equivalent to a spin 1/2 [29,30]. When working in second quantization, it is important that we define the commutation relations of the various operators. For our Pauli operators, the rules are that any two operators working on different molecules commute:

$$[B_n, B_m^\dagger] = 0, \quad [B_n, B_m] = 0 \quad (n \neq m). \tag{10}$$

In plain words this means that the operation of $B_n^\dagger$ and $B_n$ only depends on the state of molecule $n$ and not on the state of any of the other molecules. However, the operators working on one and the same molecule, behave like Fermi operators, i.e., they obey anti-commutation relations. This means that

$$B_n^\dagger B_n + B_n B_n^\dagger = 1, \quad B_n^\dagger B_n^\dagger = B_n B_n = 0. \tag{11}$$

These relations reflect the fact that a two-level molecule can at most hold one excitation. Trying to double excite it gives zero; similarly, trying to take two excitations away from a two-level molecule is impossible. This exclusion of the double excitation of a molecule is also referred to as the Pauli exclusion for the Frenkel exciton; it is important in the description of nonlinear optical response, where one in fact deals with multiple excitations of the system (Section 2.8). For later use, we note that the first relation in Eq. (11) may also be written as the commutation relation:

$$[B_n, B_n^\dagger] = 1 - 2\widehat{N}_n, \tag{12}$$

where $\widehat{N}_n = B_n^\dagger B_n$ is the operator for the number of excitations on molecule $n$. From Eqs. (10) and (12), we observe that excitons almost behave like bosons, if in some sense the term $2\widehat{N}_n$ may be considered small. We will come back to this in Section 2.4.3.

From now on, we will use the Pauli operators to denote all states and operators in a system of in interacting two-level molecules. For the states, this is simply done by operating with the creation operators on the overall ground state an appropriate number of times. If we go back to the example of the dimer, then the one-exciton state $|e_1 g_2\rangle$ may be written $B_1^\dagger |g_1 g_2\rangle$, while $|e_1 e_2\rangle = B_1^\dagger B_2^\dagger |g_1 g_2\rangle = B_2^\dagger B_1^\dagger |g_1 g_2\rangle$. Important examples of operators are the internal energy operator of an isolated molecule $n$ above its ground state, which reads $\hbar\omega_0 B_n^\dagger B_n$. The generic operator describing the excitation transfer interaction from molecule $n$ to $m$ is $J_{nm} B_m^\dagger B_n$, where $J_{nm}$ is the strength of the interaction.

Using this language, the dimer Hamiltonian, Eq. (5), is written

$$\widehat{H}_{\text{dim}} = \sum_{n=1,2} (\hbar\omega_0 + D_n) B_n^\dagger B_n + J_{1,2}(B_1^\dagger B_2 + B_2^\dagger B_1)$$
$$+ U_{1,2} B_2^\dagger B_1^\dagger B_2 B_1, \tag{13}$$

where we introduced $D_1 = D_{1,2}$, $D_2 = D_{2,1}$, and $U_{1,2} = D' - D_{1,2} - D_{2,1} = \langle e_1 e_2|\widehat{V}_{1,2}|e_1 e_2\rangle + \langle g_1 g_2|\widehat{V}_{1,2}|g_1 g_2\rangle - \langle e_1 g_2|\widehat{V}_{1,2}|e_1 g_2\rangle - \langle g_1 e_2|\widehat{V}_{1,2}|g_1 e_2\rangle$. The resulting expression is very compact, not only because we omitted the overall ground state energy (which will be set to zero from here on), but also because part of the two-exciton term in Eq. (5) is now automatically included in the first term of Eq. (13). As a consequence, terms that may really be considered two-exciton terms, namely those in which two excitations are annihilated and two are created, stand out very clearly in the current notation. Such interactions, of which the last term in Eq. (13) is a simple example, are also known as dynamic exciton–exciton interactions, in contrast to the kinematic interactions that derive from the Pauli exclusion (also see Section 2.4). We note that within the dipole approximation, the dynamic exciton–exciton interaction vanishes, unless the molecular excited states have permanent dipoles (or more accurately: the permanent dipoles of the molecular ground and excited states differ).

## 2.4. GENERAL $N$ MOLECULE SYSTEM

### 2.4.1. Hamiltonian

It is now a simple matter to generalize the foregoing to a system of $N$ two-level molecules. As before, we will neglect overlap between the relevant electronic orbitals of the different molecules. Because in this section we will be interested in general aspects of such a multi-molecule Frenkel exciton system, we will not impose any particular symmetry in their spatial arrangement. We will specialize to the case of crystalline arrangements in Section 2.5.

In second quantization, the $N$-molecule Frenkel exciton Hamiltonian reads (cf. Eq. (13)):

$$\widehat{H}_{\text{ex}} = \sum_n (\hbar\omega_0 + D_n) B_n^\dagger B_n + \sum_{n,m} J_{nm} B_n^\dagger B_m$$

$$+ \frac{1}{2} \sum_{n,m} U_{nm} B_n^\dagger B_m^\dagger B_n B_m. \quad (14)$$

As before, we have restricted ourselves to terms that conserve the total number of molecular excitations. Furthermore, we have defined the gas-condensed matter shifts as $D_n = \sum_m D_{nm}$, with

$$D_{nm} = \left( \langle e_n g_m | \widehat{V}_{nm} | e_n g_m \rangle - \langle g_n g_m | \widehat{V}_{nm} | g_n g_m \rangle \right). \quad (15)$$

Here, $\widehat{V}_{nm}$ ($n \neq m$) denotes the total Coulomb interaction between the constituents of molecule $n$ and $m$ (cf. $\widehat{V}_{1,2}$ above), and the notation of states follows a straightforward generalization of the dimer basis states. We emphasize that, since we are only dealing with pair interactions, there is no need to explicitly describe the states of all the other molecules ($\neq n, m$) in the above matrix elements, as these state are not affected by $\widehat{V}_{nm}$.

Returning to Eq. (14), the other two relevant coupling parameters are defined by:

$$J_{nm} = \langle e_n g_m | \widehat{V}_{nm} | g_n e_m \rangle = \langle g_n e_m | \widehat{V}_{nm} | e_n g_m \rangle \quad (16)$$

for the excitation transfer interaction and

$$U_{nm} = \left( \langle e_n e_m | \widehat{V}_{nm} | e_n e_m \rangle + \langle g_n g_m | \widehat{V}_{nm} | g_n g_m \rangle \right) - \langle e_n g_m | \widehat{V}_{nm} | e_n g_m \rangle$$

$$- \langle g_n e_m | \widehat{V}_{n,m} | g_n e_m \rangle \quad (17)$$

for the dynamic exciton–exciton interaction. It should be stressed that in organic materials, which often contain large molecules at relatively short distances, the point-dipole approximation (Eq. (6)) generally does not suffice to calculate the above interaction matrix elements. For the transfer interactions, a good alternative to a further multipole expansion is to use extended dipoles, i.e., the dipole is replaced by a pair of charges at a finite distance [31]. A yet better approach is to use the full electronic charge distributions of the molecular states, which should be calculated using first-principles methods [32]. For the exciton–exciton interactions, the dipole approximation often fails even qualitatively, as the permanent dipoles in the ground and the excited states vanish for centrosymmetric molecules.

### 2.4.2. Multi-Exciton Bands

The Hamiltonian equation (14) describes the excitons in the $2^N$ dimensional Hilbert space spanned by the $N$ two-level molecules. Hence, a full diagonalization in principle requires a $2^N \times 2^N$ matrix to be diagonalized. The complexity is reduced, however, as all interactions that we consider conserve the number of excitations, so that the eigenstates may be classified with respect to this number (cf. Figure 4). Thus, the overall ground state, denoted $|g\rangle$, is the state in which all molecules are in their ground state (the zero-exciton state). It should be realized that this is only true, because the typical interaction matrix elements $|J_{nm}|$ and $|U_{nm}|$ are small compared to the single-molecule excitation energy $\hbar\omega_0$ (*vide*

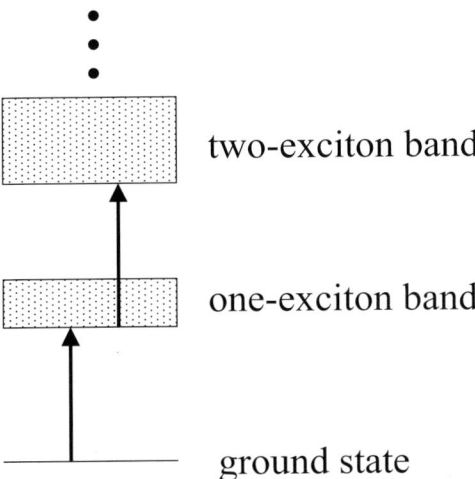

Fig. 4. Schematic energy level diagram of a system consisting of interacting two-level molecules. The excited states occur in bands, classified according to the number of excitation quanta shared by the molecules. Only transitions between adjacent exciton bands are dipole-allowed (indicated by the arrows). As a consequence, the one-exciton states suffice to describe the linear optics of the system, while its nonlinear optical response involves multi-exciton bands. Exciton–exciton interactions may give rise to bound bi-exciton states that split off below or above the continuous two-exciton band that exists for infinite systems.

*infra*), the same circumstance that allows us to neglect interactions that do not conserve the number of excitations.

The lowest set of excited states are linear combinations of singly excited states:

$$|\sigma\rangle = \sum_n \varphi_{\sigma n} B_n^\dagger |g\rangle. \tag{18}$$

These states are found by diagonalizing the $N \times N$ one-exciton Hamiltonian matrix $H_{nm} = (\hbar\omega_0 + D_n)\delta_{nm} + J_{nm}$ and in the above notation, $\varphi_{\sigma n}$ is the $n$th component of the (normalized) $\sigma$th eigenvector. The corresponding eigenvalue $\hbar\Omega_\sigma$ gives the energy of state $|\sigma\rangle$. Obviously, the one-exciton states are not affected by the exciton–exciton interactions $|U_{nm}|$. In general, the one-exciton band contains $N$ states, spread in energy over a range (the band width) of the order of the typical transfer interaction $|J|$ and centered around the energy $\hbar\omega_0$. In thermal equilibrium, the one-exciton states suffice to calculate the linear absorption spectrum. We will consider this observable more explicitly in the situation of bulk crystals and J-aggregates.

The next higher set of excited states are the two-excitons states, in which the $N$ molecules share two molecular excitation quanta. These eigenstates require an $N(N-1)/2 \times N(N-1)/2$ matrix to be diagonalized, which also contains the exciton–exciton interactions. This band is centered around twice the typical molecular excitation energy. Two-exciton states cannot be excited from the ground state through one interaction with electromagnetic fields and therefore are not visible in the linear absorption spectrum. They can be excited from the one-exciton band and, hence, they play an important role in nonlinear optical experiments (Sections 2.8 and 7.3). It should be realized that even in the absence of dynamic exciton–exciton interactions, it is in general not possible to simply express the two-exciton eigenstates in terms of the eigenvectors of the one-exciton subspace $\varphi_{\sigma n}$ obtained above. The reason is the Pauli nature of the excitons, which does not allow them to share the same site. This leads to an effective exciton–exciton interaction which is also known as the "kinematic" interaction. On the level of two-exciton states, these kinematic interactions may be dealt with by simply considering the excitons bosons and invoking a hard-core exciton–exciton repulsion, i.e., an infinitely strong repulsion for two excitations to share the same site. As the scattering matrix of such a hard-core potential can be calculated analytically, this allows one to express the two-exciton Green functions in terms of one-exciton eigenfunctions and energies [33,34]. Traditionally, approximating excitons as bosons has been common practice when dealing with Frenkel excitons in bulk crystals [27,28,35,36] and we will therefore address this approach in more detail in the next section. We note that, alternatively, for one-dimensional systems in the absence of dynamic exciton–exciton interactions, the Jordan–Wigner transformation to noninteracting fermions is very useful [37–39] (Section 7.3).

Above the two-exciton states, yet higher bands occur: three-excitons, four-excitons, etc. The actual calculation of these states gets increasingly more complicated, as they require larger and larger matrices to be diagonalized. The classification in terms of the multi-exciton bands is depicted in Figure 4. In practice, there will be a limitation to its use, as under normal conditions higher multi-exciton states (for which the average excitation density per molecule is high), will inevitably be strongly mixed with the high density of high lying molecular excited states that we neglected in our two-level picture. However, the existence of (bound) two-exciton states in semiconductor crystals [40] and molecular J-aggregates [41] has clearly been demonstrated.

### 2.4.3. Bosonization

In Section 2.3 we have seen that the Pauli operators obey the commutation relations:

$$[B_n, B_m^\dagger] = \delta_{nm}(1 - 2\widehat{N}_n), \qquad [B_n, B_m] = 0. \tag{19}$$

We note that the only difference with bosons is the term $-2\delta_{nm}\widehat{N}_n$. For excitations spread over many molecules, as is appropriate for crystals and strongly coupled aggregates of molecules, the deviation from bosonic behavior is thus expected to be proportional to the expectation value of the excitation density of the molecules, which will generally be low if low-intensity light sources are used. In particular, one expects this to hold in the regime of linear optics. This circumstance motivates the traditional treatment of excitons as bosons [27,28,35,36]. We can make this argument more rigorous by transforming the Pauli operators from the site representation to a delocalized representation:

$$B_\sigma^\dagger = \sum_n \varphi_{\sigma n} B_n^\dagger \tag{20}$$

where $\varphi_{\sigma n}$ denote the eigenvector components also introduced in Section 2.4.2. From Eq. (19), the commutation relations for these new operators are found to be

$$[B_\sigma, B_\tau^\dagger] = \delta_{\sigma\tau} - 2\sum_n \varphi_{\sigma n}^* \varphi_{\tau n} \widehat{N}_n, \qquad [B_\sigma, B_\tau] = 0. \tag{21}$$

As we have seen, the $\varphi_{\sigma n}$ denote the excitation amplitude of the $\sigma$ th one-exciton. If the excitations are spread over all $N$ molecules, as is the case in highly symmetric systems such as perfect crystals, the typical amplitude scales like $1/\sqrt{N}$ (normalization) and we thus find that for the new operators the correction to Bose behavior is of the order $\sum_n \widehat{N}_n/N$, which is the average density of excitations in the system. This confirms our above expectation that at low excitation density, excitons may be treated as bosons. Similar arguments can be made in less symmetric systems, where the excitons are not spread over all molecules, but rather over a certain delocalization range $N_{\text{del}}$.

After the transformation to the $\sigma$ basis, the Hamiltonian equation (14) reads

$$\widehat{H}_{\text{ex}} = \sum_\sigma \hbar\Omega_\sigma B_\sigma^\dagger B_\sigma + \frac{1}{2} \sum_{\sigma,\sigma',\tau,\tau'} U_{\sigma\sigma'\tau\tau'} B_\sigma^\dagger B_\tau^\dagger B_{\sigma'} B_{\tau'}, \qquad (22)$$

where $U_{\sigma\sigma'\tau\tau'}$ represents the dynamic exciton–exciton interaction in the $\sigma$ basis, which is easily expressed in terms of the $U_{nm}$ and the transformation coefficients $\varphi_{\sigma n}$. Let us now first assume that dynamic interactions are absent:

$$\widehat{H}_{\text{ex}} = \sum_\sigma \hbar\Omega_\sigma B_\sigma^\dagger B_\sigma, \qquad (23)$$

which is the situation considered in many traditional texts on Frenkel excitons. The Hamiltonian then looks like it is already in diagonal form. Generally speaking, however, this is not the case, as the kinematic interactions hidden in the general commutation relations spoil this simple picture. Only at sufficiently low excitation density we may neglect these interactions and treat the excitons as bosons. Then, indeed, Eq. (23) is in diagonal form and the multi-excitons of interest are simply described by giving the occupation number of each boson state $\sigma$. The explicit expression for such multi-boson states in the site representation is obtained by operating with the corresponding Bose creation operators an appropriate number of times on the ground state $|g\rangle$ and transforming these operators back to site representation. For instance, for a one-boson state we have $|\sigma\rangle = B_\sigma^\dagger|g\rangle = \sum_n \varphi_{\sigma n} B_n^\dagger |g\rangle$. This is exactly the one-exciton state equation (18); this exact agreement stems from the fact that for one-exciton states the commutation relations are irrelevant, in other words the kinematic interactions do not play a role. As second example, consider the two-boson state $|\sigma,\tau\rangle = B_\sigma^\dagger B_\tau^\dagger |g\rangle = \sum_{n,m} \varphi_{\sigma n} \varphi_{\tau m} B_n^\dagger B_m^\dagger |g\rangle$. It is immediately clear that this will not yield the exact two-exciton state, as this expression inevitably gives contributions where a single molecule is doubly excited. As long as the number of excited bosons per molecule is low, however, these unphysical contributions will have a small amplitude and the generated two-boson state is expected to look very much like the actual two-exciton state.

It is to be noted that for many applications, it is not necessary to go back to the explicit state representation, because most observables may be expressed in terms of correlation functions of the exciton creation and annihilation operators. For instance, the linear absorption spectrum may be expressed as a two-time correlation function of the total dipole operator [42,43], which may be re-expressed in terms of two-point correlation functions of the creation and annihilation operators. If these operators are considered Bose operators and the Hamiltonian is quadratic, the evaluation of these correlators (Green functions) is simple. Nonlinear spectroscopies require multi-time correlation functions [43] (which will vanish if the excitons are treated as bosons and the Hamiltonian contains no dynamic interactions). At this level of treatment, one takes full advantage of the power of second quantization and the machinery of Green functions, as the transformation back to explicit states is totally avoided.

In the general situation of Eq. (22), where dynamic interactions are present, it is, apart from exceptional cases, not possible to diagonalize the Hamiltonian, even if one considers the excitons to be bosons. One then has to resort to approximate techniques known from general solid state theory to treat particle–particle interactions. The most familiar technique is the mean-field approximation, of which the well-known local-field approximation of optics is an example. In this approximation, one factorizes the quartic term into forms like $\langle B_\sigma^\dagger B_{\sigma'} \rangle B_\tau^\dagger B_{\tau'}$, where the brackets denote the expectation value (other factorization contributions are possible). The Hamiltonian is now quadratic again and may always be diagonalized. As it contains some operator expectation value as parameter, the solution of the ground and excited states gives rise to a self-consistency condition containing this expectation value. This condition may be nonlinear, leading to nontrivial solutions, in which, e.g., the ground state expectation value $\langle B_\sigma^\dagger B_{\sigma'} \rangle$ is nonzero, i.e., a finite density of excitons has condensed into the ground state. This may happen only if the relevant coupling constant (the dynamic interaction matrix element) is large enough, which explains why in the introduction of Section 2.4.2 we stressed that the typical $U_{nm}$ should be sufficiently small in order for the ground state to contain no excitations. We finally note that at the level of Green functions, similar results are obtained by factorizing, for instance, a four-point correlator into two two-point ones.

Thus, the treatment of Hamiltonians like Eq. (22) is a well-known problem in solid-state physics as long as the B operators are considered Bose operators (or fermions). As we have seen, however, the approximation of excitons by bosons leads to unphysical contributions which, no matter how small, make the ad hoc approximation hard to control. As was shown by Agranovich and Toschich [44], however, a rigorous transformation exists from the Pauli operators $B_n^\dagger$ to Bose operators $\widetilde{B}_n^\dagger$. This transformation, which has also been used for investigating magnetic systems [45], can be written in various forms, the most compact of which reads:

$$B_n^\dagger = \widetilde{B}_n^\dagger \left( \sum_\nu a_\nu \widetilde{B}_n^{\dagger \nu} \widetilde{B}_n^\nu \right)^{1/2}, \tag{24}$$

where $a_\nu = (-2)^\nu/(1+\nu)!$. Invoking this transformation and its hermitian conjugate into Eq. (22), gives to lowest order in powers of $\widetilde{B}_n^{\dagger \nu} \widetilde{B}_n^\nu$ exactly the same Hamiltonian with the Pauli operators replaced by Bose operators, which is the ad hoc Bose approximation considered above. The advantage of the rigorous transformation, however, is that corrections terms to this approximation can be generated in a systematic way. For instance, the first, quadratic, term in Eq. (22) will give rise to quartic correction terms in the Hamiltonian for the bosons, which reflect the kinematic interaction at the two-particle level. It has been shown that this correction may be used to rigorously justify the hard-core boson approach for dealing with two-exciton states in the third-order nonlinear optical response functions [34]. At the same time, both the quadratic and quartic terms in Eq. (22) generate 6-boson interactions, 8-boson interactions, etc. In practice, these many-boson interactions considerably complicate working with the transformation equation (24); like any many-particle term, they can only be treated approximately, for instance in a mean-field approximation. Thus, while the transformation is formally exact, it unfortunately does not allow for exact solutions of the multi-exciton states or Green functions.[1]

As mentioned already, for one-dimensional systems an alternative to bosonization is to transform to fermions by using the Jordan–Wigner approximation. We will return to this in Section 7.

### 2.4.4. Beyond the Heitler–London Approximation

Thus far, we have only considered terms in the Hamiltonian that conserve the number of excitations. The justification of this approximation has been discussed in Section 2.2. Nevertheless, for systems of two-level molecules one type of contribution that does not obey this conservation rule is often included in the Hamiltonian. This is the contribution deriving from the two-molecule matrix elements $\langle e_n e_m | \widehat{V}_{nm} | g_n g_m \rangle = \langle g_n g_m | \widehat{V}_{nm} | e_n e_m \rangle = J_{nm}$. The first equality holds because the molecular wave functions are assumed to be real, while the equality to the transfer matrix equation (16) element holds strictly in the absence of charge transfer between molecules $m$ and $n$. Neglecting this type of matrix elements, as we have done so far, is referred to as the Heitler–London approximation. Relaxing this approximation, the Hamiltonian in second quantization reads:

$$\widehat{H}_{\text{ex}} = \sum_n (\hbar\omega_0 + D_n) B_n^\dagger B_n + \frac{1}{2} \sum_{n,m} J_{nm} \left( B_n^\dagger + B_n \right) \left( B_m^\dagger + B_m \right)$$

$$+ \frac{1}{2} \sum_{n,m} U_{nm} B_n^\dagger B_m^\dagger B_n B_m. \tag{25}$$

The new terms describe the double excitation or de-excitation of a pair of molecules caused by their Coulomb interaction (Figure 1(b)). As is seen, incorporating these extra terms can be done in a very compact way, due to the fact that their coupling constants are identical to those for the transfer interactions. This

---

[1] A generalization of the Agranovich–Toschich transformation and a constraint free bosonic representation for a system of truncated oscillators was given by A.V. Ilinskaia and K.N. Ilinski, *J. Phys. A* 29 (1996) L23.

does not mean, however, that their effect is equally large. While the transfer interactions couple states that are degenerate, the new terms couple states that differ by the order of twice an optical transition energy, $2\hbar\omega_0$, as they mix states that differ by two excitation quanta. For most molecular systems $\hbar\omega_0/J_{nm}$ is small (order 0.01–0.1) and the mixing of the multi-exciton bands may be treated perturbatively. Yet, it should be realized that with these extra interactions, the ground state has a finite density of molecular excitations, due to the fact that two-exciton states, four-exciton states, etc. mix into the ground state. This renders finding the eigenstates, even if we neglect dynamic exciton–exciton interactions, a very hard task. Only for one-dimensional systems without dynamic exciton–exciton interactions and with transfer interactions of the nearest-neighbor form, have exact results been obtained for ground state energies and first- and third-order optical response functions [46]. This was done by using techniques borrowed from spin chains, particularly the Jordan–Wigner transformation from paulions to fermions, followed by a Bogoliubov–Tyablikov transformation. The latter is a canonical transformation that mixes creation and annihilation operators,

$$a_n = \sum_m \left( u_{nm} B_m + v_{nm} B_m^\dagger \right), \tag{26}$$

and which is such that double-annihilation ($a_n a_m$) and double-excitation ($a_n^\dagger a_m^\dagger$) terms do not occur in the Hamiltonian after the transformation. The reader is referred to the literature for details [30].

In higher-dimensional systems, exact results when relaxing the Heitler–London approximation are not known. The generic approach used in the literature is to apply the Bose approximation, which, in case dynamic exciton–exciton interactions are neglected, gives rise to a Hamiltonian that is quadratic in the Bose operators. This may always be diagonalized by using a Bogoliubov–Tyablikov transformation [27,28,35,36]. For a low density of excitations (in particular in the limit where the optical response is linear), the Bose approximation is expected to be a good approximation (cf. Section 2.4.3). We will give some explicit results in Section 2.5.2. At the same time, it should be realized that the Bose approximation does not give exact expressions for the exciton eigenstates, not even those with a few excitation quanta [46].

To end this section, we make two remarks. The first calls for caution: in restricting ourselves to one molecular transition (two-level molecule), we have neglected off-resonant contributions in the Hamiltonian. By relaxing the Heitler–London approximation, we do incorporate certain off-resonant contributions, namely many-particle ones. This may be considered inconsistent. The interest in relaxing the Heitler–London approximation in two-level molecules therefore must generally be seen in the light of calculating typical corrections due to off-resonant terms. Only if the transition dipole of the selected molecular transition by far exceeds all other molecular transition dipoles, should one expect the results to have quantitative significance as well.

Finally, the physical meaning of including double (de-)excitation terms is worth mentioning. Let is assume that $J_{nm}$ is given by the interaction between the transition dipoles of molecules $m$ and $n$ (cf. Eq. (6)), then the correction to the ground state energy of a pair of molecules derives from a virtual double excitation of the pair from its ground state due to the dipolar interactions, followed by its falling back to the ground state. This is the origin of the Van der Waals interaction. Indeed, in second-order perturbation theory, the correction to the ground state energy thus obtained is $-J_{nm}^2/2\hbar\omega_0$, which has the correct attractive and $1/r_{nm}^6$ nature and is also seen to scale as $\mu^2/\hbar\omega_0$, i.e., as the molecular polarizability.

## 2.5. Frenkel Excitons in a Bulk Crystal

### 2.5.1. General Hamiltonian and One-Exciton States

We now turn to the special case of excitons in a molecular crystal, occupying an arbitrary Bravais lattice. Each unit cell of the crystal contains $S$ molecules. In practice, these molecules often are chemically identical and only differ in their orientations within the crystal. This is for instance the case for aromatic crystals, such as napthalene and anthracene. We will focus on linear optics and one-exciton states, and therefore neglect exciton–exciton interactions. The Hamiltonian equation (14) then reduces to

$$\widehat{H}_{\text{ex}} = \sum_{\mathbf{n},s}(\hbar\omega_s + D_s)B_{\mathbf{n}s}^\dagger B_{\mathbf{n}s} + \sum_{\mathbf{n},\mathbf{m},s,s'} J_{ss'}(\mathbf{n}-\mathbf{m})B_{\mathbf{n}s}^\dagger B_{\mathbf{m}s'}. \tag{27}$$

Here $\mathbf{n}$ ($\mathbf{m}$) labels the unit cell by giving its position, while $s, s' = 1, \ldots, S$ label the molecules within each unit cell. Molecules with the same label $s$ are translationally equivalent. We have used the lattice symmetry by setting the transition frequencies and dispersion shifts equal for all translationally equivalent molecules and by recognizing that the transfer interactions only depend on the separation $\mathbf{n} - \mathbf{m}$ of the two unit cells involved. Similarly, the molecular transition dipole only depends on the label $s$:

$$\boldsymbol{\mu}_{\mathbf{n}s} = \boldsymbol{\mu}_s.$$

Diagonalizing the Hamiltonian, i.e., bringing it in the form of Eq. (23), is greatly simplified by the lattice symmetry. We first make the transformation to the intermediate set of operators

$$B_{\mathbf{k}s}^\dagger = N^{-1/2} \sum_{\mathbf{n}} \exp(i\mathbf{k}\cdot\mathbf{n})B_{\mathbf{n}s}^\dagger, \tag{28}$$

with $N$ the number of unit cells in the quantization volume on which periodic boundary conditions are imposed and $\mathbf{k}$ one of the $N$ allowed wave vectors in the first Brillouin zone. This unitary transformation brings the Hamiltonian in the form

$$\widehat{H}_{\text{ex}} = \sum_{\mathbf{k},s,s'} H_{ss'}(\mathbf{k}) B_{\mathbf{k}s}^\dagger B_{\mathbf{k}s'}, \tag{29}$$

with

$$H_{ss'}(\mathbf{k}) = (\hbar\omega_s + D_s)\delta_{ss'} + \sum_{\mathbf{n}}{}' J_{ss'}(\mathbf{n}) \exp(-i\mathbf{k}\cdot\mathbf{n}). \tag{30}$$

The last term will also be denoted $J_{ss'}(\mathbf{k})$, and the prime on its summation excludes the situation with $\mathbf{n} = \mathbf{0}$ and simultaneously $s = s'$. The resulting Hamiltonian equation (29) is already diagonal in $\mathbf{k}$ space. Getting a totally diagonal form only requires the $S \times S$ matrix $H_{ss'}(\mathbf{k})$ to be diagonalized. Denoting the $s$th component of the $\alpha$th normalized eigenvector of this matrix by $\phi_{\alpha s}(\mathbf{k})$ and the corresponding eigenvector by $E_{\mathbf{k}\alpha}$ we arrive at

$$\widehat{H}_{\text{ex}} = \sum_{\mathbf{k},\alpha} E_{\mathbf{k}\alpha} B_{\mathbf{k}\alpha}^\dagger B_{\mathbf{k}\alpha}, \tag{31}$$

with

$$B_{\mathbf{k}\alpha}^\dagger = \sum_s \phi_{\alpha s}(\mathbf{k}) B_{\mathbf{k}s}^\dagger = N^{-1/2} \sum_{\mathbf{n},s} \phi_{\alpha s}(\mathbf{k}) \exp(i\mathbf{k}\cdot\mathbf{n}) B_{\mathbf{n}s}^\dagger. \tag{32}$$

When operating on the ground state of the crystal, the operator $B_{\mathbf{k}\alpha}^\dagger$ creates a one-exciton eigenstate with amplitude on the molecule $(\mathbf{n}, s)$ given by $N^{-1/2}\phi_{\alpha s} \exp(i\mathbf{k}\cdot\mathbf{n})$ and with energy $E_{\mathbf{k}\alpha}$. The eigenenergies lie on $S$ different dispersion curves, labeled by $\alpha$ and known as exciton or Davydov bands. In the next two sections, we will discuss explicitly the dispersion curves for the cases of one and two molecules per unit cell. If the molecular operators $B_{\mathbf{n}s}^\dagger$ are approximated by Bose operators, the exciton operators $B_{\mathbf{k}\alpha}^\dagger$ have Bose commutation relations as well. For one-exciton states, this is exact (within the Heitler–London approximation).

### 2.5.2. One Molecule per Unit Cell

In case we have only one molecule per unit cell (frequency $\omega_0$), the labels $s$ and $\alpha$ are suppressed and the transformation equation (28) already diagonalizes the one-exciton subspace. The only problem in determining the exciton band structure then lies in performing the lattice sum of the dipole–dipole interaction occurring in Eq. (30), $J(\mathbf{k}) = \sum_{\mathbf{n}\neq\mathbf{0}} J(\mathbf{n}) \exp(-i\mathbf{k}\cdot\mathbf{n})$. For three-dimensional lattices, this is not a simple task, due to the long-range nature of the interaction. This holds in particular for small wave vectors $\mathbf{k}$, which is the region that is of most interest for the optical absorption (see Section 2.6). A large body of literature consists

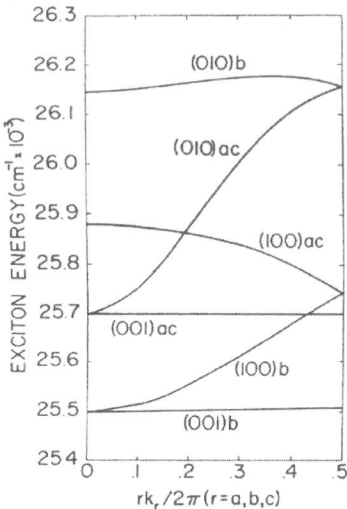

Fig. 5. Dispersion curves for the two exciton branches of the 0-0 component of the 3800 Å transition in anthracene crystals for **k** normal to the faces (001), (010), and (100). (Figure reprinted from Ref. [51] with permission from the American Institute of Physics.)

concerning this problem [47–50], the overview of which lies outside the scope of this chapter. An important property of these dipole sums is their nonanalytical dependence of $J(\mathbf{k})$ on $\mathbf{k}$ at small $|\mathbf{k}|$. In particular, $J(\mathbf{k})$ then strongly depends on the direction of $\mathbf{k}$ and only weakly on $|\mathbf{k}|$ (see Figure 5). Once the result for the summation is known, the energy of the exciton with wave vector $\mathbf{k}$ is simply given by $E_\mathbf{k} = \omega_0 + D + J(\mathbf{k})$.

Before turning our attention to the case of two molecules per unit cell, we briefly comment on generalizing the dispersion relation to include the non-Heitler–London terms $\sum_\mathbf{n} J(\mathbf{n})(B_\mathbf{n}^\dagger B_\mathbf{n}^\dagger + B_\mathbf{n} B_\mathbf{n})$ of the dipole–dipole interactions. Within the Bose approximation, this is easy to do by replacing the transformation equation (28) by a Bogoliubov–Tyablikov transformation that allows for the mixing of creation and annihilation operators. Requiring this transformation to be canonical and to bring the Hamiltonian in diagonal form, one arrives at the dispersion relation [28,35]

$$E_\mathbf{k} = \left[\left(\hbar\omega_0 + D + J(\mathbf{k})\right)^2 - J^2(\mathbf{k})\right]^{1/2}. \qquad (33)$$

If $J(\mathbf{k}) \ll \omega_0$, as is typically the case, the correction relative to the Heitler–London approximation is of the order $J^2(\mathbf{k})/(\omega_0 + D)$, as we already anticipated in Section 2.4.4. We stress again that Eq. (33) is not an exact result, as it relies on the Bose approximation. Moreover, the exciton amplitude on each molecule is, even in this approximation, not easy to calculate, as the ground state no longer is the state with all molecules of the crystal in their ground state.

### 2.5.3. Two Molecules per Unit Cell: Davydov Splitting

We consider a crystal with two molecules per unit cell. We will assume that they are chemically identical, implying that they have identical transition frequencies $\omega_0$ and dispersion shifts $D$. The transition dipoles are given by $\boldsymbol{\mu}_1$ and $\boldsymbol{\mu}_2$, for the two types of molecules $s = 1$ and 2, respectively. They only differ in orientation and have equal magnitude $\mu$. As mentioned earlier, this simple situation is relevant to aromatic crystals. To attain maximal simplicity, we will assume that the symmetry of the situation is such that $J_{1,1}(\mathbf{k}) = J_{2,2}(\mathbf{k})$ and that $J_{1,2}(\mathbf{k}) = J_{2,1}(\mathbf{k})$. For the aromatic crystals, this only holds for certain important directions of $\mathbf{k}$, like those perpendicular or parallel to the plane of symmetry of the crystal [35, p. 43]. The diagonalization of the two-dimensional matrix $H_{ss'}(\mathbf{k})$ is easily performed in this case and leads to

$$E_{\mathbf{k},\pm} = \hbar\omega_0 + D + J_{1,1}(\mathbf{k}) \pm |J_{1,2}(\mathbf{k})|, \tag{34}$$

with eigenvector components $\phi_{\pm,s} = (\pm 1)^{s+1}/\sqrt{2}$. Here the band label $\alpha$ has been replaced by $+$ and $-$. Clearly, the interaction $J_{1,2}(\mathbf{k})$ between the two types of molecules causes the exciton band of the individual sublattices for $s = 1$ and $s = 2$ to mix and split by the amount $2|J_{1,2}(\mathbf{k})|$. The mixing is very strong, as the bands for the sublattices have been taken degenerate. As we will see in Section 2.6, the absorption spectrum of the exciton system peaks at $\hbar\omega = E_{\mathbf{k}\approx 0,\alpha}$. This implies that the absorption spectrum of the system described here has two absorption bands, that are separated by $2|J_{1,2}(\mathbf{k} \approx \mathbf{0})|$. This splitting is referred to as the Davydov splitting. In the $+$ band, the molecules within each unit cell are excited exactly in phase, a collective state that is associated with a dipole per unit cell given by $(\boldsymbol{\mu}_1 + \boldsymbol{\mu}_2)/\sqrt{2}$. In the $-$ state, the molecules are excited in antiphase and the associated dipole per unit cell is $(\boldsymbol{\mu}_1 - \boldsymbol{\mu}_2)/\sqrt{2}$. Note that these two linear combinations are oriented perpendicular to each other, which implies that the two Davydov components in the absorption spectrum of aromatic type crystals have mutually perpendicular polarization. We draw attention to the fact that the above results are completely analogous to what we have seen for a simple dimer of molecules in Section 2.2. This is a natural consequence of the fact that after the transformation equation (28) to $\mathbf{k}$ space, we have in fact a dimer ($2 \times 2$ problem) for each wave vector.

### 2.5.4. Multi-Level Molecules

Thus far, we have limited ourselves to the inclusion of only one excited state per molecule. If various molecular excited states have similar energies or if one considers nonlinear optical response in a situation where the multiple of a certain transition frequency is close to one of the other transition frequencies, this restriction is not justified.

The above microscopic theory can be generalized to allow for more molecular levels. The new effect that will occur, is mixing of molecular configurations: due to the excitation transfer interaction between a level $|f\rangle$ of molecule $(\mathbf{n}, s)$ and level $|e\rangle$ of molecule $(\mathbf{m}, s')$, a mixing of the exciton bands associated with the individual molecular levels occurs [52,28]. This mixing will only be important if the corresponding transfer interaction is of the same order as or larger than the energy difference between the molecular states considered. Mixing gives rise to transfer of oscillator strength from one state to the other. Examples where the mixing plays an important role are the weak transition at 3200 Å in napthalene [53], the polarization ratios in the vibronic components of the $p$ band in anthracene [54], and the Davydov splittings in various aromatic crystals [55]. The formal set-up of the theory in second quantization, as well as consequences for the oscillator strength and the dielectric tensor, have been derived by Agranovich [28]. For the special situation of three-level molecules, several explicit results have been obtained within the context of Frenkel excitons in molecular aggregates. Thus, Knoester and Spano [56] have studied the interference of various molecular transitions in the nonlinear optical response of one-dimensional aggregates. Mukamel and co-workers [57] have reformulated the theory of excitons in three-level molecules in terms of the scattering of excitons on anharmonic potentials, which account for the energy difference between the molecular transitions considered. Juzeliūnas and Reineker [58] have calculated nonlinear spectra for a continuous density of higher excited states. Here, these results will not be described in more detail. We will come back to some important consequences of configuration mixing in the context of the dielectric theory described in Section 3.

### 2.6. Dielectric Response of Molecular Crystals

In this section, we will address the microscopic calculation of the linear dielectric response of a molecular crystal. This may be done by calculating the crystal's electric susceptibility and (or) dielectric tensor. Consider an electromagnetic wave in the crystal, with macroscopic (Maxwell) electric field given by

$$\mathbf{E}(\mathbf{n}, t) = \mathbf{E}(\mathbf{k}, \omega) \exp[i\mathbf{k} \cdot \mathbf{n} - i\omega t] + \text{c.c.}, \qquad (35)$$

where $\mathbf{k}$ and $\omega$ are the wave vector and the angular frequency of the wave, respectively, and c.c. denotes the complex conjugate. We used the position of the unit cell to indicate the position in space. The error that we make in doing so is very small, as we are typically interested in optical fields, which vary over distances of hundreds to thousands of lattice constants, allowing us to discard variations of the field inside crystal unit cells.

The electric field polarizes the crystal. As long as the response is linear, the polarization field (dipole per unit volume) takes the form

$$\mathbf{P}(\mathbf{n}, t) = \mathbf{P}(\mathbf{k}, \omega) \exp[i\mathbf{k} \cdot \mathbf{n} - i\omega t] + \text{c.c.} \tag{36}$$

The electric susceptibility tensor, $\chi(\mathbf{k}, \omega)$, of the crystal is now defined through

$$\mathbf{P}(\mathbf{k}, \omega) = \chi(\mathbf{k}, \omega) \cdot \mathbf{E}(\mathbf{k}, \omega). \tag{37}$$

Similarly, the dielectric tensor is defined through

$$\mathbf{D}(\mathbf{k}, \omega) = \varepsilon(\mathbf{k}, \omega) \cdot \mathbf{E}(\mathbf{k}, \omega), \tag{38}$$

where $\mathbf{D} = \mathbf{E} + 4\pi\mathbf{P}$ is the displacement field in the crystal. Thus, $\varepsilon(\mathbf{k}, \omega) = 1 + 4\pi\chi(\mathbf{k}, \omega)$.

In principle, the dielectric tensor contains all information about the linear optical response of the crystal. Its real part is related to the index of refraction, while the absorption spectrum is basically proportional to its imaginary part (see Section 3.2).

Above, we related the polarization and the displacement to the total Maxwell electric field. Alternatively, one may choose to relate these quantities to only the transverse part of this field, which in Fourier space is given by $\mathbf{E}^\perp(\mathbf{k}, \omega) = \mathbf{E}(\mathbf{k}, \omega) - \mathbf{E}^\|(\mathbf{k}, \omega) = (1 - \hat{\mathbf{k}}\hat{\mathbf{k}}) \cdot \mathbf{E}(\mathbf{k}, \omega)$, where $\hat{\mathbf{k}}$ is the unit vector in the direction of $\mathbf{k}$. In isotropic crystals $\mathbf{E}^\perp$ is usually associated with electromagnetic waves, while the longitudinal part, $\mathbf{E}^\|$, is associated with intermolecular Coulomb interactions. One now defines the transverse susceptibility through $\mathbf{P}(\mathbf{k}, \omega) = \chi^\perp(\mathbf{k}, \omega) \cdot \mathbf{E}^\perp(\mathbf{k}, \omega)$ and the transverse dielectric function through $\mathbf{D}(\mathbf{k}, \omega) = \varepsilon^\perp(\mathbf{k}, \omega) \cdot \mathbf{E}^\perp(\mathbf{k}, \omega)$. Again, the latter may be expressed in terms of the former. To this end, we should realize that the displacement in a system without external charges is a transverse field ($\nabla \cdot \mathbf{D} = 0$), so that $\mathbf{D} = \mathbf{D}^\perp = \mathbf{E}^\perp + 4\pi\mathbf{P}^\perp$. Using this, one arrives at $\varepsilon^\perp(\mathbf{k}, \omega) = 1 + 4\pi\chi^\perp(\mathbf{k}, \omega) - 4\pi\hat{\mathbf{k}}\hat{\mathbf{k}} \cdot \chi^\perp(\mathbf{k}, \omega)$. In this section, we will calculate the transverse susceptibility, which has as advantage that the local-field problem need not be addressed. We come back to the local field in Section 3.1.

To calculate the electric susceptibility from microscopic principles, we need the interaction $\widehat{H}_{\text{int}}$ between the electromagnetic field and the crystal. One may then use standard expressions from linear response theory (time-dependent first-order perturbation theory in $\widehat{H}_{\text{int}}$), which give the susceptibility $\chi(\mathbf{k}, \omega)$ in terms of sums over crystal eigenstates (excitons). In the Heitler–London approximation, only the one-exciton states can be reached using one interaction with the fields, so that only these states need to be calculated. The sum-over-states approach is carried out in the Schrödinger picture. An alternative method, which is very direct and elegant for our case, works via the Heisenberg picture. We will follow this approach. First, we define the interaction

$$\widehat{H}_{int} = -\sum_{\mathbf{n},s} \boldsymbol{\mu}_s \cdot \mathbf{E}^{\perp}(\mathbf{n},t)\left(B_{\mathbf{n}s}^{\dagger} + B_{\mathbf{n}s}\right), \qquad (39)$$

which is of the so-called multipolar form [59,60] and just sums the potential energies of the individual dipoles in the electric field. This form may be used for molecular systems, in which the electrons can be associated with a particular molecule. We have discarded the interaction with the possible permanent dipoles of the molecules, as these do not give rise to a response at optical frequencies.

We now need to calculate the polarization field in the presence of the electric field. Its definition as the dipole moment per unit volume allows us to write it in terms of the expectation value for the molecular dipole operators

$$\mathbf{P}(\mathbf{n},t) = \frac{1}{v_c}\sum_s \boldsymbol{\mu}_s\left(\langle B_{\mathbf{n}s}^{\dagger}(t)\rangle + \langle B_{\mathbf{n}s}(t)\rangle\right), \qquad (40)$$

where $v_c$ is the volume of a unit cell. Using the inverse of the transformation equation (32), we also have

$$\mathbf{P}(\mathbf{n},t) = \frac{1}{v_c\sqrt{N}}\sum_{\mathbf{k},s,\alpha} \boldsymbol{\mu}_s \phi_{\alpha s}^*(\mathbf{k}) \exp[-i\mathbf{k}\cdot\mathbf{n}]\langle B_{\mathbf{k}\alpha}^{\dagger}(t)\rangle + \text{c.c.} \qquad (41)$$

The expectation values $\langle B_{\mathbf{k}\alpha}^{\dagger}(t)\rangle$ may be calculated by first considering the Heisenberg equation of motion for the $B_{\mathbf{k}\alpha}^{\dagger}(t)$,

$$\frac{d}{dt}B_{\mathbf{k}\alpha}^{\dagger}(t) = \frac{i}{\hbar}\left[\widehat{H}_{ex} + \widehat{H}_{int}(t), B_{\mathbf{k}\alpha}^{\dagger}(t)\right]. \qquad (42)$$

If we treat the excitons as bosons, we have $[B_{\mathbf{k}\alpha}(t), B_{\mathbf{k}'\beta}^{\dagger}(t)] = \delta_{\mathbf{k}\mathbf{k}'}\delta_{\alpha\beta}$ and all other commutators vanish. Using this, Eqs. (31) and (39), we arrive at

$$\frac{d}{dt}B_{\mathbf{k}\alpha}^{\dagger}(t) = \frac{i}{\hbar}E_{\mathbf{k}\alpha}B_{\mathbf{k}\alpha}^{\dagger}(t) - \sum_s \sqrt{N}\boldsymbol{\mu}_s \cdot \mathbf{E}^*(\mathbf{k},\omega)\phi_{\alpha s}(\mathbf{k})e^{i\omega t}$$

$$- \sum_s \sqrt{N}\boldsymbol{\mu}_s \cdot \mathbf{E}(\mathbf{k},\omega)\phi_{\alpha s}(-\mathbf{k})e^{-i\omega t}. \qquad (43)$$

This is a linear equation, owing to the fact that we made the Bose approximation. By taking expectation values to the left and the right, we find a linear equation for $\langle B_{\mathbf{k}\alpha}^{\dagger}(t)\rangle$ in terms of the driving field. The first driving term in this equation ($\sim \exp[i\omega t]$) may result in a resonance, because the eigenfrequency $E_{\mathbf{k}\alpha}$ is positive. The second driving term is associated with a negative frequency and is so far off-resonant that we may safely neglect it. This is referred to as the rotating wave approximation [26]. After this step, straightforward solution for $\langle B_{\mathbf{k}\alpha}^{\dagger}(t)\rangle$, substitution into Eq. (41), and use of Eq. (37) gives

$$\chi^\perp(\mathbf{k}, \omega) = -\frac{1}{v_c} \sum_\alpha \frac{\mu_\alpha(\mathbf{k})\mu_\alpha^*(\mathbf{k})}{\hbar\omega - E_{\mathbf{k}\alpha} + i\hbar\gamma}, \tag{44}$$

where $\gamma$ is a small positive damping constant (which may be thought of to reflect the finite lifetime of the excitons), $\mu_\alpha(\mathbf{k}) \equiv \sum_s \mu_s \phi_{\alpha s}(\mathbf{k})$, and the vector product in the denominator is a dyadic product.

This result is identical to the one that we would have obtained by using the sum-over-states expression of standard linear response theory, independent of the Bose approximation. The Bose approximation is exact in this case, because we work in the Heitler–London approximation, where the ground state is the real vacuum state, without excitons. A similar approach may be used to account for the non-Heitler–London terms (though the Bose approximation is then no longer exact) and for more molecular levels [27,35]. We will return to some aspects of these problems in Section 3, in the context of the dielectric theory of excitons. We will there calculate the dielectric function of the crystal by starting from the linear polarizability of a single molecule and taking into account the local fields that the dipoles of the molecules in a crystal exert on each other. For linear optical response that approach is exact, as long as we work in the Heitler–London approximation.

To end this section we make a few observations concerning Eq. (44). First, as expected, we see that the transverse susceptibility has resonances at the frequencies of the excitons with the same wave vector as the exciting field. If we have $S$ different exciton bands, this implies $S$ different resonances, which will be reflected as $S$ peaks in the absorption spectrum of the crystal. The fact that the resonance frequencies are $\mathbf{k}$-dependent is referred to as spatial dispersion. It should be realized that the structure of the dispersion curves $E_{\mathbf{k}\alpha}$ occurs on the scale of $|\mathbf{k}| \sim (\text{Å})^{-1}$, which is much larger than the scale of an optical wave vector. It is for this reason that in practice one may often neglect the spatial dispersion and simply take the exciton frequencies $E_{\mathbf{k}=0,\alpha}$ as the resonances. This is referred to as the $\mathbf{k} = 0$ selection rule, which we used when introducing the Davydov splitting in Section 2.5.3. We note that accounting for spatial dispersion in the wave propagation, reflection, and absorption of crystals in general is a complicated problem, whose careful discussion requires more space than is available here [61].

## 2.7. Exciton-Polaritons

The translational symmetry of a perfect bulk crystal implies that excitons of wave vector $\mathbf{k}$ can only couple to electromagnetic waves of the same wave vector. This is the reason why the field equation (35) in linear response only gives rise to polarization fields with the same wave vector. Rigorously speaking, an exciton can couple to all field components whose wave vectors differ from the exciton's by a

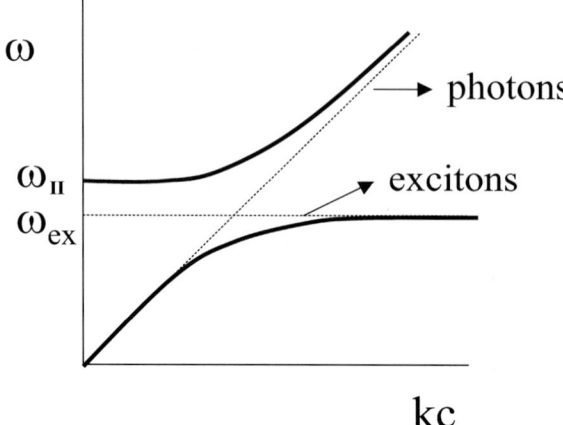

Fig. 6. Schematic dispersion diagram of exciton-polaritons in bulk crystals. The spatial dispersion of the excitons has been neglected, as has been the change of the background index of refraction when going from low to high frequencies. The frequency region between $\omega_{ex}$ and $\omega_\|$ is the stopgap, in which no electromagnetic waves propagate in the crystal. In reality, the slope of the photon-like exciton-polaritons at small $k$ is smaller than at high $k$, due to the larger value of the background index of refraction at small frequencies.

reciprocal lattice vector, but these additional field components have frequencies in the X-ray region and beyond, which makes it energetically impossible to couple efficiently. Hence, these Umklapp processes are usually neglected.

Let us now consider the situation where we have one exciton branch, in interaction with the electromagnetic field in the crystal. Quantum mechanically, the field may be considered a collection of harmonic oscillators, with a dispersion relation $\omega_\mathbf{k} = |\mathbf{k}|c$, with $c$ the velocity of light. If we apply the Bose approximation, also the excitons may be considered harmonic oscillators, with dispersion relation $E_\mathbf{k}$. Both dispersion diagrams are depicted in Figure 6, where we neglected the spatial dispersion of the excitons $E_\mathbf{k} = \hbar\omega_{ex} = \text{const}$. The coupling of the electromagnetic field and the excitons may thus be viewed as the coupling of two harmonic oscillators at each wave vector $\mathbf{k}$. This leads to two new harmonic oscillators that are mixed exciton–photon states, or in a more classical language, mixed polarization-radiation field waves. These quasi-particles are known as exciton-polaritons, in analogy to the polaritons that arise as the mixed states of optical phonons and the electromagnetic field in ionic crystals [47,62]. The microscopic theory of exciton-polaritons has been formulated independently by Hopfield [63] and Agranovich [27], and involves the mixing of exciton creation and annihilation operators and photon creation and annihilation operators into new polariton operators.

The coupling of excitons and photons leads to an avoided band crossing in the dispersion diagram, giving rise to two polariton branches (Figure 6). On the lower one, the polaritons behave like photons at small wave vector and gradually become more exciton like with increasing wave vector. For the upper branch, the situation is reversed. At the crossing point $\mathbf{k} = \omega_{\text{ex}}/c$, the polaritons have equal amount of exciton and photon character and they are then quasi-particles with a very high group velocity (in the order of $c$). The separation between both branches at the crossing point is determined by the coupling between the excitons and the radiation field, which occurs through the dipoles of the individual molecules. More precisely, the resonant splitting is proportional to the density of oscillator strength in the crystal [64]. It should also be kept in mind that the coupling is only sensitive to the projection of the dipole on the direction of the electric field. This makes the resonant splitting dependent on the direction of propagation of the polaritons through the crystal. For atomic crystals, where the excited state is three-fold degenerate with polarization in the three lattice directions, the situation is much more symmetric, though even there the crystal structure breaks the spherical symmetry.

Probably the most characteristic property of the polariton dispersion diagram is the polariton stopgap. The splitting between upper and lower branches leads to a finite splitting between the frequency $\omega_\|$ of the upper branch at $\mathbf{k} = \mathbf{0}$ and $\omega_{\text{ex}}$. As a consequence, no polariton modes exist in the interval $\omega_{\text{ex}} < \omega < \omega_\|$. As polaritons are just the quanta of electromagnetic waves dressed with the polarization field of the crystals, this implies that in this frequency region no electromagnetic waves can propagate in the crystal. Light with a frequency within this region that is incident on the crystal will be totally reflected. This explains the name "stopgap". The stopgap is indeed observed in reflection experiments [65]. In practice the reflection is not 100% throughout the stopgap region, because (i) the exciton dispersion curves have a finite width and (ii) spatial dispersion, causing the exciton dispersion to curve upward for large wave vectors, may close the stopgap.

In classical dielectric theory, the stopgap region corresponds to the frequency interval in which the dielectric function is negative. This makes the wave number $|\mathbf{k}| = \omega\sqrt{\varepsilon^\perp(\omega)}/c$ of the associated plane wave complex, leading to a damped wave that carries no energy into the crystal. In fact, the dispersion diagram of exciton-polaritons is easily obtained in a classical way by setting

$$|\mathbf{k}|^2 c^2 = \omega^2 \varepsilon^\perp(\omega) \qquad (45)$$

and using for $\varepsilon^\perp(\omega)$ the generic form found in the previous section, i.e., with a pole at the exciton frequency.

Obviously, $\omega_{\text{ex}}$, the lower boundary of the stopgap, is the frequency at which the dielectric function has a singularity. On the other hand, the upper boundary of the stopgap, $\omega_\|$, gives the frequency at which the dielectric function vanishes, $\varepsilon(\omega_\|) = 0$. It can be shown quite generally that this condition defines longitudinal waves [61].

It is noteworthy that the occurrence of stopgaps has recently resurfaced in the context of photonic crystals, where stopgaps may actually be created in the transparency region of a crystal that exists of a three-dimensional packing of dielectric units (for instance spheres) with a periodicity of the order of the wavelength of visible light. In these materials, the Umklapp processes may not be neglected and the multiple scattering involving the radiation field in various Brillouin zones gives rise to the opening of a stopgap [66].

An other interesting aspect of polaritons in infinite crystals is that they do not decay through spontaneous emission of photons, because they already fully account for the interaction between the excitons and the electromagnetic field. In large but finite crystals, polaritons decay spontaneously through emission at the crystal boundaries, in other words, the size of the crystal becomes an important factor in the spontaneous emission rate [67,68].

We finally note that, though the microscopic treatment of polaritons is based on perfect and infinite crystals, the classical definition of electromagnetic modes in a dielectric medium also allows for their introduction in quantum wells, on surfaces, and in amorphous media and liquids [69]. Their occurrence in nanostructures is discussed in the chapter by F. Bassani of this book.

### 2.8. Nonlinear Spectroscopy

In Section 2.6, we have considered the linear optical response of a bulk crystal. This is the response that arises when the intensity of the exciting fields is low enough to neglect situations where two or more excitations influence each other. Then the polarization field is linear in the amplitude of the electric field. Upon increasing the intensity of the light, nonlinear components may arise, and the polarization field may be expanded in terms of the electric field as [70]

$$\mathbf{P} = \chi^{(1)} \cdot \mathbf{E} + \chi^{(2)} : \mathbf{EE} + \chi^{(3)} \vdots \mathbf{EEE} + \cdots, \qquad (46)$$

where $\chi^{(n)}$ is the $n$th order optical susceptibility, which depends on the frequencies and wave vectors of the exciting fields. $\chi^{(n)}$ is a tensor of rank $n+1$. Familiar nonlinear optical phenomena are second harmonic generation and nonlinear absorption. In the former, an electric field of frequency $\omega$ generates a polarization of frequency $2\omega$. This second-order nonlinear effect is described by $\chi^{(2)}$ and may only occur in crystals without inversion symmetry. Nonlinear absorption is a third-order ($\chi^{(3)}$) effect and refers to the change of the absorption spectrum with increasing intensity, more specifically the change that is linear in the laser intensity. Nonlinear effects are interesting from a fundamental point of view, as they allow for a large variety of spectroscopic tools [71]. They are also of interest from a technological point of view, as they provide a means to manipulate light pulses by other light pulses (optical switching and computation [8]).

The calculation of the susceptibilities $\chi^{(n)}$ for the case of a general (multi-level) single molecule in an electric field follows standard time-dependent perturbation theory in the field–molecule interaction [70,71], also known as nonlinear response theory. This leads to sum-over-states expressions, where for the single molecule this sum runs over all molecular excited states. A well-known approach to calculate from these molecular susceptibilities the susceptibilities of the crystal, is the local-field method. In this approach one solves in a self-consistent way the polarization field in the system by allowing the individual molecular dipoles to respond to both the exciting external field and the local fields created by the surrounding dipoles [70]. This approach will be used in Section 3 to consider the linear optical response of crystals. As we mentioned in Section 2.6, the local-field approach is exact in the linear optical regime, provided we work in the Heitler–London approximation. In the nonlinear regime, however, this approach fails if multiple resonances with (multi-)exciton levels occur, i.e., when the collective excitations created by the intermolecular interactions are crucial [72–74].

In order to go beyond the local-field approximation, one should apply nonlinear response theory directly to the exciton system, i.e., use the (multi-)exciton states as basis to perform the time-dependent perturbation theory. Within the Heitler–London approximation the one-exciton states then suffice to calculate the linear response, while the one- and two-exciton states together suffice to calculate $\chi^{(2)}$ and $\chi^{(3)}$. As a consequence, to correctly describe the second- and third-order nonlinear optical response of crystals, exciton–exciton interactions should be taken into account. Here, two main types of interactions can be distinguished: dynamic and kinematic ones. Dynamic interactions are terms in the Hamiltonian in which a product of three or more exciton creation and (or) annihilation operators occurs. A simple example is the last term in Eq. (14). Even in the absence of dynamic interactions, the kinematic interactions resulting from the Pauli nature of the excitons gives rise to an intrinsic nonlinearity. Within the Heitler–London approximation it can be taken into account by making the transformation equation (24) from Pauli to Bose operators [44] and keeping the resulting effective interaction terms up to the order of interest. To describe third-order optical effects, we only need to keep terms of the form $B_n^\dagger B_n^\dagger B_n B_m$ and their hermitian conjugates [34]. These terms give rise to scattering of excitons on each other, just like the dynamic interactions do. The kinematic interactions have been studied in much detail in one-dimensional J-aggregates (Section 7), where they give rise to an energy shift of the two-exciton band-edge state relative to twice the one-exciton band-edge energy. This shift shows up as a dispersive feature in the nonlinear absorption spectrum and the pump-probe spectrum and contains information about the exciton delocalization size [38,41,75].

Dynamic exciton–exciton interactions of the form that appears in Eq. (14) are of much interest, as they may give rise to the formation of bound two-exciton states, so-called biexcitons. This only happens if the dynamic coupling $U_{nm}$ in Eq. (14) is large enough compared to the exciton band width [76,77]. Bi-excitons will show up in the nonlinear spectra as resonances at frequencies below (in the case of attraction) or above (for repulsion) twice the frequency of the linear absorption band [78–80]. Bi-exciton resonances are well-known in semiconductor crystals [40,81]. We note that in crystals with several molecules per unit cell, the kinematic interaction may also give rise to the formation of bi-excitons [82]. Finally, in charge-transfer crystals, so-called exciton strings may occur, in which a train of $n$ excitons forms a collective bound state [83].

Other types of dynamic interaction terms that are of interest for nonlinear optical response are exciton fusion and fission terms. Such terms arise from interactions involving more than one molecular transition. For instance, consider molecules with two relevant excited states, labeled $|e\rangle$ and $|f\rangle$, respectively. Then, intermolecular Coulomb interactions may give rise to terms of the form $B_{nf}^\dagger B_{me} B_{le}$, which takes away two excitations of type $|e\rangle$ and recreates one of type $|f\rangle$. Earlier we neglected such interactions, as they do not conserve the number of excitations. It is clear, however, that if the energy of state $|f\rangle$ is close to twice the energy of state $|e\rangle$, these exciton fusion terms may become important and have a noticeable effect on the nonlinear optical response. The complex conjugate of the above interaction term would be an exciton fission term in which one exciton splits up in two lower-energy ones. For one-dimensional J-aggregates, it has been shown that these type of terms give rise to three fundamentally different types of nonlinear absorption spectra, depending on the values of the parameters involved (frequency detuning between first and second molecular transition and the ratio of their transition dipoles) [56]. It should also be mentioned that exciton fusion terms are an important gateway in the process of exciton–exciton annihilation. The physics underlying this is that the higher exciton state $|f\rangle$ in practice often suffers from fast decay into the state $|e\rangle$ due to internal conversion, thereby effectively loosing one excitation quantum [84–86]. This energy loss process is an important limiting factor for the efficiency of opto-electronic devices [87]. A related loss process is the fission of the lowest singlet exciton in tetracene, which decays with a lifetime of $2 \cdot 10^{-10}$ s into two singlet excitons with an energy of 1.25 eV [88].

As an alternative to the calculation of the nonlinear susceptibilities through the sum-over-states expressions of standard nonlinear response theory, one may also apply equation of motion techniques, in analogy to the one we used to calculate the linear susceptibility in Section 2.6. In the case of nonlinear optics, one should explicitly include the quartic terms deriving from the kinematic and dynamic exciton–exciton interactions, as well as possible cubic exciton fusion and fission terms. In the right-hand side of Eq. (43), the quartic interactions give rise to additional contributions of the form $B_{\mathbf{k}\alpha}^\dagger B_{\mathbf{k}'\alpha'}^\dagger B_{\mathbf{k}''\alpha''}$. When taking the expectation value of the resulting equation to extract the equation of motion for $\langle B_{\mathbf{k}\alpha}^\dagger(t)\rangle$, these nonlinear terms should be factorized in a certain way to close the set of equations of motion. Various factorization schemes exist, for instance

$\langle B_{\mathbf{k}\alpha}^{\dagger} B_{\mathbf{k}'\alpha'}^{\dagger}\rangle \langle B_{\mathbf{k}''\alpha''}\rangle$ [89], which stresses the role of the two-exciton states, but neglects pure dephasing contributions, $\langle B_{\mathbf{k}\alpha}^{\dagger}\rangle \langle B_{\mathbf{k}'\alpha'}^{\dagger} B_{\mathbf{k}''\alpha''}\rangle$ [60], which does include these dephasing processes, but does not properly treat the two-exciton resonances, and $\langle B_{\mathbf{k}\alpha}^{\dagger}\rangle \langle B_{\mathbf{k}'\alpha'}^{\dagger}\rangle \langle B_{\mathbf{k}''\alpha''}\rangle$ [60], which is equivalent to the local-field approximation. The advantages of the first two factorization schemes can be combined into an equation of motion approach, known as the nonlinear exciton equations [90], that only neglects certain relaxation terms in variables like $\langle B_{\mathbf{k}\alpha}^{\dagger} B_{\mathbf{k}'\alpha'}^{\dagger} B_{\mathbf{k}''\alpha''}\rangle$ and leads to an accurate description of the third-order response.

To end this section, it should be noted that the proper formulation of nonlinear optical response of bulk crystals is complicated by the strong coupling between the polarization and the electromagnetic field that exists, owing to the translational symmetry. As a result of this strong coupling, the electromagnetic field really should be considered an intrinsic part of the system and not an external influence that weakly perturbs the exciton system. In other words, in order to describe nonlinear response of pure bulk crystals, one should account for the polariton modes. The classical way to include polaritons in the description of the nonlinear response, is to calculate the nonlinear susceptibilities and to substitute the polarization field equation (46) into the Maxwell equations. This automatically takes into account the proper coupling between the polarization and the electromagnetic field. The microscopic approach to take into account polariton effects re-expresses the exciton–exciton interactions in terms of polariton–polariton interactions by using the transformation between excitons and polaritons [60,91]. The nonlinear response is then described as scattering of polaritons on each other. For instance, exciton fusion terms translate into polariton fusion terms. In this approach, nonlinear optical processes in a bulk crystal are characterized by the excitation of polariton states at the boundary of the crystal and the possible mixing of these polaritons through their mutual interaction [60,91]. The advantage of the microscopic treatment over the classical one, is that in the former the scattering rates of the polaritons on imperfections and phonons may be calculated from first principles, while in the classical approach they have to be incorporated in an ad hoc fashion. In Section 4.5, we will discuss an example of the observation of polariton effects, namely the high diffusion coefficient measured in anthracene crystals using low-temperature transient grating experiments (a third-order technique). This diffusion coefficient arises from the fact that not excitons but rather polaritons (with a much higher group velocity) diffuse in the crystal.

## 2.9. Exciton–Phonon Interaction

Thus far, we have only considered electronic excitations. Of course, lattice excitations in the form of vibrations or phonons may occur as well. Even if these excitations are not directly excited by electromagnetic fields of optical frequencies, they often do play an important role in exciton systems, as they couple to the excitons and thereby change their optical response and dynamics considerably. A simple example is the formation of exciton side bands in the optical spectra, arising from the fact that the light may excite an exciton state dressed with several phonon quanta. In the case of a large collection of low-frequency phonons, this effect is seen as (homogeneous) broadening of the exciton absorption bands. The scattering of the excitons on the phonons is of crucial importance in relaxation and transport processes. We will expand upon this topic in Section 4. The effect on transport may lead to an extreme situation where the excitons, due to strong coupling to the lattice, loose their mobility. This self-trapping, also known as exciton–polaron formation, will be the topic of Section 5.

Here, we only briefly indicate what type of interaction terms may occur. The coupling of excitons and phonons arises from the fact that the various matrix elements, $D_n$, $J_{nm}$, and $U_{nm}$, in the generic Hamiltonian equation (14) depend on the distances between the molecules. As the presence of phonons alter these distances, this leads to exciton–phonon coupling. The coupling induced by the position dependence of the two-exciton interaction, $U_{nm}$, is usually not considered. The other two gives rise to what is known, respectively, as the on-site and intermolecular exciton–phonon coupling [35]. They may be written in more explicit form by expanding the matrix elements to first order in the displacements of the molecules from their equilibrium positions. Expressing the displacements in terms of phonon creation and annihilation operators, this gives rise to the following generic form of the exciton–phonon coupling in bulk crystals [35]

$$H_{\text{ex-ph}} = \sum_{\mathbf{k},\mathbf{q},r} F(\mathbf{k}-\mathbf{q};\mathbf{k};\mathbf{q}r) B^\dagger_{\mathbf{k}-\mathbf{q}} B_\mathbf{k} b^\dagger_{\mathbf{q}r} + \text{h.c.}, \qquad (47)$$

where $b^\dagger_{\mathbf{q}r}$ denotes the creation operator for a phonon of wave vector $\mathbf{q}$ in the branch labeled $r$. The coupling constant $F(\mathbf{k}-\mathbf{q};\mathbf{k};\mathbf{q}r)$ is different for the on-site and the intermolecular mechanism of coupling. It contains the first-order derivatives of the $D_n$ or the $J_{nm}$ with respect to the molecular positions [35]. The form of the interaction equation (47) reflects the conservation of quasi-momentum in the crystal, as is clear from the combinations of the wave vectors that occur. We finally notice that for certain applications it may not suffice to consider only the first-order contribution in the displacements. The next order gives the quadratic exciton–phonon coupling, which is important to describe pure dephasing processes [92].

## 3. Dielectric Theory of Frenkel Excitons

### 3.1. GENERALIZED LORENTZ LOCAL FIELD

The most general approach to the study of the optical properties of condensed media is the macroscopic electrodynamics approach making use of the concept of the dielectric tensor $\varepsilon_{ij}(\mathbf{k}, \omega)$, introduced in Section 2.6. Here, $i$ and $j$ label the Cartesian components of the tensor. Using this tensor we can calculate the refractive indexes of the normal waves. The poles of the refractive index determine the positions of absorption lines and for propagation directions for which the normal wave is transverse, these poles coincide with the poles of the dielectric tensor. The poles of the refractive index that corresponds to other (not transverse) waves can be found from the equation:

$$\sum_{i,j} \varepsilon_{ij}(\mathbf{k}, \omega) k_i k_j = 0. \qquad (48)$$

It is known [61] that all mentioned poles corresponding to transverse and other normal waves coincide with the frequencies of so-called Coulomb excitons. These excitons are the solutions of the coupled equations for matter (polarization) and electromagnetic fields in the limit $c \to \infty$. In the language of microscopic theory we can say that in this approximation the retarded interaction is neglected and only the non-retarded Coulomb interaction is taken into account. It is important to note that this approximation is the basis of the microscopic theory of excitons (see, for example, [35] and [93]). The finite velocity of light and retardation effects only need to be included in the microscopic theory when polariton effects are important.

Calculating the dielectric tensor for a specific medium is a problem of microscopic theory and for the exciton region of the spectrum is based on the various types of exciton states of the crystal (Coulomb or mechanical excitons, see [61]), which are treated as zero-order states when applying standard linear response theory. We have used this procedure in Section 2.6 to calculate $\varepsilon_{ij}^{\perp}(\mathbf{k}, \omega)$ for a crystal of two-level molecules. However, it should be stressed that we need not always know the exciton states in order to calculate the dielectric tensor of a crystal. We shall expand on this below, by considering the spectra of Frenkel singlet excited states (no charge transfer). The intermolecular interaction in this case is determined by dipoles, quadrupoles, and higher-order multipoles. It changes the frequencies of optical resonances and results in the mixing of molecular configurations.

Numerous theoretical and experimental studies have been carried out in this field, so that a whole branch of molecular optics – the optics of molecular crystals and molecular liquids – has been established. Even before Frenkel put forward his exciton concept, researchers in this branch of optics had developed a variety of exact and approximate methods for the theoretical description of optical phenomena. However, after the discovery of excitons the use of these methods became increasingly rare and many of the results obtained with them have not been sufficiently understood in the framework of exciton theory. Therefore, further development and generalization of these methods were impeded. On the other hand, since the results of pre-excitonic molecular optics were underestimated, the optical properties of crystals were treated in terms of only the exciton theory, even in those cases where this could be done much easier by using the earlier, simpler, and equally or even more clear physical concepts. The resulting situation is discussed by Agranovich [94] (also see [36, Chapter 3]). It was shown in Ref. [94] that many important results of Frenkel exciton theory, such as the Davydov splitting and the dependence of the absorption intensity on polarization, can be obtained (even in more general form) by using the local-field approach dating back to Lorentz, who used it to derive the well-known formula for the optical refractive index in isotropic media (the Lorentz–Lorenz formula).

Let us recall the derivation of this formula. According to Lorentz, the electric field $\mathbf{E}'$ acting on a molecule in an isotropic medium and causing its polarization, is not equal to the mean (macroscopic) field $\mathbf{E}$ that satisfies the phenomenological Maxwell equations, but rather is determined by

$$\mathbf{E}' = \mathbf{E} + \frac{4\pi}{3}\mathbf{P} = \frac{\varepsilon+2}{3}\mathbf{E}. \tag{49}$$

Here, $\mathbf{P}$ is the polarization field (cf. Section 2.6) and $\varepsilon$ is the dielectric constant of the medium. The factor $(\varepsilon+2)/3$ is known as the Lorentz factor. On the other hand, if we denote the polarizability of the molecule by $a$ and the number of molecules per unit volume by $N_0$, we also have $\mathbf{P} = N_0 a \mathbf{E}'$. Hence, the displacement vector $\mathbf{D}$ may be written

$$\mathbf{D} = \mathbf{E} + 4\pi\mathbf{P} = \left[1 + \frac{4\pi}{3}N_0 a(\varepsilon+2)\right]\mathbf{E}. \tag{50}$$

Since by definition the relation $\mathbf{D} = \varepsilon\mathbf{E}$ holds as well, Eq. (50) immediately gives the Lorentz–Lorenz formula,

$$\frac{\varepsilon-1}{\varepsilon+2} = \frac{4\pi}{3}N_0 a. \tag{51}$$

It should be noted that this formula, which expresses the dielectric constant of the medium in terms of the polarizability of an individual molecule, is only an approximation, even for cubic crystals. For instance, the formula does not take into account spatial dispersion. Moreover, it does not account for the contributions of higher-order multipoles to the intermolecular interactions. These additional contributions were identified in Section 2.2 as the ones responsible for the gas-condensed matter shifts $D_n$ of the molecular levels. It has been shown, however, that the Lorentz–Lorenz formula can be generalized to include these effects [94]. Following Ref. [94], we will in this section employ the local-field method to discuss the effect of mixing of molecular configurations and the spectra of impurities. In addition, we will apply it to calculate the dielectric tensor of anisotropic molecular crystals of complex structure.

Consider a crystal of which the unit cell contains $S$ identical molecules, labeled $s = 1, 2, \ldots, S$. The only difference between the $S$ molecules is their orientation with respect to the crystallographic axes. According to Born and Huang [47, Section 30], when a plane electromagnetic wave of amplitude $\mathbf{E}(\mathbf{k},\omega)$ propagates in the crystal, the electric field $\mathbf{E}^s$ acting on a molecule of type $s$ is not equal to the mean field $\mathbf{E}$, but has Cartesian components given by:

$$E_i^s = E_i + \sum_{j,s'} Q_{ij}^{ss'}(\mathbf{k}) \mu_j^{s'}. \qquad (52)$$

Here, $\mu_i^s$ is the $i$th Cartesian component of the amplitude of the dipole moment induced in the molecules of type $s$ and the coefficients $Q_{ij}^{ss'}(\mathbf{k})$ (the local-field tensor) are determined by the lattice structure only and are analytical functions of $\mathbf{k}$. For simplicity, we will assume that the molecules have no static dipole moments. If $a_{ij}^s$ denotes the polarizability tensor of the molecules of type $s$, we have

$$\mu_i^s = \sum_j a_{ij}^s E_j^s. \qquad (53)$$

At the same time, of course, the molecules may have higher-order static multipole moments. Generally speaking, these moments are different in the ground and in the excited states, which causes the interactions between the molecule and its environment to depend on the state it is in. This leads to the gas-condensed phase shifts, $D_n$, of the molecular transitions, which we already introduced in Section 2.2 for the special case of two-level molecules. We shall assume below that the only difference between the tensor $a_{ij}(\omega)$ and the respective tensor for a molecule in vacuum is determined by this frequency shift. Substitution of Eq. (53) into Eq. (52) yields:

$$E_i^s = E_i + \sum_{j,k,s'} Q_{ik}^{ss'}(\mathbf{k}) a_{kj}^{s'} E_j^{s'}. \qquad (54)$$

This equation allows one to solve for the local fields $\mathbf{E}^s$ in terms of the mean (Maxwell) field $\mathbf{E}$:

$$E_i^s(\mathbf{k},\omega) = \sum_{j,s'} A_{ij}^{ss'}(\mathbf{k},\omega) E_j(\mathbf{k},\omega), \qquad (55)$$

where the tensor $A_{ij}^{ss'}$ is the inverse of the tensor $\delta_{ss'}\delta_{ij} - \sum_k Q_{ik}^{ss'}(\mathbf{k})a_{kj}^{s'}$. Using the fact that the polarization is given by

$$P_i = \frac{1}{v_c} \sum_s \mu_i^s, \qquad (56)$$

with $v_c$ the volume of the unit cell, the dielectric tensor may be expressed in terms of the tensor $A_{ij}^{ss'}$,

$$\varepsilon_{ij} = \delta_{ij} + \frac{4\pi}{v_c} \sum_{k,s,s'} a_{ik}^s A_{kj}^{ss'}. \qquad (57)$$

## 3.2. Cubic Crystals with One Molecule per Unit Cell

If we ignore spatial dispersion, the tensor $Q_{ij}^{ss'}$ for cubic crystals with one molecule per unit cell ($S = 1$) reduces to the scalar [47]

$$Q_{ij} = \frac{4\pi}{3v_c}\delta_{ij}. \qquad (58)$$

Here, the indices $s$ and $s'$ have been dropped, because $S = 1$. If we also assume that the response of the molecules is isotropic, i.e., $a_{ij} = a\delta_{ij}$, we find $A_{ij} = A\delta_{ij}$, with $A = [1 - 4\pi a/3v_c]^{-1}$. Substituting this result into Eq. (57), we obtain $\varepsilon_{ij} = \varepsilon\delta_{ij}$, with

$$\varepsilon = 1 + \frac{4\pi}{v_c}a\left(1 - \frac{4\pi}{3v_c}a\right)^{-1}, \qquad (59)$$

which implies that $A = (\varepsilon + 2)/3$. Eq. (59) directly yields the Lorentz–Lorenz relation equation (51), with $N_0$ replaced by $1/v_c$.

Let us now consider the frequency dispersion of $\varepsilon$ by taking into account only one of the resonances of the molecular polarizability. In this approximation we have:

$$a(\omega) = \frac{F_1}{\omega_1^2 - \omega^2} \qquad (60)$$

with $F_1 = 2\mu_1^2\omega_1/\hbar = (e^2/m)f_1$. Here, $\omega_1$, $\mu_1$, and $f_1$ are the frequency, the dipole, and the oscillator strength, respectively, of the selected transition of the isolated molecule. Substituting Eq. (60) into Eq. (59) or the Lorentz–Lorenz relation yields:

$$\varepsilon(\omega) = 1 + \frac{(4\pi/v_c)F_1}{\omega_\perp^2 - \omega^2}, \qquad (61)$$

where $\omega_\perp \approx \omega_1 - (4\pi/3v_c\hbar)\mu_1^2$ denotes the frequency of the transverse exciton. Eq. (61) shows that if we take into account the local-field correction, i.e., the fact that $A \neq 1$, the oscillator strength for the transition is not changed and only the resonance frequency is shifted. Explicitly, the resonance of $\varepsilon(\omega)$ is red-shifted compared to the transition of an isolated molecule over the amount

$$\Delta\omega = \frac{4\pi}{3v_c\hbar}\mu_1^2. \qquad (62)$$

Dissipation can be taken into account by adding the imaginary term $i\gamma$ to the frequency $\omega$ in the denominator of Eq. (60). Since $\varepsilon = (n + i\kappa)^2$, with $n$ and $\kappa$ the real and imaginary parts of the refractive index, respectively, we obtain for $\gamma \to +0$

$$\int_{-\infty}^{+\infty} 2n(\omega)\kappa(\omega)\,d\omega = \frac{2\pi^2 F_1}{v_c\omega_1}. \qquad (63)$$

In this case we have for the absorption spectrum of the crystal,

$$\kappa(\omega) = \frac{1}{n(\omega)} \frac{(2\pi/v_c) F_1 \gamma(\omega)}{(\omega_\perp^2 - \omega^2)^2 + \gamma^2(\omega)}. \tag{64}$$

Here, $\gamma(\omega)$ is given by $2\omega\gamma$. The main difference between Eqs. (64) and (44) obtained through microscopic derivation, is the inclusion of antiresonant contributions, as is seen in the combination $\omega_\perp^2 - \omega^2 = (\omega_\perp - \omega)(\omega_\perp + \omega)$ appearing in the denominator above.

### 3.3. Local-Field Corrections for Impurities

Let us now consider the same crystal with a certain number of substitutional impurity molecules (the treatment can be readily modified for the case of interstitial impurities). Now the local field depends on the spatial distribution of the impurities. If we ignore fluctuations of this distribution and replace the local field by its mean value, we obtain

$$\varepsilon(\omega) = 1 + 4\pi N_0 a(\omega) \frac{\varepsilon + 2}{3} + 4\pi N_1 [\tilde{a}(\omega) - a(\omega)] \frac{\varepsilon + 2}{3} \tag{65}$$

where $\tilde{a}(\omega)$ is the polarizability of the impurity molecules, $N_0 = 1/v_c$ (as before), and $N_1$ is the concentration of impurity molecules. At small impurity concentration ($\rho \equiv N_1 v_c \ll 1$), the difference $\delta\varepsilon$ between the dielectric function of the crystal with impurities and the pure crystal is found from Eq. (65) to be

$$\delta\varepsilon(\omega) = \left(1 - \frac{4\pi N_0 a}{3}\right)^{-1} 4\pi N_1 (\tilde{a} - a) \frac{\varepsilon_0 + 2}{3}, \tag{66}$$

where $\varepsilon_0$ denotes the dielectric function for the pure crystal. Using the relation $(1 - 4\pi N_0 a/3)^{-1} = (\varepsilon_0(\omega) + 2)/3$ (cf. Section 3.1), we thus arrive at

$$\varepsilon(\omega) = \varepsilon_0(\omega) + 4\pi N_1 (\tilde{a} - a) \left(\frac{\varepsilon_0 + 2}{3}\right)^2, \tag{67}$$

which using the Lorentz–Lorenz relation may be rewritten as

$$\varepsilon(\omega) = \varepsilon_0(\omega) - \rho \frac{\varepsilon_0 + 2}{3} (\varepsilon_0 - 1) + 4\pi N_1 \tilde{a} \left(\frac{\varepsilon_0 + 2}{3}\right)^2. \tag{68}$$

In analogy to Eq. (60) we write the polarizability of the impurity molecules as,

$$\tilde{a}(\omega) = \frac{\tilde{F}_1}{\tilde{\omega}_1^2 - \omega^2 - i\tilde{\gamma}(\omega)}, \tag{69}$$

where we took into account dissipation and $\tilde{F}_1 = 2\tilde{\mu}_1^2 \tilde{\omega}_1/\hbar$. Assuming that the impurity resonance frequency $\tilde{\omega}_1$ is in the region of transparency of the host medium, we obtain for the frequencies $\omega \simeq \tilde{\omega}_1$:

$$2n(\omega)\kappa(\omega) = 8\pi N_1 \left(\frac{\varepsilon_0(\omega)+2}{3}\right)^2 \mu_1^2 \frac{\tilde{\omega}_1 \tilde{\gamma}/\hbar}{[\tilde{\omega}_1^2 - \omega^2]^2 + \tilde{\gamma}^2}. \tag{70}$$

Hence, integration in the region of the impurity absorption band yields:

$$\int 2n(\omega)\kappa(\omega)\,d\omega = 4\pi^2 N_1 \left(\frac{\varepsilon_0(\omega)+2}{3}\right)^2 \frac{\tilde{\mu}_1^2}{\hbar}. \tag{71}$$

Thus, we have found that the absorption coefficient of light due to impurities in a medium with dielectric constant $\varepsilon_0(\omega)$, as well as the integral on the left-hand side of Eq. (71) (known as the Kravets integral), are proportional to the squared Lorentz factor. This reflects a change of the observed oscillator strength of the impurities. The influence of the medium is to replace the dipole moment $\tilde{\mu}_1$ of the impurity's transition by the effective dipole moment $\tilde{\mu}_{\text{eff}} = \tilde{\mu}_1(\varepsilon_0(\tilde{\omega}_1)+2)/3$. The same enhancement factor occurs in the spontaneous emission constant of the impurity molecules [95].

If the impurity's resonance frequency $\tilde{\omega}_1$ is close to the crystal's resonance frequency $\omega_0$ and $\varepsilon_0(\tilde{\omega}_1) \gg 1$, the integrated intensity $I$ of absorption by the impurities as a function of $\tilde{\omega}_1$ increases as $(\tilde{\omega}_1 - \omega_0)^{-2}$. If we properly take into account the dependence of the exciton frequency on the wave vector, $\omega_0 = \omega_0(\mathbf{k})$, we obtain, as shown by Rashba [96], that $I \sim (\tilde{\omega}_1 - \omega_0)^{-3/2}$. A similar effect takes place, for instance, for excitons localized at the impurity. Since the experimentally measured quantity in Eq. (71) is its left-hand side, a correct introduction of the local-field correction (the Lorentz factor in this case) makes it possible to find the oscillator strength for the isolated molecule from measuring the dispersion and absorption for the molecules in solution. Of course, this can be done only if no chemical bonds arise between the molecules of the solute and the solvent, no aggregates of the impurity molecules are formed, and so on.

### 3.4. Impurity-Free Crystals: Mixing of Molecular Configurations

Now let us return to the impurity-free crystal and assume that the molecules of the crystal have the polarizability,

$$a(\omega) = a_0 + a_1(\omega), \tag{72}$$

where the function $a_1(\omega)$ is given by Eq. (60), while $a_0$ reflects the contributions to the molecular polarizability from transitions far from the resonance $\omega_1$, which can be assumed to be independent of $\omega$ in the frequency range $\omega \simeq \omega_1$. Substituting Eq. (72) into Eq. (59), we obtain

$$\varepsilon(\omega) = \varepsilon_b + \frac{(4\pi/v_c)F_1[(\varepsilon_b+2)/3]^2}{\omega_\perp^2 - \omega^2}, \tag{73}$$

where

$$\varepsilon_b = 1 + \frac{4\pi}{v_c} a_0 \left(1 - \frac{4\pi}{3v_c} a_0\right)^{-1}, \quad (74)$$

which may be considered the background dielectric constant of the crystal far from resonance (calculated under the assumption that the resonance part of molecular polarizability $a_1(\omega) = 0$). The resonance frequency $\omega_\perp$ is given by

$$\omega_\perp = \omega_1 - \frac{2\pi F_1}{v_c \omega_1} \frac{\varepsilon_b + 2}{3}. \quad (75)$$

If we introduce a weak dissipation into Eq. (72), then Eq. (63) is replaced by

$$\int 2n(\omega)\kappa(\omega)\,d\omega = \frac{2\pi^2 F_1}{v_c \omega_\perp}\left(\frac{\varepsilon_b + 2}{3}\right)^2. \quad (76)$$

Eqs. (75) and (76) demonstrate how the mixing of molecular configurations due to intermolecular (resonance dipole–dipole) interactions, leading to $\varepsilon_b > 1$, affects the frequencies and oscillator strengths of the dipole transitions.

It is important to note the difference between these equations and similar relationships, like Eq. (71), for impurity molecules in a solution or a matrix. When we consider impurity molecules in a solution the Lorentz factor on the right-hand side of Eq. (71) contains the dielectric constant of the solvent at the transition frequency of the impurity. For the impurity-free crystal and in the region of the exciton resonance, we see from Eq. (76) that this quantity in the Lorentz factor is replaced by the background dielectric constant, which is not at all equal to the squared index of refraction of light in the crystal at the resonance frequency.

Within the framework of microscopic theory, the mixing of molecular configurations of Frenkel excitons arises only if we go beyond the two-level model for the molecules. Within the two-level model this effect is absent. In this case $\varepsilon_b = 1$ and then Eq. (73) is transformed into Eq. (59), and Eq. (76) into Eq. (63).

In an alternative approach [97], it has been proposed that the background dielectric constant in the microscopic theory of Frenkel excitons can be taken into account by dividing the intermolecular resonance interaction by $\varepsilon_b$. This may be considered a renormalization of the molecular transition dipole moment and oscillator strength $\mu_1 \to \mu_1/\sqrt{\varepsilon_b}$ and $F_1 \to F_1/\varepsilon_b$. However, this approach does not even correctly account for the local-field corrections due to the long-range dipole–dipole interactions (see [36, Section 3.6]). For example, Eq. (75) shows that the shift of the frequency $\omega_\perp$ of the transverse Frenkel exciton relative to the molecular frequency, is affected by $\varepsilon_b$ through the Lorentz factor $(\varepsilon_b + 2)/3$, which increases with $\varepsilon_b$. In the renormalization approach, this shift would decrease with $\varepsilon_b$, as the effective interaction decreases. The frequency of the longitudinal Frenkel exciton, which obeys $\varepsilon(\omega_\|) = 0$, reads

$$\omega_\| = \omega_1 - \frac{2\pi F_1}{3v_c\omega_1}\frac{\varepsilon_b+2}{3} + \frac{2\pi F_1}{v_c\omega_1\varepsilon_b}\left(\frac{\varepsilon_b+2}{3}\right)^2, \tag{77}$$

and depends on $\varepsilon_b$ in an even more complicated way.

### 3.5. Anisotropic Crystals: Absorption Intensities and Davydov Splitting

We finally consider crystals with several molecules in the unit cell. We will show how the dielectric theory allows us to calculate the dielectric tensor of anisotropic crystals and to recover the results of the Frenkel exciton theory for the exciton energies and polarization intensity relations obtained in Section 2.5.3.

Let's consider a molecular crystal with $S \geqslant 1$ molecules in the unit cell. If we are interested in the optical properties of the crystal in the frequency range $\omega \simeq \omega_1$ where $\omega_1$ is the nondegenerate frequency of one of the dipole transitions of the isolated molecule, then its polarizability tensor can be written as

$$a_{ij}(\omega) = \frac{F_1 l_i l_j}{\omega_1^2 - \omega^2}, \tag{78}$$

where $l_i$ is the $i$th Cartesian component of the unit vector (**l**) parallel to the molecular transition dipole moment. Since the different molecules in the unit cell have different orientations, their polarizability tensors are

$$a_{ij}^s(\omega) = \frac{F_1 l_i^s l_j^s}{\omega_1^2 - \omega^2}, \tag{79}$$

with $s = 1, 2, \ldots, S$. Using Eqs. (52), (53), and (79), we obtain the following set of equations for the quantities $\mathbf{E}^s \cdot \mathbf{l}^s$, i.e., the projections of the local fields on the directions of the molecular transition dipoles

$$\mathbf{E}^s \cdot \mathbf{l}^s - \sum_{s'} M_{ss'}(\mathbf{k},\omega)\mathbf{E}^{s'} \cdot \mathbf{l}^{s'} = \mathbf{E} \cdot \mathbf{l}^s, \tag{80}$$

with

$$M_{ss'}(\mathbf{k},\omega) = \frac{F_1}{\omega_1^2 - \omega^2}\sum_{ij} Q_{ij}^{ss'}(\mathbf{k}) l_i^s l_j^{s'}. \tag{81}$$

Let us apply these general equations to crystals with the symmetry of the anthracene crystal, containing two molecules per unit cell. The optical properties of molecular crystals of this type were the subject of many experimental studies. For such crystals symmetry operations exist that transform molecules with $s = 1$ into molecules with $s = 2$. As a result, when $\mathbf{k} = 0$ or even when $\mathbf{k} \neq 0$ provided that the wave vector is parallel or perpendicular to the monoclinic axis of the crystal, we have

$$M_{11}(\mathbf{k}, \omega) = M_{22}(\mathbf{k}, \omega), \qquad M_{12}(\mathbf{k}, \omega) = M_{21}(\mathbf{k}, \omega) \tag{82}$$

which makes the solution of Eq. (80) less cumbersome. It can be easily seen that in this case we have

$$\mathbf{E}^s \cdot \mathbf{l}^s = \frac{1}{2}\left[\frac{L_j^+}{1 - M_{11}(\mathbf{k}, \omega) - M_{12}(\mathbf{k}, \omega)} - \frac{(-1)^s L_j^-}{1 - M_{11}(\mathbf{k}, \omega) + M_{12}(\mathbf{k}, \omega)}\right] E_j, \tag{83}$$

with

$$\mathbf{L}^\pm = \mathbf{l}^1 \pm \mathbf{l}^2. \tag{84}$$

In analogy to Eq. (57), this solution leads to the dielectric tensor

$$\varepsilon_{ij}(\mathbf{k}, \omega) = \delta_{ij} + \frac{2\pi F_1}{v_c}\left[\frac{L_i^+ L_j^+}{\Omega_+^2(\mathbf{k}) - \omega^2} + \frac{L_i^- L_j^-}{\Omega_-^2(\mathbf{k}) - \omega^2}\right], \tag{85}$$

where

$$\Omega_+^2(\mathbf{k}) = \omega_1^2 - F_1 \sum_{i,j}(Q_{ij}^{11}(\mathbf{k})l_i^1 l_j^1 - Q_{ij}^{12}(\mathbf{k})l_i^1 l_j^2), \tag{86}$$

$$\Omega_-^2(\mathbf{k}) = \omega_1^2 - F_1 \sum_{i,j}(Q_{ij}^{22}(\mathbf{k})l_i^2 l_j^2 - Q_{ij}^{21}(\mathbf{k})l_i^2 l_j^1). \tag{87}$$

Since the quantities $Q_{ij}^{ss'}(\mathbf{k})$ are analytic functions of $\mathbf{k}$, the same is true for the frequencies $\Omega_\pm^2(\mathbf{k})$. This is not surprising because the resonance frequencies of the tensor $\varepsilon_{ij}(\mathbf{k}, \omega)$ are the frequencies of the so-called mechanical excitons [61], which are analytic functions of $\mathbf{k}$, regardless of the model being used.

As the vectors $\mathbf{L}^1$ and $\mathbf{L}^2$ are orthogonal, their directions may be chosen as the $x$ and $y$ axes directions, respectively. In this coordinate system, the tensor $\varepsilon_{ij}(\mathbf{k}, \omega)$ obtains a diagonal form with the non-zero components:

$$\varepsilon_{xx}(\omega, \mathbf{k}) = 1 + \frac{2\pi}{v_c}\frac{F_{1,xx}}{\Omega_+^2(\mathbf{k}) - \omega^2}, \qquad \varepsilon_{yy}(\omega, \mathbf{k}) = 1 + \frac{2\pi}{v_c}\frac{F_{1,yy}}{\Omega_-^2(\mathbf{k}) - \omega^2},$$

$$\varepsilon_{zz}(\omega, \mathbf{k}) = 1, \tag{88}$$

with

$$F_{1,xx} = F_1|\mathbf{L}^+|^2, \qquad F_{1,yy} = F_1|\mathbf{L}^-|^2, \tag{89}$$

so that

$$F_{1,xx}/F_{1,yy} = \cot^2(\theta/2), \tag{90}$$

where $\theta$ is the angle between the direction of the two molecular dipoles, $\mathbf{l}^1$ and $\mathbf{l}^2$. These relationships show that the absorption of light propagating along the $z$-axis and with the electric vector $\mathbf{E}$ parallel to $\mathbf{L}^+$ ($\mathbf{L}^-$) has a resonance at the frequency $\Omega_+(\mathbf{k})$ ($\Omega_-(\mathbf{k})$). Thus, though we assumed that the transition of the isolated molecule at the frequency $\omega_1$ is nondegenerate, the absorption spectrum of a crystal with two molecules per unit cell should have two absorption lines, which have mutually perpendicular polarizations. The results obtained here coincide with those obtained in Section 2.5.3 within the framework of microscopic exciton theory.

The effect of mixing of molecular configurations can be considered along the same lines as was done in Section 3.4 for cubic crystals. One adds a frequency independent tensor $(a_0)_{ij}$ to the resonant part of the molecular polarizability equation (78). Mixing effects are particularly important for excitonic transitions with a small oscillator strength and also for the calculation of intensity ratios, such as the above ratio $F_{1,xx}/F_{1,yy}$ [53].

If we know the tensor $\varepsilon_{ij}(\mathbf{k}, \omega)$ we can find the position of the absorption line for arbitrary polarizations and directions of propagation of the light and also take into account spatial dispersion effects in molecular crystals of arbitrary shape. As we demonstrated, to calculate this tensor we need to know the local-field tensor $Q_{ij}^{ss'}(\mathbf{k})$, which depends only on the lattice structure, and the polarizability tensor of a single molecule. Our calculations only accounted for the dipolar terms in the local-field tensor. It is noteworthy that taking into account the higher-order multipoles only leads to changes of the resonance frequencies in the dielectric tensor [94].

The importance of the local-field method presented in this section lies in its simplicity and the possibility to generalize it to include effects of mixing of molecular configurations. Thus, while the Davydov splitting has been studied for many different crystals and was first analyzed by extending the microscopic Frenkel exciton theory to include several molecules per unit cell [35], the phenomenon can be well-understood and described in terms of a local-field analysis. It should be stressed that excitonic spectra in semiconductors cannot be understood on the basis of simple local-field arguments. The large-radius excitons in these systems, require a microscopic treatment that starts from electron band structure theory (see, for example, [93,98]).

The above should not be considered a depreciation of the microscopic theory of small-radius (Frenkel) excitons. Only within the framework of a microscopic theory can we calculate the exciton energies over the entire range of allowable values of the wave vector and consistently study effects of exciton–phonon and exciton–exciton interaction and scattering of excitons in crystals with static disorder. This is exactly what makes it possible, within the scope of small-radius exciton theory, to understand such phenomena as transfer of electronic excitation energy in crystals, the optical properties of molecular crystals with high excitation levels, nonlinear optical effects, fine-structure in absorption and luminescence spectra, and many other optical phenomena.

## 4. Diffusion of Frenkel Excitons

### 4.1. Exciton Motion and Diffusion

The transfer interaction $J_{nm}$ in Eq. (14) causes an excitation to migrate between different molecules. As mobility of excitons causes transport of the excitation energy through the system, this mobility has been in the focus of interest for many years. This interest concerns excitation energy transport in bulk crystals, but also in smaller molecular aggregates, such as the chlorophyll aggregates that occur in the photosynthetic systems of bacteria and higher plants [13].

If a molecular system is excited in one of its exciton eigenstates and no interactions occur with other degrees of freedom, the only evolution in the system will be a periodic phase change of the exciton wave function as a whole, which is not associated with spatial motion. In a more realistic situation, certainly if one deals with large systems (crystals), one excites a wave packet of excitons, which will propagate through the crystal with the group velocity. The wave packet (with some mean value of the wave vector) will scatter on static disorder (e.g., in the gas-crystal shifts $D_n$), on lattice vibrations, and on other excitons. At low excitation densities, we may neglect the latter scattering event. The scattering leads to a finite exciton mean free path. If the scattering is weak, the mean free path may be very large compared to the lattice constant and a description in terms of weakly perturbed wave packets is appropriate. This situation is referred to as the case of coherent excitons and may be described using the Boltzmann equation. In the other extreme case, the scattering is so strong and the coherence length is so short that basically the exciton loses it phase information already when propagating from one molecule to the next. This situation is referred to as the incoherent case (Förster energy transfer [99]) and is mostly described using a set of coupled rate equations for the excitation probabilities of the individual molecules.

Various methods have been developed that interpolate between the coherent and incoherent regimes (for reviews see, e.g., [100–102]). Well-known approaches use the stochastic Liouville equation, of which the Haken–Strobl–Reineker [100] model is an example, and the generalized master equation [101]. A powerful technique, which in principle deals with all aspects of the problem, uses the reduced density matrix of the exciton subsystem, which is obtained by projecting out all other degrees of freedom (the bath) from the total statistical operator [103]. This reduced density operator obeys a closed non-Markovian (integro-differential) equation with a memory kernel that includes the effects of (multiple) interactions between the excitons and the bath. In practice, one is often forced to truncate this kernel at the level of two interactions. In the Markov approximation, the resulting description is known as Redfield theory [104].

It should be realized that, independent of whether the system is in the coherent, incoherent, or intermediate regime, the motion at distances larger than the mean free path is always diffusive. Alternatively stated, on time scales large compared to the typical scattering time, the exciton motion is described by a diffusion equation. In this section, we will restrict ourselves to this diffusive regime, which presumes that the exciton life time is long compared to the scattering time, so that enough scattering events may occur before the exciton decays through spontaneous emission, internal conversion, trapping by an impurity, or any other decay channel.

In the diffusive regime, the quantity of interest is the exciton concentration $c(\mathbf{r}, t)$ as a function of position $\mathbf{r}$ and time $t$. It obeys the diffusion equation

$$\frac{\partial c}{\partial t} = D\nabla^2 c - \frac{1}{\tau_0}c + I_0(t)\kappa e^{-\kappa z}, \tag{91}$$

where $D$ is the exciton diffusion coefficient (considered isotropic here), $\tau_0$ denotes the exciton life time, $I_0(t)$ is the intensity of the pumping light incident on the sample, and $\kappa$ is the absorption coefficient in the crystal. Thus, the last term in Eq. (91) gives the number of excitons generated by the external radiation per unit volume and per unit time. We assume here that the sample is a parallel-sided slab with boundaries at $z = 0$ and $z = d$ and that the incident radiation propagates from the region $z < 0$ along the normal to the plane $z = 0$. The slab thickness $d$ should be large compared to the exciton mean free path in order for the diffusion equation to be of any use. To solve the diffusion equation we must define boundary conditions. In the steady state, the number of excitons that arrive at the boundary surface per second and per unit area is given by $D[dc/dz]_{z=0}$. In the steady state, this should equal the rate of surface annihilation, which we may write $v_a c(z = 0)$, with $v_a$ a characteristic surface annihilation velocity. The boundary condition at $z = 0$ is then written

$$D\left[\frac{dc}{dz}\right]_{z=0} = v_a c(0), \tag{92}$$

which may also be formulated as $[dc/dz]_{z=0} = c(0)/l_0$, with $l_0 = v_a/D$. It should be kept in mind that this boundary condition only determines the asymptotic behavior of the exciton concentration, i.e., for $z$ large compared to the exciton mean free path. The parameters $D$, $\tau_0$, $\kappa$, and $v$ may be taken from experiments or calculated in the framework of some microscopic theory (see, for example, Refs. [36, 102]).

A useful measure for exciton migration is the diffusion length $L = (D\tau_0)^{1/2}$. Experimental data show that for Frenkel excitons in molecular crystals at room temperature the diffusion coefficient $D \approx 10^{-3}$ cm$^2$/s and the lifetime of singlet excitons $\tau_0 \approx 10^{-8}$ s. This gives a typical diffusion length $L \approx 10^{-6}$ cm (for anthracene crystals $L \approx 5 \cdot 10^{-6}$ cm). For triplet excitons, the lifetime may be appreciably longer ($10^{-3}$–$10^{-4}$ s) and the diffusion length may be larger than for singlet excitons by more than two orders of magnitude.

In the remainder of this section, we will focus on various ways to calculate the diffusion constant from microscopic principles. We will start by considering the formal definition for quantum particles, first given by Kubo (Section 4.2). We will then consider the actual calculation in more detail for the case of coherent excitons (Section 4.3) and incoherent ones (Section 4.4). Finally, we briefly address the measurement of exciton transport properties and the effect of exciton-polariton formation on transport in molecular crystals (Section 4.5).

### 4.2. THE DIFFUSION TENSOR

According to Kubo [105], the general quantum mechanical expression for the diffusion tensor is

$$D_{ij} = \frac{1}{\beta} \int_0^{+\infty} dt\, e^{-\eta t} \int_0^{\beta} d\lambda \langle \hat{v}_i(-i\hbar\lambda)\hat{v}_j(t)\rangle, \tag{93}$$

where $\beta = 1/k_B T$, $\eta$ denotes an infinitesimal positive constant, $\hat{v}_i(t)$ is the $i$th component of the velocity operator of the migrating particle in the Heisenberg representation, and the brackets $\langle \cdots \rangle$ denote taking the statistical equilibrium average. For classical particles ($\hbar = 0$) in an isotropic medium and using a relaxation time approximation, $v_i(t) = v_i(0)\exp(-t/\tau)$, Eq. (93) leads to the well-known expression $D_{ij} = \frac{1}{3}\langle v^2 \tau \rangle \delta_{ij}$.

Let us consider the Kubo expression in somewhat more detail for the case of excitons in a molecular crystal. We will restrict ourselves to the presence of one exciton. The position of its "center of gravity" then reads

$$\mathbf{R} = \sum_n \mathbf{R}_n B_n^\dagger B_n, \tag{94}$$

where $n$ labels the molecules. In a crystal, $n$ is short for $(\mathbf{n}, s)$, the position of the unit cell and the index of the molecule within the unit cell. Furthermore, $\mathbf{R}_n$ denotes the position of the $n$th molecule. The velocity operator of the exciton may now be defined through

$$\hat{\mathbf{v}} = \frac{i}{\hbar}[\hat{H}, \mathbf{R}], \tag{95}$$

where $\widehat{H}$ is the system's total Hamiltonian, including the interactions with phonons and disorder. As we deal with one exciton only, we may neglect exciton–exciton interactions, and we have

$$\widehat{H} = \sum_n \widehat{H}_n B_n^\dagger B_n + \sum_{n,m} \widehat{J}_{nm} B_n^\dagger B_m + \sum_\kappa \omega_q b_q^\dagger b_q. \quad (96)$$

Here, $b_q^\dagger$ and $b_q$ are the creation and annihilation operators, respectively, for phonons in mode $q = (\mathbf{q}, r)$, where $\mathbf{q}$ denotes the wave vector of the phonon and $r$ is the branch label. The energy of these phonon modes is given by $\omega_q$. Furthermore, the single-molecule Hamiltonian as well as the intermolecular transfer interaction are still considered to be operators in the phonon space. The physics of the dependence of the molecular transition energy and the interactions on the phonon coordinates was introduced in Section 2.9 already, where also the linearization of these operators in the phonon coordinates was discussed.

From Eqs. (94)–(96), the velocity operator is found to be

$$\hat{v} = \frac{i}{\hbar} \sum_{n,m} \mathbf{R}_{nm} \widehat{J}_{nm} B_n^\dagger B_m, \quad (97)$$

with $\mathbf{R}_{nm} = \mathbf{R}_n - \mathbf{R}_m$. In Eq. (97) we shall ignore the dependence of $\widehat{J}_{nm}$ on the phonon operators. If the exciton–phonon interaction is weak, it is sufficient to include the dependence on the phonon operators only in the propagation of the coherences $B_n^\dagger B_m$ through the phonon dependencies in the operator $\widehat{H}_n$. Accounting for the phonon dependence of the $\widehat{J}_{nm}$ in Eq. (97) only yields small corrections to the expression for $D_{ij}$. Thus, in Eq. (97) we shall replace the operator $\widehat{J}_{nm}$ by the scalar $J_{nm}$. Then, integration of Eq. (93) over $t$ and $\lambda$ yields [106]

$$D_{ij} = \frac{1}{2\hbar^2} \sum_{n,m,n',m'} (\mathbf{R}_{nm})_i J_{nm} (\mathbf{R}_{n'm'})_j J_{n'm'}$$

$$\times \int_{-\infty}^{+\infty} dt \langle B_n^\dagger(t) B_m(t) B_{n'}^\dagger(0) B_{m'}(0) \rangle. \quad (98)$$

As can be seen from Eq. (98), the calculation of the tensor $D_{ij}$ reduces to the calculation of two-particle correlation functions. The lack of sufficiently detailed data on the exciton band structure and the exciton–phonon coupling constants considerably complicates the accurate calculation of the exciton diffusion coefficients in molecular crystals. However, the temperature dependence of this coefficient differs significantly for coherent and incoherent excitons (see below). Therefore, studying the temperature dependence of diffusion has always been an important tool to analyze the character of the energy transfer in molecular crystals. In the remainder of this section, we will focus on the main characteristics of the diffusion constant and its temperature dependence.

## 4.3. WEAK EXCITON–PHONON COUPLING: COHERENT EXCITONS

### 4.3.1. General Expressions

If the exciton–phonon coupling is sufficiently weak, the solution of the equation for the correlation function $\langle B_n^\dagger(t) B_m(t) B_{n'}^\dagger(0) B_{m'}(0) \rangle$ is equivalent to the solution of the Boltzmann equation [106]. In this coherent limit, the exciton states of the ideal lattice serve as good zeroth-order states. In other words, the wave vector still proves to be a good quantum number and the notion of excitons propagating in between scattering events as wave packets with a well-defined group velocity is useful. The kinetics of such coherent excitons under the influence of the weak exciton–phonon coupling is described by the Boltzmann equation. One may show that this equation reduces to the diffusion equation (91), if the exciton concentration changes little over lengths of the order of magnitude of the exciton mean free path.

In this case the exciton–phonon interaction can be taken into account perturbatively. The influence of this interaction on the shape and the position of the exciton band(s) is insignificant and usually can be ignored. Then, the only remaining effect of the interaction is the scattering of the excitons, as is, for instance, clear from Eq. (47). This scattering results in a change of the wave vector and the energy of the excitons. Therefore, if $\delta E_\mathbf{k}$ is the width of the energy level of the exciton with the wave vector $\mathbf{k}$ determined by the exciton–phonon interaction, the related uncertainty $\delta \mathbf{k}$ of the wave vector is given by

$$\delta E_\mathbf{k} = \hbar \mathbf{v}(\mathbf{k}) \cdot \delta \mathbf{k} \tag{99}$$

where $\mathbf{v} = (1/\hbar)(dE/d\mathbf{k})$ is the exciton group velocity. The uncertainty $\delta \mathbf{k}$ indicates that the exciton state in the crystal is realized as a wave packet rather than a plane wave. The dimensions of the wave packet, $\delta x$, $\delta y$, and $\delta z$ can be estimated from the uncertainty relations $\delta x \delta k_x \simeq 1$, etc.

The motion of the exciton wave packet causes the transport of energy. In order to find the appropriate energy diffusion coefficient we must estimate the mean free path and the mean free time of the wave packets. This situation is quite similar to that of phonon heat conductivity (see, for example, Ref. [107]).

In analogy to the mobility of electrons and holes in crystals, the diffusion coefficient for coherent excitons is determined by the relaxation time $\tau$. According to Fröhlich [108], we have

$$\frac{1}{\tau} = -\sum_{\mathbf{q},r} \frac{\Delta k_z(\mathbf{q}r)}{k_z} \left[ W_a^\mathbf{k}(\mathbf{q}r) + W_e^\mathbf{k}(\mathbf{q}r) \right], \tag{100}$$

where $\mathbf{k}$ is the exciton wave vector before the collision, $\Delta k_z(\mathbf{q}r)$ is the change of the exciton wave vector in the $z$-direction due to the collision with the phonon $\mathbf{q}r$, and $W_a^\mathbf{k}(\mathbf{q}r)$ and $W_e^\mathbf{k}(\mathbf{q}r)$ are the probability per unit time of absorption and emission, respectively, of the phonon $\mathbf{q}r$ by the exciton system. After absorption or emission, the new exciton wave vectors, accurate to an integral reciprocal lattice vector, are $\mathbf{k}' = \mathbf{k} \pm \mathbf{q}$. According to Eq. (100), we can write

$$1/\tau = 1/\tau^{ac} + 1/\tau^{op}, \tag{101}$$

where $1/\tau^{ac}$ and $1/\tau^{op}$ are the scattering rates of the excitons on acoustic and optical phonons, respectively. As is the case for electrons in semiconductors, $\tau$ is approximately equal to the mean free time of the excitons with respect to collisions with phonons. Typically, a few collisions are sufficient to reach a thermodynamic equilibrium between phonons and band excitons. Thus, if the exciton lifetime $\tau_0$ is considerably longer than the relaxation time $\tau$, as is typically the case at elevated temperatures, we can assume that the excitons are in thermodynamic equilibrium with the lattice prior to their decay. The exciton diffusion coefficient is then related to the relaxation time by

$$D = \frac{1}{3}\langle \tau v^2 \rangle \approx \frac{1}{3}\langle \tau \rangle \langle v^2 \rangle, \tag{102}$$

where $\langle v^2 \rangle$ is the (equilibrium) mean squared group velocity of the excitons, and $\langle \tau \rangle$ is the mean relaxation time. Strictly speaking, explicit expressions for wave packets should be used when calculating the quantities in Eq. (100). However, usually the absorption and emission rates vary only slightly with the exciton wave vector over the $\mathbf{k}$ interval spanned by the wave packet. Therefore, the transition probabilities can be calculated by using exciton wave functions in the form of plane waves both before and after the scattering process. We will not discuss the details of such calculations, but rather address some typical and frequently used results.

### 4.3.2. Isotropic Exciton Effective Mass and Scattering by Acoustic Phonons

Using the Fermi golden rule, the probabilities of absorption and emission of a phonon by the exciton system read

$$W_e^{\mathbf{k}}(\mathbf{q}r) = \frac{2\pi}{\hbar}|F(\mathbf{k}-\mathbf{q};\mathbf{k};\mathbf{q}r)|^2$$
$$\times (n_{\mathbf{q}r}+1)\delta\big[E(\mathbf{k}) - E(\mathbf{k}-\mathbf{q}) - \omega_r(\mathbf{q})\big], \tag{103}$$

$$W_a^{\mathbf{k}}(\mathbf{q}r) = \frac{2\pi}{\hbar}|F(\mathbf{k}+\mathbf{q};\mathbf{k};\mathbf{q}r)|^2 n_{\mathbf{q}r}\delta\big[E(\mathbf{k}) - E(\mathbf{k}+\mathbf{q}) + \omega_r(\mathbf{q})\big], \tag{104}$$

where $n_{\mathbf{q}r} = [\exp\frac{\hbar\omega_r(\mathbf{q})}{k_B T} - 1]^{-1}$ is the thermal occupation of the phonon mode $\mathbf{q}r$ and $F(\mathbf{k}\pm\mathbf{q};\mathbf{k};\mathbf{q}r)$ is the exciton–phonon coupling constant.

At sufficiently low temperatures, when the thermal energy $k_B T$ is much smaller than the exciton band width, most excitons at thermodynamic equilibrium are concentrated in the vicinity of the exciton band minimum in the wave vector space. If this minimum corresponds to $\mathbf{k} = \mathbf{0}$ and we assume for simplicity that the excitons have an isotropic effective mass $m$, we have

$$E_\mathbf{k} = E_0 + \hbar^2 \mathbf{k}^2/2m. \tag{105}$$

Moreover, under these conditions the relation $|\mathbf{k}|a \ll 1$ ($a$ is the lattice constant) is satisfied for the overwhelming majority of the excitons.

The conservation of energy for the absorption and emission processes can now be written

$$\hbar^2 |\mathbf{k}|^2/2m \pm \hbar \omega_r(\mathbf{q}) = \hbar^2 (\mathbf{k} \pm \mathbf{q})^2 / 2m. \tag{106}$$

Since we are dealing with relatively low temperatures, let us focus on the case of acoustic phonons. We then have $\omega_r(\mathbf{q}) = v_0 |\mathbf{q}|$ ($r = 1, 2, 3$), where $v_0$ is the sound velocity. For the sake of simplicity we ignore the dependence of $v_0$ on the polarization $r$ and the direction of $\mathbf{q}$. Using Eq. (106), we obtain

$$q = \mp 2k \cos \vartheta \pm 2mv_0/\hbar, \tag{107}$$

where $\vartheta$ is the angle between the vectors $\mathbf{k}$ and $\mathbf{q}$. From the Boltzmann statistics $\hbar^2 \langle |\mathbf{k}|^2 \rangle / 2m = \frac{3}{2} k_B T$, we obtain for the typical value of $|\mathbf{k}|$ at a given temperature $T$:

$$\langle |\mathbf{k}|^2 \rangle^{1/2} = \frac{1}{\hbar} [3mk_B T]^{1/2}. \tag{108}$$

Hence, we can ignore the second right-hand side term in Eq. (107) relative to the first if

$$k_B T \gg \frac{1}{3} m v_0^2. \tag{109}$$

This criterion defines the temperature region where the scattering of excitons by phonons is almost elastic. We may make this estimate more quantitative by using $v_0 \approx 10^5$ cm/s as typical velocity of sound for solids, leading to

$$T \gg \frac{0.025 m}{m_0} \text{ K}, \tag{110}$$

with $m_0$ the electron mass in vacuum.

Below we shall assume that the criterion equation (110) is satisfied and, therefore, we shall neglect the phonon energy in the argument of the delta functions in Eqs. (103) and (104), and, likewise, we shall neglect the second term on the right-hand side of Eq. (107).

Now we observe from Eq. (107) that in our model for a given value of $|\mathbf{k}|$ the exciton can interact to first approximation with the phonons with wave vectors in the range $0 \leqslant |\mathbf{q}| \leqslant 2|\mathbf{k}|$, implying that, like $|\mathbf{k}|$, the phonon wave vector is small: $|\mathbf{q}|a \ll 1$. Using the smallness of $|\mathbf{k}|$ and $|\mathbf{q}|$, we can expand $|F(\mathbf{k}+\mathbf{q},\mathbf{k};\mathbf{q}r)|^2$ in powers of these wave vectors. Keeping only the lowest-order nonzero contribution, $|F(\mathbf{k}+\mathbf{q},\mathbf{k};\mathbf{q}r)|^2$ is then a linear function of $|\mathbf{q}|$ for acoustic phonons [109]. Moreover, if we ignore the generally weak dependence of this quantity on the directions of the vectors $\mathbf{q}$ and $\mathbf{k}$ and neglect the dependence of the phonon frequency on the direction of $\mathbf{q}$ and the phonon polarization, we obtain

$$\left|F^{\mathrm{ac}}(\mathbf{k}+\mathbf{q},\mathbf{k};\mathbf{q}r)\right|^2 \approx \left|F_0^{\mathrm{ac}}\right|^2 a|\mathbf{q}|, \qquad (111)$$

where $F_0^{\mathrm{ac}}$ is a constant. Using Eq. (111), we find that (in a three-dimensional medium) the relaxation time for the exciton with wave vector $\mathbf{k}$ due to the scattering by acoustic phonons is given by

$$\frac{1}{\tau^{\mathrm{ac}}} = \frac{3|F_0^{\mathrm{ac}}|^2 a^4 m}{4\pi\hbar^3|\mathbf{k}|^3}\left(\frac{k_B T}{\hbar v_0}\right)^5 \int_0^\xi x^4 \frac{e^{-x}-e^x}{(e^x-1)(e^{-x}-1)}\, dx, \qquad (112)$$

with $\xi = 2|\mathbf{k}|\hbar v_0/k_B T$.

### 4.3.3. Temperature Dependence of the Diffusion Constant

If $\xi < 1$, the integrand in Eq. (112) can be replaced with its value for small $x$. Using Eq. (108) for the typical value for $|\mathbf{k}|$, this condition translates into

$$T > T_0 \equiv 12 m v_0^2 / k_B. \qquad (113)$$

If we now replace $|\mathbf{k}|$ with its mean value $\langle|\mathbf{k}|^2\rangle^{1/2}$ and perform the integration in Eq. (112) using the small-$x$ expansion, we obtain

$$\frac{1}{\tau^{\mathrm{ac}}} = \frac{|F_0^{\mathrm{ac}}|^2 a^4 m}{\pi\hbar^3}\left(\frac{k_B T}{\hbar v_0}\right)\langle|\mathbf{k}|^2\rangle^{1/2}. \qquad (114)$$

Thus, for $T > T_0$, we find $\tau^{\mathrm{ac}} \sim 1/T^{3/2}$ which using Eq. (102) for the diffusion constant leads to

$$D \sim 1/T^{1/2}. \qquad (115)$$

This relationship for Frenkel excitons was derived by Agranovich and Konobeev [109]; it can be seen from its derivation that it is independent of the model and, therefore, is valid also for ground state large-radius excitons as well as for electrons and holes in semiconductors.

In the region of very low temperatures when $\xi \gg 1$, that is, when the condition opposite to Eq. (113) is satisfied but the inequality (109) still holds, the main contribution to the integral in Eq. (112) comes from large $x$. Replacing the integrand with its asymptotic large-$x$ value and performing the integration then yields

$$\frac{1}{\tau^{ac}} = \frac{24|F_0^{ac}|^2 a^4 m}{5\pi\hbar^3}\langle|\mathbf{k}|^2\rangle, \tag{116}$$

which implies (Eq. (102)) that the diffusion coefficient due to scattering by acoustic phonons ceases to be temperature dependent in the region of low temperatures. Thus, with decreasing temperature the relationship $D \sim 1/\sqrt{T}$ for the diffusion coefficient reduces to $D = \text{const}$. The exciton scattering by optical phonons becomes important only at sufficiently high temperatures. The discussion of the influence of such processes on the diffusion constant can be found in Refs. [36] and [102].

To end this section, we make a few remarks. First it should be noted that the applicability of the above $D(T)$ relationships is limited by the condition of applicability of the Boltzmann equation. This condition reads $l \gg \lambda$ where $l = \langle v \rangle \tau$ and $\lambda$ is the thermal de-Broglie wavelength of the excitons. Therefore, we should expect that this condition may be satisfied only for sufficiently low temperatures. Of course, the size of the corresponding temperature range depends on the exciton–phonon interaction constants and can be found only from experimental data.

The second remark concerns the role of crystal anisotropy. The majority of the well-studied molecular crystals, such as anthracene, naphthalene, pyrene, etc., are not cubic, but very anisotropic. It is therefore natural to consider to what extent the above qualitative results for the temperature dependence of the exciton diffusion coefficient in cubic crystals are valid for anisotropic crystals. It is clear, of course, that in anisotropic crystals the exciton diffusion coefficient can exhibit anisotropy owing, for instance, to anisotropy of the exciton's effective mass. However, as long as the effective masses in various directions are of the same order of magnitude (which seems to be just the case for naphthalene crystals), the temperature dependence of the diffusion coefficient maintains the same $D \sim 1/\sqrt{T}$ character (with the exclusion of the region of low temperatures). More interesting are the class of anisotropic crystals in which the lowest exciton band corresponds to high oscillator strengths. Then the exciton energy $E_\mathbf{k}$ is known to be a nonanalytic function of $\mathbf{k}$ for small $|\mathbf{k}|$ values and this has to be taken into account when calculating the diffusion tensor. Katalnikov [110] calculated the exciton diffusion tensor in uniaxial crystals taking into account the nonanalytic energy term and found, as could be expected, that in those crystals different components of the diffusion tensor have different temperature dependence. Strong anisotropy of $D_{ij}$ in molecular crystals can lead to one- or two-dimensional exciton motion [111].

### 4.4. Strong Exciton–Phonon Coupling: Incoherent Excitons

If the exciton–phonon coupling is strong, it results in the localization of the exciton at a lattice site, so that the exciton behaves as a classical particle that can hop from cell to cell. The hopping particle is in fact a self-trapped exciton, which, due to interactions with the phonon bath, looses its phase memory on a time scale short compared to the time it takes to hop between unit cells (see Section 5.2). The resulting hopping process is known as incoherent energy transfer and, if the relevant excitation transfer interaction $J_{nm}$ is of the dipole–dipole type, it is often referred to as Förster energy transfer. To lowest order in the excitation transfer interaction, the exciton motion is then a series of uncorrelated hops (forming a Markov process), described by a random walk over the lattice sites. If $P(\mathbf{m}, t)$ is the probability that an exciton is at the lattice site $\mathbf{m}$ at the moment $t$, then the random-walk equation for $P(\mathbf{m}, t)$ has the form

$$\frac{dP(\mathbf{m}, t)}{dt} = \sum_{\mathbf{n}} \left[ W(\mathbf{n}, \mathbf{m}) P(\mathbf{n}) - W(\mathbf{m}, \mathbf{n}) P(\mathbf{m}) \right], \quad (117)$$

where $W(\mathbf{n}, \mathbf{m})$ is the probability of hopping from the site $\mathbf{n}$ to the site $\mathbf{m}$ per unit time. In the next order of approximation the hopping process is not Markovian. We then have the following integro-differential equation for $P(\mathbf{m}, t)$

$$\frac{dP(\mathbf{m}, t)}{dt} = \int_0^t \left[ \widetilde{W}(\mathbf{n}, \mathbf{m}, \tau) P(\mathbf{n}, t - \tau) \right.$$
$$\left. - \widetilde{W}(\mathbf{m}, \mathbf{n}, \tau) P(\mathbf{m}, t - \tau) \right] d\tau, \quad (118)$$

i.e., the hopping process acquires a "memory". For molecular crystals, this problem has been discussed by Kenkre and Knox [112,101]. The excellent review by Silbey [102] deals with the problem in a more general way, including exciton scattering by impurities, dispersive transport, and the Haken–Strobl–Reineker model of exciton–phonon scattering [113,100]. We also point out the recent book on transport by May and Kühn, where the problem of memory kernels is addressed in detail [103].

If we assume that the variation of the function $P(\mathbf{m}, t)$ over distances of the order of the lattice constant is small, then Eq. (117) reduces to the diffusion equation. Indeed, assuming that the vector $\mathbf{m}$ varies continuously, we obtain

$$P(\mathbf{n}) = P(\mathbf{m}) + \sum_i (\mathbf{n} - \mathbf{m})_i \frac{dP(\mathbf{m})}{dm_i}$$
$$+ \frac{1}{2} \sum_{i,j} (\mathbf{n} - \mathbf{m})_i (\mathbf{n} - \mathbf{m})_j \frac{d^2 P(\mathbf{m})}{dm_i dm_j}. \quad (119)$$

If $W(\mathbf{m}, \mathbf{n}) = W(\mathbf{n}, \mathbf{m})$ (which holds if there is no static disorder in the crystal), substitution of the above expression into Eq. (117) yields

$$\frac{\partial P(\mathbf{r}, t)}{\partial t} = D_{ij} \frac{\partial^2 P(\mathbf{r}, t)}{\partial x_i \partial x_j}, \quad (120)$$

where the diffusion coefficient tensor is

$$D_{ij} = \frac{1}{2} \sum_{\mathbf{m}} (\mathbf{n} - \mathbf{m})_i (\mathbf{n} - \mathbf{m})_j W(\mathbf{n}, \mathbf{m}). \tag{121}$$

For crystals with several molecules (labeled by $s$) per unit cell, which have symmetry operations exchanging molecules with different $s$ values, as holds for anthracene crystals, a similar procedure yields

$$D_{ij} = \frac{1}{2} \sum_{\mathbf{m},s'} (\mathbf{r_{ns}} - \mathbf{r_{ms'}})_i (\mathbf{r_{ns}} - \mathbf{r_{ms'}})_j W(\mathbf{n}s, \mathbf{m}s'). \tag{122}$$

Thus, under the given conditions, the calculation of the diffusion tensor reduces to determination of the hopping rates $W(\mathbf{n}s, \mathbf{m}s')$.

For singlet excitons the probability $W(\mathbf{n}s, \mathbf{m}s')$ can be estimated using the results of strong exciton–phonon coupling theory, as has been done by Trlifaj [114]. The intermolecular dipole–dipole interaction leads to $W(\mathbf{n}s, \mathbf{m}s') \sim 1/|\mathbf{r}_{ns} - \mathbf{r}_{ms'}|^6$, as is characteristic for Förster energy transfer [99], and in the summation equation (121) it suffices to take into account only nearest neighbors. The temperature dependence of this probability may be presented by the relation

$$D_{ij}(T) \approx D_{ij}^0 \exp(-U_a/k_B \overline{T}), \tag{123}$$

where $U_a$ is the activation energy for hopping, $\overline{T}$ is a constant for $T \ll T_D$ ($T_D$ is the Debye temperature), and $\overline{T} = T$ for $T \gg T_D$. It follows from these qualitative considerations that in the strong coupling regime the exciton diffusion constant, in contrast to the case of weak exciton–phonon coupling, increases with growing temperature.

In some molecular crystals a crossover from coherent excitons (exciton mean free path $l \gg \lambda$) to incoherent ones ($l \approx \lambda$, Ioffe–Regel criterion) takes place with increasing temperature. We then expect that upon increasing the temperature from very low values, at some threshold temperature the decreasing behavior of the diffusion constant for coherent excitons goes over into an increasing behavior.

A similar crossover phenomenon may be observed in heavily doped isotopically mixed crystals at low temperatures. In such crystals the impurity molecules are responsible for the scattering of excitons and at low temperature this scattering is almost elastic. When increasing the impurity concentration, a crossover occurs between so-called weak (at $l \gg \lambda$) and strong (at $l \approx \lambda$) Anderson localization. This crossover is analogous to the change of the electron mobility in metals upon increasing the impurity concentration and should be expected to have a strong influence on the exciton transport. A different situation arises when the electronic excitation energy of the isotopic impurity is lower than the energy of the exciton in the crystalline host. The isotopic impurities enter the lattice substitutionally and they are randomly distributed in the matrix. The electronic excitation energy transfer from one impurity to another when decreasing the impurity concentration may be used to investigate the transition from impurity band to impurity hopping transfer. Such a transition is similar to the Anderson conductor–insulator transition in semiconductors.

An extensive discussion of experiments on exciton transport in isotopically disordered crystals and numerical simulations of this phenomenon in the framework of a percolation model, may be found in the review paper by Kopelman [115]. A more recent review of this field, including the discussion of the Anderson model, may be found in the book by Pope and Swenberg [17].

### 4.5. TRANSPORT MEASUREMENTS AND DIFFUSION OF POLARITONS

Many papers have been devoted to the experimental determination of the exciton diffusion constant $D$. In most of these studies, $D$ was determined by observing how the diffusion of excitons results in their capture by impurities (sensitized fluorescence) or in bimolecular quenching of excitons (reviews of these experiments may be found in Refs. [36,17]). The interpretation of such experiments requires that not only the diffusion of the excitons to the acceptor is taken into account, but also the character of the exciton interaction with the acceptor (i.e., with the impurity molecule or with another exciton). An alternative experimental technique that does not suffer from these problems, is the picosecond transient grating (TG) method. This third-order nonlinear optical technique has been used abundantly for the study of various kinetic parameters of condensed media (liquids, semiconductors, etc.). Fayer and collaborators were the first to propose the use of TG experiments for the study of exciton transport in molecular crystals [116]. They applied the method to anthracene thin films [117,118].

In order to determine the exciton diffusion constant, one studies the decay kinetics of excitonic gratings, i.e., a spatially periodic variation in the exciton density, formed in a molecular crystal as a result of the interference of two coherent picosecond laser pulses. The periodic spatial distribution of the excitons, as well as its evolution, can be investigated by observing the diffraction of a short probe pulse sent into the crystal some delay time $t$ after creating the grating. As a result of the finite exciton lifetime and exciton diffusion, the grating amplitude decreases in time, so that the intensity $S(t)$ of the diffracted signal decreases with growing $t$. Thus, measuring $S(t)$ allows one to obtain information on the diffusion coefficient $D$ and the exciton lifetime $\tau_0$. In fact, it is easy to show that

$$S(t) = S(0)e^{-Kt}, \qquad (124)$$

with

$$K = 2\left(\frac{1}{\tau_0} + D\Delta^2\right). \qquad (125)$$

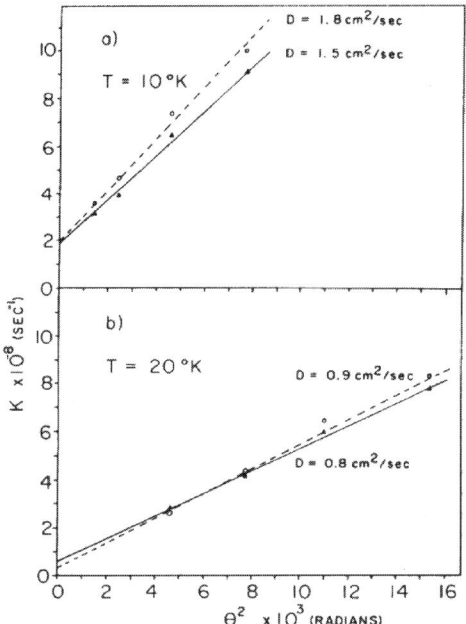

Fig. 7. The decay rate of the transient grating signal versus $\theta^2$ ($\theta$ the angle between the pump pulses) for anthracene crystals at 10 and 20 K [117]. The magnitude of the slope is proportional to the diffusion constant of the excitations in the crystal. With increasing temperature, the diffusion constant decreases. The average diffusion constant obtained from these data is about 10 times larger than the value expected for incoherent exciton motion [119]. (Figure reprinted from Ref. [117] with permission from Elsevier.)

Here, $\Delta = 2\pi/L$, where $L$ is the fringe spacing of the grating, which is given by $L = \lambda_e/(2n \sin[\theta/2])$ ($n$ is the crystal's refractive index, $\lambda_e$ the wavelength of the two excitation pulses, and $\theta$ the angle between them). By measuring the diffracted-signal decay for various values of $\theta$ and plotting the observed value of $K$ versus $\Delta^2$, the diffusion constant $D$ can be obtained from the slope, while the $\Delta = 0$ intercept is $2/\tau_0$. As the life time $\tau_0$ can also be found from other experiments (for example, from photoluminescence measurements) this provides a rigorous test of the assumption of diffusive propagation of the excitons as well.

In the above-mentioned TG experiments by the Fayer group [117,118], it was found that the diffusion constant $D$ in anthracene films at low temperature ($T = 1.8, 10, 20$ K) can reach values of the order of 1–10 cm$^2$/s (Figure 7). Such very large values of the diffusion constant too strongly contradicted the typical room temperature value of $D \sim 10^{-3}$ cm$^2$/s and attracted the attention of many investigators. It was argued [119,120] that in the interpretation of the experiments on anthracene it is necessary to take into account the fact that the lowest electronic transition in anthracene has a rather large oscillator strength, leading to a strong exciton-polariton formation. As a consequence, polaritons rather than excitons are the lowest-energy elementary excitations at low temperatures and the theoretical analysis of the decay time of the excitonic gratings should be associated with diffusion of polaritons rather than excitons.

At low temperature the polaritons are concentrated in the "bottle neck" region. Their diffusion constant in this case can be estimated as $D \approx \frac{1}{3} v_p l_p$, where $v_p$ is the group velocity of the polaritons in this region and $l_p$ is their mean free path. Since $l_p = c/(\omega \kappa)$, where $\kappa$ is the imaginary part of the refractive index $((2\omega/c)\kappa \approx 10^4$–$10^5$ cm$^{-1}$), and $v_p \approx 10^5$ cm/s, one obtains for the diffusion constant $D \geqslant 1$ cm$^2$/s, which agrees by the order of magnitude with the value obtained in the TG experiments. The discussion of the influence of bimolecular quenching and reabsorption of exciton fluorescence on the decay of the exciton gratings in organic crystals may be found in Ref. [120]. The microscopic formulation of TG and four-wave mixing experiments in molecular crystals in terms of polaritons may be found in Ref. [60].

## 5. Self-Trapping of Excitons: Spectra and Transport

### 5.1. Introductory Remarks

In Section 4 we have seen that the interaction of excitons with phonons governs the nature of the exciton motion. Coherent motion occurs for weak exciton–phonon coupling, while incoherent transport takes place for strong coupling. In the limit of strong interaction, the phonons not only affect the motion of the excitons, but they may also alter the exciton state itself considerably, to a so-called self-trapped exciton. The self-trapping (ST) of excitons is analogous to the ST of electrons and holes in ionic crystals, which also arises from the interaction of the quasi-particles with the lattice vibrations. For electrons in ionic crystals the possibility of ST was first pointed out by Landau in 1933 [121]. He showed that due to the electron–ion interaction the states of the electron "trapped by the lattice" (i.e., the states in which the lattice around an electron is strongly deformed) have an energy smaller than that of the Bloch band states in the regular lattice. The word "polaron" was later introduced by Pekar to denote such trapped states; he also developed the first consistent theoretical treatment of the self-trapped electron state by considering the model of a large-radius local state [122].

The physics of exciton ST has many features in common with electron (hole) ST. Therefore, the methods and results of the theory of electron ST have been widely used in the development of the theory of exciton ST. The effects of strong exciton–phonon coupling in organic crystals and the possibility of exciton ST were discussed in the papers by Peierls [123], Frenkel [22], and Davydov [124].

More recent discussions of exciton ST and the synthesis of almost all available approaches may be found in the reviews published as Refs. [125–128]. The reader is referred to this literature for a detailed account of exciton ST. In this section, we will restrict ourselves to a qualitative description of the most general and characteristic features of this phenomenon.

### 5.2. Self-Trapping of Frenkel Excitons

First of all, let us explain why the ST of Frenkel excitons, which in contrast to electrons (holes) have no charge, may arise. It is useful to start from the limit of a narrow exciton band and strong exciton–phonon coupling. If $\Delta$ denotes the exciton band width in the regular lattice, then $\tau_1 \approx \hbar/\Delta$ is the time scale for the transfer of the excitation from one molecule to another. The second relevant time scale is $\tau_d$, which denotes the time necessary for molecules to be displaced to new equilibrium positions upon a change in the electronic state. Such changes into a new, locally deformed, lattice configuration arise from the fact that the intermolecular interactions (like the Van der Waals interaction) are changed upon excitation of one of the molecules. In the limit of a narrow exciton band and strong exciton–lattice interaction, we have $\tau_1 \gg \tau_d$. If $E_d$ is the energy of the local deformation (the analog of the polaron energy shift), then $\tau_d \approx \hbar/E_d$. It is also clear that $\tau_d$ is longer than the characteristic period of a lattice vibration $2\pi/\omega_v$, where $\omega_v$ is a typical phonon frequency (in organic solids, $\omega_v \approx 100$ cm$^{-1}$). We thus have in the limit considered the condition $\tau_1 \gg \tau_d \geqslant 2\pi/\omega_v$. It follows that in this limit ($\Delta \ll \hbar\omega_v$) the phonons (displacements) are relatively fast, which allows them to follow the slowly moving excitons.

Thus, in the limit of narrow exciton bandwidth, the local deformation travels through the crystal following the molecular excitation. In other words, as has been metaphorically described by Frenkel, the exciton while moving through the lattice "drags with itself the entire load of atomic displacements". Thus, the already narrow exciton band is transformed into an even narrower band of "dressed" excitons. However, since the dissipative width of such states in most cases is large compared to the narrow band width, the wave vector cannot be consider a "good" quantum number. The band picture of "dressed" excitons is destroyed and localized molecular excited states (dressed with a deformation) propagate by hops from one molecule to another. It is clear that here we meet the case of incoherent excitons, with energies that, in contrast to the case of coherent excitons, do not depend on the wave vector. The diffusion constant of such excitons was addressed already in Section 4.4 [114]. It is useful to note that triplet excitons in organic solids have a rather small band width (of the order of 10 cm$^{-1}$), implying that the occurrence of low-energy ST triplet states is very likely.

We now turn to crystals with a wide exciton band, $\Delta \gg \hbar\omega_v$. Under this condition, the exciton system is fast and, thus, the usual adiabatic approximation may be used in the ST theory. This situation was first analyzed by Deigen and Pekar [129], who showed that in this case ST states of small radius $R_{ST} \approx a$ may be formed ($a$ is the lattice constant). For localization on a scale $\Delta x \approx a$, the uncertainty of the wave vector is $\Delta k \approx 1/a$, which is of the order of magnitude of the Brillouin zone. This means that when localizing, the energy of the exciton becomes of the order of the exciton band width $\Delta$. Thus, if the localization (deformation) energy $E_d > \Delta$ (strong exciton–phonon coupling), a ST state will be formed and its energy will lie below the bottom of exciton band. On the other hand, if $E_d < \Delta$, the excitons in the lowest-energy states remain coherent and ST states (if they arise at all) may appear only in the region of higher energies. More information on ST of wide-band excitons may be found in the review paper by Rashba [125].

### 5.3. Self-Trapping Barrier

In his seminal paper on self-trapping [121], Landau already noticed that to make the transition from the coherent (free) state to the ST state, the particle has to overcome a barrier. In the case of small-radius excitons (Frenkel excitons) the existence of a barrier in crystals composed of large molecules may be stipulated by purely spatial limitations. The excited molecule tries to pass into a ST state with lower energy, but the surrounding molecules, having no "wish" to be displaced or change their orientation, may prevent such passage. Thus, the coherent states are protected by a barrier, which makes them meta-stable and gives them the opportunity to exist during a finite time, even if the passage into the ST state is accompanied by a considerable energy gain. One then speaks of the coexistence of coherent and ST excitons. It should be stressed that this coexistence is not a general phenomenon and occurs only for three-dimensional (3D) systems. As was shown by Rashba [125], no barrier exists for excitons in 1D systems. In this case after initial excitation of a coherent exciton, a monotonous lowering of the total energy takes place towards the energy of the ST state. Thus, the coherent exciton states are absolutely unstable and the creation of ST states occurs without any restrictions. For polarons in 3D ionic crystals, the same physical picture was found earlier by Pekar [122].

### 5.4. Self-Trapping of Charge-Transfer Excitons

As a brief intermezzo, we mention some peculiarities of ST of charge-transfer excitons. The electron–hole interaction energy in a charge-transfer exciton is of the order of $e^2/a$, where $a$ is the distance between electron and hole. This energy may be a few eV. The change of this energy under the influence of lattice vibrations may compete with the interaction of the free electron and hole with phonons and may even be the dominant mechanism of exciton–phonon interaction. Such a situation is expected to take place for many organic crystals (see Ref. [16] and the chapter by M. Hoffman in this book). The theory of ST of charge-transfer excitons in such crystals can be found in Refs. [16,130].

Another interesting situation arises in the ST of excitons in which the electron and hole are spatially separated and are localized on different filaments (polymers or quantum wires) or on different planes (or quantum wells). In such structures the electron–hole Coulomb interaction changes when these filaments or planes are deformed. As a result a strong exciton–phonon interaction may exist, even if the individual quasi-particles (electron and hole) have very small interaction with the phonons. The theory of ST of this type of excitations may be found in Ref. [131].

### 5.5. Spectra and Mobility of ST Excitons

We return to the ST of Frenkel excitons in 3D organic structures. As a result of the Franck–Condon principle, photo-exciting a crystal from its ground state, with a regular lattice, leads to the initial creation of a coherent exciton. In order to pass into a ST state, the latter has to overcome a barrier (in a 3D lattice), which may be done either by thermoactivated tunneling or by a thermoactivated transition. A review of the relevant theory is given in Ref. [132]. It was shown that the rate of ST of thermalized excitons can be represented as

$$W(T) = \omega_v B(T) \exp[-S(T)], \qquad (126)$$

where $\omega_v$ is the characteristic phonon frequency, the pre-exponential factor $B(T)$ is always large in comparison with unity, and $S(T)$ is the temperature-dependent Hamiltonian action. At temperatures with $k_B T > \hbar \omega_v$, we have

$$S(T) = \frac{U}{k_B T}, \qquad (127)$$

where $U$ is the height of the barrier. Thus, as could be expected, the ST rate then follows an Arrhenius law.

The experimental investigations of ST in organic crystals mainly concern optical spectra, in particular time-resolved spectra. Good examples are the experiments by Matsui and co-workers reported in Ref. [133]. To discuss these spectra, let us consider the energy $E_c(\eta)$ of the crystal in its ground and excited states as functions of the coordinate $\eta$ that undergoes a strong displacement upon ST. We assume that the ground state energy has its minimum at $\eta = 0$ and that this also represents the local minimum in the excited state. As is shown in Figure 8, when the exciton–phonon interaction is taken into account, $\eta = 0$ indeed only represents a local minimum. The absolute minimum corresponds to the ST state, so that the dependence of the total energy on $\eta$ is described by an asymmetric double-well potential. As mentioned above, the coherent states are protected by a barrier; their optical spectra have to be analogous to the spectra in a regular crystal with weak

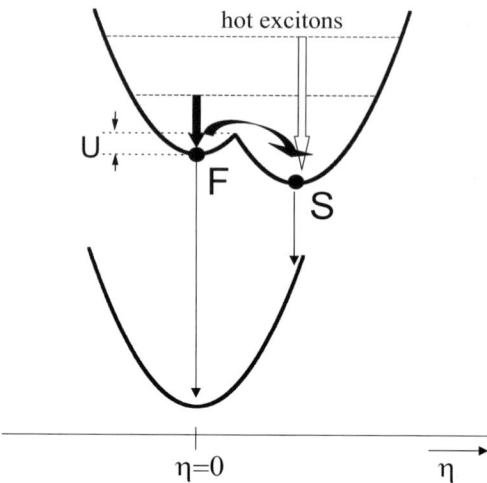

Fig. 8. Ground state potential and asymmetric double-well potential associated with the phenomenon of exciton self-trapping, as a function of the coordinate $\eta$ that undergoes a strong displacement upon self-trapping. F is the bottom of the free-exciton band, in which the lattice is not distorted ($\eta = 0$), S denotes the lowest self-trapped exciton state, and $U$ is the barrier height. The luminescence from the self-trapped state is red-shifted relative to the free-exciton luminescence. Upon photo-excitation of the system, two pathways towards the self-trapped state occur. The first possibility is that the created excitons first relax towards the bottom of the free-exciton well, after which they may further relax to the self-trapped state through tunneling or a thermoactivated process. This pathway is indicated by the filled arrows. The second possibility is that high-energy (hot) excitons relax directly to the self-trapped state, as indicated by the open arrow.

exciton–phonon coupling. We then expect the existence of narrow zero-phonon lines in the absorption and fluorescence spectra and a Davydov splitting may be observed. The existence of ST states leads to the appearance of additional broad and red shifted bands in the fluorescence spectra. As discussed already in Section 4.4, upon ST the movement of wave packets is replaced by hops. At high temperature these hops are thermoactivated leading to the Arrhenius law equation (123) for the diffusion constant. At low temperature quantum tunneling occurs, in analogy to quantum diffusion of impurities in solids (see Ref. [134] for a review).

Numerous picosecond experiments have been performed on exciton ST in organic crystals [133]. Such experiments give the possibility to study the dynamics of the ST process and allow one to determine the height of the ST barrier and the rate of the transient free-exciton luminescence. For some crystals these investigations also gave the possibility to trace the pathways of self-trapping (cf. Figure 8). An example is pyrene, which is a crystal with a rather strong exciton–phonon interaction and a barrier height $U \approx 262$ cm$^{-1}$. It was demonstrated that upon photo-generation of excitons in this crystal, even at low temperature, the process of self-trapping not always requires relaxation to the bottom of free exciton band (with $\mathbf{k} = \mathbf{0}$; black arrows in Figure 8), but sometimes takes place directly from the states with large $\mathbf{k}$ (hot excitons) avoiding the ST barrier, as is shown by the open arrow in Figure 8.

## 6. Charge-Transfer Excitons in Organic Solids

### 6.1. GENERAL CONSIDERATIONS

As we have discussed in Section 1, excitons are distinguished in two main groups: small-radius Frenkel excitons, which basically are delocalized molecular excitations, and large-radius Wannier–Mott excitons, in which electron and hole have a hydrogen-like relative motion with a radius much larger than a lattice constant. The charge-transfer exciton (CTE) occupies an intermediate place in this classification [36,135,136]. The lowest-energy CTE usually extends over two nearest-neighbor molecules and creates a so-called "donor–acceptor (D-A) complex". This is currently considered as an important intermediate state in the creation of free carriers in the photoconductivity of organic crystals, a process in which the first step is the photo-generation of a Frenkel exciton. In a CTE, the electron is localized on the acceptor and the hole on the donor. In organic crystals, such CTE localization over nearest-neighbor molecules is usually stable, because the electron–hole attraction energy is large compared to the widths of the conduction and valence bands. The localization is further stabilized by the strong tendency of the CTE to undergo self-trapping [130]. Nevertheless, such an ionic pair as a whole can be mobile; the corresponding band theory of CTEs has been discussed in many papers (see, e.g., Refs. [16,130]).

Due to the electron–hole separation in a CTE, the static dipole moment created by the positive and negative ions can assume values as large as 10–25 D. This is responsible for some of the most characteristic properties of the CTE. For example, due to its large dipole moment, the CTE contributes a large second-order nonlinear polarizability $\chi^{(2)}$ [137]. It has also been shown that the same feature can be responsible for a new type of photovoltaic effect in organic asymmetric D-A superlattices [138], for unusual intensity dependencies of nonlinear polarizabilities of D-A superlattices [139], and also for phase transitions to conducting states in a system of two-dimensional interacting CTEs [140] (see Chapter 5 below for more details). In all of these cases, it was assumed that CTEs at D-A interfaces between alternating layers of donors and acceptors are the lowest-energy electronic excited states of such an organic multi-layer structure. These states are usually populated after lattice relaxation from higher-energy Frenkel-type electronic or vibronic states.

To explain in the simplest way the structure and excitation energy of CTEs it suffices to consider a donor–acceptor pair of molecules that are neutral in their ground states. When exciting a CTE, an electron leaves the donor and transfers to the acceptor. The energy of this excited state can be estimated from the relation $E = I - A + C + P$, where $I$, $A$, $C$, and $P$ are the donor's ionization energy, the acceptor's electron affinity, the electrostatic Coulomb attraction between the electron and hole, and the polarization energy of the crystal by an infinitely separated positive and negative ion pair.

An excellent review of the optical and photoconductive properties of organic solids composed of different donor and acceptor molecules arranged in an alternating way in quasi-one-dimensional arrays, has been published by Haarer and Philpott [16]. They discussed molecular crystals with neutral ground states and with rather small ionization potentials ($I < 7$ eV) and large electron affinities ($A > 2$ eV). In such crystals the CTEs are the lowest-energy electronic excitations. As a rule these excitations are self-trapped, have very broad absorption bands, and their optical properties can be understood in the framework of a local picture, ignoring the dispersion of states in a CT exciton band. Typical examples of such crystals are anthracene-PMDA, which has the crystal structure of a mixed stack of different molecules (D-A-D-A-), and the crystal of TTF-TCNQ, which has the structure of segregated stacks of identical molecules, A-A-A-A- and D-D-D-D-.

The situation is different for molecular crystals composed of identical molecules. An excellent review of the state of the art in this field has recently been published by Petelenz [141]. Usually, the energy of CTE states in such crystals is larger than the energy of the lowest Frenkel exciton sate. However, for some quasi-one-dimensional crystals with a small distance between the nearest-neighbor molecules, like the crystal of 3,4,9,10-perylenetetracarboxylic dianhydride (PTCDA), the energy separation between the lowest-energy Frenkel and CTE states can be very small. For such crystals the nature of the lowest-energy electronic excitations is determined by the mixing of Frenkel exciton and CTE states and strongly depends on the orbital overlap between the molecules (see the chapter by M. Hoffmann in this book). This mixing occurs due to the possible virtual transformation of a Frenkel exciton into a CTE and vice versa. More explicitly, the Frenkel exciton localized at molecule $n$ can dissociate by electron transfer to molecule $m$ to give the CT state $n^+m^-$ or by hole transfer to give $n^-m^+$. The CT state can recombine by subsequent hole or electron transfer to give a Frenkel exciton located on molecule $n$ or $m$.

### 6.2. Stark Effect and Electro-Absorption of CTEs

An externally applied static electric field **F** shifts the exciton levels and these shifts are reflected in the absorption spectra. In the case of CTEs the shift is linear in the field:

$$E_-^{CT} = I - A + C + P - \boldsymbol{\mu}_p \cdot \mathbf{F}, \qquad (128)$$

where $\boldsymbol{\mu}_p$ is the static dipole moment of the CTE. If the crystal has inversion symmetry, a CTE state with opposite direction of $\boldsymbol{\mu}_p$ exists as well and this exciton acquires the energy

$$E_+^{\text{CT}} = I - A + C + P + \boldsymbol{\mu}_p \cdot \mathbf{F}. \tag{129}$$

In these expressions we neglected the corrections quadratic in the field $\mathbf{F}$. For the Frenkel exciton, which has no permanent dipole moment, such quadratic contributions are the lowest-order ones and the energy reads

$$E(\mathbf{F}) = E(\mathbf{0}) + \frac{1}{2} \sum_{i,j} a_{ij} F_i F_j, \tag{130}$$

where $a_{ij} = a_{ij}^{\text{exc}} - a_{ij}^{\text{g}}$ is the excess polarizability tensor, i.e., the difference of the molecular excited state and ground state static polarizability tensors.

The dielectric constant in the frequency region of the CT transition can be expressed by the formula

$$\varepsilon(\omega, \mathbf{F}) = \varepsilon_{0,0} + \frac{A}{E_+^{\text{CT}} - \omega - i\gamma} + \frac{A}{E_-^{\text{CT}} - \omega - i\gamma}, \tag{131}$$

where $A$ is a constant proportional to the oscillator strength and $\gamma$ is the dissipative width. It follows from this relation that up to second order in the field

$$\varepsilon(\omega, \mathbf{F}) = \varepsilon(\omega, \mathbf{0}) + \delta\varepsilon(\omega, \mathbf{F}), \tag{132}$$

with

$$\delta\varepsilon(\omega, \mathbf{F}) = \frac{1}{2}[\boldsymbol{\mu}_p \cdot \mathbf{F}]^2 \frac{d^2\varepsilon(\omega, \mathbf{0})}{d\omega^2}. \tag{133}$$

Thus, for CTEs the leading effect of a static electric field on $\varepsilon(\omega)$ is seen to be quadratic in the field and proportional to the second derivative of $\varepsilon(\omega, \mathbf{0})$.

In the frequency region of the Frenkel exciton, we have

$$\varepsilon(\omega, \mathbf{F}) = \varepsilon_{0,0} + \frac{A}{E(\mathbf{F}) - \omega - i\gamma} = \varepsilon(\omega, \mathbf{0}) + \delta\varepsilon(\omega, \mathbf{F}), \tag{134}$$

with

$$\delta\varepsilon(\omega, \mathbf{F}) = -\frac{1}{2} \sum_{i,j} a_{ij} F_i F_j \frac{d\varepsilon(\omega, \mathbf{0})}{d\omega}. \tag{135}$$

Thus, $\delta\varepsilon(\omega, \mathbf{F})$ is proportional to $\sum_{i,j} a_{ij} F_i F_j$ and to the first derivative of $\varepsilon(\omega, \mathbf{0})$.

For the interpretation of experiments it is important to know the corrections to the real and imaginary parts of the refractive index, $n$ and $\kappa$, respectively, where

$\kappa$ measured as a function of frequency gives the absorption spectrum. These corrections can be found easily from the relation $\varepsilon = (n+i\kappa)^2$. Thus, the corrections to $n$ and $\kappa$, which we denote as $\delta n$ and $\delta \kappa$, read

$$\delta\kappa = \frac{n_0(\delta\varepsilon)'' - \kappa_0(\delta\varepsilon)'}{2(n_0^2 + \kappa_0^2)}, \tag{136}$$

and

$$\delta n = \frac{\kappa_0(\delta\varepsilon)'' + n_0(\delta\varepsilon)'}{2(n_0^2 + \kappa_0^2)}, \tag{137}$$

where $(\delta\varepsilon)'$ and $(\delta\varepsilon)''$ are the real and imaginary parts of $\delta\varepsilon$, respectively ($\delta\varepsilon = (\delta\varepsilon)' + i(\delta\varepsilon)''$), while $n_0(\omega)$ and $\kappa_0(\omega)$ are the real and imaginary parts, respectively, of the refractive index in the absence of the static electric field.

If in the spectral region under consideration the absorption is not very strong, so that $n_0(\omega) \gg \kappa_0(\omega)$, the expressions for $\delta\kappa$ and $\delta n$ are reduced to

$$\delta\kappa = \frac{(\delta\varepsilon)''}{2n_0}, \quad \delta n = \frac{(\delta\varepsilon)'}{2n_0}. \tag{138}$$

The formula for $\delta\kappa$ forms the basis of electro-absorption spectroscopy, which is an experimental technique in which one measures the change of the absorption spectrum induced by a slowly varying external electric field. (Using a slowly varying field, as opposed to a static field, makes it experimentally simpler to extract the change of the absorption spectrum by focusing on the component of the spectrum that varies in time according to the applied slow frequency.) As we demonstrated, for Frenkel excitons the change of the absorption spectrum is proportional to the first-derivative signal, which is an antisymmetric function of $\omega$ relative to the resonance frequency. On the other hand, a second-derivative signal, which is symmetric around the resonance frequency, is commonly associated with CTEs (see, for example, [142–144]). This may be made explicit by considering the simple generic form for $\varepsilon''(\omega, \mathbf{0})$ close to a resonance frequency $\omega_0$:

$$\varepsilon''(\omega, \mathbf{0}) = \frac{g}{(\omega - \omega_0)^2 + \gamma^2}. \tag{139}$$

Using this expression and Eqs. (135) and (138), we find for a Frenkel exciton resonance

$$\delta\varepsilon(\omega, \mathbf{F}) = \frac{1}{2} \sum_{i,j} a_{ij} F_i F_j \frac{2g(\omega - \omega_0)}{[(\omega - \omega_0)^2 + \gamma^2]^2}, \tag{140}$$

which indeed is seen to be an antisymmetric function relative to the resonance frequency. On the other hand, if $\omega_0$ is a CTE resonance, we find from Eqs. (133)

and (138)

$$\delta\kappa(\omega, \mathbf{F}) = \frac{1}{2}[\boldsymbol{\mu}_p \cdot \mathbf{F}]^2 \left[ -\frac{2g}{[(\omega-\omega_0)^2 + \gamma^2]^2} + \frac{4g(\omega-\omega_0)^2}{[(\omega-\omega_0)^2 + \gamma^2]^3} \right], \quad (141)$$

which is symmetric relative to the resonance frequency.

It should be noted that the above classification of the electro-absorption spectrum is valid only approximately, because first of all Eq. (138) is correct only in the case of weak absorption and, second, the Frenkel and CT exciton states usually mix. We finally mention that the change of the refractive index $\delta n$ is of the same order as $\delta\kappa$; new experimental techniques are required to measure this change, however. Good candidates for such methods have been proposed by Warman and coworkers [145]. The success of such measurements could be the basis of electro-refraction spectroscopy, complimentary to the existing electro-absorption spectroscopy.

## 7. Molecular Aggregates: Low-Dimensional Exciton Systems

### 7.1. Introductory Remarks

In 1936, Jelley [146] and Scheibe [147] independently discovered that upon increasing the concentration of a solution of the dye pseudo-isocyanine (PIC, Figure 9), the absorption spectrum strongly changed. The relatively broad absorption band of PIC monomers at 525 nm disappeared and was replaced by a much narrower absorption band around 570 nm. It was soon realized that this narrow band, which is now generally known as the J (Jelley) band (sometimes also as the S (Scheibe) band), was a manifestation of Frenkel exciton states that existed on large groups of aggregated PIC molecules [148]. Polarized experiments in streaming solutions, performed by Scheibe, suggested that these aggregates have a thread-like structure [149]. Ever since these initial discoveries, a strong interest in the optical properties of molecular aggregates of these and other so-called polymethine cyanine dyes has persisted. The class of molecular aggregates with a red-shifted absorption band are referred to as J-aggregates. The narrow absorption band with a large intensity (oscillator strength) is the most characteristic property of these systems. A good example of the growing of the aggregate J band at the expense of the single-molecule absorption band is given in Figure 10 for the dye 5,5′,6,6′-tetrachloro-1,1′-diethyl-3,3′-di(4-sulfobutyl)-benzimidazolocarbocyanine (TDBC).

One of the driving forces to study cyanine dyes and their aggregates, is their abundant application as photosensitizers in photographic emulsions. Even to date, this remains a topic of much study [12]. However, also on a more fundamental level, J-aggregates have aroused much interest. In particular the collective optical

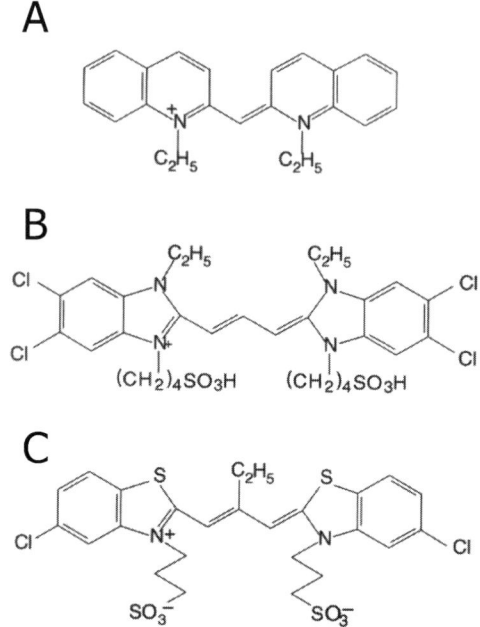

Fig. 9. Three examples of cyanine dyes that are known to form quasi-one-dimensional molecular J-aggregates. Shown are (a) 1,1′-diethyl-2,2′-cyanine (PIC), (b) 5,5′,6,6′-tetrachloro-1,1′-diethyl-3,3′-di(4-sulfobutyl)-benzimidazolo-carbocyanine (TDBC), and (c) 3,3′-bis(sulfopropyl)-5,5′-dichloro-9-ethylthiacarbocyanine (THIATS).

and nonlinear optical response of the molecules that form the aggregate, caused by the delocalized Frenkel excitons, has been the key word in this research. The above mentioned red-shift of the absorption spectrum and the narrowness of the J-band are examples of such collective properties. The narrowness is ascribed to exchange narrowing (or motional narrowing) of disorder by the delocalized exciton states [150,151]. Other collective properties that have attracted attention more recently are the ultrafast cooperative spontaneous emission ("exciton superradiance") [1,2,152] and the possibility of size-enhanced nonlinear optical response [6,7,153–157]. During the past 15 years, these optical properties have been studied with a growing number of optical techniques, including fluorescence excitation, hole burning, photon-echoes, and pump-probe experiments (see, e.g., [25, 41]). The motivation in performing these experiments is the intrinsic interest in the collective properties and the interest in unraveling the static and dynamic properties of the exciton states underlying them. Finally, the field of molecular aggregates recently has received a renewed interest, following the discovery that the photosynthetic systems of bacteria and higher plants often contain highly orga-

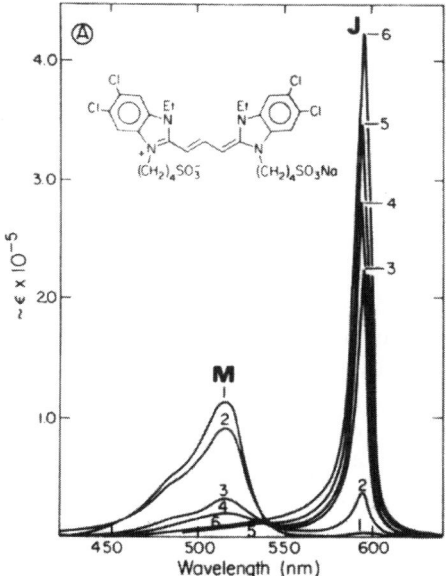

Fig. 10. Changing of the absorption spectrum of a solution of TDBC molecules (Figure 9(b)) upon increasing the solute concentration (from curve 1 to 6). Clearly is demonstrated that the absorption peak changes from the broad monomeric absorption peak at about 520 nm to the narrow and red-shifted J-band resulting from aggregated molecules. (Figure reprinted from Ref. [11] with permission from Elsevier.)

nized light-harvesting (LH) systems that are basically aggregates of chlorophyll molecules [13]. We will address these types of aggregates in Section 8.

Molecular aggregates differ from bulk crystals in several respects. Most important is the difference in structure. Aggregates generally do not have a three-dimensional structure, but instead the electrostatic forces that favor their self-assembly may lead to one-dimensional geometries (molecular chains or bundles), two-dimensional ones (molecular monolayers, such as may be formed using the Langmuir–Boldgett technique or by adsorption on silver-halide microcrystals in photographic applications), or even more complex, curved, geometries, such as rings or cylinders. The latter in particular occur as natural LH complexes, but may also be produced synthetically. Sometimes, the finite dimensions of these aggregates, such as the diameter of rings or cylinders, plays an important role. Finally, aggregates often occur in strongly disordered host media, such as liquid solutions, glasses, or protein scaffolds. The configurational randomness in this environment leads to disorder in the electronic Hamiltonian of the aggregate, which causes localization of the exciton states and has important consequences for the optical properties of the aggregates.

It is of interest to note that bulk crystals and aggregates traditionally belong to different communities. Crystal properties are part of condensed matter physics, in which delocalized electronic states are the rule and a limitation of the extent of these states, due to disorder, is referred to as localization. On the other hand, the field of molecular aggregates has developed within the physical chemistry of dye molecules, where one takes a different point of view, emphasizing the delocalizing effect of the intermolecular interactions on the electronic states. Thus, the simultaneous occurrence of the terms localization length and delocalization length is explained from the fact that two different fields meet.

In this section, we will address the optical and nonlinear optical response of Frenkel excitons in molecular aggregates. The role of disorder and multi-exciton states (see Figure 4) will be stressed. For explicitness, we will mostly restrict ourselves to one-dimensional aggregates. There is abundant evidence that upon self-assembly in solution, many cyanine J-aggregates, in particular those of PIC, are indeed (quasi-)linear [149,158,159]. As a consequence most of the theory has been developed using one-dimensional models. It has appeared that these models suffice to understand the salient optical properties the aggregates.

### 7.2. Linear Optics of One-Dimensional J-Aggregates

#### 7.2.1. Homogeneous Aggregates

We consider a linear aggregate, i.e., a chain of $N$ equidistant two-level molecules, with their transition dipoles of magnitude $\mu$ all oriented parallel to each other (Figure 11). Now the Frenkel exciton Hamiltonian in the Heitler–London approximation takes the form (cf. Eq. (14))

$$H = \sum_{n=1}^{N} \omega_n B_n^\dagger B_n + \sum_{n,m=1}^{N} J_{nm} B_n^\dagger B_m, \qquad (142)$$

where we have included the gas-condensed phase shift in the molecular transition frequency $\omega_n$ (this also includes all interactions with the host molecules) and we

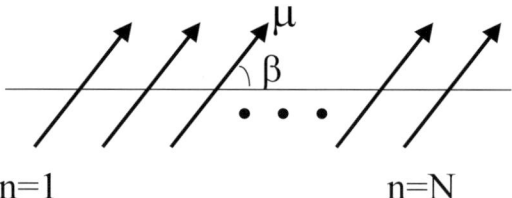

Fig. 11. Schematic picture of a linear aggregate of $N$ equidistant two-level molecules with parallel transition dipoles of magnitude $\mu$. The angle between the dipoles and the chain is $\beta$. For $\beta < 54.7°$, the intermolecular dipole–dipole interaction is negative and the aggregate is a J-aggregate. Otherwise it is an H-aggregate.

have neglected dynamic exciton–exciton interactions. The latter may safely be done, as such interactions do not influence the linear optical response. We will impose open boundary conditions on the chain.

We will assume that the total aggregate is short compared to an optical wavelength. This is not necessarily true for cyanine J-aggregates, but the relevant optical length scale, imposed by the localization length of the excitons, usually does obey this condition. We may then write the (transition) dipole operator of the aggregate as

$$\mathbf{M} = \mu \sum_n (B_n^\dagger + B_n), \quad (143)$$

where $\mu$ is the transition dipole of the individual molecules. From the form of this dipole operator, it is clear that in linear optics one can only probe properties of one-exciton states (cf. Section 2.4.2).

It is useful to start our discussion of collective linear optical properties by considering homogeneous aggregates, in which all transition frequencies are taken equal, $\omega_n = \omega_0$, and the transfer interaction $J_{nm}$ is just a function of the distance between the molecules $n$ and $m$. To maximize the simplicity, we will first assume that we only have transfer interactions between neighboring molecules and we will denote the interaction strength by $J$. In one-dimensional systems, approximating the long-range dipole–dipole interaction equation (6) by a nearest-neighbor one is not too bad an approximation, certainly if one is mostly interested in the essential physics. For quantitative fits to experiment, the long-range nature may even in linear systems be of importance [160].

Keeping only the nearest-neighbor interactions, one finds for the one-exciton eigenstates:

$$|k\rangle = \sqrt{\frac{2}{N+1}} \sin\left(\frac{\pi k n}{N+1}\right) B_n^\dagger |g\rangle, \quad (144)$$

with the energy

$$\Omega_k = \omega_0 + 2J \cos\left(\frac{\pi k}{N+1}\right). \quad (145)$$

Here, $k = 1, 2, \ldots, N$ denotes the quantum number of the state. Clearly, all states are delocalized standing waves of excitation on the chain, with the $k$th state having $k - 1$ nodes. For $N = 2$, the above solution reduces to the states $|\pm\rangle$ of the dimer, separated by $2|J|$. With growing $N$, the one-exciton band obtains a width $4|J|$, centered around the molecular frequency $\omega_0$. The oscillator strength between the ground state and the one-exciton $k$ state is given by [161]

$$\mu_{k,g}^2 = |\langle k|\mathbf{M}|g\rangle|^2 = \frac{1-(-1)^k}{2} \frac{2\mu^2}{N+1} \cot^2 \frac{\pi k}{2(N+1)}. \quad (146)$$

Analysis of this result shows that almost the entire oscillator strength between the ground state and the one-exciton band resides in the transition to the $k=1$ state: $\mu_{k=1,g}^2 = 0.81(N+1)\mu^2$ for $N \gg 1$, which is 81% of the total. This is not strange, as the $k=1$ state is the only state in which the wave function contains no nodes, so that the dipoles of the individual molecules oscillate maximally in phase in this state. All states with $k=$ even have no oscillator strength, as their wave functions are odd with respect to the chain center. The $k=3$ state contains, for $N \gg 1$, 9% of the oscillator strength to the one-exciton band. The oscillator strength of the $k=1$ state, being of the order of $N\mu^2$, is generally referred to as a "giant" oscillator strength. States having giant oscillator strengths, proportional to the volume, also occur in other exciton systems with dimensions small compared to an optical wavelength, e.g., in semiconductor microcrystallites [153,162]. Obviously, such states dominate the optical response.

Thus, the absorption spectrum of the ordered chain is dominated by a peak at the position of the $k=1$ state, which for $N \gg 1$ occurs to a good approximation at $\omega_0 + 2J$. For $J > 0$, one thus expects that the absorption spectrum of the aggregate is blue-shifted relative to the monomer absorption spectrum, while for $J < 0$ we expect a red-shift. The former case is referred to as an H-aggregate, while the latter case is called a J-aggregate. The optical properties of H-aggregates are often harder to describe, as the state with most oscillator strength lies at the top of the exciton band, allowing for very fast relaxation to lower exciton states after its excitation. This makes the H absorption bands very broad and exciton–phonon coupling is an essential ingredient in the proper description of its properties [163]. By contrast, the absorption of J-aggregates occurs at the bottom of the band, giving sharp absorption peaks (also see Section 7.2.2), that allow for a more detailed study and description. We will be mostly interested in J-aggregates, i.e., $J < 0$. The sharp absorption band of J-aggregates is known as the J-band.

The frequency shift $2J$ of the J-band relative to the monomer absorption spectrum (in the same solution), gives the magnitude $J$ of the transfer interaction. For J-aggregates of cyanine dyes, this interaction strength is typically of the order $J \approx -1000$ cm$^{-1}$. Some caution is in place here, as this way of estimating $J$ does not account for the gas-condensed phase shifts induced by the surrounding aggregate molecules. Still, independent calculation of the transfer interactions, using extended dipole models, confirms the typical magnitude of the interactions for cyanine dyes to be of the order quoted above [164].

Next, we address the radiative emission of J-aggregates. It turns out that this process can be considerably faster than the single-molecule spontaneous emission [2,152]. From the above, the explanation is clear. In a typical fluorescence experiment, used to measure the radiative emission, one excites off-resonantly in the blue wing of the J-band. If the excitation rapidly relaxes to the lowest ($k=1$) one-exciton state, the spontaneous emission will occur with a rate that is of the order of $N$ times the single-molecule spontaneous emission rate. This is due to

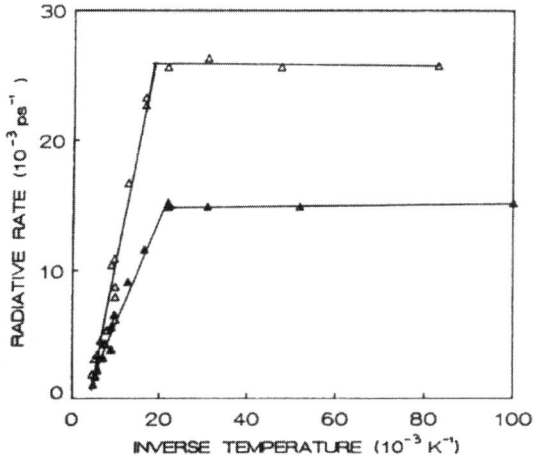

Fig. 12. Temperature dependence of the inverse fluorescence lifetime of PIC J-aggregates for the red absorption site (solid symbols) and the blue site (open symbols). The straight lines are guides to the eye, not theoretical fits. (Figure reprinted from Ref. [152] with permission from Elsevier.)

the fact that the oscillator strength of the $k = 1$ state according to Eq. (146) is of the order of $N$ times the oscillator strength of a single molecule. For this reason, the $k = 1$ state is often called a superradiant state and the process of fast emission is also referred to as "exciton superradiance" or "cooperative spontaneous emission". Physically, the effect results from the fact that the $k = 1$ state has the dipoles of the individual molecules oscillating almost perfectly in phase.

The cooperative spontaneous emission is only observed at low temperatures. With increasing temperature, excitons will not necessarily all relax to the bottom of the band, leading to an, on the average, lower oscillator strength per excited state [165,166]. This lowers the observed emission rate, as is indeed seen in experiment [152,165,167] (Figure 12). It should be stressed that, even at low temperatures, the interpretation of fluorescence experiments in terms of a superradiant rate generally is more complicated than sketched above, due to the fact that the kinetics of the populations of the various $k$ states depends sensitively on the competition between the relaxation and emission processes [168]. It also is of interest to note that the superradiant enhancement of the oscillator strength of the $k = 1$ state starts to break down when the chain length approaches the relevant optical wavelength $\lambda$, simply due to the fact that the dipole approximation for the entire aggregate breaks down. For perfectly ordered chains that are very large compared to the wavelength, we have to introduce the polariton concept, which for this one-dimensional case reveals one radiatively stable branch and one branch that decays very fast, with a rate of the order $\lambda/a$ times the single-molecule emission rate (*a* the lattice constant) [169]. In practice, however, J-aggregates suffer from too much disorder to enter this regime.

Finally, we briefly address the effect of going beyond the nearest-neighbor approximation for the transfer interactions. If we assume point-dipole interactions, we have $J_{nm} = J/|n-m|^3$, where $J$ still parametrizes the nearest-neighbor interactions. Obviously, the existence of long-range interactions tends to delocalize the one-exciton states even more. It turns out, however, that for chains with $N \gg 1$, the wave functions of these one-exciton states are still very well described by Eq. (144), thus preserving the essential features, like the existence of the dominant $k=1$ state. This has been found using numerical simulations [160] as well as analytical estimates [170]. The oscillator strength of the dominant state increases even somewhat (to 83% of the total). The long-range interactions cause the total exciton band to take an asymmetric position around $\omega_0$. In particular, the bottom of the band, i.e., the frequency of the $k=1$ state, moves to $\omega_0 + 2.4J$, while the top lies at $\omega_0 - 1.8J$ (remember that $J < 0$).

### 7.2.2. Disordered Aggregates: Exchange Narrowing and Localization

As stated already, disorder, induced by, for instance, configurational randomness in the host medium has important effects on the electronic states of the aggregate. A frequently used method to include such effects, uses in Eq. (142) $\omega_n = \omega_0 + \delta\omega_n$, where the $\delta\omega_n$ are taken randomly and without correlation from the Gaussian distribution

$$P(\delta\omega) = \frac{1}{\sqrt{2\pi}\sigma} \exp\left(-\frac{(\delta\omega)^2}{2\sigma^2}\right). \tag{147}$$

This is referred to as static diagonal or frequency disorder. The parameter $\sigma$ is a measure of the strength of this disorder. The effect of intermolecular correlations in the disorder can also be included, but we will not address this problem here [151,171–173]. Also, we will not discuss the effect of off-diagonal disorder (interaction disorder) [160,174].

The addition of the disorder contribution, $H_{\text{dis}} = \sum_n \delta\omega_n B_n^\dagger B_n$ leads to shifts of the homogeneous exciton states found in Section 7.2.1. In addition, the disorder will mix those states. To lowest order in the disorder, the shifts are given by

$$\delta\Omega_k = \langle k|H_{\text{dis}}|k\rangle = \frac{2}{N+1} \sum_n \sin^2\left(\frac{\pi k n}{N+1}\right) \delta\omega_n. \tag{148}$$

Using the Gaussian and uncorrelated nature of the $\delta\omega_n$, one finds that each of these shifts is a Gaussian random variable as well, with a standard deviation, i.e., a typical value, given by [151,175,171]

$$\sigma_{kk} = \sigma\sqrt{\frac{3}{2(N+1)}}. \tag{149}$$

In this result, we see reflected the effect of "exchange narrowing": the delocalized exciton states average over $N$ uncorrelated Gaussian variables, which reduces the fluctuations in their energy by a factor of the order of $\sqrt{N}$. The disorder-induced coupling between the different $k$ states, $\langle k|H_{\text{dis}}|k'\rangle$, can be calculated in a similar way [171,175]. It turns out that its typical magnitude is also exchange narrowed and is given by ($k \neq k'$):

$$\sigma_{kk'} = \sigma \sqrt{\frac{1}{N+1}}. \tag{150}$$

As long as the mixing of the exciton states due to their disorder-induced coupling is negligible, it is easy to account for the disorder. This is the case if the typical size of the coupling is small compared to the energy difference between the homogeneous states. Using Eq. (150) and the fact that the smallest energy separation occurs at the bottom of the exciton band and is given by $\Omega_{k=2} - \Omega_{k=1} \approx 3\pi^2|J|/N^2$ (Eq. (145) with $N \gg 1$), this yields the criterion

$$\sigma \ll 3\pi^2|J|/N^{3/2} \tag{151}$$

which clearly is seen to strongly depend on the size of the chain. This limit, in which the exciton states all maintain their delocalized homogeneous wave function, is referred to as the exchange narrowing (or motional narrowing) limit. In this limit of weak disorder, the absorption spectrum of an ensemble of aggregates is dominated by a Gaussian peak centered at $\Omega_{k=1}$ and with a standard deviation $\sigma\sqrt{\frac{3}{2(N+1)}}$. Thus, the tendency of the delocalized exciton states to average over the disorder offsets of the individual molecules, leads to the narrow absorption lines of J-aggregates observed in experiment.

From Eq. (151), we see that the size of the chain plays in important role when neglecting mixing between the homogeneous exciton states. In fact, as J-aggregates of cyanine dyes may be thousands of molecules long, Eq. (151) should never be expected to hold for these systems (the typical value of $\sigma$ is in the order of 10's to 100's of cm$^{-1}$ while $J$ is in the order of 1000 cm$^{-1}$). Thus, the excitons are expected to strongly mix and this mixing leads to new eigenstates that are localized on a relatively small part of the chain. This effect of localization of one-particle states in a disordered system is well-known in the field of disordered conductors (Anderson localization [176]) and is particularly strong in one-dimensional systems. The typical localization length depends on the energy of the state (stronger localization at the exciton band edges) and on the ratio $\sigma/|J|$: $J$ tends to delocalize the excitons, while $\sigma$ tends to localize them.

If the excitons are localized, one cannot expect the absorption line width to scale like $\sigma/\sqrt{N}$ anymore, and similarly, one cannot expect the low-temperature spontaneous emission rate to scale proportional to $N$. Instead, one should expect $N$ in these expressions to be replaced by the typical exciton delocalization length

$N_{\text{del}}$ at the bottom of the exciton band (where the optically dominant states reside). The reason is that only $N_{\text{del}}$ molecules are coherently coupled and have dipoles that can oscillate in phase. The effective replacement of $N$ by $N_{\text{del}}$ was noted by Knapp already in his seminal paper on exchange narrowing [151].

It has turned out that in the case of weak localization ($N \gg N_{\text{del}} \gg 1$), it is possible to understand the low-lying exciton states, i.e., the ones that dominate the optical response, from a simple self-consistent picture [175]. The idea is that the chain segment on which an exciton state is localized, may be considered a weakly disordered chain of effective length $N_{\text{del}}$, on which the effect of disorder on the exciton state may be treated as in the exchange narrowing limit. Using this picture in a self-consistent way, leads to a simple derivation of various scaling relations [175,170,177,178], as we will demonstrate in the following.

Consider a segment with effective chain length $N_{\text{del}}$, which is determined by the disorder and is as yet unknown. On this segment, the typical disorder-induced coupling between the effective exciton states scales like $\sigma_{kk'} = \sigma/\sqrt{N_{\text{del}} + 1}$. The typical energy difference between the effective states at the bottom of the band, is given by $\Delta = 3\pi^2|J|/(N_{\text{del}} + 1)^2$. If the coupling is larger than the energy separation, the states will mix and the delocalization length will further decrease. Thus, in order for the segment length to be self-consistent, $N_{\text{del}}$ should be determined by equating $\sigma_{kk'} = \Delta$. This leads to the scaling relation

$$N_{\text{del}} + 1 = (3\pi^2)^{2/3}\left(\frac{\sigma}{|J|}\right)^{-2/3} \approx 9.6\left(\frac{\sigma}{|J|}\right)^{-2/3}. \tag{152}$$

Similarly, the half-width at half maximum $W$ of the absorption line (the J-band) is given by the effective exchange narrowed value

$$W = \sigma\sqrt{\frac{3}{2(N_{\text{del}} + 1)}}. \tag{153}$$

Substituting Eq. (152) into Eq. (153), leads to

$$W = \sqrt{\frac{3}{2}}\frac{1}{(3\pi^2)^{1/3}}|J|\left(\frac{\sigma}{|J|}\right)^{4/3} \approx 0.4|J|\left(\frac{\sigma}{|J|}\right)^{4/3}. \tag{154}$$

This scaling relation for $W$ is in fact well-known and may be derived by various other methods [179,180]. It has also been confirmed by numerical simulations [160,179,181]. The power 4/3 is not affected by adding long-range dipole–dipole interactions, while even the prefactor 0.4 is only slightly changed [160] (note that we use the half-width for $W$, where some other authors use the full width).

The scaling relation equation (152) is less well-known. As the low-temperature spontaneous emission rate is expected to scale proportional to $N_{\text{del}} + 1$, this relation should describe the scaling of this rate with $\sigma/|J|$ [175]. Indeed, the power $-2/3$ and even the prefactor 9.6 agrees rather well with numerical simulations of this quantity [160]. It turns out that this prefactor is sensitive to the inclusion of long-range interactions [160], implying that a quantitative fit to experiment does require such interactions to be considered.

The fact that the scaling relations, obtained using the above self-consistent arguments, agree well with those obtained by other techniques, means that the picture of effective chain segments of length $N_{\text{del}}$ works well for the low-energy exciton states. Indeed, it has been confirmed by numerical simulations that the lowest exciton states of a disordered chain have a structure that resembles the homogeneous states on small segments of the chain [170,178]. It should be kept in mind, however, that the thus obtained segment length $N_{\text{del}}$ is a *typical* value, which in practice undergoes rather big fluctuations [178].

The above scaling relations may and have been used to extract information concerning the exciton delocalization length and (or) the disorder strength from experimental observables. For instance, the low-temperature spontaneous emission rate, $N_{\text{del}}\gamma_{\text{mol}}$, with $\gamma_{\text{mol}}$ the single-molecule emission rate, has been used by Wiersma and co-workers to find $N_{\text{del}} \approx 50$ for the case of PIC aggregates in a 1.5 K glassy host [152,158,165]. It should be noted that application of this method requires an accurate measurement of the fluorescence quantum yield [182] and also requires that the vibration assisted intraband exciton relaxation is fast compared to the superradiant emission rate [168].

The second observable that has been used to measure the delocalization length, is the absorption line width, in combination with the exchange narrowing relation equation (153). A problem with this method is that, generally, one does not have independent information on the microscopic parameter $\sigma$. By lack of better information, one then takes this parameter from the inhomogeneous width of dilute solutions, in which aggregation does not occur. This is a crude procedure, as a molecule in a chain has quite a different environment than an isolated molecule in solution. For PIC aggregates at cryogenic temperatures, this approach yields values of $N_{\text{del}}$ of hundreds of molecules, up to 1600 [2], which is a big overestimation of the above quoted (and now generally accepted as correct) value of 50.

If one chooses to use $W$ to estimate $N_{\text{del}}$, it is better to do this without relying on $\sigma$. The self-consistent arguments given above allow one to extract $N_{\text{del}}$ from $W$ if $J$ is known. This may be done by using Eqs. (154) and (152), or alternatively, by directly equating $W$ to the splitting $3\pi^2|J|/(N_{\text{del}}+1)^2$. Both methods in principle work well [177]. Although also the experimental determination of $J$ meets with some uncertainties (Section 7.2.1), it is generally better known than $\sigma$, making these methods more reliable.

So far, we have only discussed the emission rate and the absorption line width. Of course, the absorption spectrum contains more information, in particular its detailed shape and position. In order to determine these aspects, more extensive calculations are necessary. Approximate semi-analytical techniques may and have been used to calculate the absorption spectrum in detail. An example is the co-

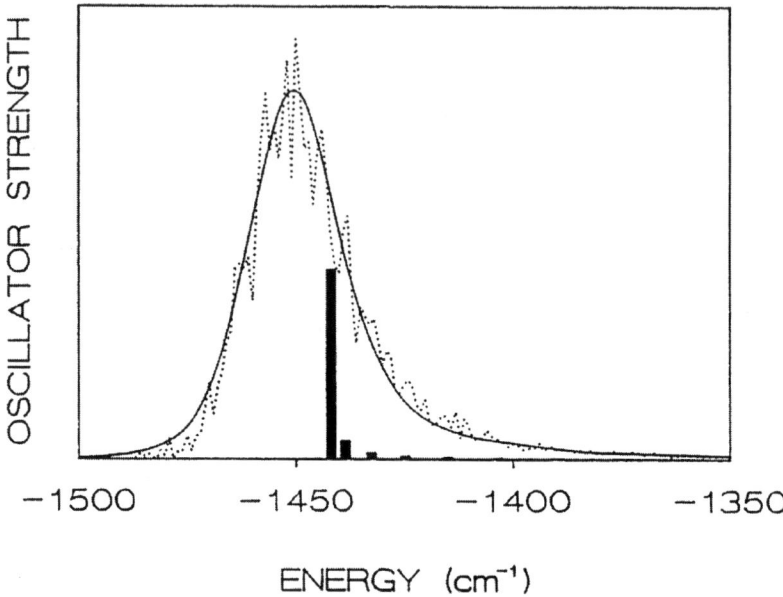

Fig. 13. Measured (solid) and simulated (dashed) absorption spectrum of the 570 nm J-band of PIC aggregates at 1.5 K. All dipole–dipole interactions are included in the simulations and the energy is measured relative to the monomer energy $\omega_0$. The bars give the positions and oscillator strengths for a homogeneous aggregate of the same length (250 molecules) with the same interactions. (Figure courtesy of H. Fidder, also cf. Ref. [184].)

herent potential approximation [180]. The most popular technique, however, is straightforward numerical simulation of the line shape [179,160], in which the $N \times N$ one-exciton Hamiltonian is numerically diagonalized for many random realizations of the disorder (both diagonal and off-diagonal disorder may be considered [160]). In practice, the length of the chain should be much larger than the typical exciton delocalization length in order for the results to be reliable.

As an example, we show in Figure 13 the simulation of the J-band shape of PIC aggregates at $T = 1.5$ K, using 500 chains of 250 molecules with long-range dipole–dipole interactions. The solid line is the experiment, the dotted line the numerical fit. The only fitting parameter is $\sigma/|J|$, which was obtained to be 0.11 from these simulations. The value of $J$ for PIC is $-600$ cm$^{-1}$, implying a value of $\sigma \approx 65$ cm$^{-1}$. The typical delocalization length at the position of the top of the J-band, calculated using the participation ratio [179,160], is in the order of 50 molecules, which agrees with the observed enhancement of the spontaneous emission rate. The calculated line shape is seen to nicely fit the observed one. In particular the asymmetry of having a more pronounced high-energy than low-energy tail, which is characteristic for J-aggregates, is well-reproduced. This asymmetry can be understood in a perturbative picture from the fact that higher $k$ states obtain oscillator strength as a consequence of their disorder-induced mixing with the superradiant $k = 1$ state. Such mixing does not occur at the low-energy side of the $k = 1$ state, as this is the lowest state in the homogeneous chain.

We observe that, although the simulated spectrum obviously fits the observed one well, it is still quite noisy. This may be improved by simulating more disorder realizations, which is a time-consuming solution. A better way to considerably improve the statistics of the simulated spectra, is to use a rigorous smoothing technique, in which one performs the average over the mean value of the molecular frequencies within each chain analytically [183]. Since fluctuations in this average value for the simulated (finite) chains are the main source of noise, this smoothing technique is very powerful.

### 7.3. Nonlinear Optics of One-Dimensional J-Aggregates

*7.3.1. Interest*

The nonlinear optical response of J-aggregates has drawn considerable attention since the late 1980s. Initially, most interest was aroused by the fact that the occurrence of superradiant transitions could give rise to strong, so-called size-enhanced, nonlinear optical susceptibilities. This would be of great interest for the creation of nonlinear optical devices, like all-optical switches. To understand the effect of size-enhancement, let us consider the nonlinear absorption spectrum, which is the imaginary part of the third-order susceptibility $\chi^{(3)}(\omega; \omega, \omega, -\omega)$ [70,71]. Any third-order response function contains a product of four transition dipoles, three to account for the interactions with the exciting fields and one to generate the signal. In the case of exciton systems, these dipoles either connect the ground state with the one-exciton band, or the one-exciton band with the two-exciton band. We have seen already that one transition from the ground state to the one-exciton band exists (the $k = 1$ transition) which has a dipole squared of the order of $N\mu^2$ where $\mu$ is the single-molecule transition dipole and $N$ the number of molecules in the chain. Thus, the contribution to $\chi^{(3)}$ where all transitions occur between the ground state and the $k = 1$ one-exciton state, has a magnitude that scales proportional to $N^2$, which is factor of $N$ stronger than the nonlinear response of $N$ noninteracting molecules. This contribution – which comes with a minus sign – reflects the saturation of the one-exciton absorption. In addition, positive two-photon absorption contributions to the two-exciton band exist that involve two transition dipoles between the ground state and the one-exciton band as well as two between the one-exciton band and the two-exciton band. Also among the latter dipoles, superradiant ones occur (see Section 7.3.2), so that these contributions also scale proportional to $N^2$.

There are a number of limitations to the applicability of this size-enhanced scaling. First, it is a purely resonant phenomenon. Out of resonance with the J-band (i.e., the absorption band), $\chi^{(3)}$ simply scales propertional to $N$ [154,74]. Second, as in the case of exchange narrowing and spontaneous emission, one does not expect these size scaling relations to be valid under realistic conditions, as the exciton delocalization size will generally be limited by disorder, in stead of by the system size. Handwaving replacement of $N$ by $N_{del}$ would yield an enhancement by a factor $N_{del}^2$. Numerical simulations have shown that the peak value of the absorption saturation per molecule in the sample, scales like $N_{del}^\delta$, with $\delta = 2.36$ [156]. Thus, the delocalization may indeed lead to strong nonlinear response. Experiments on PIC J-aggregates have confirmed that they do exhibit large nonlinear absorption coefficients [6,7,157]. A direct check of the size scaling has recently been performed by creating cyanine dyes with different substituents and in different solutions, which enables one to vary the disorder, and thus the delocalization length, in a controlled way [185].

In addition to the application driven interest of size-enhancement, the nonlinear optical response of J-aggregates also has aroused much interest as spectroscopic technique for studying the dynamics of the exciton states and for probing two-exciton (and higher) states. The latter may, for instance, be done using the two-photon contribution to the nonlinear absorption mentioned above or the pump-probe (transient absorption) spectrum. In general, any third-order optical technique is sensitive to the two-exciton states. In the remainder of this section, we will address the calculation of the two-exciton states in our aggregate model, their role in the pump-probe spectrum, and the possibility which probing the two-exciton states gives to determine the exciton delocalization length.

*7.3.2. Incorporating the Two-Exciton States*

Calculating the two-exciton states in general requires the diagonalization of the Hamiltonian in the two-exciton subspace, which is a matrix of size $N(N-1)/2 \times N(N-1)/2$. In the case of disorder, where one has to consider many random realizations, this is almost prohibitively time-consuming for any realistic value of $N$. It should be realized that the problem in calculating the multi-exciton states for the Frenkel exciton Hamiltonian in the absence of dynamic exciton–exciton interactions, like Eq. (142), resides in properly accounting for the Pauli exclusion (kinematic interaction – cf. Section 2.4.2), which forbids two excitations to occupy the same molecule. Although in general dynamic exciton–exciton interactions affect the two-exciton Hamiltonian and may lead to the formation of bi-excitons, such bound states have thus far not been unambiguously observed in J-aggregates. This motivates us to focus on the kinematic interactions only.

Two methods exist to account for this interaction without diagonalizing such big matrices. The first method is the hard-core boson approach [33,34,44], which has already been explained in Sections 2.4.2 and 2.4.3. It may be used for systems of any dimension and with arbitrary long-range excitation transfer interactions. It is limited to the inclusion of two-exciton states and relies on the Heitler–London approximation. In fact, this method does not directly calculate these states, but rather calculates the Green's function of the scattering of two bosonic one-excitons on an imposed hard-core potential that mimics the Pauli exclusion. Third-order optical observables can be formulated in terms of this Green's function, whose calculation requires the diagonalization and inversion of $N \times N$ matrices. For more detail on this method, we refer to the literature [33,34].

The second method relies on the so-called Jordan–Wigner transformation [37–39], which only works for one-dimensional systems, but has the advantage that it allows for the calculation of all $2^N$ multi-exciton states (even if one relaxes the Heitler–London approximation [46]). Another advantage is that it gives simple expressions for these states, which allows for direct physical insight. This method has been used in many simulations [39,74,75,156], and we will briefly explain it below.

The Jordan–Wigner transformation transforms the Pauli excitations to non-interacting fermions. This works for one-dimensional systems, if only nearest-neighbor interactions are included. Adding long-range interactions, gives rise to weakly interacting fermions. For excitons, the transformation was first used by Chesnut and Suna [37]. It reads

$$c_n^\dagger = B_n^\dagger \exp\left(-\pi i \sum_{m=1}^{n-1} B_m^\dagger B_m\right), \quad c_n = \exp\left(\pi i \sum_{m=1}^{n-1} B_m^\dagger B_m\right) B_n. \quad (155)$$

It can be checked in a straightforward way that the new creation and annihilation operators indeed obey Fermi anti-commutation relations. Moreover, it follows that the Frenkel exciton Hamiltonian equation (142) (keeping only nearest-neighbor interactions $J$), transforms to:

$$H = \sum_{n=1}^{N} \omega_n c_n^\dagger c_n + J \sum_{n=1}^{N-1} (c_n^\dagger c_{n+1} + c_{n+1}^\dagger c_n) \equiv \sum_{n=1}^{N} A_{nm} c_n^\dagger c_m. \quad (156)$$

The advantage of the new Hamiltonian is not so much the nearest-neighbor interaction, that we have restricted ourselves to, but rather the fact that it reflects a system of non-interacting fermions, as is recognized from the fact that only terms that are quadratic in the operators occur. All multi-fermion states can now be obtained by diagonalizing the tridiagonal $N \times N$ matrix $A$ implicitly defined in Eq. (156). Let the eigenvalues of this matrix be denoted $\Omega_p$ and the normalized eigenvectors $\phi_{pn}$, then the diagonalized Hamiltonian reads

$$H = \sum_{p=1}^{N} \Omega_p \eta_p^\dagger \eta_p, \quad (157)$$

where the new fermion eigen-operators are given by $\eta_p^\dagger = \sum_n \phi_{pn} c_n^\dagger$.

An arbitrary $\nu$-exciton state may now be created by operating on the ground state $|g\rangle$ (still the state with all molecules in the ground state), with $\nu$ $\eta_p^\dagger$ operators. As the latter are Fermi operators, each mode $p$ only can be excited once. Thus the energy of a $\nu$-exciton state is given by the sum of $\nu$ different one-fermion (or simply: one-exciton) energies $\Omega_p$. The wave function in the molecular representation is obtained by translating the $\eta_p^\dagger$ operators, via the $c_n^\dagger$ back to the operators $B_n^\dagger$. This yields Slater determinants of the eigenvectors $\phi_{pn}$ as the expansion coefficients in the molecular representation [39]. It should be stressed that all this holds in the presence of arbitrary disorder in the molecular energies (and the nearest-neighbor interactions).

The $N \times N$ matrix $A$ which is to be diagonalized, is identical to the one-exciton Hamiltonian in the molecular representation. Thus, one immediately finds that the one-exciton states calculated via the Jordan–Wigner transformation are identical to the ones obtained directly within the paulion picture. The advantage lies in the fact that the same diagonalization now suffices to also obtain all multi-exciton states.

It is useful to briefly consider the special case of the homogeneous chain. Then, the eigenvalues of the matrix $A$ are the one-exciton frequencies $\Omega_k$ given in Eq. (145) (the fermion label $p$ is now replaced by $k$) and the $\phi_{kn}$ are given by the $\sin(\pi k n/(N+1))$ one-exciton eigenfunctions. Using these results, we find that the dominant transition from the superradiant $k = 1$ one-exciton state to the two-exciton band occurs to the state $\eta_{k=1}^\dagger \eta_{k=2}^\dagger |g\rangle$, in which the fermion states of quantum label $k = 1$ and $k = 2$ are occupied. This transition is superradiant as well, with an oscillator strength equal to $1.27(N+1)\mu^2$ (which is 70% of the total oscillator strength from the $k = 1$ state to the two-exciton band) [74]. Thus, for third-order experiments carried out resonantly with the J-band, one expects an effective three-level picture, in which one only accounts for the ground state $|g\rangle$, the one-exciton state $\eta_{k=1}^\dagger |g\rangle$, and the two-exciton state $\eta_{k=1}^\dagger \eta_{k=2}^\dagger |g\rangle$, to describe the essential results.

### 7.3.3. Pump-Probe Spectrum of J-Aggregates

In the pump-probe technique, one measures the linear absorption spectrum of a weak probe beam after the system has been optically excited by a pump-pulse. The pump-probe spectrum is defined as the difference of this absorption spectrum and the normal linear absorption spectrum, obtained without pumping the system [71].

For molecular aggregates, the pump pulse (if weak enough) can only excite the system into one of its one-exciton states. Three different effects now contribute to the pump-probe spectrum. (i) After pumping, less aggregates reside in the ground state, causing the probe absorption into the one-exciton states to diminish. This effect, known as bleaching, gives negative contributions to the pump-probe spectrum at all allowed one-exciton frequencies. (ii) The probe pulse will lead to stimulated emission of the one-exciton state excited by the pump-pulse. This, too, gives a negative contribution. (iii) The probe pulse gives rise to extra absorption contributions from the excited one-exciton state into the two-exciton band. These so-called induced absorption contributions are positive and their spectral positions reflect the frequencies of dipole-allowed transitions between the pumped one-exciton state and the two-exciton band.

Based on the effective three-level picture found at the end of Section 7.3.2 for the third-order response, one expects that for homogeneous aggregates only two transitions dominate if the experiment is performed in resonance with the center of the J-band. The $k = 1$ one-exciton state dominates the bleaching and the stimulated emission. On the other hand, the two-exciton state that dominates the induced absorption from the $k = 1$ state, is the state $\eta^\dagger_{k=1}\eta^\dagger_{k=2}|g\rangle$. Thus, the dominant negative peak in the pump-probe spectrum is expected to occur at $\omega = \Omega_{k=1}$, while the dominant positive peak occurs at $\omega = \Omega_{k=2}$ (here $\omega$ denotes the frequency of the probe pulse). As was first pointed out by Juzeliūnas [38], within the fermion picture for the excitons one finds that the positive contribution is shifted relative to the negative one over a frequency difference

$$\Delta\omega = \Omega_{k=2} - \Omega_{k=1} \approx -\frac{3\pi^2 J}{(N+1)^2}, \tag{158}$$

which follows from Eq. (145) with $N \gg 1$. For J-aggregates ($J < 0$), $\Delta\omega$ is positive, i.e., the induced absorption is blue shifted compared to the one-exciton bleaching and stimulated emission dip (see Figure 14). This shift is simply a consequence of the Pauli exclusion of the double excitation of a single molecule and may be referred to as the *Pauli exclusion gap*. In simple terms, the second exciton effectively has less space than the first one, as it may not reside on the same mole-

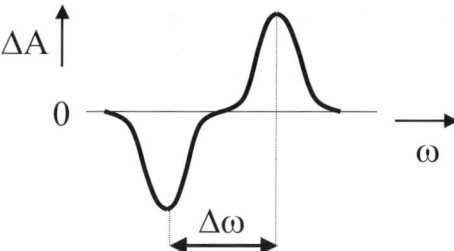

Fig. 14. Schematic picture of the pump-probe spectrum for a J-aggregate pumped in the center of the J-band, with $\omega$ the frequency of the probe beam. The negative peak reflects the bleaching and stimulated emission of the dominant one-exciton transition, while the positive peak is the induced absorption resulting from the dominant one- to two-exciton transition.

cule. For J-aggregates, this forces the second exciton into a higher-energy state than the first exciton. Although the transformation to non-interacting fermions can only be made in one dimension, the above rationale holds in all dimensions, and one generally expects for J-aggregates the induced absorption peak to be blue shifted relative to the bleaching and stimulated emission dip.

The blue-shifted induced absorption peak has been seen in many types of J-aggregates [41,186–190], including also the ring-shaped aggregates of chlorophyll molecules occurring in bacterial light-harvesting systems [191]. It seems to be an almost generic feature for J-aggregates, implying that often the Pauli exclusion shift dominates the nonlinear effects. The first strong evidence that indeed the unbound two-exciton state may be observed in J-aggregates, was found for PIC aggregates at 1.5 K [41]. The experiment clearly yielded a blue-shifted induced absorption and the theoretical description of the spectrum based on the Pauli exclusion effect, perfectly agreed with the experimental data, without using free parameters (Figure 15).

Mostly, the pump-probe spectrum has been used not so much to confirm the existence of the two-exciton state, but rather to extract information on the delocalization length [172,188–191]. From Eq. (158) we see that the blue shift $\Delta \omega$ in the pump-probe spectrum is related to the size $N$ of the chain. As before, it is not reasonable to expect that the physical chain size will show up in the optical spectra. Rather, one expects Eq. (158) to hold with $N$ replaced by the typical delocalization length at the position of the J-band, $N_{\text{del}}$. In other words, one expects the length scale obtained from the pump-probe spectrum through

$$N_{\text{del}}^{\text{pp}} = \sqrt{\frac{3\pi^2 |J|}{\Delta \omega} - 1} \tag{159}$$

to be a good measure of the actual exciton delocalization length. This expectation is supported by the hidden structure of the exciton states mentioned in Section 7.2.2 [170,178]. Knoester and Bakalis have assessed the validity of this expectation by performing numerical simulations of pump-probe spectra for chains of 250 molecules over a range of the degree of disorder $\sigma/|J|$ [75]. This study showed that as long as static disorder dominates the line width, an almost perfect linear scaling exists between $N_{\text{del}}^{\text{pp}}$ and the delocalization length $N_{\text{del}}$ obtained from the participation ratio at the center of the J-band: $N_{\text{del}}^{\text{pp}} = 1.3 N_{\text{del}} + 1.7$ (Figure 16). This confirms that at low temperature, the pump-probe spectrum indeed serves as a useful tool to measure the exciton delocalization length. The scaling breaks down when the homogeneous linewidth of the exciton transitions approaches the detuning $\Delta \omega$ [75,192], as is also demonstrated in Figure 16.

It is quite remarkable that the simple Hamiltonian equation (142), without dynamic exciton–exciton interactions and accounting for only one molecular excited state, has been so successful in describing the pump-probe spectrum of J-aggregates. Theoretical work on the inclusion of dynamic interactions in linear

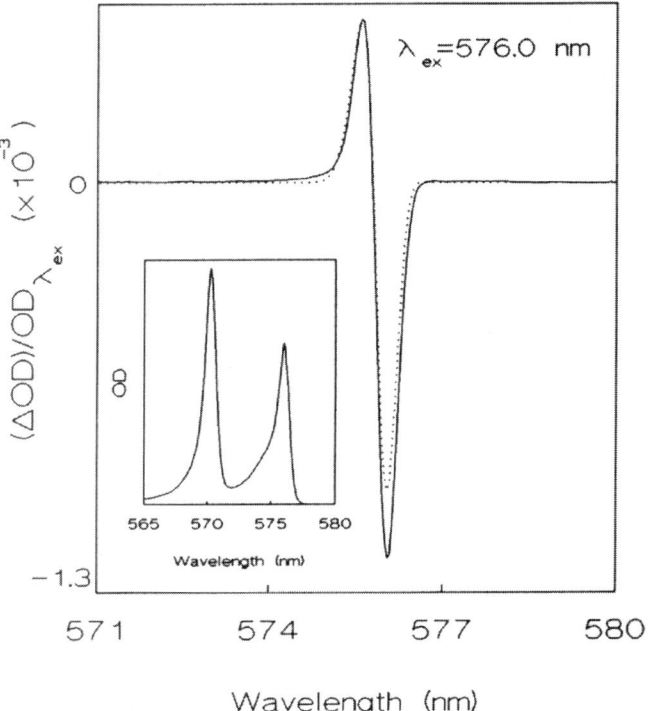

Fig. 15. Measured pump-probe spectrum (solid line) of PIC aggregates in an ethylene glycol/water glass at 1.5 K after pumping at 576 nm. The dotted line is the theoretical curve, based on the effective three-level picture discussed at the end of Section 7.3.2. Apart from the vertical scale, no free parameters were used in the theory; all parameters were taken from linear optical experiments. The insert shows the linear absorption spectrum. (Figure reprinted from Ref. [41] with permission from the American Institute of Physics.)

aggregates and the associated possible formation of bound bi-exciton states has been reported in Refs. [77,80,58]. Yet, we know of very few experimental reports where such bound states have been suggested in J-aggregates [157,193]. Spano has shown that only for a small range of exciton–exciton coupling strengths, the pump-probe spectrum in the frequency region of the J-band is appreciably affected by the extra interaction [80]. These coupling strengths may in practice rarely be realized. Also the inclusion of a second molecular excited state $|f\rangle$, dipole-allowed from the first exited state $|e\rangle$, has been studied theoretically in the context of linear aggregates. A most interesting phenomenon in this model, which includes a fusion interaction of the form $B_{nf}^{\dagger} B_{me} B_{ne}$, is the possible interference between single-molecule and collective nonlinear optical response, with the appearance and disappearance of two-photon absorption (or induced absorption)

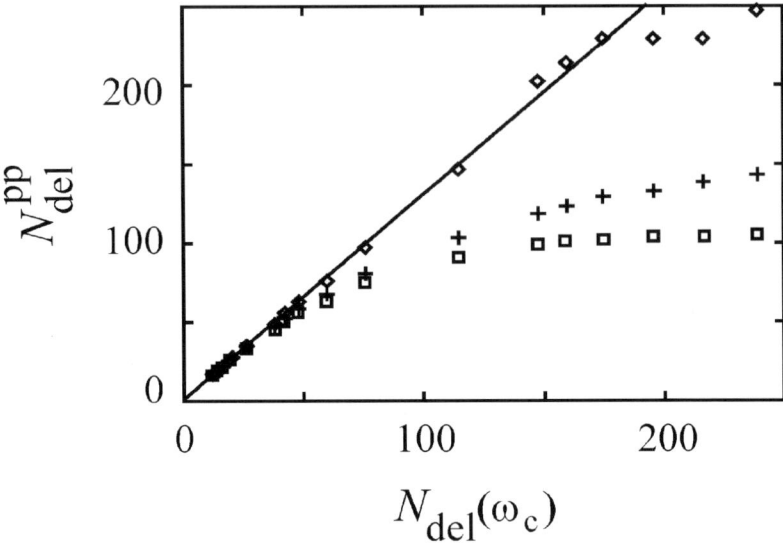

Fig. 16. The exciton delocalization length as obtained from the simulated pump-probe spectrum through Eq. (159), in comparison with the real delocalization length as calculated from the participation ratio at the center of the simulated J-band. In the simulations, chains of $N = 250$ molecules were used, with a diagonal disorder strength varying between $\sigma = 0.001|J|$ and $0.3|J|$. Data points represent homogeneous line widths of 0 ($\Diamond$), $5 \times 10^{-4}|J|$ (+), and $8 \times 10^{-4}|J|$ ($\Box$). The solid line is the best fit through the data with vanishing homogeneous line width and has a slope of 1.3. (Figure reprinted from Ref. [75] with permission from the American Chemical Society.)

peaks as a function of the detuning and the relative dipole strengths of the two intra-molecular levels [56]. The same three-level model has been used to model exciton–exciton annihilation [85], an effect that plays a crucial role in molecular aggregates at high laser intensities, limiting the effective exciton life time [194, 195] and affecting the nonlinear optical response [196].

### 7.4. AGGREGATES OF OTHER GEOMETRIES

Above, we have considered the optical properties of linear J-aggregates. Two-dimensional cyanine aggregates are also well-known. They are formed by adsorption on silver-halide microcrystals in photographic emulsions [11,12]. The properties of such aggregates have been studied with similar techniques and models as we used above for the one-dimensional case. In general, the J-band is broader than for the one-dimensional J-aggregates formed in solution (see, e.g., Refs. [197,198]). This also holds for two-dimensional cyanine aggregates formed via the Langmuir–Blodgett technique [199–201]. In such two-dimensional aggregates the nearest-neighbor approximation for the excitation transfer interactions can generally not be made and often the point-dipole approximation is very poor as well. Better results may be obtained by using extended dipoles [31,198]. Finally, the end-surface of a molecular crystal may be considered a two-dimensional aggregate, because the frequency of the molecules in this layer is detuned from the inner layers, as a result of different gas-condensed phase shifts for molecules on the surface layer [202]. Observing the fluorescence from the end-surface of anthracene crystals, this has allowed for the first observation of superradiance of two-dimensional exciton-polaritons [203].

Recently, the Daehne group has succeeded in preparing self-assembled cyanine aggregates with interesting geometries, by making use of the special interactions between amphiphilic side groups and the solvent. This has for instance resulted in the preparation of cylindrical cyanine aggregates [204], which mimic the cylindrical light-harvesting systems that occur in green bacteria [205,206] (Section 8). As a result of the circular geometry, a mono-wall cylindrical aggregate has three superradiant transitions, two of which are degenerate [207,208]. As has been confirmed by experiment [209], this leads to the formation of two perpendicularly polarized absorption bands, as opposed to the single J-band observed for linear J-aggregates. Finally, an interesting class of aggregates are the columnar H-aggregates prepared from discotic molecules, with an absorption peak that is blue-shifted relative to the monomer transition [210].

## 8. Excitons in Biological Systems

Molecular electronic excitations play an important role in the photosynthetic systems of bacteria and higher plants. In such organisms, light-harvesting (LH) systems occur, whose task it is to absorb the light of the sun and to transport the excitation energy to the photosynthetic reaction center [13]. Chlorophyll molecules are the main building blocks of these LH systems. They play the role of both absorber and transport medium. Various types of (bacterio)chlorophyll molecules exist, which may absorb in different parts of the electromagnetic spectrum. The efficiency of these systems is extremely high: more than 90% of the excitations created by the sun light lead to a charge separation in the reaction center. As chlorophyll molecules have a spontaneous emission time of the order of 1 ns, this implies that the transport has to take less than 100 ps. Over the past decades, much work has been devoted to understanding the mechanism of this efficient transport. Here the main question is whether the transport should be described as incoherent, i.e., Förster transport from molecule to molecule, or whether coherent excitons, delocalized over several or many molecules play a role (cf. Section 4). As biological systems generally have a complicated and not very regular configuration, one might expect that coherence plays only a minor role. It is of interest, however, that during the past decade the structure of several LH systems has been

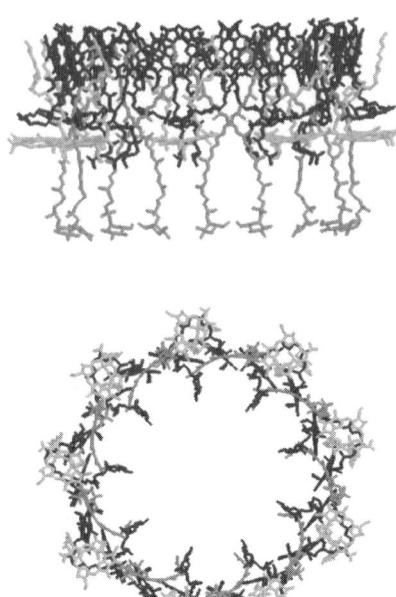

Fig. 17. Side and top views of the pigment structure of the light-harvesting complex LH2 of the purple bacterium Rhodopseudomonas acidophola. Two rings of bacteriochlorophyll molecules exist. The upper one contains 18 molecules, the lower one 9. The linear molecules are carotenoids. (Figure courtesy of R.J. Cogdell.)

unraveled in detail and that some of these systems, especially those occurring in bacteria, exhibit a surprising amount of symmetry.

Probably best known are the circular light-harvesting systems LH1 and LH2 of purple bacteria, which are systems with a 16-fold and 8- or 9-fold symmetry, respectively [211,212]. For Rhodopseudomonas acidophila, LH2 has a 9-fold symmetry (Figure 17). It contains two rings stacked on top of each other, both with a diameter of about 5 nm, one of which (the B800 ring, absorbing at about 800 nm) contains 9 bacteriochlorophyll molecules and the other (the B850 ring, absorbing at about 850 nm) contains 18 [211]. The transition dipoles of all molecules lie (almost) in the plane of the rings and rotate around the ring with the position of the molecule. If the rings have no electronic disorder, this situation would give rise to Frenkel exciton states of the Bloch form, with only the states with wavenumber $\pm 2\pi/N$ ($N$ the number of molecules in the ring) being dipole allowed [213–215]. The random conformations of the surrounding protein scaffold, however, breaks the electronic symmetry and may lead to localization. It is generally accepted that the transfer interactions between the molecules in the B800 ring are too weak to overcome the disorder strength and to delocalize the excitations in it. This has indeed nicely been demonstrated by fluorescence excitation experiments carried out on single LH2 complexes [216]. For the B850 ring, the situation is different, as the intermolecular distances between neighboring molecules are small (about 9 Å), leading to interactions of the order of a few hundred wave numbers. Similar single-complex experiments have indicated that the absorption spectrum of each of these rings is typically dominated by two perpendicularly polarized transitions, which are separated by about 100 cm$^{-1}$ [215]. These transitions may be understood as originating from the $\pm 2\pi/N$ transitions, with the splitting being caused by disorder and (or) a deformation of the ring [215,217]. This strongly suggests that the primary excitations, created immediately after the absorption of light, are strongly delocalized (over almost the entire ring). At the same time, pump-probe [191] and superradiant emission [218] experiments indicate that the excitations after vibronic relaxation extend over only a few molecules. Much work has also been done on the energy transfer from the B800 ring to the B850 ring [219,220], which occurs through rather weak interactions and may be understood in terms of incoherent hopping transport. It has been demonstrated by Sumi and coworkers that the involvement of many molecules requires an adaption of the usual Förster method to calculate the relevant transfer rate [221,222].

A second type of highly symmetric LH system are the rod elements in the chlorosomes of green bacteria [205]. For Chloroflexus aurantiacus, these are cylindrical aggregates of bacteriochlorophyll molecules, with a diameter of about 5 nm and a length of up to a few hundred nm [206]. These aggregates may be thought of as 6 helices of chlorophyll molecules intertwined on the cylinder surface. In the limit of long homogeneous cylinders, such aggregates exhibit three superradiant states, two of which are degenerate [208,223]. These transitions are too broad to separate in ensemble absorption experiments. Additional information may be obtained using polarized spectroscopic techniques, such as linear and circular dichroism. It is of interest that such helical molecular aggregates contain a circular dichroism (CD) signal, even if the individual molecules have no rotational activity [208,223]. This is yet another interesting example of collective optical response in molecular aggregates. For the possible development of bio-mimicking synthetic complexes, it is of much interest that cylindrical aggregates have also been prepared synthetically, through self-assembly from TDBC cyanine dyes with amphiphilic substituents [204] (cf. Section 7.4). As cyanine dyes have stronger transition dipoles than chlorophyll molecules, such synthetic complexes may at some time surpass the natural ones in performance.

We finally note that other LH complexes in bacteria (for instance, the FMO protein [224]) and higher plants (photosystems I and II [225,226]) have a much lower degree of symmetry. Although the exciton delocalization is not very strong in these systems, effects of several molecules participating in the excitation can still be observed in some of these systems [227].

## 9. Concluding Remarks

In this chapter we have described the basic theory and several key experiments of Frenkel and charge-transfer excitons in organic solids. While in some respects, this field may be considered classical in the sense that it has gone through a history of almost three-quarters of a century, it still constitutes an extremely important field to date. Most of the optical properties as well as the energy transport properties in molecular solids result from the static and dynamic properties of these neutral excitations. Moreover, the scope of this field has undergone a strong development from the study of three-dimensional organic bulk solids, such as the classical aromatic crystals, to include nanoscale systems, such as molecular J-aggregates and biological light-harvesting systems. The increasing interest in energy transport, storage, and manipulation on molecular length scales, both in synthetic and in natural systems, should be expected to further strengthen this development.

## Acknowledgement

We would like to thank Professor R.J. Silbey for stimulating discussions.

Jasper Knoester gratefully acknowledges the hospitality and support from the Massachusetts Institute of Technology while working on this manuscript.

## References

1. M. Kasha, in: B. di Bartolo (Ed.), Spectroscopy of the Excited State, Vol. B12, Plenum, New York, 1976.
2. S. de Boer, K. Vink, D.A. Wiersma, *Chem. Phys. Lett.* 137 (1987) 99.
3. N.C. Greenham, S.C. Moratti, D.D.C. Bradley, R.H. Friend, A.B. Holmes, *Nature* 365 (1993) 628.
4. U. Lemmer, A. Haugeneder, C. Kallinger, J. Feldmann, in: G. Hadziioannou, P.F. van Hutten (Eds.), Semiconducting Polymers, Wiley-VCH, Weinheim, 2000.
5. B.I. Greene, J. Orenstein, S. Schmitt-Rink, *Science* 247 (1990) 679.
6. V.L. Bogdanov, E.N. Viktorova, S.V. Kulya, A.S. Spiro, *Pis'ma Zh. Èksper. Teoret. Fiz.* 53 (1990) 100; *JETP Lett.* 53 (1991) 105.
7. Y. Wang, *J. Opt. Soc. Amer. B* 8 (1991) 981.
8. H.M. Gibbs, Optical Bistability: Controling Light with Light, Academic Press, New York, 1985.
9. D.W. Snoke, J.P. Wolfe, A. Mysyrowicz, *Phys. Rev. Lett.* 64 (1990) 2543.
10. J.C. Kim, J.P. Wolfe, *Phys. Rev. B* 57 (1998) 9861.
11. A.H. Herz, *Adv. Colloid Interface Sci.* 8 (1977) 237.
12. T. Tani, Photographic Sensitivity, Oxford University Press, Oxford, 1995.
13. H. van Amerongen, L. Valkunas, R. van Grondelle, Photosynthetic Excitons, World Scientific, Singapore, 2000.
14. R.S. Knox, Theory of Excitons, Academic Press, New York, 1963.

15. H. Haken, Quantum Field Theory of Solids: An Introduction, North-Holland, Amsterdam, 1976.
16. D. Haarer, M.R. Philpott, in: V.M. Agranovich, R.M. Hochstrasser (Eds.), Spectroscopy and Excitation Dynamics of Condensed Molecular Systems, North-Holland, Amsterdam, 1983, pp. 27–82.
17. M. Pope, C.E. Swenberg, Electronic Processes in Organic Crystals and Polymers, Oxford University Press, Oxford, 1999.
18. A. Kronenberger, P. Pringsheim, Z. Phys. 40 (1926) 75.
19. I.V. Obreimov, J. Russia Soc. Phys. Chem. 59 (1927) 548 (in Russian).
20. I.V. Obreimov, W.J. de Haas, Proc. Acad. Sci. Amsterdam 31 (1928) 353; Proc. Acad. Sci. Amsterdam 32 (1929) 1324.
21. Ya.I. Frenkel, Phys. Rev. 37 (1931) 17, ibid 1276.
22. Ya.I. Frenkel, Phys. Z. Soviet Union 9 (1936) 158.
23. I.V. Obreimov, A.F. Prikhotko, Sov. Phys. 1 (1932) 203, Sov. Phys. 9 (1934) 34, Sov. Phys. 9 (1936) 48.
24. V.L. Broude, E.I. Rashba, E.F. Sheka, Spectroscopy of Molecular Excitons, Springer, Berlin, 1985.
25. T. Kobayashi (Ed.), J-Aggregates, World Scientific, Singapore, 1996.
26. M. Sargent III, M.O. Scully, W.E. Lamb Jr., Laser Physics, Addison-Wesley, London, 1974.
27. V.M. Agranovich, Zh. Èxper. Teoret. Fiz. 37 (1959) 430; English translation: Sov. Phys. JETP 37 (1960) 307.
28. V.M. Agranovich, Fiz. Tverd. Tela 3 (1961) 811; English translation: Sov. Phys. Solid State 3 (1961) 592;
V.M. Agranovich, The Theory of Excitons, Nauka, Moscow, 1968 (in Russian).
29. R.P. Feynman, F.L. Vernon Jr., R.W. Hellwarth, J. Appl. Phys. 28 (1957) 49.
30. S.V. Tyablikov, Methods in the Quantum Theory of Magnetism, Plenum, New York, 1967.
31. V. Czikkely, H.D. Försterling, H. Kuhn, Chem. Phys. Lett. 6 (1970) 207.
32. B.P. Krueger, G.D. Scholes, G. Fleming, J. Phys. Chem. B 102 (1998) 5378.
33. J.A. Leegwater, S. Mukamel, Phys. Rev. A 46 (1992) 452.
34. G. Juzeliūnas, J. Knoester, J. Chem. Phys. 112 (2000) 2325.
35. A.S. Davydov, Theory of Molecular Excitons, Plenum, New York, 1971.
36. V.M. Agranovich, M.D. Galanin, in: V.M. Agranovich, A.A. Maradudin (Eds.), Electronic Excitation Energy Transfer in Condensed Matter, North-Holland, Amsterdam, 1982.
37. D.B. Chesnut, A. Suna, J. Chem. Phys. 39 (1963) 146.
38. G. Juzeliūnas, Z. Phys. D 8 (1988) 379.
39. F.C. Spano, Phys. Rev. Lett. 67 (1991) 3424.
40. Vu Duy Phach, A. Bivas, B. Hönerlage, J.B. Grun, Phys. Status Solidi B 86 (1978) 159.
41. H. Fidder, J. Knoester, D.A. Wiersma, J. Chem. Phys. 98 (1993) 6564.
42. R. Kubo, M. Toda, N. Hashitsume, Statistical Physics II, Springer, Berlin, 1992.
43. P.N. Butcher, D. Cotter, The Elements of Nonlinear Optics, Cambridge University Press, Cambridge, 1990, pp. 56–61.
44. V.M. Agranovich, B.S. Toshich, Zh. Èksper. Teoret. Fiz. 53 (1967) 149; Sov. Phys. JETP 26 (1968) 104.
45. M.I. Kaganov, in: A.S. Borovik-Romanov, S.K. Sinha (Eds.), Spin Waves and Magnetic Excitations, North-Holland, Amsterdam, 1988.
46. L.D. Bakalis, J. Knoester, J. Chem. Phys. 106 (1997) 6964.
47. M. Born, K. Huang, Dynamical Theory of Crystal Lattices, Clarendon, Oxford, 1954.
48. M.H. Cohen, F. Keffer, Phys. Rev. 99 (1955) 1128.
49. M.R. Philpott, J. Chem. Phys. 60 (1974) 1410.
50. M. Orrit, P. Kottis, Adv. Chem. Phys. LXXIV (1988) 1.
51. M.R. Philpott, J. Chem. Phys. 54 (1971) 111.

52. D.P. Craig, *J. Chem. Soc.* (1955) 2302.
53. D.P. Craig, S.H. Walmsley, Excitons in Molecular Crystals, Benjamin, New York, 1968.
54. R.J. Silbey, J. Jortner, S. Rice, *J. Chem. Phys.* 42 (1965) 1515.
55. M.R. Philpott, *J. Chem. Phys.* 50 (1969) 5117.
56. J. Knoester, F.C. Spano, *Phys. Rev. Lett.* 74 (1995) 2780.
57. V. Chernyak, W.M. Zhang, S. Mukamel, *J. Chem. Phys.* 109 (1998) 9587.
58. G. Juzeliūnas, P. Reineker, *J. Chem. Phys.* 109 (1998) 6919.
59. D.P. Craig, T. Thiranumachandran, Molecular Quantum Electrodynamics, Academic Press, London, 1984.
60. J. Knoester, S. Mukamel, *Phys. Rep.* 205 (1991) 1.
61. V.M. Agranovich, V.L. Ginzburg, Crystal Optics with Spatial Dispersion and Excitons, Springer, Berlin, 1984.
62. C. Kittel, Quantum Theory of Solids, Wiley, New York, 1987, p. 42.
63. J.J. Hopfield, *Phys. Rev.* 112 (1958) 1555.
64. S.H. Stevenson, M.A. Connolly, G.J. Small, *Chem. Phys.* 128 (1988) 157.
65. J.J. Hopfield, D.G. Thomas, *Phys. Rev.* 132 (1963) 563.
66. J.D. Joannopoulos, R.D. Meade, J.N. Winn, Photonic Crystals: Molding the Flow of Light, Princeton University Press, Princeton, 1995.
67. J. Knoester, *Phys. Rev. Lett.* 68 (1992) 654.
68. V.M. Agranovich, D.M. Basko, O. Dubovskii, *J. Chem. Phys.* 106 (1997) 3896.
69. V.N. Denisov, B.N. Mavrin, V.B. Podobedov, *Phys. Rep.* 151 (1987) 1.
70. N. Bloembergen, Nonlinear Optics, Benjamin, New York, 1965.
71. S. Mukamel, Principles of Nonlinear Optical Spectroscopy, Oxford University Press, Oxford, 1995.
72. F.C. Spano, S. Mukamel, *Phys. Rev. Lett.* 66 (1991) 1197.
73. F.C. Spano, *J. Chem. Phys.* 96 (1992) 8109.
74. J. Knoester, *Phys. Rev. A* 47 (1993) 2083.
75. L.D. Bakalis, J. Knoester, *J. Phys. Chem. B* 103 (1999) 6620.
76. N.A. Efremov, E.P. Kaminskaya, *Fiz. Tverd. Tela* 15 (1973) 3338.
77. G. Vektaris, *J. Chem. Phys.* 101 (1994) 3031.
78. V.M. Aganovich, N.A. Efremov, E.P. Kaminskaya, *Opt. Commun.* 3 (1971) 3.
79. F.C. Spano, V.M. Agranovich, S. Mukamel, *J. Chem. Phys.* 95 (1991) 1400.
80. F.C. Spano, *Chem. Phys. Lett.* 234 (1995) 29.
81. J. Shah, Ultrafast Spectroscopy in Semiconductors and Semiconductor Nanostructures, Springer, Berlin, 1999.
82. V.M. Agranovich, O.A. Dubovsky, A.M. Kamchatkov, *Synth. Met.* 116 (2001) 293.
83. M. Kuwata-Gonokami, N. Peyghambarian, K. Meissner, B. Fluegel, Y. Sato, K. Ema, R. Shimano, S. Mazumdar, F. Guo, T. Tokihiro, H. Ezaki, E. Hanamura, *Nature* 367 (1994) 47.
84. L. Valkunas, G. Trinkunas, V. Liuolia, in: D.L. Andrews, A.A. Demidov (Eds.), Resonance Energy Transfer, Wiley, New York, 1999.
85. V.A. Malyshev, H. Glaeske, K.-H. Feller, *Chem. Phys. Lett.* 305 (1999) 117.
86. I.V. Ryzhov, G.G. Kozlov, V.A. Malyshev, J. Knoester, *J. Chem. Phys.* 114 (2001) 5322.
87. G.J. Denton, N. Tessler, N.T. Harrison, R.H. Friend, *Phys. Rev. Lett.* 78 (1997) 733.
88. M. Pope, N.E. Geacintov, F. Vogel, *Mol. Cryst. Liq. Cryst. Sci. Technol.* 6 (1969) 83.
89. F.C. Spano, S. Mukamel, *J. Chem. Phys.* 91 (1989) 683.
90. V. Chernyak, W.M. Zhang, S. Mukamel, *J. Chem. Phys.* 109 (1998) 9587.
91. L.N. Ovander, *Uspekhi Fiz. Nauk* 86 (1965) 3; English translation: *Sov. Phys.-Usp.* 8 (1965) 337.
92. J.L. Skinner, D. Hsu, *J. Phys. Chem.* 90 (1986) 4931.
93. P.Y. Yu, M. Cardona, Fundamentals of Semiconductors, Springer, Berlin, Heidelberg, 1996.

94. V.M. Agranovich, *Uspekhi Fiz. Nauk* 112 (1974) 143, English translation: *Sov. Phys.-Usp.* 17 (1974) 103.
95. J. Knoester, S. Mukamel, *Phys. Rev. A* 40 (1989) 7065.
96. E.I. Rashba, in: Physics of Impurity Centers in Crystals, Proc. Internat. Seminar, Tallin, p. 427; V.L. Broude, E.I. Rashba, E.F. Sheka, Spectroscopy of Molecular Excitons, Springer, Heidelberg, 1984.
97. A.S. Davydov, Theory of Molecular Excitons, Plenum Press, 1971, p. 65; A.S. Davydov, E.N. Myasnikov, *Sov. Phys.-Dokladi* 173 (1967) 1040; English translation: *Sov. Phys. Dokl.* 12 (1967) 346.
98. H. Haug, S.W. Koch, Quantum Theory of the Optical and Electronic Properties of Semiconductors, World Scientific, Singapore, 1990.
99. Th. Förster, *Ann. Phys. (Leipzig)* 2 (1948) 55.
100. P. Reineker, in: G. Höhler (Ed.), Exciton Dynamics in Molecular Crystals and Aggregates, Springer, Berlin, 1982.
101. V.M. Kenkre, in: G. Höhler (Ed.), Exciton Dynamics in Molecular Crystals and Aggregates, Springer, Berlin, 1982.
102. R. Silbey, in: V.M. Agranovich, R.M. Hochstrasser (Eds.), Spectroscopy and Excitation Dynamics of Condensed Molecular Systems, North-Holland, Amsterdam, 1983, pp. 1–26.
103. V. May, O. Kühn, Charge and Energy Transfer Dynamics in Molecular Systems, Wiley-VCH, Berlin, 2000.
104. A. Redfield, *Adv. Magn. Res.* 1 (1965) 1.
105. R. Kubo, *J. Phys. Soc. Japan* 12 (1957) 570.
106. D.N. Zubarev, *Uspekhi Fiz. Nauk* 71 (1960) 71; English translation: *Sov. Phys.-Usp.* 3 (1960) 320.
107. R. Peierls, Quantum Theory of Solids, Clarendon Press, Oxford, 1955.
108. H. Fröhlich, *Proc. Roy. Soc.* 160 (1937) 230.
109. V.M. Agranovich, Yu.V. Konobeev, *Opt. Spektrosk.* 6 (1959) 242; English translation: *Opt. Spectrosc.* 6 (1959) 115; *Fiz. Tverd. Tela* 5 (1963) 1373; English translation: *Sov. Phys.-Solid State* 5 (1963) 999; *Phys. Stat. Sol.* 27 (1968) 435.
110. V.V. Katal'nikov, *Fiz. Tekh. Poluprovodn.* 2 (1968) 1392; English translation: *Sov. Phys.-Semicond.* 2 (1969) 1165; *Fiz. Tverd. Tela* 16 (1974) 3498; English translation: *Sov. Phys.-Solid State* 16 (1974) 2271;
V.V. Katal'nikov, O.S. Rudenko, *Ukrain. Fiz. Zh.* 22 (1977) 1362.
111. V.M. Agranovich, The Theory of Excitons, Nauka, Moscow, 1968 (in Russian).
112. V.M. Kenkre, R.S. Knox, *Phys. Rev. B* 9 (1974) 5279;
V.M. Kenkre, *Phys. Rev. B* 11 (1975) 1741.
113. H. Haken, G. Strobl, *Z. Phys.* 262 (1973) 135.
114. M. Trlifaj, *J. Czech. Phys.* 6 (1956) 533.
115. R. Kopelman, in: V.M. Agranovich, R.M. Hochstrasser (Eds.), Spectroscopy and Excitation Dynamics of Condensed Molecular Systems, North-Holland, Amsterdam, 1983, pp. 139–184.
116. M.D. Fayer, in: V.M. Agranovich, R.M. Hochstrasser (Eds.), Spectroscopy and Excitation Dynamics of Condensed Molecular Systems, North-Holland, Amsterdam, 1983, pp. 185–248.
117. T.S. Rose, R. Rigini, M.D. Fayer, *Chem. Phys. Lett.* 106 (1984) 13.
118. T.S. Rose, V.J. Newel, J.S. Meth, M.D. Fayer, *Chem. Phys. Lett.* 145 (1988) 475.
119. V.M. Agranovich, A.M. Ratner, M.Kh. Salieva, *Solid State Comm.* 63 (1986) 329;
V.M. Agranovich, T.A. Leskova, *Solid State Comm.* 68 (1988) 1029.
120. V.M. Agranovich, A.M. Ratner, M.Kh. Salieva, *Chem. Phys.* 128 (1988) 23.
121. L.D. Landau, *Phys. Z. Soviet.* 3 (1933) 664.
122. S.I. Pekar, Untersuchungen über die Electronentheorie der Kristallen, Academie-Verlag, Berlin, 1954.

123. R. Peierls, *Ann. Phys.* 13 (1932) 905.
124. A.S. Davydov, Theory of Molecular Excitons, McGraw-Hill, New York, 1962.
125. E.I. Rashba, in: E.I. Rashba, M.D. Sturge (Eds.), Excitons, North-Holland, Amsterdam, 1982, p. 543.
126. N. Schwentner, E.E. Koch, J. Jortner, Electronic Excitations in Condensed Rare Gases, in: Springer Tracts in Modern Physics, Vol. 107, Springer, Berlin, 1985.
127. M. Ueta, H. Kanzaki, K. Kabayashi, Y. Toyozawa, E. Hanamura, Excitonic Processes in Solids, in: Springer Series in Solid State Sciences, Vol. 60, Springer, Berlin, 1985.
128. G. Zimmerer, in: Excited State Spectroscopy in Solids, XCVI Corso, Societe Italiana di Fisica, Bologna, 1987, p. 37.
129. M.F. Deigen, S.I. Pekar, *Zh. Èksper. Teoret. Fiz.* 21 (1951) 803.
130. V.M. Agranovich, A.A. Zakhidov, *Chem. Phys. Lett.* 50 (1977) 278.
131. V.M. Agranovich, B.P. Antonyuk, A.G. Mal'shukov, *Pis'ma Zh. Èksper. Teoret. Fiz.* 23 (1976) 492; English translation: *JETP Lett.* 23 (1976) 448;
    V.M. Agranovich, B.P. Antonyuk, E.P. Ivanova, A.G. Mal'shukov, *Zh. Èksper. Teoret. Fiz.* 72 (1977) 614; English translation: *Sov. Phys. JETP* 45 (1977) 322.
132. A.S. Ioselevich, E.I. Rashba, in: Yu. Kagan, A.J. Legget (Eds.), Quantum Tunneling in Condensed Media, North-Holland, Amsterdam, 1992, p. 347.
133. K. Mizuno, A. Matsui, *J. Lumin.* 38 (1987) 323;
    A. Matsui, K. Mizuno, N. Tamai, I. Yamazaki, *J. Lumin.* 40–41 (1988) 455;
    A. Matsui, K. Mizuno, *Proc. SPIE* 910 (1988) 131; *Proc. SPIE* 1054 (1989) 224;
    M. Furukawa, K. Mizuno, A. Matsui, N. Tamai, I. Yamazaki, *Chem. Phys.* 138 (1989) 423;
    A. Matsui, *J. Opt. Soc. Amer. B* 7 (1990) 1615.
134. Yu. Kagan, N.V. Prokof'ev, in: Yu. Kagan, A.J. Legget (Eds.), Quantum Tunneling in Condensed Media, North-Holland, Amsterdam, 1992, p. 37.
135. E.A. Silinsh, Organic Molecular Crystals: Their Electronic States, Springer-Verlag, Berlin, 1980.
136. M. Pope, C.E. Swenberg, Electronic Processes in Organic Crystals, Clarendon Press, Oxford, 1982.
137. D.S. Chemla, J. Zyss, Nonlinear Optical Properties of Organic Molecules and Crystals, Academic Press, Orlando, 1987.
138. V.M. Agranovich, G.C. La Rocca, F. Bassani, *Pis'ma Zh. Èksper. Teoret. Fiz.* 62 (1995) 407; English translation: *JETP Lett.* 62 (1995) 418.
139. V.M. Agranovich, G.C. La Rocca, F. Bassani, *Chem. Phys. Lett.* 247 (1995) 355.
140. V.M. Agranovich, K.N. Ilinski, *Phys. Lett. A* 191 (1994) 309.
141. P. Petelenz, in: V.M. Agranovich, G.C. La Rocca (Eds.), Organic Nanostructures: Science and Applications, IOS Press, Amsterdam, 2002.
142. L. Sebastian, G. Weiser, H. Bässler, *Chem. Phys.* 61 (1981) 125.
143. L. Sebastian, G. Weiser, G. Peter, H. Bässler, *Chem. Phys.* 75 (1983) 103.
144. G. Weiser, *Phys. Rev. B* 45 (1992) 14076.
145. G.H. Gelinck, J.J. Piet, J. Warman, *Synth. Met.* 101 (1999) 553.
146. E.E. Jelley, *Nature* 138 (1936) 1009, *Nature* 139 (1937) 631.
147. G. Scheibe, *Angew. Chem.* 49 (1936) 563, *Angew. Chem.* 50 (1937) 212.
148. J. Franck, E. Teller, *J. Chem. Phys.* 6 (1938) 861.
149. G. Scheibe, in: W. Först (Ed.), Optische Anregungen organischer Systeme, Verlag Chemie, Weinheim, 1966, p. 109.
150. I.I. Abram, R.M. Hochstrasser, *J. Chem. Phys.* 72 (1980) 3617.
151. E.W. Knapp, *Chem. Phys.* 85 (1984) 73.
152. S. de Boer, D.A. Wiersma, *Chem. Phys. Lett.* 165 (1990) 45.
153. E. Hanamura, *Phys. Rev. B* 37 (1988) 1273.

154. F.C. Spano, S. Mukamel, *Phys. Rev. A* 40 (1989) 5783.
155. H. Ishihara, K. Cho, *Phys. Rev. B* 42 (1990) 1724.
156. J. Knoester, *Chem. Phys. Lett.* 203 (1993) 371.
157. R.V. Markov, A.I. Plekhanov, V.V. Shelkovnikov, J. Knoester, *Phys. Status Solidi B* 221 (2000) 529.
158. H. Fidder, Collective Optical Response of Molecular Aggregates, PhD Thesis, University of Groningen, 1993.
159. H. von Berlepsch, C. Böttcher, L. Dähne, *J. Phys. Chem. B* 104 (2000) 8792.
160. H. Fidder, J. Knoester, D.A. Wiersma, *J. Chem. Phys.* 95 (1991) 7880.
161. R.M. Hochstrasser, J.D. Whiteman, *J. Chem. Phys.* 56 (1972) 5945.
162. T. Itoh, M. Furumiya, T. Ikehara, C. Gourdon, *Solid State Commun.* 73 (1990) 271.
163. I.G. Scheblykin, O.Y. Sliusarenko, L.S. Lepnev, A.G. Vitukhnovsky, M. Van der Auweraer, *J. Phys. Chem. B* 105 (2001) 4636.
164. H. Kuhn, C. Kuhn, in: T. Kobayashi (Ed.), J-Aggregates, World Scientific, Singapore, 1996.
165. H. Fidder, J. Knoester, D.A. Wiersma, *Chem. Phys. Lett.* 171 (1990) 529.
166. F.C. Spano, J.R. Kuklinski, S. Mukamel, *Phys. Rev. Lett.* 65 (1990) 211.
167. V.F. Kamalov, I.A. Struganova, K. Yoshihara, *J. Phys. Chem.* 100 (1996) 8640.
168. M. Bednarz, V.A. Malyshev, J. Knoester, *J. Chem. Phys.* 117 (2002) 6200.
169. V.M. Agranovich, O.A. Dubovsky, *Pis'ma Zh. Èksper. Teoret.* 3 (1966) 345; English translation: *JETP Lett.* 3 (1966) 223.
170. V.A. Malyshev, P. Moreno, *Phys. Rev. B* 51 (1995) 14587.
171. J. Knoester, *J. Chem. Phys.* 99 (1993) 8466.
172. J.R. Durrant, J. Knoester, D.A. Wiersma, *Chem. Phys. Lett.* 222 (1994) 450.
173. M. Wang, G.R. Fleming, *J. Chem. Phys.* 113 (2000) 2823.
174. V.A. Malyshev, F. Domínguez-Adame, *Chem. Phys. Lett.* 313 (1999) 255.
175. V.A. Malyshev, *Opt. Spectrosk.* 71 (1991) 873, *Opt. Spectrosc.* 71 (1992) 505; *J. Lumin.* 55 (1993) 225.
176. P.W. Anderson, *Phys. Rev.* 109 (1958) 1492.
177. L.D. Bakalis, J. Knoester, *J. Lumin.* 87–89 (2000) 66.
178. A.V. Malyshev, V.A. Malyshev, *Phys. Rev. B* 63 (2001) 195111.
179. M. Schreiber, Y. Toyozawa, *J. Phys. Soc. Japan* 51 (1982) 1528, 1537.
180. A. Boukahil, D.L. Huber, *J. Lumin.* 45 (1990) 13.
181. J. Köhler, A.M. Jayannavar, P. Reineker, *Z. Phys. B* 75 (1989) 451.
182. H. Fidder, D.A. Wiersma, *Phys. Status Solidi B* 188 (1995) 285.
183. D.V. Makhov, V.V. Egorov, A.A. Bagatur'yants, M.V. Alfimov, *Chem. Phys. Lett.* 246 (1995) 371.
184. H. Fidder, J. Terpstra, D.A. Wiersma, *J. Chem. Phys.* 94 (1991) 6895.
185. R.V. Markov, A.I. Plekhanov, Z.M. Ivanova, V.V. Shelkovnikov, J. Knoester, in: F. Charra, et al. (Eds.), Organic Nanophotonics, Kluwer, Amsterdam, 2003, pp. 279–290.
186. R. Gadonas, R. Danielius, A. Piskarskas, S. Rentsch, *Izv. Akad. Nauk SSSR Fiz.* 47 (1983) 2445; English translation: *Bull. Akad. Sci. USSR Phys.* 47 (1983) 151.
187. A.E. Johnson, S. Kumazaki, K. Yoshihara, *Chem. Phys. Lett.* 211 (1993) 511.
188. K. Minoshima, M. Taiji, K. Misawa, T. Kobayashi, *Chem. Phys. Lett.* 218 (1994) 67.
189. M. van Burgel, D.A. Wiersma, K. Duppen, *J. Chem. Phys.* 102 (1995) 20.
190. J. Moll, S. Daehne, J.R. Durrant, D.A. Wiersma, *J. Chem. Phys.* 102 (1995) 6362.
191. T. Pullerits, M. Chachisvilis, V. Sündstrom, *J. Phys. Chem.* 100 (1996) 10787.
192. T. Meier, V. Chernyak, S. Mukamel, *J. Phys. Chem. B* 101 (1997) 7332.
193. R. Gagel, R. Gadonas, A. Laubereau, *Chem. Phys. Lett.* 217 (1994) 228.
194. H. Stiel, S. Daehne, K. Teuchner, *J. Lumin.* 39 (1988) 351.
195. V. Sündstrom, T. Gillbro, R. Gadonas, A. Piskarskas, *J. Chem. Phys.* 89 (1988) 2754.

196. E. Gaizaukas, K.-H. Feller, R. Gadonas, *Opt. Commun.* 118 (1995) 360.
197. A.A. Muenter, D.V. Brumbaugh, J. Apolito, L.A. Horn, F.C. Spano, S. Mukamel, *J. Phys. Chem.* 96 (1992) 2783.
198. L.D. Bakalis, I. Rubtsov, J. Knoester, *J. Chem. Phys.* 117 (2002) 5393.
199. S. Kirstein, H. Möhwald, M. Shimomura, *Chem Phys. Lett.* 154 (1989) 303.
200. A. Nabetani, A. Tomioka, H. Tamaru, K. Miyano, *J. Chem. Phys.* 102 (1995) 5109.
201. A.G. Vitukhnovsky, A.N. Lobanov, A.V. Pimenov, Y. Yonezawa, N. Kometani, K. Asumi, J. Yano, *J. Lumin.* 87–89 (2000) 260.
202. M.R. Philpott, J.M. Turlet, *J. Chem. Phys.* 64 (1976) 3852.
203. Y. Aviksoo, Y. Lippmaa, T. Reinot, *Opt. Spectrosk.* 62 (1987) 706; English translation: *Opt. Spectrosc.* 62 (1987) 419.
204. S. Kirstein, H. von Berlepsch, C. Böttcher, C. Burger, A. Ouart, G. Reck, S. Daehne, *CHEMPHYSCHEM* 3 (2000) 146.
205. L.A. Staehelin, J.R. Golecki, G. Drews, *Biochim. Biophys. Acta* 589 (1980) 30.
206. A.R. Holzwarth, K. Schaffner, *Photosynth. Res.* 41 (1994) 225.
207. M. Bednarz, J. Knoester, *J. Phys. Chem. B* 105 (2001) 12913.
208. C. Didraga, J. Klugkist, J. Knoester, *J. Phys. Chem. B* 106 (2002) 11474.
209. C. Spitz, J. Knoester, A. Ouart, S. Daehne, *Chem. Phys.* 275 (2002) 271.
210. C. Ecoffet, D. Markovitsi, P. Millié, J.P. Lemaistre, *Chem. Phys.* 177 (1993) 629.
211. G. McDermott, S.M. Prince, A.A. Freer, A.M. Hawthornthwaitelawless, M.Z. Papiz, R.J. Cogdell, N.W. Isaacs, *Nature* 374 (1995) 517.
212. J. Koepke, X. Hu, C. Muenke, K. Schulten, H. Michel, *Structure* 4 (1996) 581.
213. V.I. Novoderezhkin, A.P. Razjivin, *FEBS* 330 (1993) 5.
214. X. Hu, K. Schulten, *Phys. Today* 50 (1997) 28.
215. A.M. van Oijen, M. Ketelaars, J. Köhler, T.J. Aartsma, J. Schmidt, *Science* 285 (1999) 400.
216. A.M. van Oijen, M. Ketelaars, J. Köhler, T.J. Aartsma, J. Schmidt, *Biophys. J.* 78 (2000) 1570.
217. M.V. Mostovoy, J. Knoester, *J. Phys. Chem. B* 104 (2000) 12355.
218. R. Monshouwer, M. Abrahamsson, F. van Mourik, R. van Grondelle, *J. Phys. Chem. B* 101 (1997) 7241.
219. A.P. Shreve, J.K. Trautman, H.A. Frank, T.G. Owens, A.C. Albrecht, *Biochim. Biophys. Acta* 1058 (1991) 280.
220. R. Jimenez, S.N. Dikshit, S.E. Bradforth, G.R. Fleming, *J. Phys. Chem.* 100 (1996) 6825.
221. H. Sumi, *J. Phys. Chem. B* 103 (1999) 252.
222. K. Mukai, S. Abe, H. Sumi, *J. Phys. Chem. B* 103 (1999) 6096.
223. O.J.G. Somsen, R. van Grondelle, H. van Amerongen, *Biophys. J.* 71 (1996) 1934.
224. J.M. Olson, *Biochim. Biophys. Acta* 594 (1980) 33.
225. P. Jordan, P. Fromme, H.T. Witt, O. Klukas, W. Sänger, N. Krauss, *Nature* 411 (2001) 909.
226. A. Zouni, H.T. Witt, J. Kern, P. Fromme, N. Krauss, W. Sänger, P. Orth, *Nature* 409 (2001) 739.
227. M. Wendling, M.A. Przyjalgowski, D. Glen, S.I.E. Vulto, T.J. Aartsma, R. van Grondelle, H. van Amerongen, *Photosynth. Res.* 71 (2002) 99.

## Chapter 2
# Wannier–Mott Excitons in Semiconductors

### G.C. LA ROCCA
*Scuola Normale Superiore and I.N.F.M., Piazza dei Cavalieri 7, 56126 Pisa, Italy*

1. Introduction . . . . . . . . . . . . . . . . . . . . . . . . . . . . . . . . . . . 97
2. Wannier–Mott Excitons in Bulk Semiconductors . . . . . . . . . . . . . . . . . . 98
3. Quantum Confined Wannier–Mott Excitons . . . . . . . . . . . . . . . . . . . 106
4. Quantum Well Exciton Optical Nonlinearities . . . . . . . . . . . . . . . . . . 115
   References . . . . . . . . . . . . . . . . . . . . . . . . . . . . . . . . . . . 126

## 1. Introduction

In this second introductory chapter, the concept of Wannier–Mott exciton in bulk and low-dimensional semiconductors will be reviewed, including non-linear effects in quantum wells, with all the emphasis on a physically motivated discussion rather than on formal derivations. With no pretense of completeness, the presentation is tailored to the needs of readers not already familiar with semiconductor physics [1] and the choice of material has a bias towards the general topic of this volume. The impressive achievements in inorganic semiconductor quantum confined structures [2], and in exciton based optoelectronics in particular, are an omen of success for the rapidly developing field of optically active organic based low-dimensional structures.

The low temperature optical properties of pure semiconductor crystals and of their heterostructures are dominated, just below the band edge, by exciton absorption lines typically arranged in a hydrogenic series. The excitons in semiconductors have a Bohr radius $a_0$, related to the relative motion of electron and hole, which is much larger than the lattice constant. Thus, they are well described as Wannier–Mott, or weak binding, excitons [3], as opposed to tightly bound Frenkel excitons [4]. Apart from a small reduced mass, the electron–hole attraction is screened by the large dielectric constant $\varepsilon \approx 10$ and the exciton binding energy $E_b$, i.e., the difference between the $1S$ exciton line and the electronic band structure gap, is only a tiny fraction of the Rydberg, not more than a few tens of meV.

The envelope function approach is well suited to describe Wannier–Mott excitons, as discussed in Section 2 for the bulk and in Section 3 for quantum confined structures [5]. In the most relevant case of a quantum well thickness $L$ of the order of the Bohr radius, the motion of the electron and of the hole along the growth direction are separately quantized and the binding energy and oscillator strength of the quantum well exciton are significantly increased compared to the bulk.

At high excitation density, non-linear effects come into play and the picture of excitons as ideal bosons is to be abandoned [6]. In Section 4, the most simple and basic considerations are discussed in the complex and still rapidly developing field of many-body and correlation effects in semiconductor optics [7]. In the intermediate density regime where the excitonic resonances still persist, the nonlinearities scale as $n/n_S$ where $n$ is the two-dimensional exciton density and $n_S$ the corresponding saturation density $n_S \approx 1/(\pi a_{2D}^2)$, being $a_{2D}$ the two-dimensional exciton radius. Moreover, for appropriate conditions, biexcitons, i.e., molecular states of two excitons bound together, may form and affect the optical properties [8]. Finally, the possibility of Bose–Einstein condensation of excitons will be considered [9].

## 2. Wannier–Mott Excitons in Bulk Semiconductors

### 2.1. Band Structure of Direct Gap Semiconductors

The electronic states of inorganic semiconductor crystals are well described by a band structure calculated assuming an appropriate crystal potential $V$ in a rigid lattice in the one-electron approximation, i.e., including electron–electron interactions only on average with no correlation effects. The corresponding single particle Hamiltonian is

$$H = \frac{p^2}{2m_0} + V(\mathbf{r}) \qquad (1)$$

where $m_0$ is the electron mass, $\mathbf{p}$ the momentum operator, and $V$ the mean field potential having the full symmetry of the crystal, in particular the translational symmetry.

The single particle eigenstates have energies $E_{n\mathbf{k}}$ and are Bloch states characterized by a wavevector $\mathbf{k}$ within the first Brillouin zone and by a band index $n$

$$\psi_{n\mathbf{k}}(\mathbf{r}) = \frac{1}{\sqrt{N}} e^{i\mathbf{k}\cdot\mathbf{r}} u_{n\mathbf{k}}(\mathbf{r}), \qquad (2)$$

where $u_{n\mathbf{k}}(\mathbf{r})$ is periodic on the unit cell of volume $\Omega$ and the total volume of the crystal is $V = N\Omega$. As an example, the band structure of GaAs is shown in Figure 1. The highest occupied valence band and the lowest empty conduction band have each large bandwidths and small effective masses, corresponding to well delocalized electronic states.

The energy band gap $E_g$ between the bottom of the conduction band and the top of the valence band falls in the visible or near infrared spectral region. As the wavevector of light $2\pi/\lambda$ is much smaller than the size of the Brillouin zone

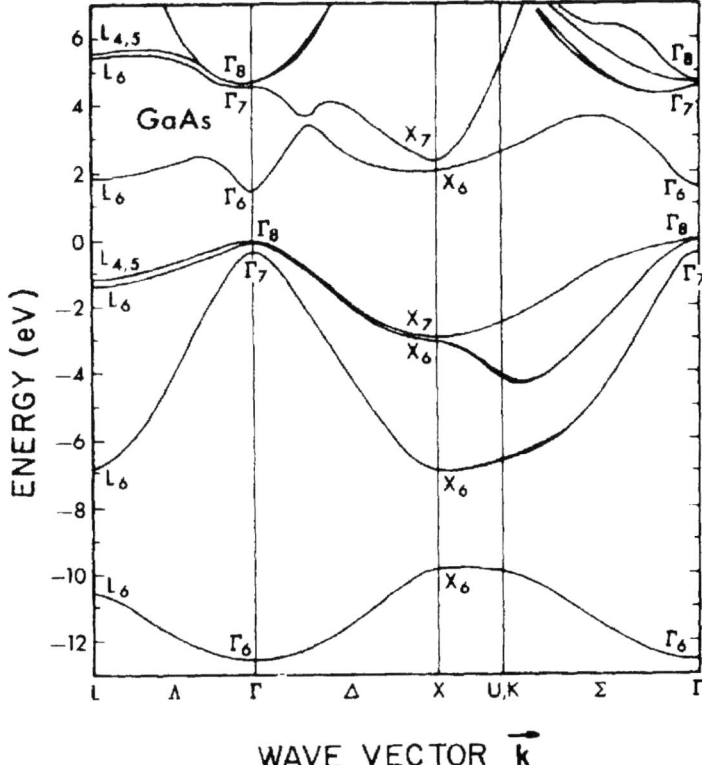

Fig. 1. The band structure of GaAs (from J.R. Chelikowsky and M.L. Cohen, *Phys. Rev. B* 14 (1976) 556). The top of the valence band ($\Gamma_8$) and the bottom of the conduction band ($\Gamma_6$) occur at $\vec{k} = 0$ ($\Gamma$ point).

$\pi/l$, being $l$ the lattice constant, only "vertical" transitions are allowed between electronic states having the same **k**. A direct gap semiconductor has the top of the valence band and the bottom of the conduction band at the same point of the Brillouin zone (typically the $\Gamma$ point at **k** = 0), and, if the symmetry of the Bloch functions correspond to an allowed transition, it exhibits a well pronounced absorption edge at the band gap, becoming opaque at frequencies $\hbar\omega > E_g$ with an absorption coefficient $\alpha \approx 10^4$ cm$^{-1}$. As an example, the absorption edge of InSb is shown in Figure 2 [10].

### 2.2. Interband Transitions and Optical Constants

The quantum mechanical transition probability from an initial valence band state to a final conduction band state under an electromagnetic field described by the

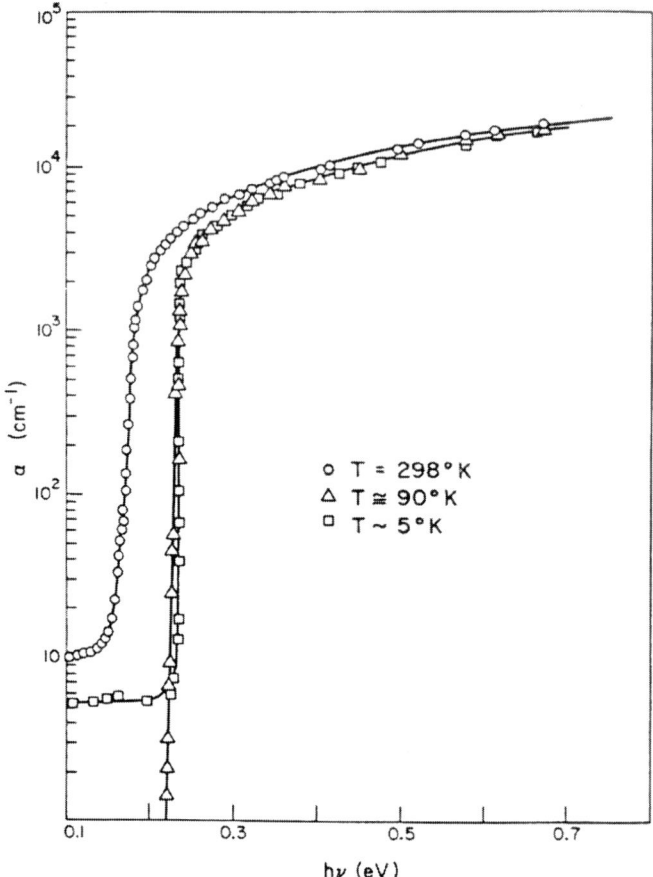

Fig. 2. The absorption edge of InSb (from G.W. Gobeli, H.Y. Fan, in: Semiconductor Research, Purdue University, 1956).

vector potential $\mathbf{A} = A_0 \hat{e}(e^{-i\omega t} + \text{c.c.})$ is given by the Fermi Golden rule

$$\mathcal{P} = \frac{2\pi}{\hbar} \left| \frac{eA_0}{m_0 c} \langle \psi_{c\mathbf{k}} | \hat{e} \cdot \mathbf{p} | \psi_{v\mathbf{k}} \rangle \right|^2 \delta(E_{c\mathbf{k}} - E_{v\mathbf{k}} - \hbar\omega). \qquad (3)$$

In the above expression, contrary to the case of localized molecular excitations [4], the electromagnetic field responsible for the interband transition is just the macroscopic Maxwell field, because for the case of delocalized electronic states local field effects are negligible. The same holds true for the large radius Wannier–Mott excitons discussed later on.

The imaginary part $\varepsilon_2(\omega)$ of the optical dielectric constant can be directly calculated summing on all possible transitions from an occupied to an empty state (then, via Kramers–Kronig analysis $\varepsilon_1(\omega)$, and all other optical constants can be obtained) [3]

$$\varepsilon_2(\omega) = \left(\frac{2\pi e}{m_0 \omega}\right)^2 \sum_{c,v} \int \frac{2 d\mathbf{k}}{(2\pi)^3} |\hat{e} \cdot \mathbf{M}_{cv}(\mathbf{k})|^2 \delta(E_{c\mathbf{k}} - E_{v\mathbf{k}} - \hbar\omega), \quad (4)$$

where the integral is over the Brillouin zone and a factor 2 for spin has been included; the transition matrix element is

$$\mathbf{M}_{cv}(\mathbf{k}) = \langle \psi_{c\mathbf{k}} | \mathbf{p} | \psi_{v\mathbf{k}} \rangle = \langle \psi_{c\mathbf{k}} | \frac{im_0}{\hbar} [H, \mathbf{r}] | \psi_{v\mathbf{k}} \rangle$$

$$= \frac{im_0}{\hbar}(E_{c\mathbf{k}} - E_{v\mathbf{k}}) \langle \psi_{c\mathbf{k}} | \mathbf{r} | \psi_{v\mathbf{k}} \rangle$$

$$= \frac{im_0}{e\hbar}(E_{c\mathbf{k}} - E_{v\mathbf{k}}) \mathbf{d}_{cv}(\mathbf{k}), \quad (5)$$

and the corresponding dimensionless oscillator strength $f$ is

$$f = \frac{2}{m_0 \hbar \omega} |\hat{e} \cdot \mathbf{M}|^2. \quad (6)$$

For a direct gap semiconductor with an allowed transition at the absorption edge, it is a good approximation for $\hbar\omega \approx E_g$ to consider the transition dipole matrix element $\mathbf{M}$ as a constant. Then, the integral in Eq. (4) reduces to the joint density of states (*JDOS*) which vanishes for $\hbar\omega < E_g$ and for $\hbar\omega \gtrsim E_g$ can be evaluated near the absorption edge assuming a parabolic dispersion of the conduction and valence band states

$$JDOS(\omega) = \int \frac{2 d\mathbf{k}}{(2\pi)^3} \delta(E_{c\mathbf{k}} - E_{v\mathbf{k}} - \hbar\omega) \propto \sqrt{\hbar\omega - E_g}. \quad (7)$$

This result correctly describes the gross features of the absorption edge for allowed interband transitions in bulk semiconductors.

### 2.3. Electron–Hole Correlation

The discussion above neglects the attractive interaction between the electron promoted into the conduction band and the hole left behind in the valence band. However, their motion is correlated and, in particular, they can be bound into Wannier–Mott excitonic states with an energy below that of the single particle energy gap. The exciton radius $a_0$ is large and the binding energy correspondingly small, therefore the exciton absorption lines are typically prominent only at low temperatures, as shown for example in Figure 3.

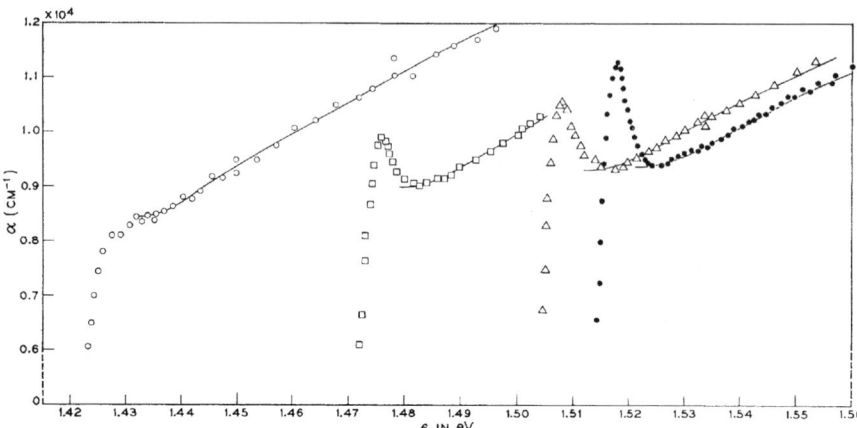

Fig. 3. Excitonic absorption in bulk GaAs: at room temperature (open circles) the $1S$ exciton peak is no longer observed (from M.D. Sturge, *Phys. Rev.* 127 (1962) 768).

The basic exciton physics can be discussed considering the case of non-degenerate conduction and valence bands with extrema at $\mathbf{k} = 0$ and a dipole allowed interband transition matrix element. Spin degrees of freedom will be neglected, for the time being, and the exciton state written as

$$|\Phi_{j\mathbf{Q}}\rangle = \int \frac{V\,d\mathbf{k}}{(2\pi)^3} A_j(\mathbf{k})\, c^+_{\mathbf{k}+m_e\mathbf{Q}/M}\, v_{\mathbf{k}-m_h\mathbf{Q}/M} |G\rangle, \qquad (8)$$

where $|G\rangle$ is the semiconductor ground state, $c^+$ creates an electron in the conduction band while $v$ destructs one in the valence band. The exciton center of mass wavevector $\mathbf{Q}$ for resonantly generated excitons is given by the wavevector of light, and even in other cases is typically much smaller than the size of the Brillouin zone. The function $A_j(\mathbf{k})$ describes the correlated relative motion, labelled by $j$ (e.g., $j = 1S, 2S, 2P\ldots$); $A_j(\mathbf{k})$ has a range around $\mathbf{k} = 0$ of the order of $\pi/a_0$, much smaller than the size of the Brillouin zone. Thus, in the above equation the parabolic approximation is used for the electron and hole dispersion, with effective masses, respectively, $m_e$ and $m_h$, being $M = m_e + m_h$ the total mass of the exciton.

### 2.4. Effective Mass Approximation

Using the hole representation for the valence band, the exciton state can also be described by a two particle wavefunction $\Phi(\mathbf{r}_e, \mathbf{r}_h)$

$$\Phi_{j\mathbf{Q}}(\mathbf{r}_e, \mathbf{r}_h) = \sqrt{\frac{\Omega}{N}}\, F_j(\mathbf{r}_e - \mathbf{r}_h)\, e^{i\mathbf{Q}\cdot(m_e\mathbf{r}_e + m_h\mathbf{r}_h)/(m_e+m_h)}\, u_{c0}(\mathbf{r}_e) u^*_{v0}(\mathbf{r}_h), \qquad (9)$$

where the cell periodic functions $u$ are approximated by those at $\mathbf{k} = 0$. The envelope function $F_j(\mathbf{r} = \mathbf{r}_e - \mathbf{r}_h)$ of the relative motion is

$$F_j(\mathbf{r}) = \sqrt{V} \int \frac{d\mathbf{k}}{(2\pi)^3} e^{i\mathbf{k}\cdot\mathbf{r}} A_j(\mathbf{k}), \tag{10}$$

and, within the effective mass approximation [3], it satisfies the hydrogen-like Schrödinger equation

$$\left( -\frac{\hbar^2 \nabla^2}{2\mu} - \frac{e^2}{\varepsilon r} \right) F_{nL}(\mathbf{r}) = E_n F_{nL}(\mathbf{r}), \tag{11}$$

where the reduced mass given by $\mu = (m_e m_h)/(m_e + m_h)$ is usually only a fraction of $m_0$, and the dielectric constant $\varepsilon$ screening the electron–hole Coulomb attraction is typically large ($\varepsilon \approx 10$), its choice being discussed below.

On the basis of the above equation, the Wannier–Mott exciton turns out to be a "rescaled hydrogen atom" with binding energies depending only on the principal quantum number $n$

$$E_n = -\frac{R^*}{n^2}, \quad R^* = \frac{\mu e^4}{2\varepsilon^2 \hbar^2} = \frac{\hbar^2}{2\mu a_0^2} = \frac{e^2}{2\varepsilon a_0}, \tag{12}$$

the effective Rydberg $R^*$ being typically of the order of 10 meV and the exciton Bohr radius $a_0$ larger than 1 nm, which justifies *a posteriori* the effective mass approximation. The exciton energy levels are given by

$$E_{n\mathbf{Q}} = E_g + \frac{\hbar^2 Q^2}{2M} - \frac{R^*}{n^2}; \tag{13}$$

and, e.g., the lowest bound state $n = 1$ has the $1S$ envelope function

$$F_{1S}(\mathbf{r}) = \frac{1}{\sqrt{\pi a_0^3}} e^{-r/a_0}. \tag{14}$$

### 2.5. Effective Screening

As for the dielectric constant $\varepsilon$ screening the electron–hole attraction, it takes into account both the background contribution of electronic origin and the lattice contribution. The latter, however, is fully effective only when the binding energy is smaller than the longitudinal optical phonon energy (and, correspondingly, the exciton radius larger than the polaron radius); in this case, that applies for instance to weakly polar semiconductors as GaAs, the static dielectric constant should be used, together with effective masses and band gap values including the electron–phonon interaction, i.e., the experimentally measured ones.

In other cases, for instance, for ionic semiconductors with comparatively small radius excitons like ZnSe, a potential interpolating from the full static screening at large electron–hole distances to the high frequency electronic screening only at small distances could be used [11], or the phonons and the electron–phonon interaction should be explicitly included in the dynamics.

## 2.6. ELECTRON–HOLE EXCHANGE INTERACTION

So far, spin degrees of freedom have been ignored. In the case of negligible spin orbit coupling, it is convenient to consider separately spin triplet and spin singlet excitons, only the latter being optically active. Then, for spin singlet excitons only, the electron–hole exchange interaction is also present and adds to the effective mass Hamiltonian of Eq. (11) for the envelope function a term of the form

$$\left(J_{\text{exch}}^{\text{an}} + J_{\text{exch}}^{\text{nan}}(\widehat{Q})\right)\delta(\mathbf{r}); \tag{15}$$

the exchange interaction comprises an analytical part (independent of $\widehat{Q}$) that gives the tiny splitting between the singlet and triplet excitons, and a non-analytical part. In isotropic media, such as most semiconductors of interest, the latter contribution splits the longitudinal from the transverse excitons [3], having polarization along $\widehat{Q}$ or perpendicular to it, respectively. This contribution to the energy of a longitudinal Wannier–Mott $S$ excitons is $J_{\text{exch}}^{\text{nan}}|F_{nS}(\mathbf{r}=0)|^2 \approx d^2/(na_0)^3$, being $d$ the interband transition electric dipole, and is much smaller than that of the electron–hole Coulomb attraction. Moreover, the non-analytical part of the exchange interaction corresponds to the long range unretarded dipole–dipole interaction, and should not be included in the exciton energy appearing in the dielectric function given below (Eq. (17)) [12].

## 2.7. OSCILLATOR STRENGTH

Optical transitions are only allowed to $\mathbf{Q} \simeq 0$ spin singlet excitons, and, when the interband dipole matrix element at the band edge does not vanish, only to $S$ states.

The dimensionless oscillator strength is obtained as

$$f_n = 2\frac{2}{m_0\hbar\omega}\left|\langle\Phi_{nS,\mathbf{Q}=0}|\hat{e}\cdot\mathbf{p}|G\rangle\right|^2$$

$$= 2\frac{2}{m_0\hbar\omega}\left|\langle\psi_{c0}|\hat{e}\cdot\mathbf{p}|\psi_{v0}\rangle\right|^2\left|\int\frac{V\,d\mathbf{k}}{(2\pi)^3}A_{nS}(\mathbf{k})\right|^2$$

$$= 2\frac{2}{m_0\hbar\omega}\left|\langle\psi_{c0}|\hat{e}\cdot\mathbf{p}|\psi_{v0}\rangle\right|^2 V\left|F_{nS}(\mathbf{r}=0)\right|^2, \tag{16}$$

the factor $|F_{nS}(\mathbf{r}=0)|^2$ tells that the transition rate is proportional to the probability to find electron and hole in the same place, in particular $f \propto (l/a_0)^3$, and the oscillator strength per unit volume scales as $f_n/V \sim (na_0)^{-3}$ (e.g., $f_2/f_1 = 1/8$).

In Eq. (16) the factor 2 appropriate for pure singlet states in the absence of spin–orbit interaction is included. In the presence of spin orbit coupling, the appropriate factor is twice the spin singlet component in the exciton state (and the same factor should also appear in the exchange interaction of Eq. (15)) [15].

The bound state optical resonances can be described by the following complex dielectric constant

$$\varepsilon(\omega) = \varepsilon_b + \frac{4\pi e^2}{m_0 V} \sum_n \frac{f_n}{(E_{n\mathbf{Q}=0}/\hbar)^2 - \omega^2 - i\gamma_n\omega}, \quad (17)$$

where $E_{n\mathbf{Q}=0}$ is the energy of the transverse exciton, $\varepsilon_b$ is a background dielectric constant and $\gamma_n$ a phenomenological linewidth due, e.g., to phonon scattering. Such dielectric constant can be used in Maxwell equations to obtain the corresponding polaritons [13].

### 2.8. Sommerfeld Enhancement

Apart from the hydrogenic series of lines appearing below the band gap energy and corresponding to the optical transitions to the $nS$ bound exciton states, the electron–hole Coulomb attraction also modifies the absorption band for energies just above the absorption edge as also unbound electron–hole pairs are not uncorrelated. Such effect leads to an enhanced value of $\varepsilon_2(\omega)$ for $\hbar\omega \gtrsim E_g$ given by the

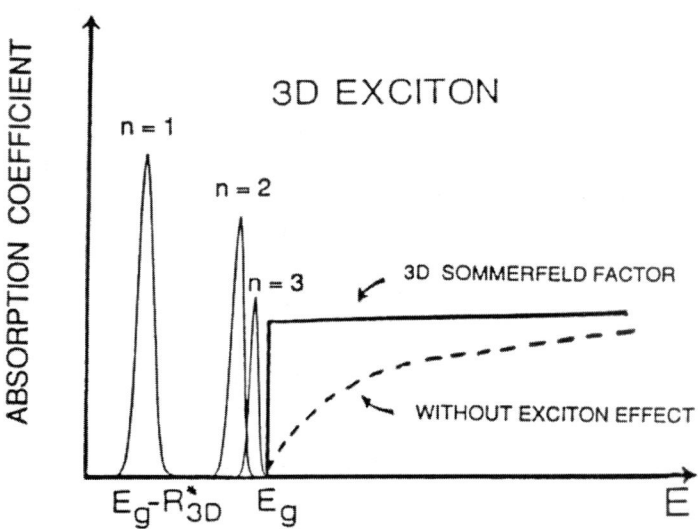

Fig. 4. Sketch of excitonic effects on the absorption edge, the dashed line showing the uncorrelated interband absorption.

Sommerfeld factor [3]

$$S(\omega) = \frac{\pi x e^{\pi x}}{\sinh(\pi x)}, \quad x(\omega) = \sqrt{\frac{R^*}{\hbar\omega - E_g}}, \quad (18)$$

as sketched in Figure 4.

### 2.9. Valence Band Degeneracy

The description of the bulk exciton properties given above is based on the effective mass equation of Eq. (11), appropriate for non-degenerate valence and conduction bands. In most semiconductors of interest, the uppermost valence band is degenerate (as described by the Luttinger matrix [1]) and the exciton wavefunction is a linear superposition containing the various degenerate bands, as a consequence the kinetic energy term becomes a matrix and the single equation (11) a set of coupled equations [3]. Yet, due to the smallness of the reduced mass the diagonal terms are dominant and the resulting exciton spectrum is again nearly hydrogenic [14].

## 3. Quantum Confined Wannier–Mott Excitons

The Wannier–Mott exciton resonances in an intrinsic bulk semiconductor have properties determined by the electronic band structure as discussed above, and thus they can only be changed with stringent limitations by choosing a different material. Starting from the seminal work of Esaki and Tsu around 1970, however, the possibility of tailoring the electronic properties of man-made layered materials has led to the impressive development of the field of semiconductor nanostructures [2]. Band structure engineering allows to change with great flexibility the electronic and optical properties of quantum confined semiconductor systems.

### 3.1. Quantum Well Confinement

The basic example of an artificial low-dimensional structure is an epitaxially grown quantum well in which a central semiconductor layer (the well) is sandwiched between semiconductors having a larger band gap (the barriers) in such a way that the profile of the bottom of the conduction band along the growth direction confines the electrons within the well layer, and similarly for the holes in the valence band (i.e., a type I quantum well). If the coherence length of the electrons (and holes) is larger than the quantum well thickness $L_w$, the regime of quantum confinement is realized in which their motion along the growth axis is discretized, while the one along the well plane remains free, giving rise to a two-dimensional

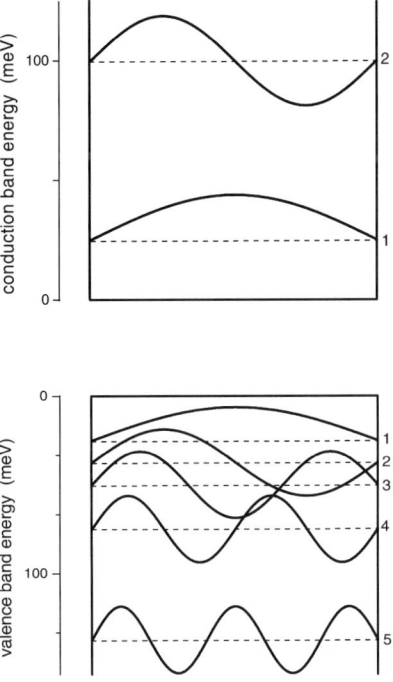

Fig. 5. Schematic quantum well band profile and confined wavefunctions.

subband structure as sketched in Figure 5. As the size of a quantum well is typically much larger than the lattice constant $l$, an effective mass description of the quantum confined electronic states is valid, taking into account the appropriate boundary conditions at the well/barrier interfaces [3].

The confined electron wave function is written as

$$\psi_{c,m\mathbf{k}_\parallel}(\mathbf{r}_\parallel, z) = \frac{1}{\sqrt{N_{2D}}} e^{i\mathbf{k}_\parallel \cdot \mathbf{r}_\parallel} F_{c,m}(z) \sqrt{l}\, u_{c0}(\mathbf{r}), \qquad (19)$$

where $m$ is the subband index, $z$ is along the growth direction and $\mathbf{r}_\parallel = (x, y)$ along the quantum well plane, $\mathbf{k}_\parallel$ is the wavevector of the two-dimensional free motion, $N_{2D}$ the number of unit cells along the plane and $F_{c,m}(z)$ the quantum confinement envelope function. Assuming infinite barriers, the confinement energies and envelope functions are

$$E_{c,m\mathbf{k}_\parallel} = \frac{\hbar^2 \mathbf{k}_\parallel^2}{2 m_e} + \frac{\hbar^2}{2 m_e}\left(\frac{\pi m}{L_w}\right)^2, \qquad F_{c,m}(z) = \sqrt{\frac{2}{L_w}} \sin\left(\frac{m\pi z}{L_w}\right), \qquad (20)$$

and similarly for hole states. Thus, for $m_e \simeq 0.1\, m_0$ and $L_w \simeq 5$ nm, the scale of the confinement energy is $E_{m=1,\mathbf{k}_\parallel=0} \simeq 140$ meV.

In the realistic case of finite barriers, the envelope function leaks out of the well into the barrier region and can be obtained matching at the interfaces the values of $F(z)$ and $\frac{1}{m_e(z)}\frac{\partial}{\partial z}F(z)$, where the discontinuity of the effective mass is accounted for and probability current conservation ensured [3]. At least one bound state is always obtained, and for sufficiently wide or deep wells the lowest bound states are qualitatively similar to the infinite barrier ones of Eq. (20).

### 3.2. Intersubband Transitions

In such a quantum well, undoped and made of a semiconductor with an optically allowed direct gap $E_g$, the absorption edge (disregarding excitonic effects) corresponds to the promotion of an electron from the highest confined valence subband to the lowest confined conduction subband at an energy given by

$$E_{g2D} \simeq E_g + \frac{\hbar^2}{2\mu}\left(\frac{\pi}{L_w}\right)^2, \qquad (21)$$

being $\mu$ the reduced mass in the well.

In the case of a single quantum well for light propagating along the growth direction, the quantity of direct physical meaning to describe the optical properties is the dimensionless absorption probability [15] defined as

$$w(\omega) = \frac{\text{absorbed energy per unit time and unit surface}}{\text{incident energy per unit time and unit surface}}, \qquad (22)$$

which, similarly to Eq. (4), is calculated summing over all possible transitions

$$w(\omega) = \frac{(2\pi e)^2}{n m_0^2 \omega c} \sum_{c,v,m} \int \frac{2\,d\mathbf{k}_\|}{(2\pi)^2} |\hat{e}\cdot \mathbf{M}_{cv,m}(\mathbf{k}_\|)|^2 \delta(E_{c,m\mathbf{k}_\|} - E_{v,m\mathbf{k}_\|} - \hbar\omega) \qquad (23)$$

where n is the index of refraction, the integral is over the two-dimensional Brillouin zone, and the transition matrix element is

$$\begin{aligned}\mathbf{M}_{cv,m} &= \langle\psi_{c,m\mathbf{k}_\|}|\mathbf{p}|\psi_{v,m\mathbf{k}_\|}\rangle \\ &\simeq \int_\Omega d\mathbf{r}\, u_{c0}^*(\mathbf{r})\,\mathbf{p}\,u_{v0}(\mathbf{r}) \int dz\, F_{c,m}^*(z)\,F_{v,m}(z) \\ &\simeq \frac{i m_0 E_{g2D}}{e\hbar}\mathbf{d}_{cv},\end{aligned} \qquad (24)$$

approximately independent of $\mathbf{k}_\|$; in the above equations, the selection rule $\Delta m = 0$ appropriate for the infinite barrier case has been taken into account. Then, at the two-dimensional absorption edge given by Eq. (21) (and similarly at each higher subband edge corresponding to $m > 1$) the integral in Eq. (23) reduces to the two-dimensional joint density of states which vanishes for $\hbar\omega < E_{g2D}$ and for $\hbar\omega \geqslant E_{g2D}$ is a constant, i.e., it is a step function as sketched in Figure 6.

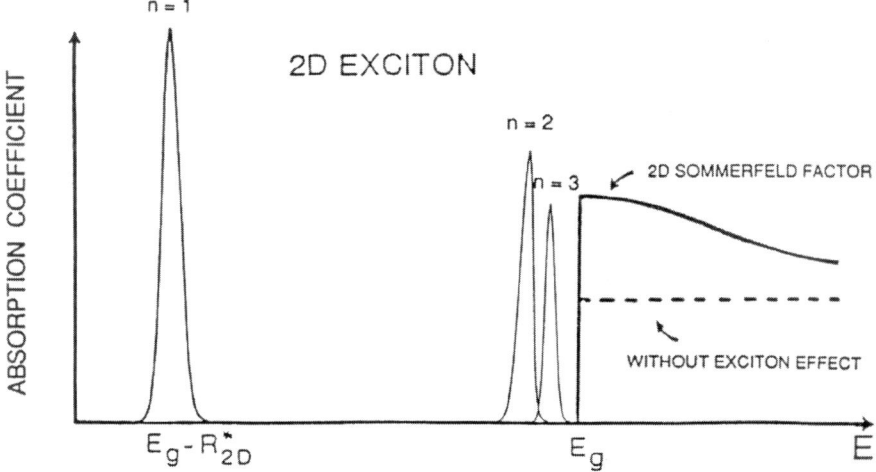

Fig. 6. Sketch of a quantum well absorption edge, the dashed line showing the uncorrelated inter-subband absorption.

### 3.3. QUANTUM WELL EXCITON STATES

When excitonic effects are considered, two different regimes may be realized. In very large quantum wells ($L_w \gg a_0$), the single particle confinement energies (Eq. (20)) are small compared to the effective Rydberg $R^*$ (Eq. (12)), the electron and hole relative motion is unchanged with respect to the bulk case, while the exciton center of mass motion itself is confined along the growth direction. In narrow wells ($L_w \lesssim a_0$), the electron and hole motion along the growth direction is separately quantized (Eq. (20)) with a confinement energy large compared to $R^*$, and mainly the dynamics along the quantum well plane is affected by the electron hole attraction giving rise to quasi two-dimensional excitons.

The latter case is more interesting as quantum confinement can be used to tailor the exciton properties, and in particular to enhance its binding energy and oscillator strength. The lowest optically active quantum well exciton can be described by the two particle wavefunction

$$\Phi_j(\mathbf{r}_e, \mathbf{r}_h) = \frac{l^2}{\sqrt{N_{2D}}} F_j^{2D}(\mathbf{r}_{e\parallel} - \mathbf{r}_{h\parallel}) F_{c,1}(z_e) F_{v,1}^*(z_h) u_{c0}(\mathbf{r}_e) u_{v0}^*(\mathbf{r}_h), \quad (25)$$

where $F_j^{2D}$ is the two-dimensional exciton envelope function, the quantum number $j$ refers to the in plane relative motion and only the lowest pair (i.e., $m = 1$ in Eq. (20)) of electron and hole subbands has been included.

The effective mass equation [5] satisfied by $F^{2D}$ can be written as

$$\left(-\frac{\hbar^2 \nabla_\parallel^2}{2\mu} - \frac{e^2}{\varepsilon}\int dz_e \int dz_h \frac{|F_{c,1}(z_e) F_{v,1}(z_h)|^2}{\sqrt{\mathbf{r}_\parallel^2 + (z_e - z_h)^2}}\right) F_{nL}(\mathbf{r}) = E_n F_{nL}(\mathbf{r}), \tag{26}$$

the electron and hole confinement along $z$ tending to increase their attraction with respect to the bulk case.

In the strictly two-dimensional limit (i.e., for infinite barriers with $L_w \to 0$, and $|F_{c,1}(z)|^2 = |F_{v,1}(z)|^2 = \delta(z)$), the effective mass equation satisfied by $F^{2D}$ can be solved exactly [16] giving

$$E_n^{2D} = -\frac{R^*}{(n - 1/2)^2}, \tag{27}$$

the binding energy of the lowest exciton in the strictly two-dimensional limit is therefore $R_{2D}^* = 4 R^*$ (see also discussion below). The exciton energy levels, including also the in plane center of mass translational energy, are given by

$$E_{n\mathbf{Q}_\parallel} = E_{g2D} + \frac{\hbar^2 \mathbf{Q}_\parallel^2}{2M} - \frac{R^*}{(n - 1/2)^2}, \tag{28}$$

and, e.g., the lowest bound state $n = 1$ has the $1S$ envelope function

$$F_{1S}^{2D}(\mathbf{r}) = \frac{2\sqrt{2}}{a_0 \sqrt{\pi}} e^{-2r/a_0}, \tag{29}$$

where the two-dimensional exciton radius is $a_{2D} = a_0/2$.

### 3.4. QUANTUM WELL EXCITON ABSORPTION

The dimensionless oscillator strength is obtained from Eq. (25) as

$$f_n^{2D} = 2\frac{2}{m_0 \hbar \omega} |\hat{e} \cdot \mathbf{M}_{cv,1}|^2 (N_{2D} l^2) |F_{nS}^{2D}(\mathbf{r} = 0)|^2, \tag{30}$$

where $\mathbf{M}_{cv,m=1}$ is given in Eq. (24) and, in the strictly two-dimensional limit, the oscillator strength per unit area scales as $f_n^{2D}/(N_{2D} l^2) \sim (n - 1/2)^{-3} a_0^{-2}$.

Also for quantum wells, the electron–hole interaction leads to an enhancement of the absorption probability even for unbound electron–hole pairs at energies $\hbar\omega \gtrsim E_{g2D}$, by an amount given by the two-dimensional Sommerfeld factor [16]

$$S_{2D}(\omega) = \frac{2}{1 + e^{-\pi x}}, \quad x(\omega) = \sqrt{\frac{4R^*}{\hbar\omega - E_{g2D}}}, \tag{31}$$

as sketched in Figure 6.

The dimensionless absorption probability of Eq. (22) integrated over a quantum well exciton line is directly related to the oscillator strength per unit area of Eq. (30)

$$\int w(\omega)\,d\omega = \frac{2\pi^2 e^2}{nm_0 c}\frac{f_n^{2D}}{N_{2D}l^2}, \qquad (32)$$

and, for a multiple quantum well system of period $L_w + L_b$, being $L_b$ the barrier width, the absorption coefficient is $\alpha(\omega) = w(\omega)/(L_w + L_b)$.

A single quantum well, because of the broken translational symmetry along the growth axis, should in general be described by a non-local dielectric response [15], it is however useful the introduction of an effective complex dielectric constant for a quantum well exciton resonance given by

$$\varepsilon_n(\omega) = \varepsilon_b + \frac{4\pi e^2}{m_0\, N_{2D}\, l^2\, L}\frac{f_n^{2D}}{(E_{n\mathbf{Q}_\parallel=0}/\hbar)^2 - \omega^2 - i\gamma_n\omega}, \qquad (33)$$

where the characteristic length $L$ is the well thickness $L_w$ (for a multiple quantum well, instead, $d = L_w + L_b$), $\varepsilon_b$ the background dielectric constant and $\gamma_n$ the quantum well exciton phenomenological linewidth discussed below. The above expression can replace the non-local theory provided that the width of the structure $L$ is small compared to the wavelength inside the structure (i.e., $\sqrt{\varepsilon_n(\omega)}(\omega/c)\,L \ll 1$), condition that reduces to [20]

$$\left|\varepsilon_n(\omega) - \varepsilon_b\right|^2 \left(\frac{L\omega}{2\pi c}\right)^2 \ll 1, \qquad (34)$$

which is usually satisfied even near resonance for typical thicknesses $L$ and broadenings $\gamma$. The expression given by Eq. (33) is useful, for instance, to calculate the normal incidence reflection and transmission coefficients of a multilayered structure via a standard transfer matrix approach in which each layer is considered as a homogeneous medium. In the case of oblique incidence, due account of the polarization dependence should also be taken (the effective medium being uniaxial).

### 3.5. Exciton Linewidth

Differently than for the bulk case, the linewidth of a quantum well exciton is significantly affected by inhomogeneous broadening due to structural imperfections, in particular well width fluctuations, which are more severe in narrow wells, the contribution $\gamma_{\text{inhom}}$ of thickness fluctuations scaling as $L_w^{-3}$. The disorder is the dominant broadening mechanism at low temperatures, and much care has been devoted to the growth of high quality quantum narrow wells in which the residual inhomogeneous broadening is of the order of 1 meV or less.

With increasing temperature, the linewidth is affected by phonon scattering according to

$$\gamma_{\text{phon}}(T) = aT + \frac{b}{e^{\hbar\omega_{\text{LO}}/(K_B T)} - 1}, \qquad (35)$$

where both the acoustic and longitudinal optic phonon scattering are included, typical values for GaAs, e.g., being $a \simeq 5$ μeV/K and $b \simeq 10$ meV, the LO phonon scattering ($\hbar\omega_{\text{LO}} = 36$ meV) being the dominant source of broadening at room temperature [21].

### 3.6. THEORETICAL REFINEMENTS

For a more detailed and realistic description of quantum well excitons a number of additional effects should be considered. First of all, the valence band degeneracy of typical semiconductors such as GaAs is split by the confinement potential giving rise to two sets of hole states, the "heavy" hole and "light" hole states [2,5], corresponding to larger and smaller effective masses for the motion along the growth direction, respectively. In a first approximation, they may be separately considered, the heavy hole ones having a lower confinement energy and being typically the lowest available hole states. From the symmetry of the electronic bands, it follows that the heavy hole states have a transition dipole matrix element polarized in the quantum well plane, whereas light hole states are optically active also for a polarization along the growth direction.

The second important effect is barrier penetration because the confining potential is not infinitely high: electron and hole wavefunctions extend into the barrier regions and, thus, the strictly two-dimensional limit is only approximately realized. As a consequence, both the exciton binding energy and oscillator strength depend on the quantum well thickness. For narrow wells, they tend to increase with decreasing well thickness until barrier penetration becomes large and they decrease again. In Figure 7 the dependence of the binding energy on well thickness for a GaAs quantum well is shown, in which case the optimal thickness is around 5 nm [17].

An extreme case that goes beyond the validity of the usual envelope function description of quantum confinement is that of excitons bound to a monolayer impurity plane (e.g., InAs in GaAs) [18]; such two-dimensional excitons have a large oscillator strength.

More refined calculations [19] for quantum well excitons also consider the effects of valence band mixing (i.e., between heavy and light holes), Coulomb coupling among different subbands and dielectric constant discontinuities; the obtained binding energies may even exceed $4R^*$ and the oscillator strength per unit area is typically $f_n^{2D}/(N_{2D} l^2) \approx 5 \cdot 10^{-2}$ nm$^{-2}$, as shown in Figure 8. In general, quantum confinement strengthens the effects of the electron–hole attraction and

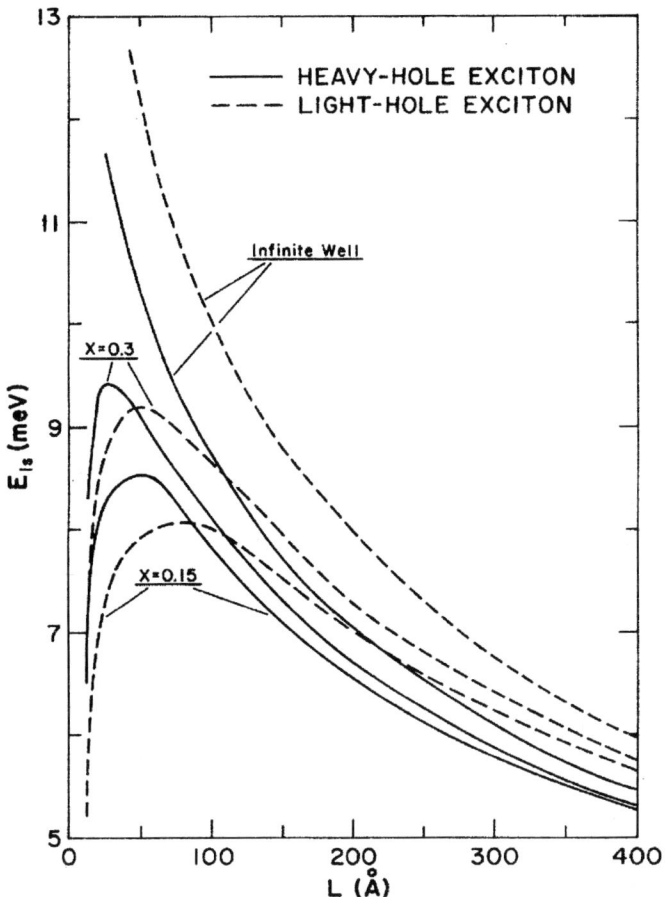

Fig. 7. Well thickness dependence of the 1S exciton binding energy in a GaAs/Al$_x$Ga$_{1-x}$As quantum well (from Ref. [17]).

the excitonic features in quantum wells are more prominent than in the bulk, and clearly visible even at room temperature as shown in Figure 9.

### 3.7. Quantum Wires and Dots

Quantum confined structures with lower dimensionality than quantum wells, i.e., one-dimensional quantum wires [22] and zero-dimensional quantum dots [23] have also been realized. In the latter case, atomic like spectra are obtained which

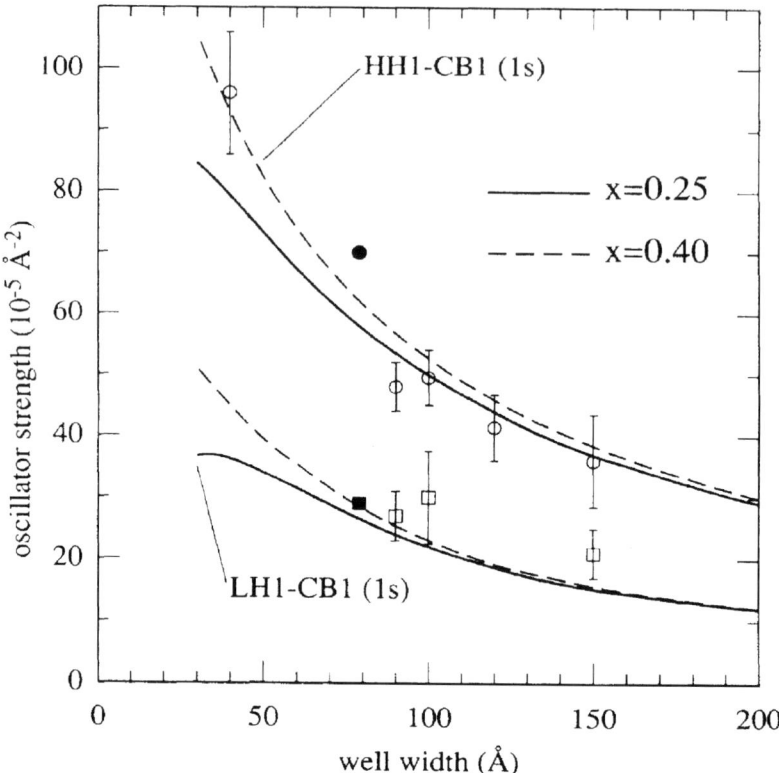

Fig. 8. Well thickness dependence of the lowest heavy hole and light hole exciton oscillator strength in a GaAs/Al$_x$Ga$_{1-x}$As quantum well (from Ref. [19]).

do depend on the electron–hole interaction; however, the exciton concept as a travelling excitation is no longer applicable.

In the case of quantum wires, there still is one degree of translational motion and quasi-one-dimensional excitons play a significant role. For their analysis, it is important a realistic description of the wire transversal dimensions as the strictly one-dimensional limit (differently from the two-dimensional one) is unphysical as it predicts an infinitely bound fundamental exciton state.

Both for the case of quantum dots and quantum wires, the effects of structural imperfections are very severe and give rise to a typically large inhomogeneous linewidth for the optical excitations of ensembles of wires or dots. In view of the study of organic multilayers and heterostructures, the case of quantum well excitons is more relevant and we will concentrate here on two-dimensional confinement.

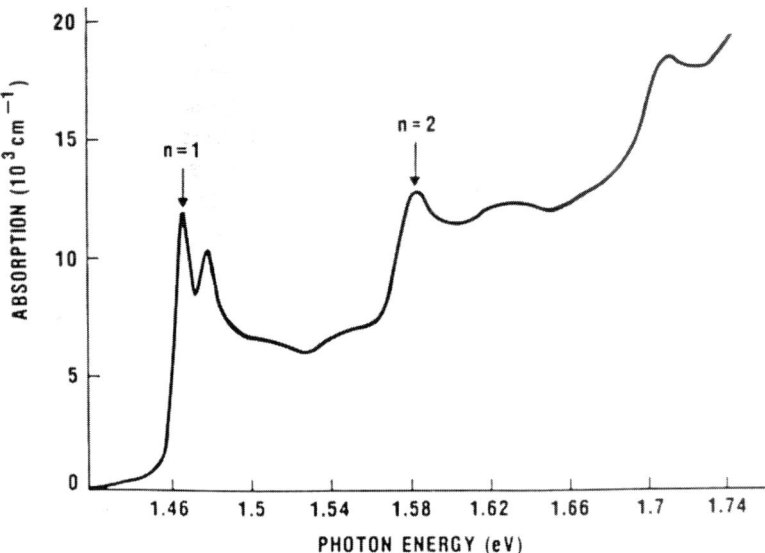

Fig. 9. Multiple quantum well exciton lines at room temperature, at the first intersubband edge ($n = 1$) both heavy and light hole $1S$ excitons are resolved (from Ref. [6]).

## 4. Quantum Well Exciton Optical Nonlinearities

At low excitation densities, the optical response of Wannier–Mott excitons is linear, as considered above, and they can be approximately described as ideal bosons. At very high excitation densities, the exciton gas gives way to an electron–hole plasma showing a different phenomenology, such as band gap renormalization and gain [24]. In the present section, we discuss the intermediate density regime in which excitonic resonances are still dominating the near edge optical response, but nonlinear effects play a significant role. The most relevant case of two-dimensional excitons will be considered in the density regime $n \lesssim n_S$, the saturation density being $n_S \approx 1/(\pi a_{2D}^2)$.

### 4.1. Real and Virtual Excitations

A useful distinction among different nonlinear optical regimes can be made according to the excitation energy [6,21], in a typical pump and probe configuration.

For a pump laser frequency well below the exciton line, i.e., in the transparency region, only virtual excitations can be driven by the laser field; the corresponding nonlinearities are comparatively weak, but very fast as the induced excitation directly follows the laser pulse. In such a case, excitons are totally coherent having a well defined wavevector and phase.

For the opposite case of a pump laser frequency well above the exciton line in the electron–hole continuum, large excitonic nonlinearities turn on after the generated electrons and holes have relaxed via inelastic scattering dissipating the excess energy and have reached the bottom of their bands, this process taking typically a few hundred femtoseconds, to eventually form incoherent excitons.

In the case in which the excitons are resonantly pumped, the initial coherence is lost after the exciton dephasing time, which at low temperatures and low exciton densities is of the order of a picosecond, and the exciton nonlinearities are then due to an incoherent population of excitons with a lifetime of the order of a nanosecond.

The excitonic nonlinearities will be discussed in the following on the basis of the semiconductor Bloch equations [25] with a phenomenological dephasing rate, while the ultrafast dynamics during which coherence is destroyed is very involved and would require a quantum kinetic approach [7,26].

## 4.2. Exciton Statistics

Strictly speaking, excitons are not bosons, they are composite objects made of two fermionic particles: one electron and one hole. For the sake of simplicity, we neglect for the time being spin degrees of freedom and consider only $1S$ excitons with a vanishing center of mass wavevector. Then, similarly to Eq. (8), the two-dimensional exciton creation operator can be written as

$$B^\dagger = \sum_{\mathbf{k}} A^{2D}(\mathbf{k}) c_{\mathbf{k}}^+ h_{-\mathbf{k}}^+ \quad \text{with} \quad \sum_{\mathbf{k}} |A^{2D}(\mathbf{k})|^2 = 1, \tag{36}$$

where the hole creation operator is introduced ($h_{-\mathbf{k}}^+ = v_{\mathbf{k}}$) and the summation is over the $N_{2D}$ wavevectors within the two-dimensional Brillouin zone (here, and in the following, all wave vectors are two-dimensional and only the lowest confined subbands are considered).

Similarly to Eq. (10), the function $A^{2D}(\mathbf{k})$ is given according to Eq. (29) by

$$A^{2D}(\mathbf{k}) = \frac{1}{\sqrt{N_{2D} l^2}} \int d\mathbf{r}\, e^{-i\mathbf{k}\cdot\mathbf{r}} F_{1S}^{2D}(\mathbf{r}) = \frac{\frac{a_0}{l}\sqrt{\frac{2\pi}{N_{2D}}}}{[1 + (a_0 k/2)^2]^{3/2}}. \tag{37}$$

Thus, from the fermionic commutation rules of the electron and hole operators it follows

$$[B, B^\dagger] = \sum_{\mathbf{k}} |A^{2D}(\mathbf{k})|^2 (1 - \hat{n}_e(\mathbf{k}) - \hat{n}_h(-\mathbf{k})), \tag{38}$$

where the electron and hole number operators are $\hat{n}_e(\mathbf{k}) = c^+(\mathbf{k})c(\mathbf{k})$ and $\hat{n}_h(\mathbf{k}) = h^+(\mathbf{k})h(\mathbf{k}) = 1 - v^+(-\mathbf{k})v(-\mathbf{k})$. Evaluating on a state containing $N$ excitons the expectation values of the operators appearing in Eq. (38), it follows to the lowest order in $N$

$$[B, B^\dagger] = 1 - 2N \sum_{\mathbf{k}} |A^{2D}(\mathbf{k})|^4 = 1 - \frac{8\pi}{5} a_0^2 n, \qquad (39)$$

where $n$ is the two-dimensional exciton density $n = N/(N_{2D}l^2)$.

The equation above shows how the commutation rules for excitons depart from the bosonic ones with an increasing exciton density, and it should be compared with the analogous results for Frenkel excitons given by Eq. (21) of the chapter by J. Knoester and V.M. Agranovich in this book; in the case of Wannier–Mott excitons, as soon as the density $n$ is of the order of the saturation density $n_S = 1/(\pi a_{2D}^2)$ the ideal boson approximation fails completely.

### 4.3. COULOMB INTERACTION AND FERMIONIC HAMILTONIAN

The density dependent modification of the bosonic commutation rules described by Eq. (39) is only due to the composite nature of the excitons and to the underlying fermionic statistics of electrons and holes. Such "kinematic" nonlinearities correspond to the phase space filling effects common to atom-like two-level systems, however exciton nonlinearities are also due to dynamical effects stemming from the Coulomb interaction.

A theoretical description of both kinds of nonlinearities can be given in terms of electron and hole fermionic operators starting from the Hamiltonian

$$H_T = H_{SP} + H_{Coul} + H_L, \qquad (40)$$

where the single particle Hamiltonian describing parabolic electron and hole sub-bands is

$$H_{SP} = E_{g2D} + \sum_{\mathbf{q}} \frac{\hbar^2 q^2}{2m_e} c_{\mathbf{q}}^+ c_{\mathbf{q}} + \sum_{\mathbf{q}} \frac{\hbar^2 q^2}{2m_h} h_{\mathbf{q}}^+ h_{\mathbf{q}}. \qquad (41)$$

The Coulomb interaction, i.e., electron–electron and hole–hole repulsion plus electron–hole attraction, is given by

$$H_{Coul} = \frac{1}{2} \sum_{\mathbf{q} \neq 0, \mathbf{k}, \mathbf{k}'} V_{\mathbf{q}} \big( c_{\mathbf{k}+\mathbf{q}}^+ c_{\mathbf{k}'-\mathbf{q}}^+ c_{\mathbf{k}'} c_{\mathbf{k}} + h_{\mathbf{k}+\mathbf{q}}^+ h_{\mathbf{k}'-\mathbf{q}}^+ h_{\mathbf{k}'} h_{\mathbf{k}}$$
$$- 2 c_{\mathbf{k}+\mathbf{q}}^+ h_{\mathbf{k}'-\mathbf{q}}^+ h_{\mathbf{k}'} c_{\mathbf{k}} \big), \qquad (42)$$

with

$$V_{\mathbf{q}} = \frac{2\pi e^2}{\varepsilon q} \frac{1}{N_{2D}l^2}, \qquad (43)$$

being $\varepsilon$ the static dielectric constant. Finally, the interaction of the interband polarization with the laser electric field is written as

$$H_L = -\left(\mathcal{E}(t) \cdot \mathbf{d}_{cv}\right) \sum_{\mathbf{q}} c_{\mathbf{q}}^+ h_{-\mathbf{q}}^+ + \text{h.c.} \tag{44}$$

### 4.4. Mean Field Optical Response

On the basis of the Hamiltonian of Eq. (40), the Heisenberg equations of motion of the electron and hole populations ($n_e(\mathbf{k}) = \langle c^+(\mathbf{k})c(\mathbf{k})\rangle$ and $n_h(\mathbf{k}) = \langle h^+(\mathbf{k})h(\mathbf{k})\rangle$) and of the interband polarization ($P(\mathbf{q}) = \langle h(-q)c(\mathbf{q})\rangle$) are easily written, however they do not form a closed system because the Coulomb interaction couples them to those of four-particle terms and the hierarchy of such equations must be somehow truncated [25]. In the first approximation, all expectation values of four-particle operators are factorized and expressed as products of population and polarization terms ("random phase" approximation), the resulting equations of motion corresponding to a Hartree–Fock treatment.

Screening and correlation effects are here neglected as they are less important in the two-dimensional case [6], and excitation induced dephasing, as discussed later on, will be included along with electron–phonon scattering in the phenomenological linewidth.

The optical response will be obtained from the interband polarization, which in this meanfield approximation obeys the equation of motion

$$i\hbar \frac{\partial}{\partial t} P(\mathbf{k}, t) = H_0 P(\mathbf{k}, t) + H_1 P(\mathbf{k}, t)$$
$$- \left(1 - n_e(\mathbf{k}) - n_h(-\mathbf{k})\right)\mathcal{E}(t) \cdot \mathbf{d}_{cv}, \tag{45}$$

with

$$H_0 P(\mathbf{k}, t) = \left(E_{g2D} + \frac{\hbar^2 \mathbf{k}^2}{2\mu}\right) P(\mathbf{k}, t) - \sum_{\mathbf{k}' \neq \mathbf{k}} V_{\mathbf{k}-\mathbf{k}'} P(\mathbf{k}', t), \tag{46}$$

and

$$H_1 P(\mathbf{k}, t) = -P(\mathbf{k}, t) \sum_{\mathbf{k}' \neq \mathbf{k}} V_{\mathbf{k}-\mathbf{k}'} \left(n_e(\mathbf{k}') + n_h(-\mathbf{k}')\right)$$
$$+ \left(n_e(\mathbf{k}) + n_h(-\mathbf{k})\right) \sum_{\mathbf{k}' \neq \mathbf{k}} V_{\mathbf{k}-\mathbf{k}'} P(\mathbf{k}', t). \tag{47}$$

Equations of motion for the populations can be similarly derived and together with Eq. (45) amount to the so-called semiconductor Bloch equations: their numerical solutions describe well a variety of nonlinear effects, including wave mixing [25,7].

In the totally coherent case of virtual excitations, the population and polarization values satisfy the relations $n_e(\mathbf{k}) = n_h(-\mathbf{k}) = n(\mathbf{k})$ and $|P(\mathbf{k})|^2 = n(\mathbf{k})(1 - n(\mathbf{k}))$.

However, we here assume that electron and hole populations are given parameters, whereas the driving electric field appearing in Eq. (45) is that of the probe laser. In particular, we consider present a given density $n$ of incoherent $1S$ excitons at the bottom of their band ("cold" excitons), for which case one has [6]

$$n_e(\mathbf{k}) = n_h(-\mathbf{k}) = n|A^{2D}(\mathbf{k})|^2 N_{2D} l^2, \tag{48}$$

which corresponds to an occupation of the fermionic electron and hole states given by their weight in the exciton wavefunction of Eq. (36).

The interband polarization equation of motion comprises the homogeneous terms $H_0 P$, the linear one which corresponds to the strictly two-dimensional limit of the exciton effective mass equation (Eq. (26)) written in $\mathbf{k}$ space, and $H_1 P$, the nonlinear one proportional to the excitation density $n$, plus the inhomogeneous driving term diminished by Pauli blocking.

We will solve Eq. (45) perturbatively to the lowest order in the excitation density, expanding the solution of the inhomogeneous equation in terms of those of the homogeneous one [7], i.e., the bound and unbound exciton.

Assuming a monochromatic probe laser $\mathcal{E}(t) = \mathbf{E} e^{-i\omega t} + \text{c.c.}$ nearly resonant with the lowest exciton state, one can simply write the solution in terms of the $1S$ exciton wavefunction given by Eq. (37) slightly perturbed by $H_1$

$$P(\mathbf{k}, t) = P(\mathbf{k}) e^{-i\omega t} \simeq \alpha \left( A^{2D}(\mathbf{k}) + \delta A^{2D}(\mathbf{k}) \right) e^{-i\omega t}. \tag{49}$$

Before analysing the nonlinear perturbative effects, we recall that the optical susceptibility $\chi(\omega)$ is related to the driven Maxwell polarization

$$\mathbf{P}(\omega) = \frac{1}{N_{2D} l^2 L} \sum_{\mathbf{k}} \mathbf{d}_{cv}^* P(\mathbf{k}), \tag{50}$$

simply by $\mathbf{P}_i(\omega) = \chi_{ij}(\omega) \mathbf{E}_j$. In the linear regime, projecting the equation of motion Eq. (45) on the $1S$ exciton state, it follows

$$\alpha = \frac{\mathbf{d}_{cv} \cdot \mathbf{E}}{E_{1S} - \hbar\omega} \sum_{\mathbf{k}} A^{2D}(\mathbf{k})^* = \frac{\mathbf{d}_{cv} \cdot \mathbf{E}}{E_{1S} - \hbar\omega} \sqrt{N_{2D} l^2} \, F_{1S}^{2D}(0)^*, \tag{51}$$

thus, in the isotropic case, including the phenomenological linewidth $\gamma$, the linear susceptibility is

$$\chi_0(\omega) = \frac{|F_{1S}^{2D}(0)|^2}{L} \frac{|\mathbf{d}_{cv}|^2}{E_{1S} - \hbar\omega - i\hbar\gamma}, \tag{52}$$

which agrees with Eq. (33) for $\hbar\omega \simeq E_{1S} = E_{1\mathbf{Q}_\parallel = 0}$ and gives the same oscillator strength $f_1^{2D}$ of the $1S$ two-dimensional exciton.

## 4.5. EXCITONIC OPTICAL NONLINEARITIES

The nonlinear changes of resonant frequency and oscillator strength can be evaluated from Eqs. (45), (48) to first order in the excitation density $n$ as follows [6].

The correction to the 1S exciton energy is simply the expectation value of $H_1$ (Eq. (47)) on the zeroth order wavefunction $A^{2D}(\mathbf{k})$ (Eq. (37)) which gives

$$\delta E_{1S} \simeq 1.93\,\pi(a_0/2)^2(4R^*)n, \tag{53}$$

this blue-shift is a manifestation of the short range hard-core repulsion among excitons (in the three-dimensional case, it is compensated by screening and correlation effects).

The correction to the oscillator strength comprises two contributions: the first from the Pauli blocking in the inhomogeneous term of the driving electric field (last term in Eq. (45)), the second from the first order perturbation of the exciton wavefunction $\delta A^{2D}$. As a matter of fact, the corrected oscillator strength is proportional to

$$\delta f_1^{2D} \propto \sum_{\mathbf{q}} \left(A^{2D}(\mathbf{q}) + \delta A^{2D}(\mathbf{q})\right)$$
$$\times \sum_{\mathbf{k}} (1 - n_e(\mathbf{k}) - n_h(-\mathbf{k}))\left(A^{2D}(\mathbf{k}) + \delta A^{2D}(\mathbf{k})\right)^*, \tag{54}$$

and to first order in the excitation density $n$, it follows

$$\frac{\delta f_1^{2D}}{f_1^{2D}} = -\frac{\sum_{\mathbf{k}}(n_e(\mathbf{k}) - n_h(-\mathbf{k}))A^{2D}(\mathbf{k})^*}{\sum_{\mathbf{k}} A^{2D}(\mathbf{k})^*} + \left(\frac{\sum_{\mathbf{k}} \delta A^{2D}(\mathbf{k})}{\sum_{\mathbf{k}} A^{2D}(\mathbf{k})} + \text{c.c.}\right), \tag{55}$$

with

$$\delta A^{2D}(\mathbf{k}) = \sum_{n \neq 1S} \frac{\langle n | H_1 | 1S \rangle}{E_{1S} - E_n} A_n^{2D}(\mathbf{k}), \tag{56}$$

where the sum is over all zeroth order bound and unbound exciton states, i.e., the eigenstates of $H_0$, with wavefunctions $A_n^{2D}$ and energies $E_n$. Finally, using Eq. (48), the correction of the oscillator strength is [6,27]

$$\frac{\delta f_1^{2D}}{f_1^{2D}} = -\frac{n}{n_S} \simeq -(2.29 + 1.70)\pi(a_0/2)^2 n; \tag{57}$$

the corresponding saturation density at which exciton bleaching takes place being $n_S \approx 1/(\pi a_{2D}^2)$, as anticipated (the first term 2.29 is the contribution of the Pauli blocking). In fact, the above perturbative results substantiate the intuitive picture of "hard disc" excitons, the Coulomb interaction contribution being of the same order as that of phase space filling [28]. Of course, the numerical factors appearing in Eqs. (53) and (57) are only a rough estimate because the strictly

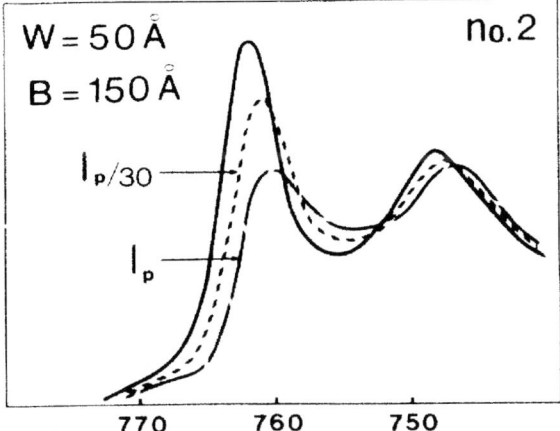

Fig. 10. Quantum well exciton blue shift and bleaching with increasing excitation density (from Ref. [29]).

two-dimensional limit has been used to describe the quantum well confinement and no account has been taken of the spin degrees of freedom and of the exciton fine structure (e.g., heavy hole excitons in III–V quantum wells have four nearly degenerate states, of which only two are optically active) [21].

The exciton blue-shift and the reduction of the oscillator strength with increasing excitation density as determined experimentally are in good agreement with the theoretical predictions scaling as $n a_0^2$ [6]. As an example [29], the $1S$ heavy hole exciton absorption in a narrow GaAs quantum well as a function of increasing excitation density is shown in Figure 10, while Figure 11 illustrates the predicted behaviour $(\delta E_{1s}/E_{1s}) = C \, \delta f_1^{2D}/f_1^{2D}$, the proportionality constant from the simple estimates above being $C \simeq 0.5$.

Considering the excitonic optical resonance described by Eq. (52), an additional, important, nonlinear effect concerns the increase of the linewidth $\gamma$ with growing excitation density $n$, i.e., collisional broadening. Its theoretical description goes beyond the unscreened Hartree–Fock meanfield formalism described above and calls for a higher order truncation scheme of the infinite hierarchy of many-body equations of motion. The following simple expression to lowest order in $n$ has been found to describe well the experimental observations [30,31] as shown in Figure 12

$$\gamma(n) \simeq \gamma_0 + 0.48\pi (a_0/2)^2 (4R^*) n. \qquad (58)$$

Significantly larger values of such excitation induced dephasing have also been reported [32]. In any case, this broadening can be seen as an imaginary self-energy contribution accompanying the energy shift of Eq. (53) and being of the same

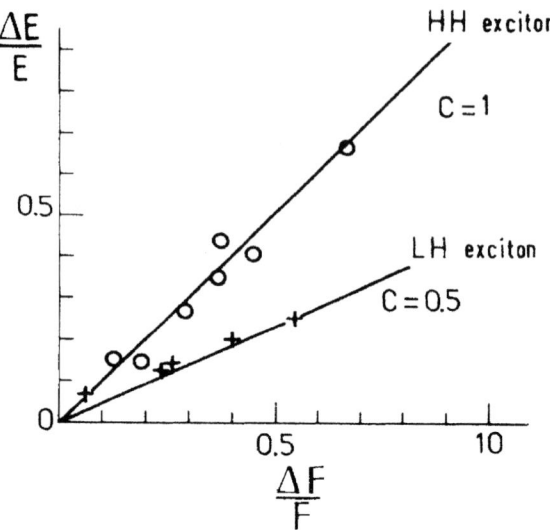

Fig. 11. Experimentally observed proportionality between the blue shift and the oscillator strength reduction: $(\delta E_{1s}/E_{1s}) = C\,\delta f_1^{2D}/f_1^{2D}$ (from Ref. [29]).

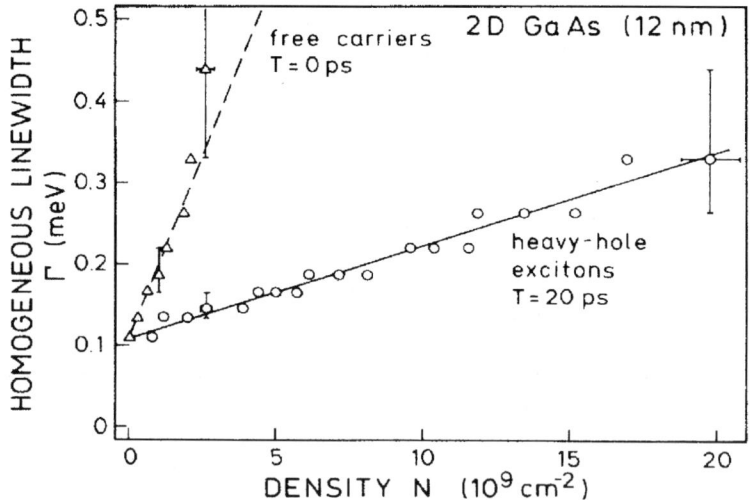

Fig. 12. Density dependence of the homogeneous linewidth of quantum well excitons due to collisional broadening (from Ref. [30]).

order of magnitude. Such a simple description of excitation induced dephasing has limitations, especially in the context of coherent optical processes [7].

## 4.6. BIEXCITONS

In this section so far, spin degrees of freedom have been ignored: when they are included, in most semiconductors of interest, it is useful to make a distinction between optically active excitons generated by right and by left circularly polarized light. While the interaction between excitons having the same polarization is repulsive, that between excitons of opposite polarization is attractive and may lead to the formation of bound states of two excitons (i.e., biexcitons) [8], in analogy to the formation of hydrogen molecules. In a simple adiabatic picture, the internal structure of each exciton is not modified, but only their relative motion is correlated and described by a molecular wavefunction having typically a radius $a_m$ a few times larger than the exciton one $a_0$. The energy of a biexciton formed by two excitons each of energy $E_x$ is written as $E_{xx} = 2 E_x - B^*$, where the biexciton binding energy $B^*$ is small compared to the exciton one $R^*$. However, their ratio is significantly enhanced in the quantum well case with respect to the bulk: variational calculations predict a value of $(B_{2D}^*/R_{2D}^*) \approx 0.2$, in good agreement with experiments [8,33–35].

Biexcitons are involved in three basic optical processes [36]: two-photon absorption from the ground state (one-photon absorption from the ground state being forbidden), one-photon absorption induced by the presence of excitons and radiative decay into one-photon plus one exciton. All of them are enhanced by the so-called giant oscillator strength of the exciton–biexciton transition which, in the two-dimensional case, is proportional to $a_m^2$. In the two-photon transition the sum of the energies of the two absorbed photons must equal the biexciton energy $E_{xx}$, and the transition rate is resonantly enhanced when the energy of one photon matches the exciton energy $E_x$. The induced absorption and the biexciton luminescence are reverse processes of each other, in which the photon absorbed or emitted, respectively, has an energy equal to $E_{xx} - E_x = E_x - B^*$, i.e., lower than the exciton energy by an amount corresponding to the biexciton binding energy [24].

From the theoretical standpoint of Eq. (40), the description of biexciton related nonlinearities would require a proper treatment of higher order correlations including at least the four-particle terms [7]. Many experiments have proved the importance of biexcitonic contributions in coherent nonlinear optical processes [21], a particularly striking example being the observation of exciton–biexciton oscillations in pump and probe or four-wave-mixing experiments with a beating period corresponding to the biexciton binding energy, as shown in Figure 13.

## 4.7. BOSE–EINSTEIN CONDENSATION AND INDIRECT EXCITONS

As discussed above, at densities well below the saturation regime excitons (and similarly biexcitons) behave as weakly interacting bosons, the question therefore

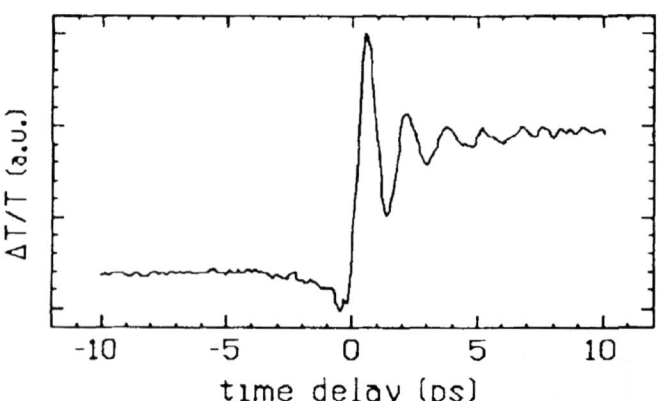

Fig. 13. Transient differential transmission signal for pump and probe of opposite circular polarizations showing beatings corresponding to the quantum well biexciton binding energy (from S. Bar-Ad, I. Bar-Joseph, *Phys. Rev. Lett.* 68 (1992) 349).

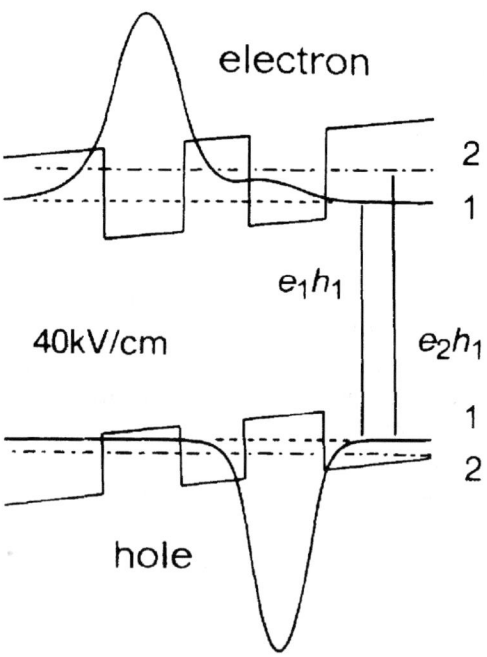

Fig. 14. Potential profile and confined wavefunctions in a coupled quantum well under an electric field, the indirect exciton corresponds to the $e_1 h_1$ transition (from Ref. [43]).

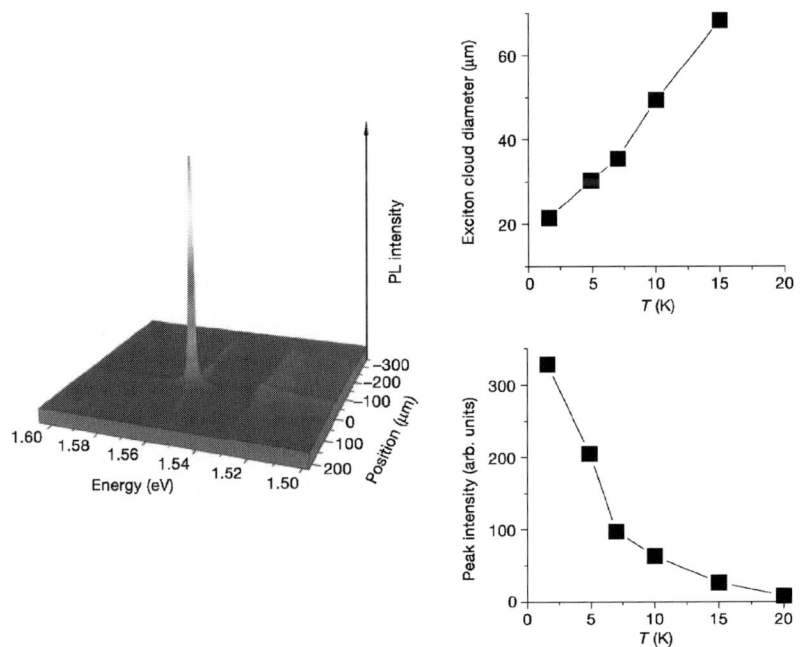

Fig. 15. Spectrally and spatially resolved luminescence from indirect excitons condensed in a trap and temperature dependence of the size of the exciton cloud and of the peak intensity of its luminescence (from Ref. [44]).

arises of their quantum statistical properties and, in particular, the possibility of their Bose–Einstein condensation (BEC) [8,37,38,9]. A short lifetime is detrimental to the realization of large concentrations and low temperatures, thus long-lived forbidden exciton states like the "yellow" paraexciton of $Cu_2O$ have been considered as good candidates for BEC [39,40].

As for quantum well excitons, a very interesting configuration is that of two coupled wells in which electrons and holes are spatially separated under the action of an external static electric field along the growth direction (quantum confined Stark effect [41]). This system is very promising for achieving quantum degeneracy and, possibly, a phase transition to a macroscopically coherent state [42–45]. The main advantage in using such "indirect" excitons is the fact that the electron–hole separation (see Figure 14) leads to a suppressed oscillator strength and, thus, an enhanced lifetime by a few orders of magnitude; moreover, the cooling of excitons due to acoustic phonon emission is also enhanced by a few orders of magnitude compared to the bulk case due to the relaxation of the momentum conservation law along the growth direction. The indirect excitons may in addition be confined laterally in the potential fluctuations due, e.g., to variations of the quantum well thickness; in such traps a large concentration of excitons can indeed be collected as shown by spatially resolved luminescence measurements (see Figure 15) [44].

Indirect excitons are also characterized by a repulsive interaction due to their static electric dipoles all oriented along the growth direction: at low densities this repulsion is beneficial to BEC, while at higher densities gives rise to a strong blue-shift and a screening of the external static electric field [46]. Such a blue shift is analogous to the one experienced by charge transfer excitons at a donor–acceptor interface [47]. As indirect excitons have a low oscillator strength, their large density dependent blue-shift does not lead directly to large all-optical nonlinearities; however, more complicated schemes of coupled inorganic quantum wells are conceivable [48] in which direct and indirect excitons are hybridized to take advantage of the charge transfer character of the indirect ones and of the large oscillator strength of the direct ones, in analogy to the case of hybrid Frenkel–Wannier–Mott excitons in organic-inorganic heterostructures [49].

## References

1. Y. Yu, M. Cardona, Fundamentals of Semiconductors, Springer, Berlin, 1996.
2. C. Weisbuch, B. Vinter, Quantum Semiconductor Structures, Academic Press, San Diego, 1991.
3. F. Bassani, G. Pastori Parravicini, Electronic States and Optical Transitions in Solids, Pergamon Press, Oxford, 1975.
4. J. Knoester, V.M. Agranovich, in: V.M. Agranovich, G.F. Bassani (Eds.), Electronic Excitations in Organic Based Nanostructures, in: Thin Films and Nanostructures, Vol. 31, Elsevier, San Diego, 2003, Chapter 1.
5. G. Bastard, Wave Mechanics Applied to Semiconductor Structures, Les Editions de Physique CNRS, Paris, 1988.
6. S. Schmitt-Rink, D.S. Chemla, D.A.B. Miller, *Phys. Rev. B* 32 (1985) 6601; *Adv. Phys.* 38 (1989) 89.
7. D.S. Chemla, in: E. Garmire, A. Kost (Eds.), in: Semiconductors and Semimetals, Vol. 58, Academic Press, San Diego, 1999;
D.S. Chemla, J. Shah, *Nature* 411 (2001) 549.
8. A.L. Ivanov, H. Haug, L.V. Keldysh, *Phys. Rep.* 296 (1998) 237.
9. S.A. Moskalenko, D.W. Snoke, Bose–Einstein Condensation of Excitons and Biexcitons, Cambridge University Press, Cambridge, 2000.
10. E.J. Johnson, in: Semiconductors and Semimetals, Vol. 3, Academic, New York, 1967, p. 153.
11. A. Antonelli, J. Cen, K.K. Bajaj, *Semicond. Sci. Technol.* 11 (1996) 74.
12. V.M. Agranovich, V.L. Ginzburg, Crystal Optics with Spatial Dispersion and Excitons, Springer, Berlin, 1984.
13. G.F. Bassani, in: V.M. Agranovich, G.F. Bassani (Eds.), Electronic Excitations in Organic Based Nanostructures, in: Thin Films and Nanostructures, Vol. 31, Elsevier, San Diego, 2003, Chapter 3.
14. A. Baldereschi, N.O. Lipari, *Phys. Rev. B* 3 (1971) 439.
15. L.C. Andreani, in: E. Burstein, C. Weisbuch (Eds.), Confined Electrons and Photons, Plenum Press, New York, 1995, p. 57.
16. M. Shinada, S. Sugano, *J. Phys. Soc. Japan* 21 (1966) 1936.

17. R.L. Green, K.K. Bajaj, D.E. Phelps, *Phys. Rev. B* 29 (1984) 1807.
18. R. Cingolani, O. Brandt, L. Tapfer, G. Scamarcio, G.C. La Rocca, K. Ploog, *Phys. Rev. B* 42 (1990) 3209;
    R. Iotti, L.C. Andreani, M. Di Ventra, *Phys. Rev. B* 57 (1998) 15072;
    A.R. Goni, M. Stroh, C. Thomsen, F. Heinrichsdorff, V. Tuerck, A. Krost, D. Bimberg, *Appl. Phys. Lett.* 72 (1998) 1433.
19. L.C. Andreani, A. Pasquarello, *Phys. Rev. B* 42 (1990) 8928.
20. E.L. Ivchenko, *Sov. Phys. Solid State* 33 (1991) 1344.
21. J. Shah, Ultrafast Spectroscopy of Semiconductors and Semiconductor Nanostructures, Springer, Berlin, 1999.
22. R. Cingolani, R. Rinaldi, *Riv. Nuovo Cimento* 16 (1993) 1.
23. U. Woggon, Optical Properties of Semiconductor Quantum Dots, Springer, Berlin, 1997;
    S. Gaponenko, Optical Properties of Semiconductor Nanocrystals, Cambridge University Press, Cambridge, 1998.
24. R. Cingolani, K. Ploog, *Adv. Phys.* 40 (1991) 535.
25. H. Haug, S.W. Koch, Quantum Theory of the Optical and Electronic Properties of Semiconductors, World Scientific, Singapore, 1994.
26. H. Haug, A.-P. Jauho, Quantum Kinetics in Transport and Optics of Semiconductors, in: Springer Series in Solid State Sciences, Vol. 123, Springer, Berlin, 1996;
    F. Rossi, T. Kuhn, *Rev. Mod. Phys.* 74 (2002) 895.
27. R. Zimmermann, *Phys. Stat. Sol. B* 146 (1988) 371.
28. B.I. Green, J. Orenstein, S. Schmitt-Rink, *Science* 247 (1990) 679.
29. D. Hulin, A. Mysyrowicz, A. Antonetti, A. Migus, W.T. Masselink, H. Morkoc, H.M. Gibbs, N. Peyghambarian, *Phys. Rev. B* 33 (1986) 4389.
30. A. Honold, L. Schultheis, J. Kuhl, C.W. Tu, *Phys. Rev. B* 40 (1989) 6442.
31. R. Eccleston, B.F. Feuerbacher, J. Kuhl, W.W. Rühle, K. Ploog, *Phys. Rev. B* 45 (1992) 11403.
32. B. Deveaud, F. Clerot, N. Roy, K. Satzke, B. Sermage, D.S. Katzer, *Phys. Rev. Lett.* 67 (1991) 2355;
    C. Ciuti, V. Savona, C. Piermarocchi, A. Quattropani, P. Schwendimann, *Phys. Rev. B* 58 (1998) 7926.
33. D.A. Kleinman, *Phys. Rev. B* 28 (1983) 871.
34. J.-J. Liu, X.-J. Kong, Y. Liu, *J. Appl. Phys.* 84 (1998) 2638.
35. D. Birkedal, J. Singh, V.G. Lyssenko, J. Erland, J.M. Hvam, *Phys. Rev. Lett.* 76 (1996) 672.
36. M. Ueta, H. Kanzaki, K. Kobayashi, Y. Toyozawa, E. Hanamura, in: Excitonic Processes in Solids, in: Springer Series in Solid State Sciences, Vol. 60, Springer, Berlin, 1986.
37. L.V. Keldysh, A.N. Kozlov, *Sov. Phys. JETP* 27 (1968) 521.
38. A. Griffin, D.W. Snoke, S. Stringari (Eds.), Bose–Einstein Condensation, Cambridge University Press, Cambridge, 1995.
39. D.W. Snoke, J.P. Wolfe, A. Mysyrowicz, *Phys. Rev. Lett.* 59 (1987) 827; *Phys. Rev. Lett.* 64 (1990) 2543.
40. G.M. Kavoulakis, *Phys. Rev. B* 65 (2002) 035204;
    S. Denev, D.W. Snoke, *Phys. Rev. B* 65 (2002) 085211;
    A. Jolk, M. Joergen, C. Klingshirn, *Phys. Rev. B* 65 (2002) 245209, and references therein.
41. D.A.B. Miller, in: E. Burstein, C. Weisbuch (Eds.), Confined Electrons and Photons, Plenum Press, New York, 1995, p. 675.
42. Yu.E. Lozovik, V.I. Yudzon, *Sov. Phys. JETP* 44 (1976) 389.
43. T. Fukuzawa, E.E. Mendez, J.M. Hong, *Phys. Rev. Lett.* 64 (1990) 3066.
44. L.V. Butov, C.W. Lai, A.L. Ivanov, A.C. Gossard, D.S. Chemla, *Nature* 417 (2002) 47.
45. L.V. Butov, A.C. Gossard, D.S. Chemla, *Nature* 418 (2002) 751;
    D. Snoke, S. Denev, Y. Liu, L. Pfeiffer, K. West, *Nature* 418 (2002) 754.

46. V. Negoita, D.W. Snoke, K. Eberl, *Phys. Rev. B* 61 (2000) 2779.
47. V.M. Agranovich, G.C. La Rocca, in: V.M. Agranovich, G.F. Bassani (Eds.), Electronic Excitations in Organic Based Nanostructures, in: Thin Films and Nanostructures, Vol. 31, Elsevier, San Diego, 2003, Chapter 6.
48. C. Ciuti, G.C. La Rocca, *Phys. Rev. B* 58 (1998) 4599.
49. V.M. Agranovich, V.I. Yudson, P. Reineker, in: V.M. Agranovich, G.F. Bassani (Eds.), Electronic Excitations in Organic Based Nanostructures, in: Thin Films and Nanostructures, Vol. 31, Elsevier, San Diego, 2003, Chapter 7.

## Chapter 3

# Polaritons

FRANCO BASSANI

*Scuola Normale Superiore and I.N.F.M., Piazza dei Cavalieri 7, 56126 Pisa, Italy*

1. Introduction and General Concepts . . . . . . . . . . . . . . . . . . . . . . . . . . 129
2. Classical Theory of Polaritons . . . . . . . . . . . . . . . . . . . . . . . . . . . . 132
3. Quantum Theory of Polaritons . . . . . . . . . . . . . . . . . . . . . . . . . . . 135
4. Real Space Density Matrix Approach . . . . . . . . . . . . . . . . . . . . . . . 137
5. Experiments on Polaritons . . . . . . . . . . . . . . . . . . . . . . . . . . . . . 140
6. Surface Polaritons . . . . . . . . . . . . . . . . . . . . . . . . . . . . . . . . . . 141
7. Quantum Well Polaritons . . . . . . . . . . . . . . . . . . . . . . . . . . . . . . 147
8. Quantum Wire Polaritons . . . . . . . . . . . . . . . . . . . . . . . . . . . . . . 161
9. Exciton-Polaritons in Microcavities . . . . . . . . . . . . . . . . . . . . . . . . 166
    Acknowledgements . . . . . . . . . . . . . . . . . . . . . . . . . . . . . . . . . 180
    References . . . . . . . . . . . . . . . . . . . . . . . . . . . . . . . . . . . . . . 180

We review the theory of polaritons, referring in particular to excitonic polaritons, and also mention some experimental evidence. The semiclassical theory, which makes use of a model dielectric function incorporated in Maxwell's equations, is developed and compared with the full quantum theory, which implies the quantization of the electromagnetic field and of the polarization field.

The polariton concept is extended to the case of surfaces and interfaces, where localized exciton modes are present.

The case of reduced dimensionality nanostructures, such as Quantum Wells (Q.W.) and Quantum Well Wires (Q.W.W.), is also considered, and resonant and surface-like polariton modes are described.

The extension of the polariton concept to the case of microcavities is finally discussed.

## 1. Introduction and General Concepts

We consider the electronic excitation in solids, due to the electron bound to its hole and forming an exciton, as described in the chapter by G.C. La Rocca. The electron–hole interaction is responsible for the relative motion and produces states characterized by their binding energies. In addition, since the total momentum of the excitation $\hbar \vec{k}_{\text{ex}}$ is a good quantum number, the exciton energies also depend on

$\hbar \vec{k}_{\text{ex}}$. We recall that in the case of inorganic semiconductors one obtains Wannier–Mott excitons, characterized by a large effective Bohr radius (compared with the lattice constant), while in the case of organic materials or large gap insulators one obtains Frenkel–Peierls excitons, characterized by a small Bohr radius [1].

The excitons always carry a transition dipole moment $\vec{\mu}_{\text{ex}}$, related to the dynamic polarization of the medium, so that we can have longitudinal excitons ($\vec{\mu}_{\text{ex}} \parallel \vec{E}$) and transverse excitons ($\vec{\mu}_{\text{ex}} \perp \vec{E}$), always separated in cubic crystals. We recall that the energy separation between longitudinal and transverse modes at $\vec{k}_{\text{ex}} = 0$ ($\Delta_{LT}$) is related to the oscillator strength per unit volume $f_{\text{ex}}/V$ by [1,2]

$$\Delta_{LT} \approx \frac{2\pi}{\varepsilon_\infty} \frac{\hbar e^2}{m\omega_T} \frac{f_{\text{ex}}}{V}, \qquad (1)$$

where $\varepsilon_\infty$ is the value of the dielectric function for $\omega \gg \omega_T$, and the oscillator strength is related to the transition dipole moment $\vec{\mu}_{\text{ex}}$ by

$$\frac{f_{\text{ex}}}{V} \simeq \frac{2m\omega_T}{e^2 \hbar} |\mu_{\text{ex}}|^2. \qquad (2)$$

The transition dipole moment $\vec{\mu}_{\text{ex}}$, expressed in terms of the conduction and valence band extrema Bloch functions $u_e$ and $u_v$ and of the exciton radius $a_0$, is in the case of Wannier–Mott excitons

$$\vec{\mu}_{\text{ex}} = \langle \psi_{\text{ex}} | e\vec{r} | 0 \rangle \simeq \langle u_v | e\vec{r} | u_c \rangle |F_{\text{ex}}(0)| \simeq \vec{\mu}_{vc} \left( \frac{1}{\pi a_0^3} \right)^{1/2}, \qquad (3)$$

where $|F_{\text{ex}}(0)|$ is the value of the envelope function at null distance between electron and hole. The last expression (3) holds also for Frenkel excitons, with the lattice constant $d$ instead of the Bohr radius $a_0$, so that the ratio of the oscillator strengths $f_{\text{ex}}^{\text{W.M.}}/f_{\text{ex}}^{\text{F}}$ is generally of order $10^{-2}$.

The dependence of the exciton energy on $\vec{k}_{\text{ex}}$ is quite different for Wannier–Mott excitons and for Frenkel–Peierls excitons. In the former case it is mostly given by the exciton center of mass motion, and is about the same for longitudinal and transverse excitons, while in the latter case a relevant contribution is given by the long range electron–hole exchange term, which in cubic crystals in the dipole approximation has positive (negative) sign for transverse (longitudinal) excitons [2]. The higher multipoles in some crystals may also significantly contribute to the dependence of the exciton energy on $\vec{k}_{\text{ex}}$.

The polaritonic concept originates from considering the electromagnetic radiation of frequency close to that of a transverse exciton. In a medium the radiation is associated to a pseudomomentum $\frac{\hbar \omega}{c} n$, where $n = \sqrt{\varepsilon}$ is the refractive index. The total momentum is defined modulus $\hbar \vec{h}$, $\vec{h}$ being a reciprocal lattice vector, but in the case of the optical properties the relevant momentum is confined to the first Brillouin Zone in correspondence to $\vec{h} = 0$ [3]. Since in proximity of an exciton mode the index of refraction has a strong dependence on frequency, one can see immediately that frequency and wave vector are not linearly related, but an anomalous dispersion occurs, which produces radiation modes strongly influenced by the excitonic polarization modes.

Such dressed electromagnetic modes can be obtained directly by including the polarization $\vec{P}(\omega, \vec{k})$ in Maxwell equations, with $\vec{D} = \varepsilon \vec{E} = \vec{E} + 4\pi \vec{P}$, and solving the eigenvalue equations. For the transverse modes (polaritons) we have

$$\omega_T(k) = \frac{c}{\sqrt{\varepsilon(\omega_T, k)}} |\vec{k}|, \tag{4}$$

while for the longitudinal modes the eigenvalue equation is

$$\varepsilon(\omega_L, k) = 0. \tag{5}$$

The polariton modes have then a mixed character because they contain a polarization and an electric field contribution. They were first introduced by K. Huang and K.B. Tolpygo [4] to account for the polarization associated to the optical vibrational modes in dipolar lattices.

In the late fifties U. Fano [6], J.J. Hopfield [7], and V.M. Agranovich [8] demonstrated that the polariton modes can be expressed as quantum field particles in crystal media, analogous to the photons of the electromagnetic field in vacuum. The name "polaritons" was introduced by Hopfield and Agranovich to stress their polarization origin (optical phonons or excitons), and is commonly used to denote such mixed modes and the related field particles. Hopfield and Agranovich also proved that the eigenvalue equation obtained from the quantum theory is the same as that obtained from the classical theory with the appropriate expression for $\varepsilon(\omega, 0)$. Such proof was later extended to include spatial dispersion [9], and the quantum mechanical exact eigenfunctions were also obtained [10].

In the sixties, the polariton idea has become basic in linear and nonlinear optics [5]. Experimental evidence of the "polariton" concept described above is abundant [11], and also the prediction of an "additional optical wave", introduced by Pekar [12] as a consequence of spatial dispersion, has been amply confirmed [11].

The polariton concept has been extended to the specific optical modes bound to surfaces and interfaces [13]. In this case the polariton modes are travelling waves depending on the wave vector in the direction parallel to the surface plane $\vec{k}_\parallel$, and are evanescent waves in the direction perpendicular to the surface.

The recent study of the optical properties of reduced dimension nanostructures has required the introduction of polaritons in quantum wells (Q.W.) and in quantum well wires (Q.W.W.), with the derivation of resonant modes and surface-like modes as functions of the appropriate in-plane wave vector $\vec{k}_\parallel$ or of the linear wave vector $k$ [14,15].

The recent developments with semiconductor microcavities have also led to the introduction of specific polariton modes appropriate to microcavities [16]. In this case one can consider the specific nature of the relevant material contained in the microcavities.

In the following sections we will present the concepts described above, giving the general theory and referring to some relevant experiments.

## 2. Classical Theory of Polaritons

To obtain the transverse optical modes in a medium we must solve Eq. (4), with an appropriate expression for the dielectric function $\varepsilon(\omega, \vec{k})$ in the vicinity of an excitonic transverse resonance frequency $\omega_{ex}$. Using the expression obtained with the Lorentz oscillator model, which is, for a cubic crystal, in terms of the plasma frequency $\omega_p$ and of the oscillator strength $f_{ex}$

$$\varepsilon(k, \omega) = \varepsilon_\infty + \frac{\omega_p^2 f_{ex}}{\omega_{ex}^2(k) - \omega^2 - i\omega\gamma_{ex}}, \tag{6}$$

where $\varepsilon_\infty$ is a constant, we obtain the following equation for the polariton dispersion modes

$$\frac{c^2 k^2}{\omega^2} = \varepsilon_\infty + \frac{\omega_p^2 f_{ex}}{\omega_{ex}^2(\vec{k}) - \omega^2 - i\omega\gamma_{ex}}. \tag{7}$$

In the above equations the plasma frequency is expressed in terms of the electron density $N/V$ as $\omega_p = \sqrt{\frac{4\pi e^2}{m} \frac{N}{V}}$, $\omega_{ex}(\vec{k})$ is the exciton transverse mode frequency, $f_{ex}$ is the oscillator strength, $\varepsilon_\infty$ is the background dielectric constant corresponding to $\omega \gg \omega_{ex}$, and $\gamma_{ex}(\omega)$ is a damping frequency which accounts for dissipation, mostly due to exciton–phonon interaction and to disorder, with a small contribution due to electron–hole recombination.

The solution of Eq. (7) gives two modes for every $\vec{k}$ value, a lower polariton branch (L.P.) and an upper polariton branch (U.P.). The upper polariton starts at a frequency which coincides with the longitudinal exciton value $\omega_L$ at $k = 0$, and at high frequency goes asymptotically to the linear optical behavior $\omega = \frac{c}{\sqrt{\varepsilon_\infty}} k$. The value of the longitudinal mode at $k = 0$ can be obtained immediately from Eqs. (5) and (6) with $\omega = \omega_L$, $\omega_0 = \omega_{ex}(k=0)$ and $\gamma_{ex} = 0$,

$$\omega_L = \omega_0 \left(1 + \frac{\omega_p^2 f_{ex}}{\omega_0^2 \varepsilon_\infty}\right)^{1/2}. \tag{8}$$

The lower polariton branch frequency goes asymptotically to $\omega_{ex}(\vec{k})$ for large $k$ values, and linearly to zero as $k \to 0$, with velocity

$$\frac{c}{\sqrt{\varepsilon_\infty + \frac{\omega_p^2 f_{ex}}{\omega_0^2}}}.$$

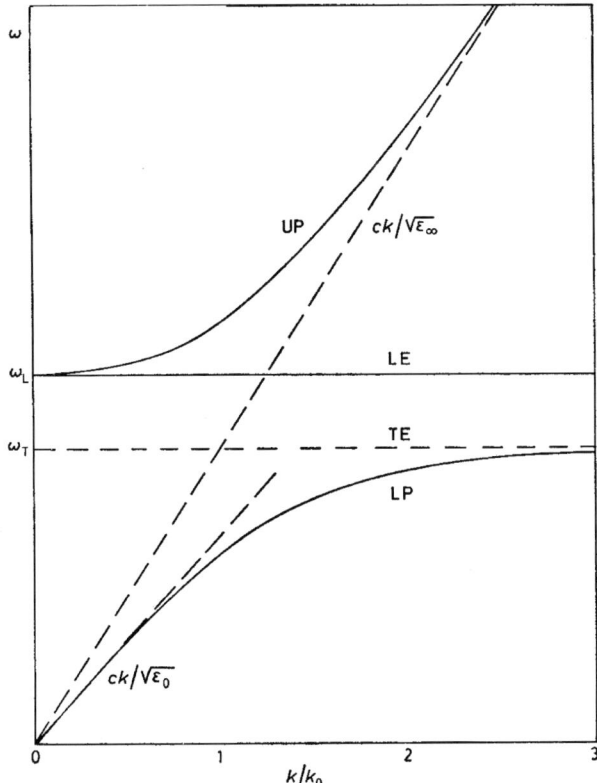

Fig. 1. Schematic dispersive behavior of excitons and related L. and U. polaritons, neglecting spatial dispersion. The dashed lines give the independent dispersion of electromagnetic waves and excitons. The value $k_0$ corresponds to their crossing point. Only transverse excitons interact with the radiation field.

The polariton modes so defined are analogous to the electromagnetic modes in vacuum, but the presence of the electronic excitations in the medium changes their dispersive behavior. They contain both crystal polarization and crystal electromagnetic waves. Their schematic behavior is displayed in Figure 1 when the $\vec{k}$ dependence in $\omega(k)$ is not considered, and in Figure 2 in the case of Wannier–Mott excitons with $\vec{k}$ dependence. In the case of Frenkel–Peierls excitons the dispersive behavior is usually less pronounced because of the larger values of the effective mass and is rather different for lower polariton and longitudinal exciton because of the larger value of the electron–hole exchange [9].

In the former case the strong dispersive behavior of the exciton energy ($\frac{\hbar^2 k^2}{2M}$) makes the lower polariton branch bend strongly upwards so that for $\omega > \omega_L$ an

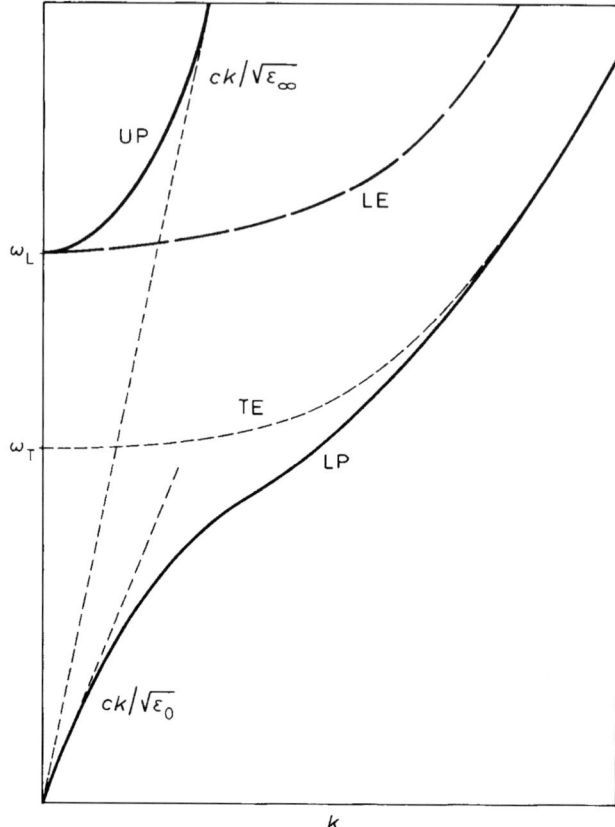

Fig. 2. Schematic dispersive behavior of Upper and Lower Polaritons from Wannier–Mott excitons, including spatial dispersion. The $T$–$L$ splitting is $\Delta_{LT} = \frac{4\pi e^2}{\hbar} X_{0n}^2 \frac{N}{V}$ (from Ref. [9]).

additional propagating wave must exist, as first pointed out by Pekar [17]. The simultaneous propagation of two waves at the same frequency poses a problem about their relative amplitudes because Maxwell boundary conditions are not sufficient, and the problem of additional boundary conditions (A.B.C.) arises.

In the case of those Frenkel excitons, where one can consider the total mass to be infinite ($M \simeq \infty$), the wave vector dependence is much less important. It can be obtained in the dipole approximation from the long range electron–hole exchange contribution, and it has opposite sign for transverse and longitudinal excitons; it was computed by Heller and Marcus [18] for cubic crystals and by L. Pirozzi et al. [2] for anisotropic materials. In layered molecular crystals with strong exciton dipole moment the exciton band width can be rather large. In these cases the approximation of infinite exciton effective mass is not fulfilled at least in layer directions where the distance between molecules is small and where the resonance intermolecular interaction is large [19].

## 3. Quantum Theory of Polaritons

It has been shown that the polariton modes can be interpreted as quantum particles of the electromagnetic fields, analogous to the "photons" in vacuum [6–10]. This is obtained from the quantum treatment of radiation and excitons. The radiation can be expressed in terms of creation ($a_k^+$) and destruction operators ($a_k$) of photons of energy $\hbar v k$ (with $v = c/\sqrt{\varepsilon_\infty}$ in the medium). Also the excitons, as shown by G. La Rocca in Chapter 2 of this book, can be expressed in terms of creation ($b_k^+$) and destruction ($b_k$) operators for excitation particles (electron–hole pairs), which obey the boson commutation rules for small exciton densities. Their mutual interaction can be expressed in terms of the dipole matrix elements from the ground to the exciton states (3) or, equivalently, in terms of the oscillator strength (2), through the parameter

$$\beta = f_{\text{ex}} \frac{\omega_p^2}{4\pi \omega_{\text{ex}}^2}.$$

The resulting Hamiltonian consists of free photons and free excitons, and of the photon–exciton interaction. For cubic crystals it is [9]:

$$H = \sum_k \left[ \hbar v k \left( a_k^+ a_k + \frac{1}{2} \right) + \hbar \omega_k \left( b_k^+ b_k + \frac{1}{2} \right) \right.$$
$$+ \sum_k [i C_k (a_k^+ + a_{-k})(b_k - b_{-k}^+)$$
$$\left. + D_k (a_k^+ + a_{-k})(a_k + a_{-k}^+) ] \right], \tag{9a}$$

where the coupling constants are:

$$C_k = \hbar \omega_0 \left( \frac{\pi \beta \omega_k}{v k \varepsilon_\infty} \right)^{1/2}, \tag{9b}$$

and

$$D_k = \hbar \omega_0 \frac{\pi \beta \omega_0}{v k \varepsilon_\infty}. \tag{9c}$$

The first interaction term originates from the contribution $-\frac{e}{mc} \vec{A} \cdot \vec{p}$ and the second from the $\frac{e^2}{2mc^2} \vec{A} \cdot \vec{A}$ interaction. If we neglect spatial dispersion ($\omega_k \simeq \omega_0$) in the exciton frequencies, they are related because $D_k = C_k^2/\hbar \omega_0$. In general $D_k$ and

$C_k$ are much smaller than $\hbar\omega_0$ for large values of $k$, but they diverge as $k \to 0$ (infrared divergence of quantum electrodynamics). This makes it impossible to treat the interaction perturbatively when $k < \beta\omega_0/v$.

The above Hamiltonian (9) is quadratic and it couples only photons and excitons of momentum $\vec{k}$ and $-\vec{k}$, so that it can be made diagonal by a Bogoliubov–Tyablikov-like linear transformation, introducing new creation $\alpha^+_{k_1}$, $\alpha^+_{k_2}$, and destruction operators $\alpha_{k_1}$, $\alpha_{k_2}$, linearly related to the previous ones. The new diagonal Hamiltonian for the two polariton modes $i = 1, 2$ becomes [9]

$$H = \sum_{k,i} \alpha^+_{ki} \alpha_{ki} \hbar\Omega_k, \qquad (10)$$

and $\Omega_k$ turns out to be given by the same equation (7) as in the classical case, with $\gamma_{ex} = 0$. The explicit expressions of the transformations are obtained from the Bose commutation rules of the $\alpha$ operators and from

$$[\alpha_{ki}, H] = \hbar\Omega_{ik}\alpha_{ki}, \quad i = 1, 2. \qquad (11)$$

They turn out to be [7,9,11]:

$$\begin{bmatrix} \alpha_{k_1} \\ \alpha_{k_2} \\ \alpha^+_{-k_1} \\ \alpha^+_{-k_2} \end{bmatrix} = \begin{bmatrix} W_1 & X_1 & Y_1 & Z_1 \\ W_2 & X_2 & Y_2 & Z_2 \\ Y_1^* & Z_1^* & W_1^* & X_1^* \\ Y_2^* & Z_2^* & W_2^* & X_2^* \end{bmatrix} \begin{bmatrix} a_k \\ b_k \\ a^+_{-k} \\ b^+_{-k} \end{bmatrix}, \qquad (12)$$

with

$$W_1 = X_2 = \frac{(\omega_k^2 - \Omega_{1k}^2)(vk + \Omega_{ik})}{2(vk\Omega_{1k})^{1/2}\{(\omega_k^2 - \Omega_{1k}^2)^2 + 4\pi\beta\omega_k^2\omega_0^2/\varepsilon_\infty\}^{1/2}},$$

$$X_1 = W_2 = \frac{-2iC_k vk/\hbar}{(\omega_k - \Omega_{1k})(vk + \Omega_{1k})} W_1,$$

$$Y_1 = Z_2 = \frac{\Omega_{1k} - vk}{\Omega_{1k} + vk} W_1,$$

$$Z_1 = Y_2 = \frac{\omega_k - \Omega_{1k}}{\omega_k + \Omega_{1k}} X_1. \qquad (13)$$

The fact that the quantum theory gives the same results for the $\Omega_{ik}$ as the classical theory is not surprising, but the quantum theory describes the polariton modes as particles, so that polariton statistics can be made and polariton squeezing obtained [20]. The quantum theory is also appropriate for all values of $\vec{k}$, even outside the first Brillouin Zone, which is important for investigating linear and nonlinear dynamical processes.

With the quantum theory one can also find the eigenfunctions, provided the new vacuum is determined, because the ground state of Hamiltonian (10) with interaction does not coincide with the ground state of the Hamiltonian without interaction (zero photons and excitons). The new vacuum of the polaritons $|0'\rangle$ is defined from

$$\alpha_{ki}|0'\rangle = 0, \tag{14}$$

and turns out to be given by [10,11]:

$$|0'\rangle = \prod_k \frac{1}{N_k} \exp\left[\frac{1}{2}\sum_{i,j} G_{i,j}(k)\alpha_{ik}^+\alpha_{j,-k}^+\right]|0\rangle, \tag{15a}$$

with

$$G_{11}(k) = G_{22}(k) = -\frac{\Omega_{1k} + \Omega_{2k} - \omega_k - vk}{\Omega_{1k} + \Omega_{2k} - \omega_k + vk},$$

$$G_{12}(k) = G_{21}(k) = -i\left(\frac{vk\omega_k\varepsilon_\infty}{\pi\beta\omega_0^2}\right)^{1/2} G_{11}(k), \tag{15b}$$

and the normalization constant

$$N_k^2 = |W_1|^2 + |X_1|^2, \tag{15c}$$

with $W_1$ and $X_1$ given by (13).

The polariton vacuum $|0'\rangle$ is quite different from the independent particle vacuum $|0\rangle$. The polariton states are obtained by applying to the polariton vacuum $|0'\rangle$ powers of the operator $\alpha_{ik}^+$, which create a number of polariton states. Even in a one-polariton state any number of photons and excitons are contained. The linear combination of a one-photon and one-exciton state often used is a very crude approximation, which is valid only for large values of $k$. As an example we report in Figure 3 the one-photon and one-exciton component, the two-photon and one-exciton component, and the three-photon components for the lower polariton of CuCl.

## 4. Real Space Density Matrix Approach

Polariton states and the excitation spectra can be also discussed when more exciton states are involved and may influence one another, and when coherence effects between the electric field and electron and hole need to be taken into account. In these cases an alternative approach to the one described above can be used, particularly appropriate for studying the dynamics of polaritons and nonlinear properties. It consists in extending the density matrix method and the Bloch equations of atomic physics to solid state problems. This approach has been developed by Stahl and Baslev [21], who consider the Heisenberg equation for the

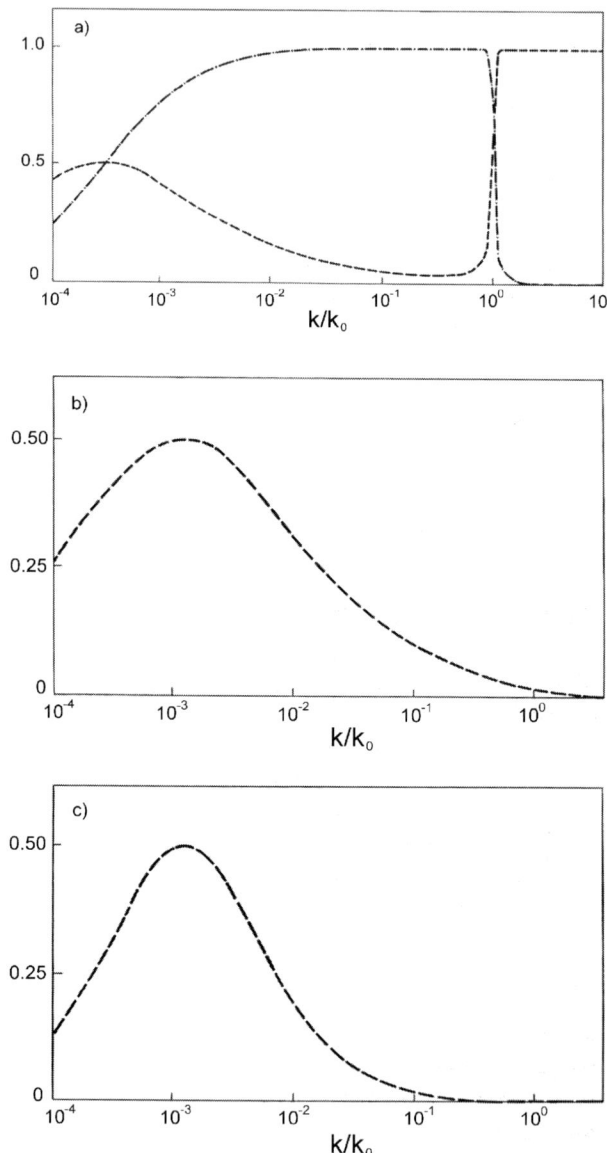

Fig. 3. Components of the lower polariton state, computed for the 1s exciton-polariton of CuCl (from Ref. [10]): (a) one-photon (— · — · — · —) and one-exciton (— — — —) components; (b) two-photon and one-exciton component; (c) three-photon component.

density matrix

$$i\hbar \frac{\partial \rho}{\partial t} = [H, \rho] + i\hbar \left(\frac{\partial \rho}{\partial t}\right)_{\text{irr.}}, \quad (16)$$

with inclusion of an irreversible decay contribution, due to electron–electron and electron–phonon interaction. In the case of two bands this leads to a Schrödinger-like equation for a coherent wave function $Y(\vec{r}, \vec{R})$, which gives the probability amplitude of an electron–hole pair, including the coupling to the electric field, and a broadening value $\Gamma$ to account for irreversible decay processes, i.e.,

$$\left[E_g - \hbar\omega - i\Gamma - \frac{\hbar^2}{2M}\nabla_R^2 - \frac{\hbar^2}{2\mu}\nabla_r^2 + V(r)\right] Y(\vec{r}, \vec{R}) = M(\vec{r}) E(\vec{R}), \quad (17)$$

where $\vec{r}$ is the relative coordinate and $\vec{R}$ the center of mass coordinate; $\vec{E}$ denotes the electric field and $V(r)$ the electron–hole attraction. This couples the exciton to the electric field $E(\vec{R})$ through the transition dipole density $M(\vec{r})$ (Fourier transform of the dipole matrix element between Bloch functions at the same $k$). When more bands are considered coupled equations of type (17) result. The polarization is obtained from the coherent wave function as

$$P(R) = \int d\vec{r}\, M(\vec{r}) Y(\vec{r}, \vec{R}). \quad (18)$$

This can be computed from the simultaneous resolution of (17) and of Maxwell equations to obtain the electric field. In this approach the polariton spectrum is obtained from the peaks in the imaginary part of the susceptibility. As an example, the results of such calculations are given in Figure 4 for GaAs, where we have separated Heavy Hole and Light Hole exciton states [21,22].

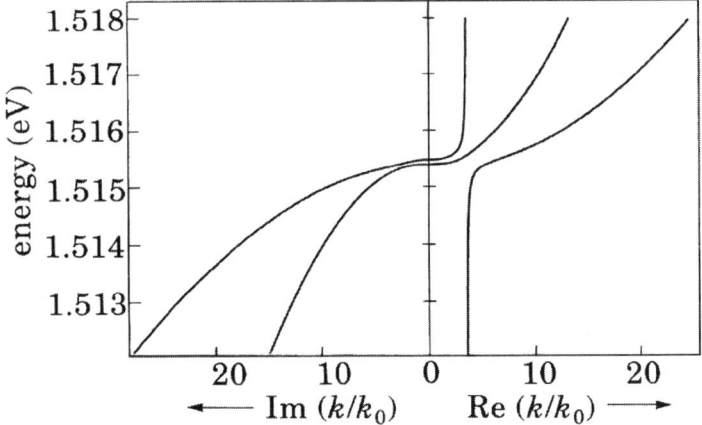

Fig. 4. Calculated dispersive behavior of polaritons in GaAs (from Ref. [22]).

An essentially equivalent procedure is given by K. Cho in terms of the current densities [21].

## 5. Experiments on Polaritons

Experimental investigation of the "polaritons" described above is abundant. As an example we give in Figure 5 the lowest excitation of CuCl, with the dispersive behavior obtained from nonlinear optical experiments (two-photon absorption and hyper-Raman scattering) [23]. As a further indication of the polariton nature of the excitation in CuCl we can find the dispersive behavior of Longitudinal exciton and of the Lower and Upper Polariton in CuCl, from the two-photon transition to biexciton states [23]. In Figure 6 we show the reabsorption dips in the biexciton luminescence produced by resonant two-photon absorption from the ground state to the biexcitonic state with an external laser. We also plot the intensity minima

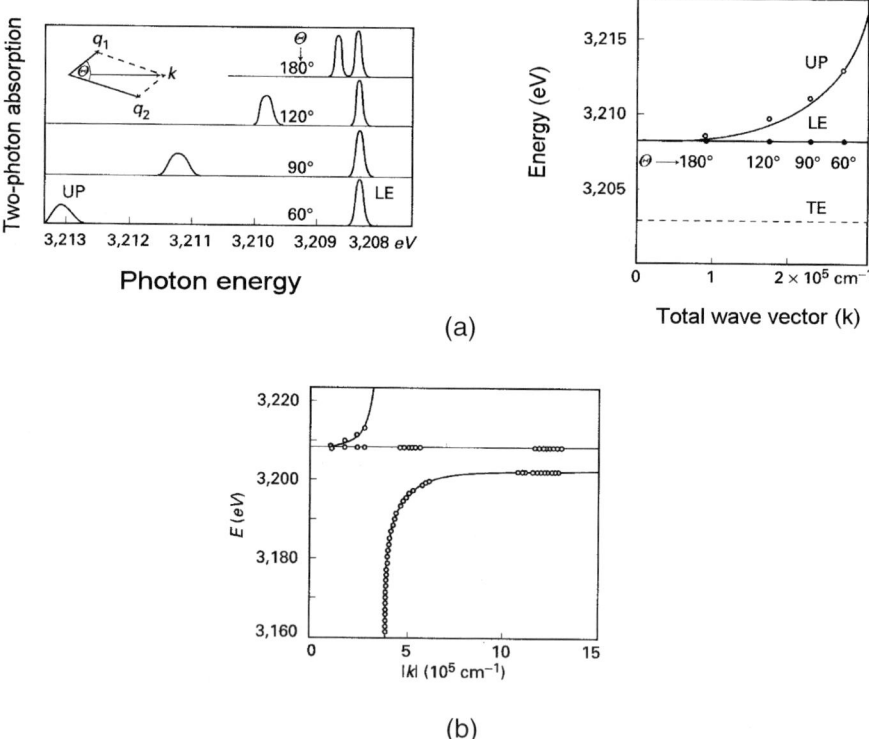

Fig. 5. Polariton dispersion of the 1s exciton-polaritons in CuCl for small values of $k$, as obtained from (a) two-photon absorption at different angles and (b) hyper-Raman scattering (from Ref. [23]).

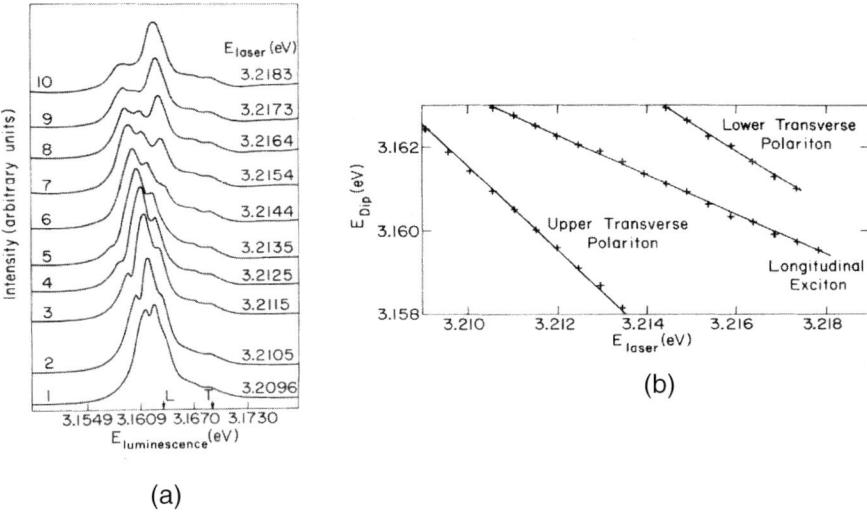

Fig. 6. Resonant two-photon transitions to biexciton states in CuCl (from Pribram et al. [23]): (a) dips in the luminescence from biexciton states due to the reabsorption in the presence of laser light; (b) plot of the two-photon absorption with respect to laser energy.

with respect to laser energy, which give the dispersion of the three intermediate states, L.P., longitudinal exciton, and U.P. We also report in Figure 7 the polariton dispersive behavior of the lowest exciton in GaAs, as obtained from the reflectivity of a thin slab (6000 Å) grown by molecular beam epitaxy (M.B.E.), where the additional wave due to the $k$-dispersion of the polaritons produces additional interference dips [24]. We also present in Figure 8 the comparison of the experimental reflectance spectrum with that calculated [25] using the Stahl and Baslev real-space density matrix approach described above [21]. In Figure 9 we finally show the direct observation of the additional wave in CdS obtained by Lebedev et al. [26]. Other experimental evidence is reviewed in Ref. [11]. The polariton concept is therefore amply demonstrated in bulk crystals and thin films.

## 6. Surface Polaritons

The crystal boundaries lead to surface states in the spectra of all excitations, and therefore also surface excitons interacting with photons must result in surface polaritons [13]. These are mixed modes, localized at the surface, with frequencies in the region between $\omega_T$ and $\omega_L$, where no bulk polaritons exist in the absence of spatial dispersion. They do not couple directly to free electromagnetic waves because in their frequency range it is not possible to satisfy simultaneously the energy and momentum conservation laws, as is generally true for independent eigenmodes.

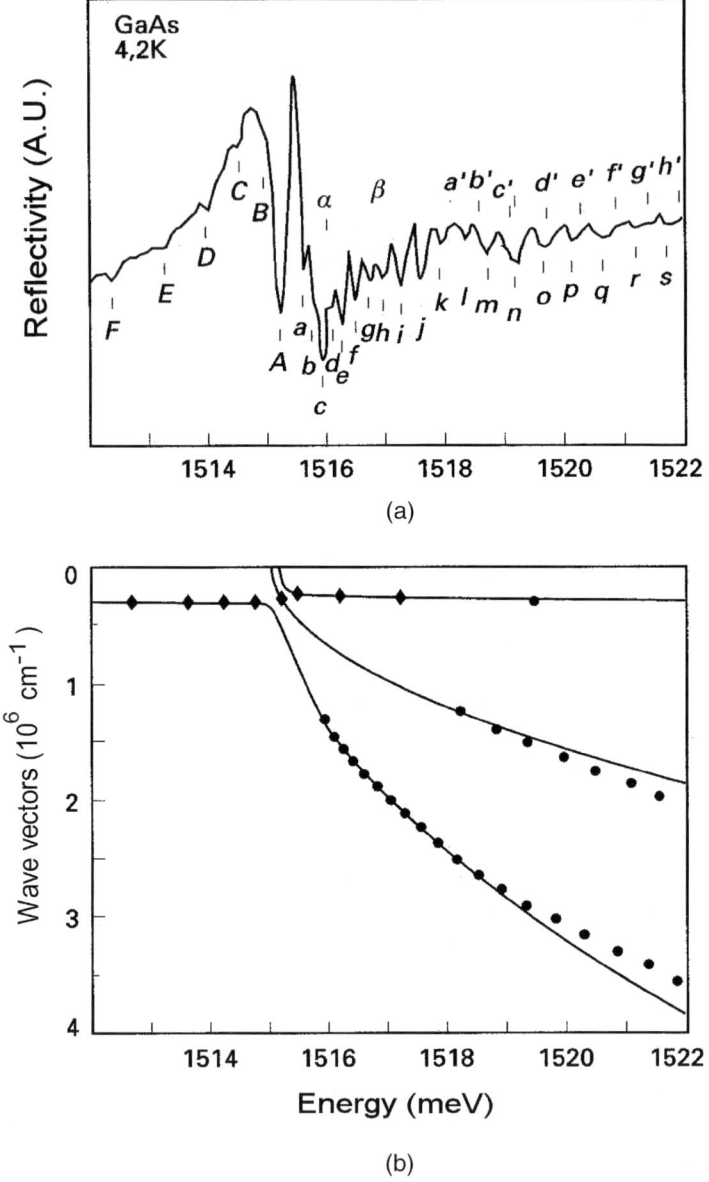

Fig. 7. (a) Reflectance of a thin film (6000 Å) of GaAs, with evidence of the dips produced by lower polariton interference (capital letters), and those with the additional waves (from Ref. [24]). (b) Corresponding dispersive behavior of exciton polaritons in GaAs. Observe the double L.P. curves due to the heavy and light hole mass (from Ref. [24]).

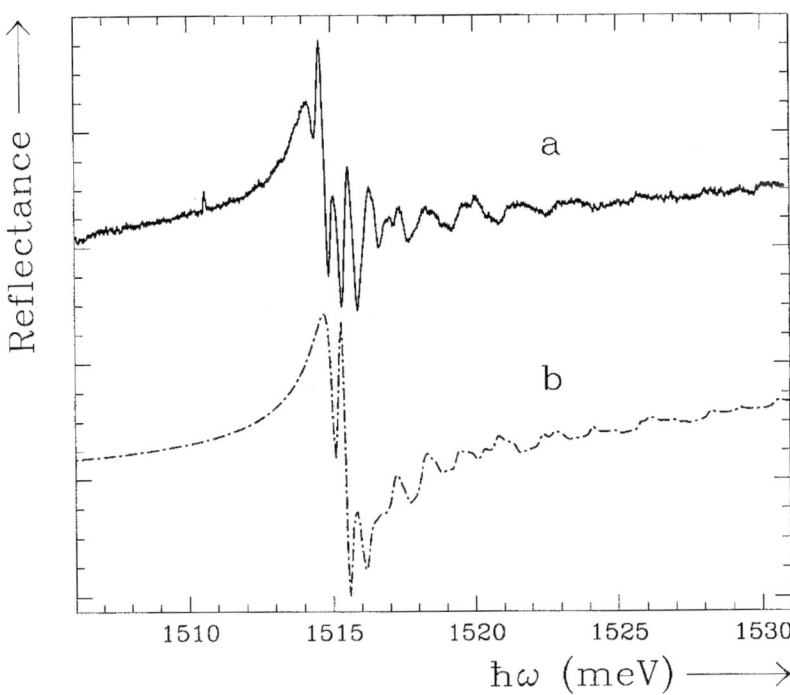

Fig. 8. Comparison of (a) theoretical and (b) experimental reflectance spectra in a GaAs thin film of about 600 nm thickness (from Ref. [25]).

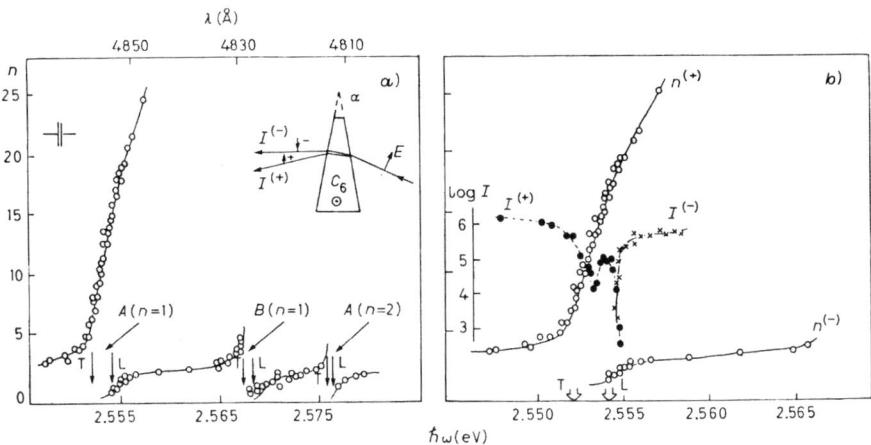

Fig. 9. (a) Dispersion of the real part of the index of refraction in a CdS crystal with the electric field polarized in the laser plane. (b) Real part of the index of refraction and transmitted intensities of the LP($I^+$) and UP($I^-$). From Ref. [26].

To understand in a simple way how all this happens one can use the phenomenological classical approach, as described in Section 2 for bulk polaritons.

One considers the dielectric function near an exciton state in the form

$$\varepsilon(\omega, k) = \varepsilon_\infty \left( 1 + \frac{\omega_L^2 - \omega_T^2}{\omega_T^2 + \alpha k^2 - \omega^2 - i\gamma\omega} \right), \tag{19}$$

which includes spatial dispersion ($\alpha \simeq \hbar \omega_T / M$ for Wannier excitons), and damping ($\gamma \neq 0$).

The concept of surface polaritons can be presented in the simplest way by neglecting spatial dispersion and damping [27]. Considering the surface plane $x$–$y$, where periodic translation symmetry is preserved, the good quantum number is the two-dimensional wave vector $\vec{q}(q_x, q_y)$ in the surface plane. Let us direct $\vec{q}$ along the $x$-axis $(q_x, 0)$, and consider the Maxwell equation for the $y$ component of the magnetic field (T.M. mode). We obtain

$$\frac{\partial^2 H_y}{\partial z^2} - \left( q_x^2 - \frac{\varepsilon \omega^2}{c^2} \right) H_y = 0, \tag{20}$$

which gives a $z$-dependence which decays exponentially away from the surface, for $z < 0$ as

$$e^{k_2 z} = \exp\left( z \sqrt{q_x^2 - \frac{\omega^2}{c^2} \varepsilon} \right),$$

and for $z > 0$, in vacuum, as

$$e^{-k_1 z} = \exp\left( -z \sqrt{q_x^2 - \frac{\omega^2}{c^2}} \right).$$

Such a $z$-dependence also holds for the electric field. The electric field is in the $xz$-plane, and for $z > 0$ (in vacuum), we obtain $E_z^{(1)} = \frac{iq_x}{k_1} E_x^{(1)}$, while for $z < 0$ (in the medium) we obtain $E_z^{(2)} = \frac{-iq_x}{k_2} E_x^{(2)}$.

The eigenvalue dispersive equation for the surface polaritons is obtained by imposing the boundary conditions at the surface ($E_x^{(1)} = E_x^{(2)}$ and $E_{z=0}^{(1)} = \varepsilon E_{z=0}^{(2)}$), which gives

$$\frac{iq_x}{\sqrt{q_x^2 - \frac{\omega^2}{c^2}}} = \frac{-iq_x \varepsilon}{\sqrt{q_x^2 - \frac{\omega^2}{c^2} \varepsilon}}. \tag{21}$$

The above equation gives the dispersion relation of surface polaritons

$$\omega = cq_x \sqrt{\frac{\varepsilon(\omega) + 1}{\varepsilon(\omega)}}. \tag{22}$$

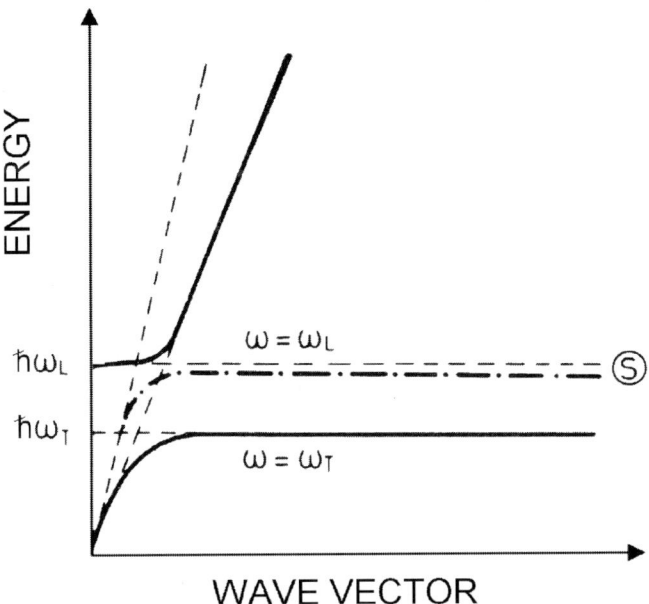

Fig. 10. Schematic behavior of the surface polariton dispersion $\omega(q_x)$ when spatial dispersion and damping are neglected. The dash-dotted line gives the surface state as a function of $q_x$. The continuous lines denote upper and lower bulk polaritons.

From Eq. (21) it follows that the only acceptable solutions require $\varepsilon(\omega)$ to be negative, which implies, from expression (19), that $\omega$ be in the interval between $\omega_T$ and $\omega_L$, where no bulk solutions exist in the absence of spatial dispersion. Substituting expression (19) into (22) we obtain the surface polariton dispersion dependence. This gives for $\omega$ a monotonically increasing function of $q_x$, which is schematically shown in Figure 10 in the limit $\gamma \to 0$, $\alpha \to 0$. The frequency range of the surface polariton $\omega_S$ is above $\omega_T$ (the lowest state is in correspondence to $\varepsilon(\omega) = -\infty$) and below $\omega_L$ (the highest state is in correspondence to $\varepsilon(\omega) = -1$).

The surface polariton is therefore incompatible with all polariton modes of the crystal or of the vacuum, and consequently cannot decay into them. All surface polaritons are therefore nonradiative, and cannot be excited by shining light into the surface. A similar situation occurs in the case of an interface, with the dielectric constant $\varepsilon_1$ used instead of 1 in expressions (21) and (22). When $\varepsilon_1$ approaches $\varepsilon_\infty$, $\omega_S \to \omega_L$.

To consider spatial dispersion in the dielectric function, $\alpha \neq 0$ in Eq. (19), makes the calculation of surface polaritons more complicated; in this case the usual boundary conditions are not sufficient to determine eigenvalues analogous to (22) because of the additional ray. Consequently additional boundary condi-

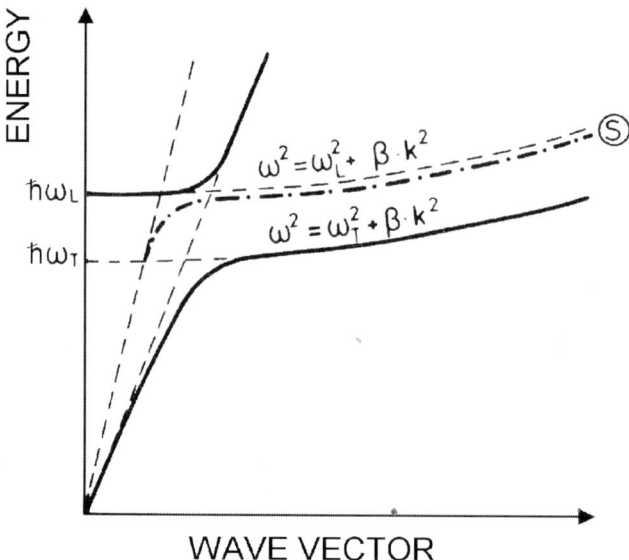

Fig. 11. Schematic behavior of surface polariton modes with consideration of spatial dispersion (from Ref. [27]). Symbols as in Figure 10.

tions (A.B.C.) are required, as shown by Agranovich in a specific chapter of the book dedicated to surface polaritons [28]. The surface polariton modes have a behavior which is schematically indicated in Figure 11. In this case Maradudin and Mills [29] have shown that the admixture of bulk modes and surface modes of the same energy but different $\vec{q}$ values results in an additional damping mechanism.

To observe surface exciton polaritons is an experimentally difficult task even in the case when spatial dispersion is considered, because, though their energy can overlap that of bulk polaritons or vacuum photons, the corresponding in-plane wave vector $\vec{q}$ is always larger. One must use techniques which generate sufficiently large wave vectors parallel to the surface. We can mention the method of attenuated total reflection (A.T.R.), the use of periodic surface structures (gratings), or inelastic scattering of light or electrons.

The most effective technique is A.T.R., which is obtained when light is totally reflected inside a prism of larger index of refraction $n_p$, applied to the surface. If $\alpha$ is the angle of incidence, larger than the angle of total reflection, an evanescent wave with wave vector

$$q = k_\| = \frac{n_p \omega}{c} \sin \alpha \qquad (23)$$

enters the crystal. If the prism is close to the surface, and the value given by (23) is sufficiently large, this evanescent wave can couple to the surface mode and excite it. This effect decreases the reflectance in correspondence to the appropriate

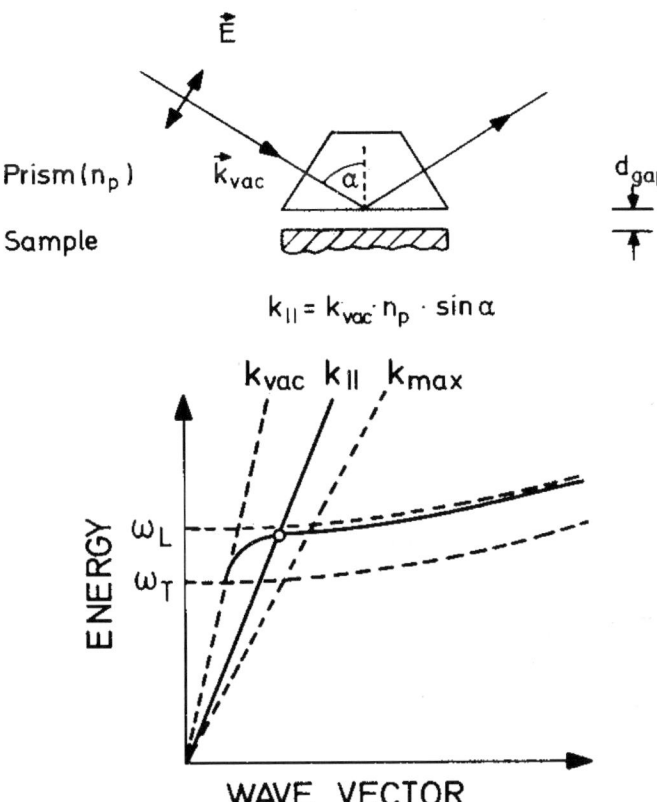

Fig. 12. Schematic view of the experimental set-up for observing A.T.R. The top scheme shows the experimental setup. The bottom scheme indicates the condition for exciting the surface polaritons. From Ref. [27].

value of $\omega$ and $\vec{q}$, and produces a dip in the reflected intensity at the appropriate frequency and angle. The scheme is shown in Figure 12, and experimental results for the $Z_3$ exciton of CuCl (1s exciton of Figures 5 and 6) at 77 K are reported in Figure 13 [30]. One can notice the dip in the reflected intensity for light polarized parallel to the plane of incidence at a frequency between $\omega_T$ and $\omega_L$, and observe how its energy increases with increasing angle of incidence, thus allowing to measure the dispersion of surface polaritons.

## 7. Quantum Well Polaritons

The polariton concept also applies to electromagnetic waves and electronic excitations in nanostructures, provided the translational symmetry is preserved in some

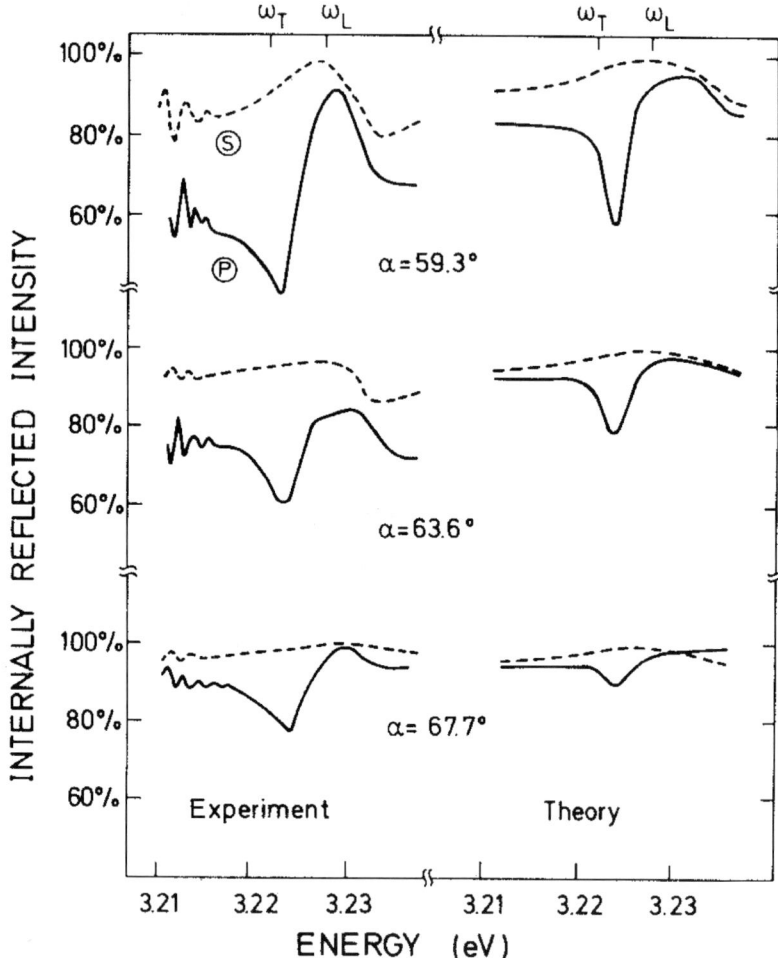

Fig. 13. A.T.R. due to surface polariton excitation in the $Z_3$ exciton of CuCl (from Ref. [30]). The dashed line refers to reflection with light polarized perpendicular to the plane of incidence and the continuous line to parallel polarization. The surface polariton dip is clearly visible.

directions. We show in Figure 14 the scheme of how this can be obtained. The simplest of such nanostructures is the Quantum Well (Q.W.), consisting of a layer which contains a number of planes of a material, confined by a different material chosen to have a larger energy gap. As the surface, it has a two-dimensional translational symmetry, but is limited by potential barriers in both directions, which produces discrete localized states, with a two-dimensional subband associated to

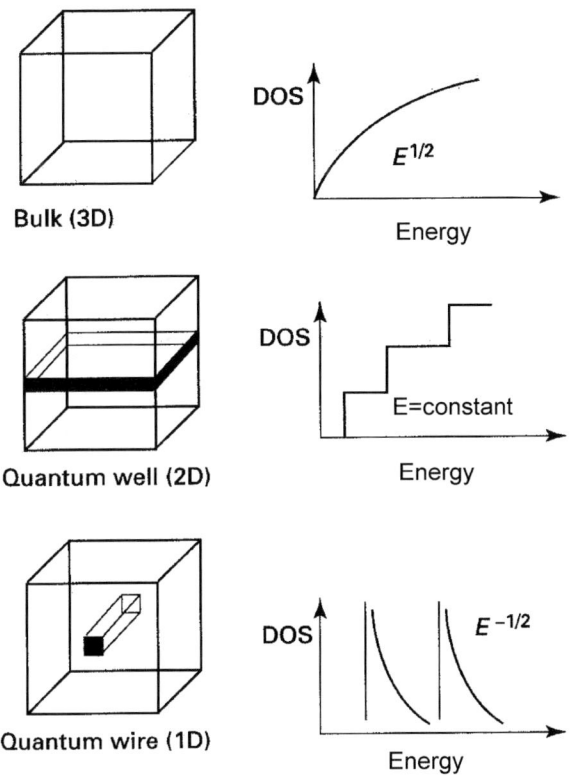

Fig. 14. Schematic view of bulk material, Q.W. with two-dimensional translational symmetry, Q.W.W. with one-dimensional symmetry. The different densities of states are also indicated.

each of them. The electron–hole interaction produces excitons associated to each couple of subbands [31].

The typical material for inorganic Q.W. is GaAs in $Ga_{1-x}Al_xAs$, because GaAs and AlAs have different energy gaps (1.42 eV and 2.4 eV, respectively, at room temperature), and they have the same lattice parameter so that internal strains are not produced. The aluminium concentration fixes the value of the gap in the mixed crystal and hence the value of the potential barriers. Other types of inorganic Q.W. have been obtained, including the nitrates $GaN/Ga_{1-x}Al_xN$, which have excitons in the ultraviolet spectral region. Organic Quantum Wells have also been produced [32]. Typical are the T.C.N.Q., but all organic materials can be grown to form nanostructures, with the advantage that lattice matching is not required because the molecular binding in this case is long range, being mostly due to van der Waals interaction.

Table I. Symmetry classification of excitons in Q.W. (groups $D_{2d}$). We consider only the lowest exciton states originating from the lowest H.H., L.H., or from the split off (S.O.) valence subbands. Other states can be found in Ref. [33]

| VB  | CB  | $\Gamma_V$ | $\Gamma_c$ | Exciton | One photon | g factor |
|-----|-----|------------|------------|---------|------------|----------|
| HH1 | CB1 | $\Gamma_6$ | $\Gamma_6$ | $\Gamma_1$ $\Gamma_2$ $^2\Gamma_5$ | $x, y$ | 1 |
| LH1 | CB1 | $\Gamma_7$ | $\Gamma_6$ | $\Gamma_3$ $\Gamma_4$ $^2\Gamma_5$ | $z$ $x, y$ | 4/3 1/3 |
| SO1 | CB1 | $\Gamma_7$ | $\Gamma_6$ | $\Gamma_3$ $\Gamma_4$ $^2\Gamma_5$ | $z$ $x, y$ | 2/3 2/3 |

The excitons in Q.W. have a larger binding energy than in the bulk because the electron–hole attraction has a stronger effect (the purely two-dimensional binding energy of the lowest exciton is 4 Ry*) [31]. However, the excitation energy is larger than in the bulk because the confinement lowers the valence subbands and increases the energy of the conduction subbands. The reduced symmetry splits the top valence band ($\Gamma_8^{(4)}$ in zincblend material), into two bands $\Gamma_6$ and $\Gamma_7$, called for convenience Heavy Hole and Light Hole bands. Consequently the excitons are subdivided into Heavy Hole (H.H.) and Light Hole (L.H.) excitons, and so are the polaritons.

The calculations of the exciton energies and wave functions with their $\vec{k}_\parallel$-dependence, as functions of the materials and of the Q.W. thickness have been carried out in great detail and are described in the chapter by G.L. La Rocca of this book. The reduced symmetry produces an internal structure of the exciton, depending on the direction of the dipole moment with respect to the growth axis. One has a longitudinal exciton (L.) and a transverse exciton (T.) with dipoles in the plane, and a Z exciton with dipole moment oriented in the growth axis. The exciton symmetries are reported in Table I, where it can be observed that the T and Z excitons can be excited in the case of the L.H. lowest exciton, while only the T excitons are excited in the case of the H.H. lowest exciton [33].

We wish here to show how the mixing of the electromagnetic waves produces also in this case exciton polaritons, as we have seen for surface polaritons. However, in this case, as first pointed out by V. Agranovich and O. Dubovsky for the purely two-dimensional limiting case [34], all possible values of $\vec{k}_\parallel$, including zero, are allowed, and for all values with $|\vec{k}_\parallel| < \frac{\omega}{n} c = k_0$, the polaritons are degenerate with the photons in the barrier and can decay into them. Only the polaritons with $|\vec{k}_\parallel| > k_0$ behave as surface-like waves outside the Q.W., due to the conservation of the parallel component of $\vec{k}$ on the border, which gives outside the well

$$|\vec{k}| = k_0 = \sqrt{k_\parallel^2 + k_z^2}, \tag{24}$$

i.e., evanescent waves in the $z$ direction for $k_\parallel > k_0$. This explains the fact that excitons in Q.W. are excited by the impinging radiation, and photoluminescence is observed in Q.W. polaritons, while it is absent for surface polaritons. In the above mentioned paper [34], also the 1D case (quantum wire) has been considered. It was shown that in 1D the lifetime $\tau$ is of the order of $(2\pi a/\lambda)\tau_0$, where $\tau_0$ is the molecular lifetime and $a$ the lattice constant (for Wannier–Mott excitons it would correspond to the Bohr radius), while in 2D, $\tau \approx (2\pi a/\lambda)^2 \tau_0$. In 1D the effect is weaker, in the 2D case the "superradiant" decay has been observed for the first time in anthracene [35].

## 7.1. Electron–Hole Exchange Effects

We first wish to consider the electron–hole exchange contribution, which is generally neglected in the calculation of Q.W. excitons, but is essential to compute the internal structure of the excitons. Since the electron–hole exchange originates from the Coulomb interaction between electrons, this corresponds to solving Maxwell equation with electron–electron interaction but in the limit of $c \to \infty$, i.e., without retardation.

The $k$-dependent electron–hole exchange term appearing in the effective mass equation is

$$J(\vec{k}) = g(\Gamma)\langle \psi_{ck_c} \psi_{vk_v'} | \frac{e^2}{r_{12}} | \psi_{vk_v} \psi_{ck_c'} \rangle, \tag{25}$$

where $\vec{k} = \vec{k}_c - \vec{k}_v = \vec{k}_c' - \vec{k}_v'$, and $g(\Gamma)$ is a numerical factor which gives the singlet component in every specific symmetry determined state ($g = 4/3$ for $\Gamma_8$ valence band excitons, and $g = 2/3$ for the $\Gamma_7$ split off valence band). Expanding the term $1/r_{12}$ of Eq. (25) we can separate $J(\vec{k})$ into a short range contribution, which is independent of $\vec{k}$, and a long range contribution, which depends on $\vec{k}$. The short range contribution gives the oscillator strength and the $T - L$ splitting, while the long range contribution gives a spatial dispersive behavior which amounts to the total $\vec{k}$ dependence in the case of Frenkel excitons with total mass $M_\parallel \to \infty$. In the case of Q.W. the symmetry is $D_{2d}$ and the values of $g(\Gamma)$ can be obtained for all exciton states (see Table I). A separation occurs between the transverse excitons with dipole on the plane (T excitons) and those with dipole along the growth axis (Z excitons). One can observe in Table I relevant exciton symmetries and selection rules for the point groups $D_{2d}$, appropriate to Q.W. originating from materials with cubic symmetry, when one neglects the lack of inversion of the zincblend symmetry on the well plane.

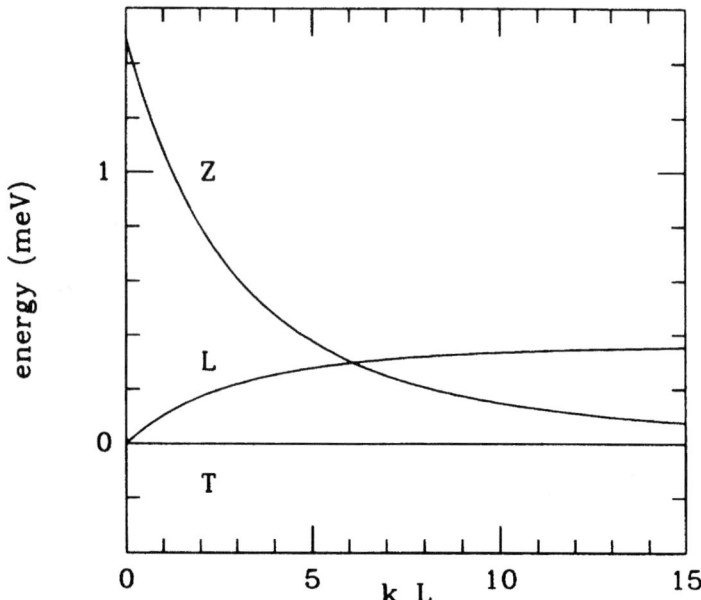

Fig. 15. Internal structure of the LH-CB exciton of a Q.W. of GaAs/AlGaAs (60 Å wide), computed from the calculation of the electron–hole exchange (from Ref. [36]).

All the above considerations are implemented by carrying out the calculations of the exchange contribution (25) in Q.W.s of different thicknesses. In this case one can use the expansions at the band extrema in terms of the bulk Bloch functions $u_c(\vec{r})$ and $u_v(\vec{r})$, with envelope confining functions $c(z)$ and $v(z)$. For the conduction subband we have

$$\psi_c(\vec{k}_\parallel, \vec{r}) = \left(\frac{V}{S}\right)^{1/2} e^{i\vec{k}_\parallel \cdot \vec{\rho}} u_c(0, \vec{r}) c(z), \tag{26}$$

where $V$ denotes the volume and $S$ the surface. A similar expression holds for the valence subband. The resulting contributions from Eq. (25) give a short range exchange and a long range exchange greatly increased with respect to those of the bulk, due to the larger exciton envelope function value at the origin $F_{ex}(0)$ and to the large overlap of the confining functions in the $z$ direction [36]. The short range exchange gives a local field correction, while the long range contribution is obtained by screening the electron–hole interaction with the $\varepsilon_\infty$ dielectric constant [37].

We report in Figure 15 the computed splittings of the T, L, and Z modes in the L.H. exciton of a GaAs/GaAlAs Q.W, as a function of $\vec{k}_\parallel$. The computed value of the splitting at $\vec{k}_\parallel = 0$ between the Z exciton and the T exciton is 1.2 meV,

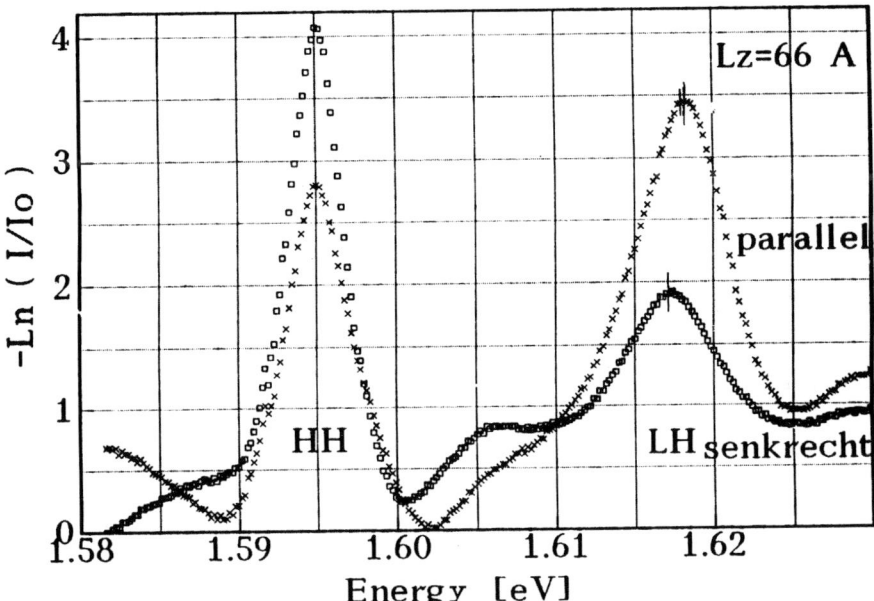

Fig. 16. Experimental absorption on the Q.W. polariton for an angle of incidence of 66°, with polarization parallel and perpendicular to the plane of incidence. The shift of the parallel polarized peak is observed for *LH* excitons as expected (from Ref. [38] and private communication by D. Frölich).

in agreement with the experimental observation obtained by Fröhlich et al. [38], who measured the transparency for perpendicular and for oblique impinging light, polarized in the plane of incidence. Their experimental results are reported in Figure 16, and show for oblique impinging light, with polarization parallel to the plane of incidence, the presence of the Z Light-Hole exciton peak, displaced by about 1.2 meV with respect to the T Light-Hole exciton peak. This occurs only for L.H. exciton polaritons, where the Z polariton is optically active, and not in the H.H. exciton polariton case, where the Z polariton is not coupled to the radiation, as can be seen in Table I.

### 7.2. Retardation Effects and Q.W. Polaritons

To obtain the polaritons one can use the quantum picture of Section 3 or the semiclassical approach of Section 2. We follow the latter scheme and solve Maxwell equations including retardation [36,39]. The susceptibility near any exciton resonance is non-local and can be expressed as follows

$$\chi_{Q.W.}(\vec{k}_\parallel, \omega, z, z') = \chi_{Q.W.}(\omega, \vec{k}_\parallel)\rho(z)\rho(z')$$

$$= \frac{\mu_{cv}^2}{\hbar} |F_{ex}^{Q.W.}(0)|^2 \rho(z)\rho(z')$$

$$\times \left[ \frac{1}{\omega_{ex}(\vec{k}_\parallel) - \omega + i\varepsilon} + \frac{1}{-\omega_{ex}(\vec{k}_\parallel) - \omega + i\varepsilon} \right], \quad (27)$$

where $\rho = c(z)v^*(z)$ is the product of the confinement functions of electron and hole. We observe that $\omega_{ex}(\vec{k}_\parallel)$ does not include the long range exchange described above because its effects appear in the solution of Maxwell equations, while the short range exchange appears in the form of a local field effect, as shown by Cho [37].

The susceptibility (27), substituted into Maxwell equations, originates the following integro-differential equation for the electric field $\vec{E}(\omega, k_\parallel, z)$:

$$\vec{\nabla} \wedge \vec{\nabla} \wedge \vec{E} - \frac{\omega^2}{c^2} \left[ \varepsilon_\infty \vec{E}(\omega, k_\parallel, z) + 4\pi \int_{-\infty}^{+\infty} dz' \chi(\omega, \vec{k}_\parallel, z, z') \vec{E}(\omega, k_\parallel, z') \right]$$
$$= 0. \quad (28)$$

The solutions of the above equation (28) in the inhomogeneous system are obtained by matching at the boundaries the solutions inside the Q.W. with the free electromagnetic waves in the barriers. As mentioned before, two types of solutions can be found: the resonant polaritons when $k_\parallel < k_0 = \frac{\omega}{c}\sqrt{\varepsilon_\infty}$ and the surface-like polaritons when $k_\parallel > k_0 = \frac{\omega}{c}\sqrt{\varepsilon_\infty}$.

To find the resonances, we look for the general solutions for $k_\parallel < k_0$, considering the scattering of waves by the Q.W. and expressing the scattering coefficients in the Breit–Wigner form

$$\alpha = \frac{\omega - \bar{\omega} - i\Gamma}{\omega - \bar{\omega} + i\Gamma} \exp(ik_z L), \quad (29)$$

where the condition $\omega = \bar{\omega}(\vec{k}_\parallel, \omega)$ gives the values of the resonances of the optical modes and the corresponding $\Gamma(\vec{k}_\parallel, \omega)$ gives the linewidths. We obtain for the three optically active modes [36,39]:

$$\bar{\omega}_T(\vec{k}_\parallel, \omega) = \omega(k_\parallel) - \frac{4\pi \mu_{cv}^2 |F_{ex}(0)|^2}{\varepsilon_\infty \hbar} k_0^2 P(k_\parallel), \quad (30)$$

$$\Gamma_T(\vec{k}_\parallel, \omega) = \frac{2\pi \mu_{cv}^2 |F_{ex}(0)|^2 Q^2(k_z)}{\varepsilon_\infty \hbar} \frac{k_0^2}{k_z}, \quad (31)$$

$$\bar{\omega}_L(\vec{k}_\parallel, \omega) = \omega(\vec{k}_\parallel) - \frac{4\pi \mu_{cv}^2 |F_{ex}(0)|^2}{\varepsilon_\infty \hbar} k_z^2 P(k_z), \quad (32)$$

$$\Gamma_L(\vec{k}_\parallel, \omega) = \frac{2\pi \mu_{cv}^2 |F_{ex}(0)|^2 Q^2(k_z) k_z}{\varepsilon_\infty \hbar}, \quad (33)$$

$$\bar{\omega}_Z(\vec{k}_\|,\omega) = \omega(\vec{k}_\|) + \frac{4\pi\mu_{cv}^2|F_{ex}(0)|^2}{\varepsilon_\infty \hbar}\left(\int dz\, \rho^2(z) - k_\|^2 P(k_z)\right) \quad (34)$$

$$\Gamma_Z(\vec{k}_\|,\omega) = \frac{2\pi\mu_{cv}^2|F_{ex}(0)|^2 Q^2(k_z)}{\varepsilon_\infty \hbar}\frac{k_\|^2}{k_z}. \quad (35)$$

In the above expressions

$$Q(k_z) = \int_{-L/2}^{+L/2} \rho(z)\cos(k_z z)\, dz, \quad (36)$$

and

$$P(k_z) = -\int_{-L/2}^{+L/2} dz \int_{-L/2}^{+L/2} dz' \frac{1}{2k_z}\sin(k_z|z-z'|)\rho(z)\rho(z'). \quad (37)$$

For $\vec{k}_\| > k_0$ surface-like modes are found by matching at the boundaries the evanescent waves of the barriers with the solutions inside the Q.W. The surface-like modes are stationary, with zero radiative linewidths, and propagate only in the quantum well plane. They are given by the same expressions (30), (32), and (34) of the previous case, with $\Gamma = 0$, provided the quantity $P(k_z)$ is replaced by $\bar{P}(k_z)$, obtained by using the imaginary wave vector $ik_z$, instead of $k_z$, i.e.,

$$\bar{P}(k_z) = \int_{-L/2}^{+L/2} dz \int_{-L/2}^{+L/2} dz' \frac{1}{2k_z}\exp[-k_z(z-z')]\rho(z)\rho(z'). \quad (38)$$

The modes described above are the polaritons in the quantum wells, totally analogous to bulk polaritons and surface polaritons, except for their radiative linewidths for $\vec{k}_\| < k_0$. For the purpose of illustration we show in Figure 17 the dispersion of polaritons, computed for a typical quantum well GaAs/GaAlAs. In Figure 18 we show the computed dispersive behavior and the lifetime broadening of the $Z_3$ exciton polariton in a Q.W. CuCl/Ca$_2$F for small values of the in plane $k$ vector, and in Figure 19 the computed dispersive behavior as $k_\|$ greatly increases.

A more extended discussion of Q.W. polaritons can be found in a recent review article by L.C. Andreani [40].

### 7.3. Q.W. Polariton Lifetimes

While in bulk material polaritons would have an infinite radiative lifetime in an unlimited crystal, in confined structures we have seen that polaritons have a radiative broadening when $\vec{k}_\| < k_0$. When light impinges on a Q.W. only resonant modes are produced because this condition holds (the radiation is close to normal incidence because of the large refractive index of the barrier material). Photoluminescence therefore results, with the decay lifetime $1/\Gamma$ given above in Eqs. (31), (33), (35). As pointed out by Agranovich and Dubovsky for the two-dimensional

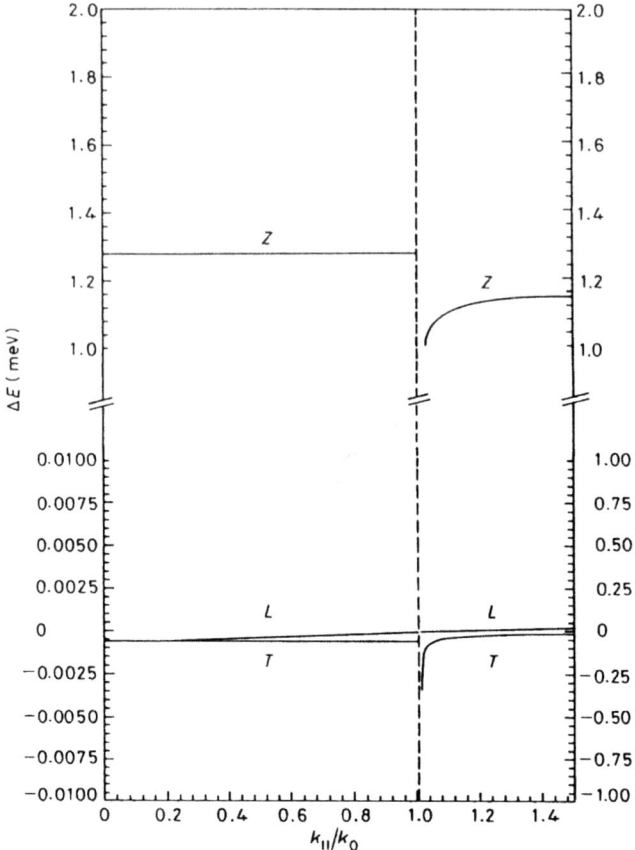

Fig. 17. Polariton modes for a Q.W. of GaAs/AlGaAs (from Ref. [39]).

limit [34] such a lifetime at $k_\parallel = 0$ is orders of magnitude smaller than in bulk crystals.

An equivalent alternative way of considering lifetimes is to compute the decay rates of the quantum well polaritons into the photons of the barrier with the same $\vec{k}_\parallel$ and all possible values of $k_z$, characterized by the one-dimensional density of states

$$\rho(\vec{k}_\parallel, \omega) = \frac{V}{\pi S}\left(\frac{n}{\hbar c}\right)^2 \frac{\hbar \omega}{\sqrt{k_0^2 - k_\parallel^2}} \theta(k_0 - k_\parallel), \tag{39}$$

where $\theta(x)$ is the step function (0 for $x < 0$, 1 for $x > 0$), and $n$ the refractive index. From Fermi's golden rule one obtains the decay rates. They can be expressed for any polarization direction $\vec{\varepsilon}$ in terms of the exciton oscillator strengths per unit

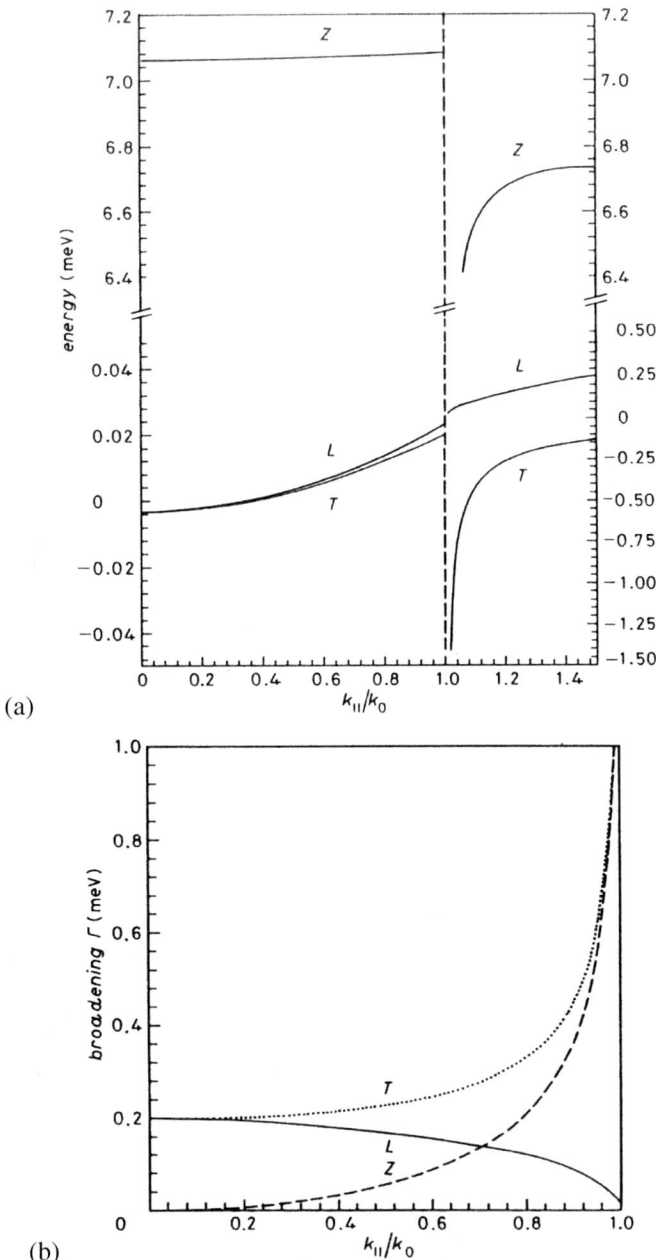

Fig. 18. (a) Dispersive behavior of polaritons in the lowest exciton resonance ($Z_3$) of CuCl/Ca$_2$F Q.W. (from Ref. [39]). The lower curves (b) give the computed radiative linewidths for $k_\| < k_0$.

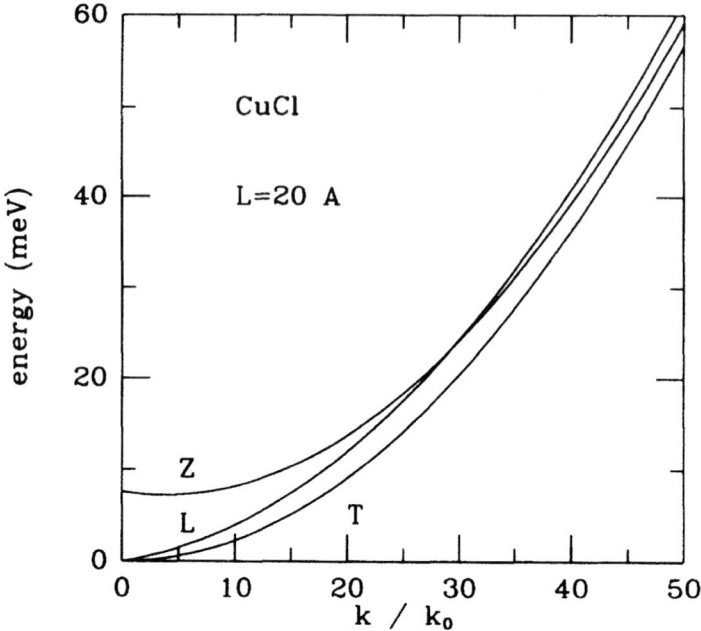

Fig. 19. Dispersive behavior of CuCl/Ca$_2$F Q.W. polaritons for large values of the in-plane wave vector $k$ (from Ref. [39]).

area

$$f_\varepsilon = \frac{1}{S} \frac{2}{m_0 \hbar \omega_0} |\langle \psi_0 | \vec{\varepsilon} \cdot \vec{p} | \psi_{ex} \rangle|^2, \quad (40)$$

and for $k_\parallel = 0$, one obtains [41]

$$\Gamma_0 = \left( \frac{2\pi}{n} \right) \frac{e^2}{m_0 c} f_\varepsilon. \quad (41)$$

For the H.H. polaritons the L and T modes have the same decay rates and the Z-mode does not decay because $f_z = 0$, as shown in Table I. For L.H. polaritons $f_z = 4 f_{xy}$ and also the Z-mode decays. An estimate has been obtained for Q.W. of the type GaAs/AlGaAs of about 100 Å width. From a computed oscillator strength $f_{xy} \simeq 5 \cdot 10^{-4}$ Å$^{-2}$ a radiative lifetime $\tau_0$ of about 12 picoseconds has been obtained [15,41].

The experimental data usually give a larger lifetime, with a linear temperature dependence at low temperature [42]. This is due to the fast thermalization of the polariton modes produced by inelastic scattering with the acoustic phonons, with only a fraction of the polaritons occupying the sates with $k < k_0$, which can decay

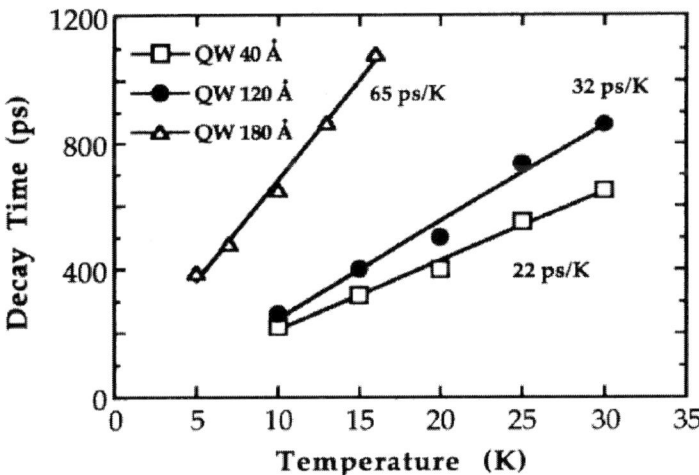

Fig. 20. P.L. decay times as functions of temperature for GaAs/AlGaAs L.H. polaritons in Q.W. of various thicknesses (from Ref. [42]). Observe the linear dependence and the increase with Q.W. thickness.

radiatively with lifetime $\tau_0$. Such a thermalization gives for the lifetime of the H.H. exciton polariton [41]

$$\tau(T) = \frac{3Mk_BT}{\hbar^2 k_0^2}\tau_0, \qquad (42)$$

in fair agreement with experimental results [42], as shown in Figure 20.

Experimentally, one can also distinguish between the "fast" lifetimes for $k < k_0$ [43] at very low temperatures, and the "slow" lifetimes, linear with $T$ for thermalized polaritons, according to Eq. (42) [42].

Another evidence of the radiative lifetime is obtained from its influence on the normal-incidence reflectivity, which has been computed with the procedure suggested by K. Cho [44], using linear response theory with the appropriate nonlocal susceptibility. For a Q.W. with a barrier of infinite length on one side and a barrier of thickness $D$ on the side where one measures the reflectivity, one finds [41]:

$$R(\omega) = \left(\frac{n-1}{n+1}\right)^2 \\ - \frac{8n(n-1)}{(n+1)^2} \cdot \frac{(\gamma+\Gamma_0)\cos k_0\ell + (\omega_0-\omega)\sin k_0\ell}{(\omega_0-\omega)^2 + (\gamma+\Gamma_0)^2}\Gamma_0, \qquad (43)$$

where $\ell = L + 2D$, $\omega_0$ is the resonance frequency and $\gamma$ is the nonradiative width. We observe that the correction to the background reflectivity strongly depends on the radiative decay width $\Gamma_0$.

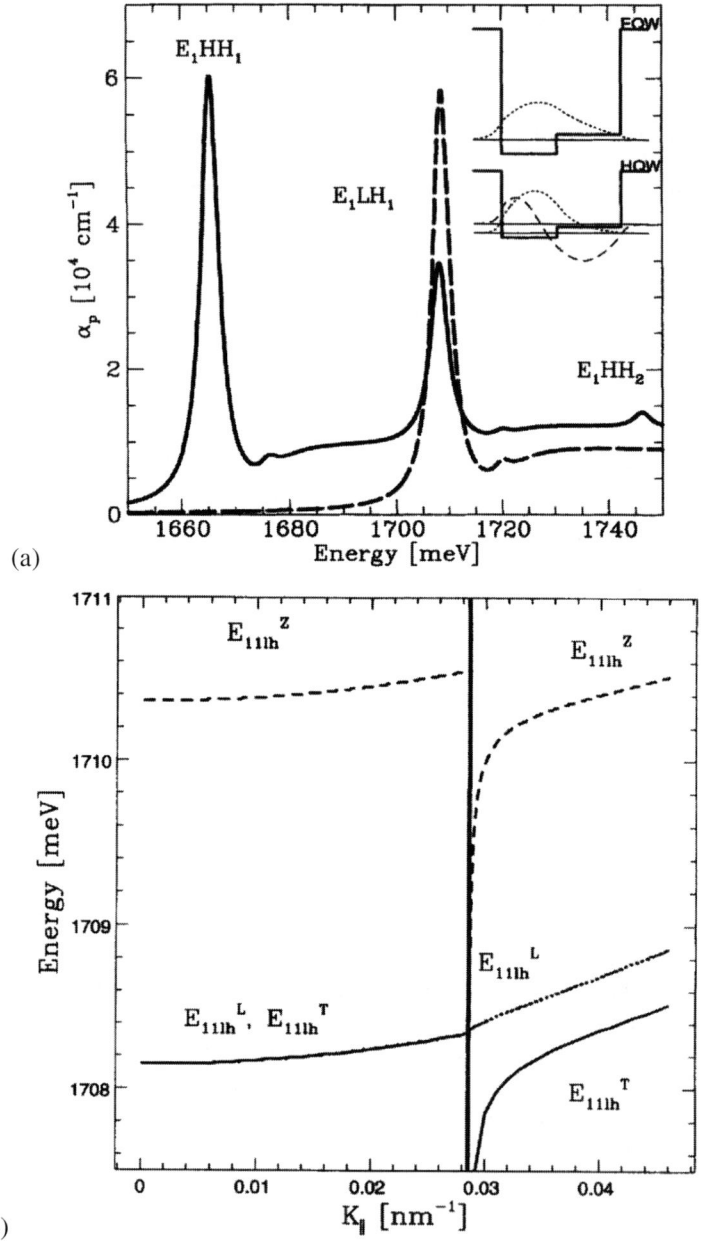

Fig. 21. (a) Absorption of an A.Q.W. at normal incidence (continuous line) and polarized along the $z$ axis (dashed line). In the inset the electron and hole confining potentials. (b) Dispersion curves of the L.H. polaritons for the A.Q.W. of (a) (from Ref. [46]).

## 7.4. Asymmetric and Double Quantum Wells

Polariton states can be computed also in more complicated two-dimensional structures, such as Asymmetric Quantum Wells (A.Q.W.) or Double Quantum Wells (D.Q.W.), with a confining potential of different depth in different parts of the well or two coupled wells of different thicknesses respectively. In this case it is convenient to use a simplified procedure suggested by V.M. Agranovich [45], based on the fact that the wavelength of the radiation is large compared to the width of the well, and consequently one can use for all fields inside the Q.W. the expansion

$$F(z) = F(0) + \left(\frac{\partial F}{\partial z}\right)_0 z, \tag{44}$$

and

$$F(\vec{r}, t) = F(z) e^{i \vec{k}_\parallel \cdot \vec{\rho}} e^{-i\omega t}. \tag{45}$$

The phase conservation imposes a constant $k_\parallel$ at all the boundaries, and all other boundary conditions give eigenvalue equations for the parallel and perpendicular components of the fields in terms of the perpendicular and parallel dielectric functions $\varepsilon_\perp$ and $\varepsilon_\parallel$, and of the background dielectric function $\varepsilon_b$ [46]. The reflected and transmitted fields can be found in terms of the incident ones, and the reflection coefficients $R_S$ and $R_P$ for polarization perpendicular and parallel to the plane of incidence can be computed when the material parameters are known. The complex poles of $R_S$ and $R_P$ give the values and the dispersions of the polariton states and their radiative linewidths. Also in this case the results for $k_\parallel > k_0$ are obtained from those at $k_\parallel < k_0$ by replacing $k_z$ with $ik_z$, the linewidths being equal to zero [46,47]. We give in Figure 21 the computed absorption spectra of an A.Q.W. of GaAs/GaAlAs and the computed dispersion curves of the polariton states.

## 7.5. Detection of Q.W. and A.Q.W. Polaritons

While the resonant Q.W. polaritons ($k_\parallel < k_0$) can be observed from the peaks in the absorption and in the reflectivity, as shown by Fröhlich et al. [23] and by Berz et al. [47], the surface-like Q.W. polaritons can be excited only by producing high values of $k_\parallel$ with surface gratings or with total reflection as shown in Figure 12. As in the case of surface polaritons, one expects dips corresponding to attenuated total reflections, an example of which is shown in Figure 22, based on a detailed calculation on a CuCl Q.W. polariton [48].

# 8. Quantum Wire Polaritons

The exciton polariton concept can be extended to the Quantum Well Wires (Q.W.W.), where the translational symmetry is preserved only in one direction.

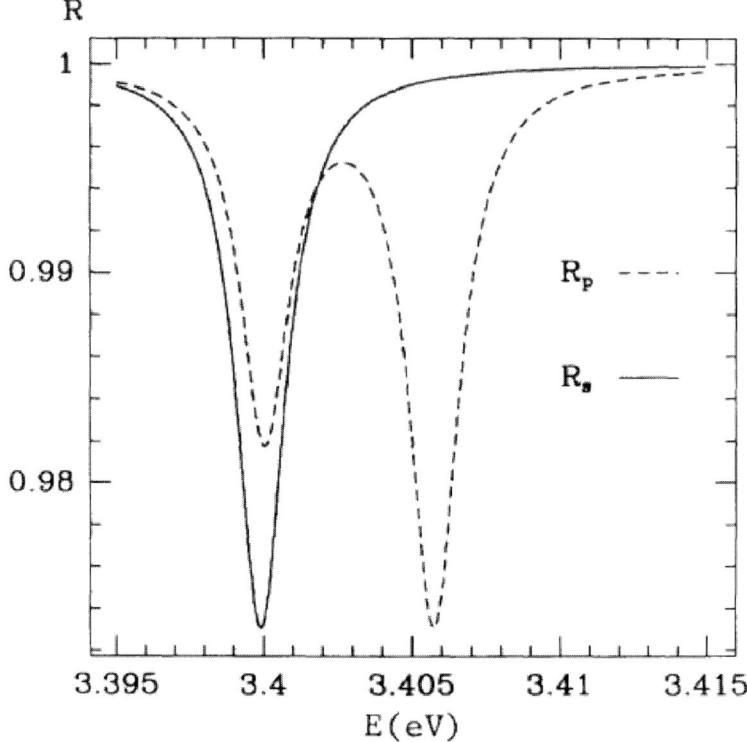

Fig. 22. Attenuated total reflectivity of a Q.W. of CuCl at nearly grazing incidence, with light polarized in the plane of incidence (– – – $R_p$), and perpendicular to it (— $R_s$). One can observe the appearence of the L.H. polariton in the former case. From Ref. [48].

This of course is also true for those organic crystals (like polyacetylene) that are linear molecular chains, so separated one from another that one can neglect the interaction among them. In this case the $\vec{k}$ vector is also one-dimensional, and one obtains one-dimensional polariton modes.

As in the Q.W. case, we can find the polariton modes by solving Maxwell equations with the appropriate boundary conditions and with the appropriate nonlocal susceptibility. One starts with the exciton energies and wave functions, which are computed by introducing a confining potential in two directions and the Coulombic electron–hole attraction in three dimensions (or an effective attraction in one dimension, since the Coulombic attraction in one dimension would give a divergent binding energy) [49]. Also in this case the cylindrical symmetry separates the degeneracy of the valence band, so that one can use a two-band model, and a trial envelope function of the form

$$\psi^{\text{exc.}}_{k,M_e,M_h}(\vec{\rho}_e, \vec{\rho}_h, x_e, x_h)$$
$$= \sum_{m_e,n_e} f_{n_e,M_e k/2}(\rho_e, x_e) f_{n_h,M_h,k/2}(\rho_h, x_h) F^{\text{ex}}_{n_e,n_h,M_e,M_h}(x_e - x_h), \quad (46)$$

where $f_n$ denote the wire confinement functions and $M$ the angular momenta (total momentum $M = M_e + M_h$). The trial wave function $F^{\text{ex}}(x_e - x_h)$ requires a rather complicated calculation [50], but in the approximation of an infinitely large confining potential the confinement functions in (46) take the simple form

$$f_{n,M,k}(\rho, x) = N e^{iM\Phi} e^{ikx} J_M(\kappa_n \rho) \theta(R - \rho), \quad (47)$$

where $N$ is the normalization, $R$ is the wire radius and $\kappa_n = \rho_{0n}/R$, $\rho_{0n}$ being the $n$th zero of the Bessel function. The oscillator strength of the optical transition (per unit length) is

$$\frac{f_0}{L} = \frac{2m\omega_0}{\hbar e^2} \mu_{vc}^2 \left| \int d\rho_e \, dx_e \, \psi_{\text{ex}M}(\rho_e = \rho_h, x_e = x_h) \right|^2, \quad (48)$$

so that only $M = 0$ excitons are allowed in the dipole approximation because the others have null exciton wave function at the origin of the relative coordinates. Other selection rules can be obtained from symmetry, as in the case of quantum wells. An enhancement of the binding energy and of the oscillator strength with respect to the case of quantum wells of comparable widths arises because confinement is increased and electrons and holes are forced to stay closer.

The nonlocal susceptibility of the Q.W.W. can be expressed in a form analogous to (27) as

$$\chi(\omega, k, M, \rho, \rho') = \frac{1}{\hbar} \sum_{c,v} \mu_{cv}^2 \left[ \frac{1}{\omega_{\text{ex},n} - \omega + i\varepsilon} + \frac{1}{-\omega_{\text{ex},n} - \omega + i\varepsilon} \right]$$
$$\times f_{\text{ex}}(\rho) f_{\text{ex}}(\rho') |F_{\text{ex}}(0)|^2, \quad (49)$$

where $f_{\text{ex}}(\rho)$ is the product of the confinement functions of electron and hole $f_{\text{ex}}(\rho_e) f_{\text{ex}}(\rho_h)$ in the wire, and $F_{\text{ex}}(0)$ is the relative motion envelop wave function at the origin. This can be inserted into Maxwell equations, as in the case of quantum wells, and an analogous integro-differential equation of the Fredholm type is obtained, whose solutions inside the wire can be matched at the border with the travelling waves in the barrier, using for the ratio of outgoing to incoming waves the scattering coefficients of the Breit–Wigner form (29). As shown by Tassone and Bassani [50], one can take advantage of the cylindrical symmetry by expressing the electromagnetic fields in terms of the angular momentum $M$ and the wave vector in the wire direction $k$ as

$$\vec{E}(k, x) = \vec{E}_M(k, \rho) e^{iM\phi} e^{ikx}, \quad (50)$$

and solving the coupled integral equations for the components inside the wire. Only electromagnetic waves of angular momentum 0 or ±1 are optically active, i.e., have nonvanishing polarization.

For $M = 0$ the excitonic polarization lies along the wire axis and the corresponding mode is longitudinal (L), while for $M = \pm 1$ the excitonic polarization lies in the orthogonal plane and the two degenerate modes are transverse (W modes). Also in this case one obtains for $M = 0$ resonant modes for $k < k_0 = \frac{\omega}{c}\sqrt{\varepsilon_b}$, and surface-like modes for $k > k_0 = \frac{\omega}{c}\sqrt{\varepsilon_b}$. We give the expressions for the longitudinal mode for the case $k < k_0$:

$$\bar{\omega}_L(k,\omega) = \omega_{\text{ex}}(k) - 4\pi \frac{\mu_{cv}^2|F_{\text{ex}}(0)|^2}{\varepsilon_\infty \hbar} k^2 P, \tag{51}$$

and its radiative broadening

$$\Gamma_L = \frac{2\pi^2 \mu_{cv}^2 |F_{\text{ex}}(0)|^2 Q^2 k^2}{\varepsilon_\infty \hbar}, \tag{52}$$

where

$$P = \iint d\rho\, d\rho'\, \rho\rho'\, G(\rho,\rho') f_{\text{ex}}(\rho) f_{\text{ex}}(\rho'), \tag{53}$$

and

$$Q = \int \rho\, d\rho\, J_0(k\rho) f_{\text{ex}}(\rho), \tag{54}$$

the Green function being expressed in terms of the regular Bessel function $J_0$ and of the singular Bessel function $Y_0$ as:

$$G(\rho,\rho') = \begin{cases} -\pi/2 J_0(k\rho_<), & \text{for } \rho_< = \min(\rho,\rho'); \\ -\pi/2 Y_0(k\rho_>), & \text{for } \rho_> = \max(\rho,\rho'). \end{cases} \tag{55}$$

In the case of $k > k_0$ the expressions (51) are the same, but the Bessel functions $J_0$ and $Y_0$ in the Green's function (55) are replaced by the modified Bessel functions and $\Gamma_L = 0$ because asymptotically decaying fields in the barrier must be matched to the solutions inside the wires.

For the $M = \pm 1$ case the same approach gives [50]:

$$\bar{\omega}_W(k,\omega) = \omega_{\text{ex}}(k) - \frac{2\pi \mu_{cv}^2 |F_{\text{ex}}(0)|^2 k_0^2 - P_1}{\varepsilon_\infty \hbar}, \tag{56}$$

and

$$\Gamma_W(k_\parallel,\omega) = \frac{\pi^2 \mu_{cv}^2 |F_{\text{ex}}(0)|^2 Q^2 (k_0^2 + k^2)}{\varepsilon_\infty \hbar}, \tag{57}$$

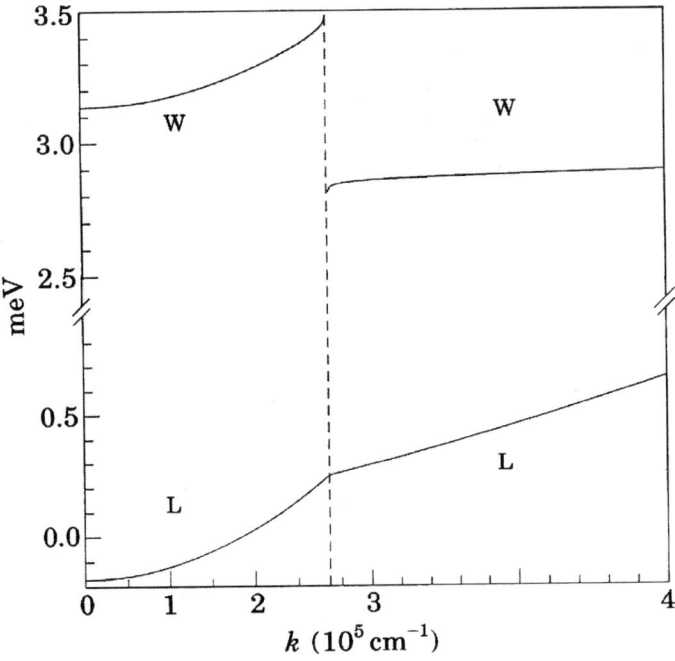

Fig. 23. Polariton dispersion curves in a cylindrical Q.W.W. of 40 Å in GaAs/GaAlAs. The broken line indicates the photon dispersion $\omega = vk$. The difference between the longitudinal and the wire mode is due to the radiative correction with retardation (from Ref. [50]).

where

$$P_1 = \iint \rho\rho' \, d\rho \, d\rho' \left( G_\Phi^{(0)} - i G_\Phi^{(0)} \right) f_{\text{ex}}(\rho) f_{\text{ex}}(\rho')$$
$$- \frac{1}{k_0^2} \int \rho \, d\rho \, f_{\text{ex}}^2(\rho), \tag{58}$$

where the last term produces a resonance shift. The broadening in this case never vanishes and the mode can decay for all values of $k$.

The dispersion character of $W$ and $L$ polariton modes of a typical Q.W.W. is exemplified in Figure 23.

### 8.1. Q.W.W. Polariton Lifetimes

The radiative lifetimes of quantum wire polaritons can be inferred from the broadenings (52) and (57). They can also be computed, as in the case of quantum wells, from the transition probability rate of one-dimensional excitons to the two-dimensional photons of the barrier with the same $k$ and with all possible $\vec{k}_\rho$ in the plane perpendicular to the wire direction. Considering the higher value of the oscillator strength in quantum wire excitons compared to quantum well excitons one would expect a shorter radiative lifetime, but the photon density of states is also different, so that the lifetime is usually larger in quantum wires. Also, the temperature-dependence due to thermalization is in this case proportional to $\sqrt{T}$ [51].

An additional effect to be considered in the study of lifetimes is the loss of coherence, when scattering by phonons and disorder reduces the coherence length $\ell_c$, with a spread in wavevectors $\delta k \simeq \ell_c^{-1}$ and an additional homogeneous broadening [51]. A full discussion of radiative lifetimes can be found in the review articles by L.C. Andreani [15,40].

## 9. Exciton-Polaritons in Microcavities

In recent years a new type of heterostructure has been considered, which appears very promising for enhancing optical effects through increased coupling between excitons and photons. It is the microcavity, typically constituted by a material of length comparable to the wavelength of light in the medium, with totally reflecting barriers in specific directions [16]. The scheme is shown in Figure 24, where the optical barriers on two sides of the cavity are indicated as Distributed Bragg Reflectors (D.B.R.), multiple repeats of alternating layers of high and low index of refraction, each of thickness $\lambda/4$, where $\lambda = \lambda_{\text{vacuum}}/n$ is the chosen wavelength

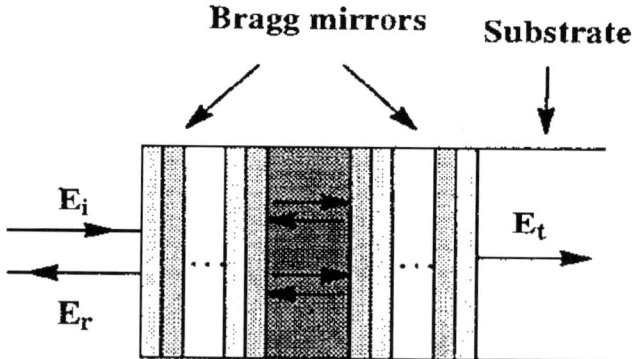

Fig. 24. Scheme of a planar microcavity, where the light is confined by Distributed Bragg Reflectors (D.B.R.).

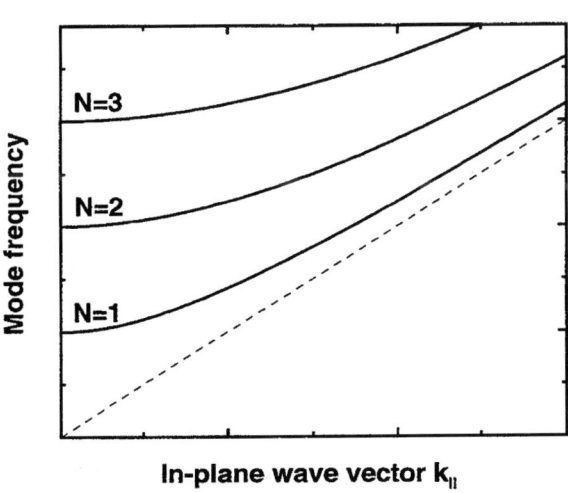

Fig. 25. Schematic behavior of the resonant modes in a planar cavity (from Ref. [52]).

of the layer material. Such D.B.R. are preferred to metal coated surfaces mainly because surface quality allows coherence to be better preserved in the reflections. A comparison of the properties of cavities with metallic mirrors and with D.B.R. is given by V. Savona [52].

The cavity resonant modes, in the ideal situation of perfectly reflecting mirrors, are obtained from the Fabry–Perot condition

$$k_z L_c = L_c \sqrt{\frac{\omega^2 \varepsilon_c}{c^2} - k_\parallel^2} = n\pi, \qquad (59)$$

where $\varepsilon_c$ is the constant dielectric function of the medium and $k_\parallel$ is the in-plane wave vector. A schematic behavior of the lowest Fabry–Perot cavity modes is shown in Figure 25. We can observe that such modes, which are free propagating waves in the plane, can exist only for $\omega^2 > \frac{c^2 k_\parallel^2}{\varepsilon_c}$, otherwise $k_z$ is imaginary and in the $z$ direction we have evanescent waves.

In realistic situations the reflectance of the D.B.R. and the cavity resonant modes are not the ideal ones; they must be found from the solutions of Maxwell equations with the appropriate boundary conditions. In our case we can express the space-dependence of monochromatic fields as

$$\vec{E}_{k_\parallel}(\vec{\rho}, z) = \vec{e}_{k_\parallel} U_{k_\parallel, \omega}(z) e^{i\vec{k}_\parallel \cdot \vec{\rho}}, \qquad (60)$$

which gives for the mode function $U_{k_\parallel, \omega}(z)$

$$\frac{d^2 U_{k_\parallel,\omega}(z)}{dz^2} + \left(\frac{\omega^2}{c^2}\varepsilon(z) - k_\parallel^2\right) U_{k_\parallel,\omega}(z) = 0. \tag{61}$$

In each layer with constant $\varepsilon$ the above equation gives two waves propagating in opposite directions, with $k_z = (\frac{\omega^2}{c^2}\varepsilon - k_\parallel^2)^{1/2}$,

$$U_{k_\parallel,\omega}(z) = E_\ell(k_\parallel) e^{-ik_z z} + E_r(k_\parallel) e^{ik_z z}. \tag{62}$$

Their complex amplitudes have to be determined by imposing the boundary conditions at each interface. This task is made very simple, also for complicated systems, by the transfer matrix approach [53], which consists in connecting the two amplitudes of incoming waves to those of outgoing waves with a matrix of order two, whose elements are determined from the Maxwell boundary conditions at each interface, and considering the products of such matrices at each layer. One obtains in this way the reflectivity of the D.B.R. and the values of the electric field in the microcavity (M.C.) spacer and out of it. Typical of the D.B.R. is a stop band of nearly total reflectance centered around the frequency $\omega_s = v k_0$, where $v$ is the light velocity in the medium and $k_0$ the vacuum wave vector of light. At normal incidence the reflectance $R = |r|^2$ is about constant in the stop band; and for $N$ layer pairs is given by [52]

$$R \simeq 1 - \frac{4 n_{\text{ext}}}{n_{\text{cav}}} \left(\frac{n_2}{n_1}\right)^{2N} \quad (n_2 < n_1), \tag{63}$$

with the phase given by

$$\Phi = \frac{n_1 L_{DBR}}{c}(\omega - \omega_S), \tag{64}$$

where $\omega_S$ is the frequency at the center of the stop band, and

$$L_{DBR} = \frac{\lambda}{2} \frac{n_1 n_2}{n_{\text{cav}}(n_1 - n_2)}, \tag{65}$$

is an effective penetration thickness ($\frac{\lambda}{4}$ = layer thickness). For the purpose of exemplification we show in Figure 26 the computed reflectivity of a D.B.R. of 20 layer pairs.

The penetration length (65) and the subsequent phase (64) modify the cavity mode frequency $\omega_m$ with respect to the Fabry–Perot frequency $\omega_c$ of Eq. (59) in the following way [49]:

$$\omega_m = \frac{L_c \omega_c + L_{DBR} \omega_S}{L_c + L_{DBR}}. \tag{66}$$

This means that when the cavity is designed in a way that $\omega_S = \omega_c$, the cavity mode coincides with the Fabry–Perot condition (59) and its frequency is at the center of the stop band. For the purpose of exemplification we show in Figure 27

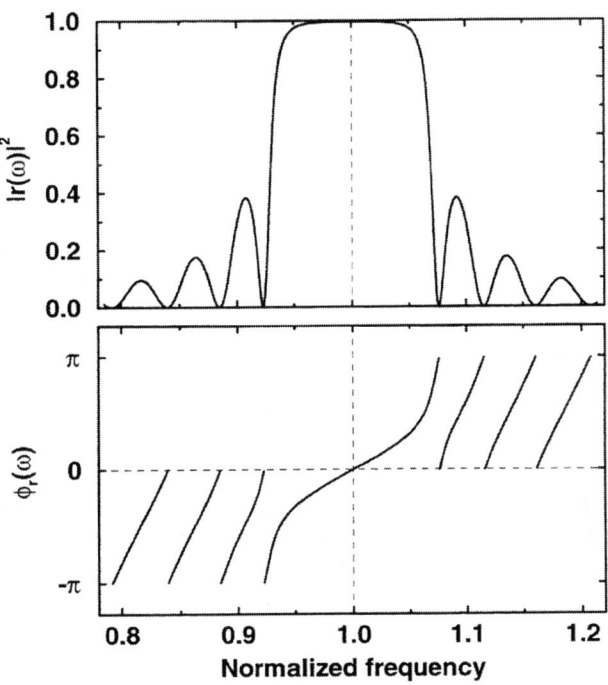

Fig. 26. Reflectivity at normal incidence of a D.B.R. of 20 layer pairs with $n_1 = 3.6$ and $n_2 = 3$. One can observe the oscillating behavior about the stop band (from Ref. [52]).

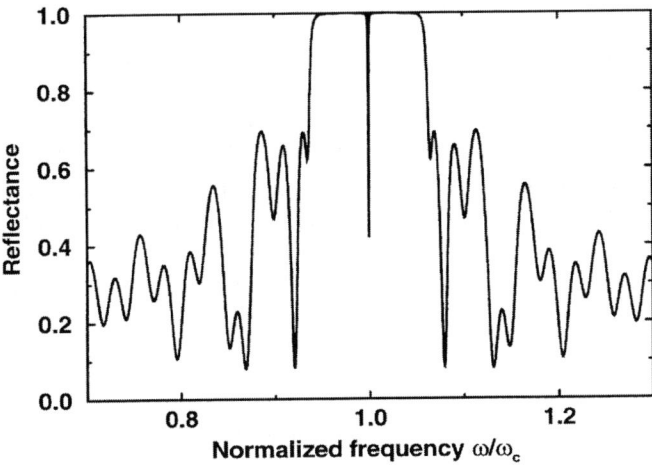

Fig. 27. Reflectivity of a λ M.C. with the cavity resonance mode at the center of the stop band (from Ref. [52]).

the reflectance of a microcavity, with the dip due to the cavity resonance at $\omega = \omega_c$. The width of the cavity mode has also been computed and, for $R$ close to 1, is given by [54]

$$\gamma_c \simeq \frac{(1-R)c}{n_c(L_c + L_{DBR})}. \tag{67}$$

The calculations can be performed for all values of $k_\parallel$, and the linewidth can be made much smaller than for metallic reflectors, which gives a good value for the finess (ratio of the separation between cavity modes and linewidths).

### 9.1. POLARITONS IN Q.W. IMPLANTED M.C.

The material inside the cavity spacer contains excitons and, if the optical mode frequency of the cavity is tuned to an exciton resonance frequency ($n = 1$ exciton in particular) the two states interact. The splitting at resonance, in analogy with the case of atoms in a microwave cavity, is called the Rabi splitting. When the Rabi splitting is large in comparison to the natural linewidths of the cavity mode and of the exciton, the strong coupling regime holds and two separate modes are produced. In the opposite case of weak coupling regime the damping prevails over the light-matter interaction, so that only the radiative decay rates are modified. A detailed experimental and theoretical study of the strong coupling regime can be found in the review article by Skolnick, Fisher and Whittaker [55], and a general analysis of cavity polaritons is in the review article of Savona et al. [16].

The simplest way to consider the coupling between the cavity mode and the excitonic state to obtain the cavity polaritons is to vary the separation between the two frequencies and obtain the two modes as a mixture of cavity mode and exciton. A simple Hamiltonian treatment can be made at $k_\parallel = 0$, with a coupling characterized by an energy which gives the vacuum Rabi splitting $\hbar\Omega$ at resonance. This is related to the oscillator strength, and in the case of a Q.W. embedded in the M.C. is given by [54]

$$\Omega \simeq 2\left(2\pi \frac{e^2 f_{ex}}{\varepsilon_c m_c (L_c + L_{DBR})}\right)^{1/2}. \tag{68}$$

The matrix Hamiltonian is

$$H = \begin{vmatrix} E_e & \frac{\hbar\Omega}{2} \\ \frac{\hbar\Omega}{2} & E_c \end{vmatrix}, \tag{69}$$

and its eigenvalues give the energies of upper and lower cavity polaritons:

$$E_\pm = \frac{E_e + E_c}{2} \pm \frac{1}{2}\sqrt{(E_e - E_c)^2 + (\hbar\Omega)^2}. \tag{70}$$

The Hamiltonian treatment cannot consider the linewidths. To take this into account the master equation for the density matrix, with an irreversible contribution, must be considered, using Eq. (16) of Section 4. Close to resonance one can simplify the procedure by modifying the Hamiltonian (69), with the addition of an imaginary contribution $-i\hbar\gamma$ to $E_e$ and $E_c$ in order to account for the linewidths of exciton and cavity modes, the latter given by (67). At resonance one obtains:

$$\omega^{\pm} = \omega_{\text{ex}} - \frac{i}{2}(\gamma_{\text{ex}} + \gamma_m) \pm \frac{1}{2}\sqrt{\Omega^2 - (\gamma_{\text{ex}} - \gamma_m)^2}, \quad (71)$$

which shows that the condition for strong coupling is $|\Omega| > |\gamma_{\text{ex}} - \gamma_m|$. One can visualize in Figure 28 the dependence of the frequency and linewidth of the polariton microcavity at resonance on the reflectance of the D.B.R. The weak and

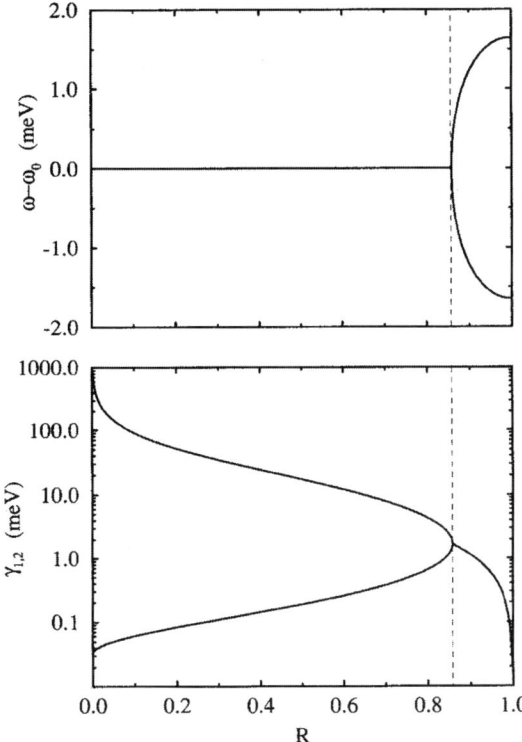

Fig. 28. Microcavity polariton frequencies and linewidths at resonance, in the weak and strong coupling regimes, as functions of the reflectance. The upper and lower polariton modes separate in strong coupling, while the linewidths decrease. In strong coupling at resonance the linewidths of U.P. and L.P. coincide (from Ref. [54]).

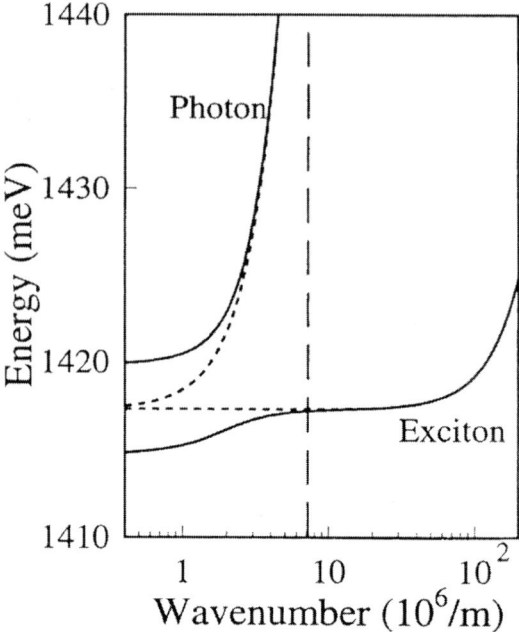

Fig. 29. Calculated polariton dispersion as function of $k_\parallel$. The broken curves show the uncoupled exciton and photon dispersion curves (taken to be degenerate at $k_\parallel = 0$). The broken vertical line separates resonant and surface-like polaritons (from Ref. [55]).

strong coupling regimes are indicated, and the vacuum Rabi splitting corresponds to $R = 1$, in which case the broadening is given by the exciton linewidth.

In a similar way one can compute the cavity polariton dispersion, taking into account the fact that the cavity modes and the exciton modes interact only for the same value of the in-plane wave vector $\vec{k}_\parallel$ [56,57]. Since the dispersions of excitons and photons are very different, the coupling decreases as $k_\parallel$ increases and the upper polariton becomes more photon-like while the lower polariton is more exciton-like, as in the case of bulk polaritons. An example of such Q.W. exciton-cavity polaritons has been computed in the strong coupling regime and is shown in Figure 29 [55,56]. Such calculations have been extended to the case when more Q.W. are present in one cavity or more cavities are coupled [58].

The polariton dispersion described above can be measured in angle tuning experiments, as in the case of surface polaritons, by observing the dips in total reflectance. As an example, we show in Figure 30 the results of experimental data of Baxter et al. [59]. One can notice the strong shift of the cavity mode with $k_\parallel$ and a large Rabi splitting, corresponding to an anticrossing at $k_\parallel \simeq 4 \cdot 10^6$ m$^{-1}$.

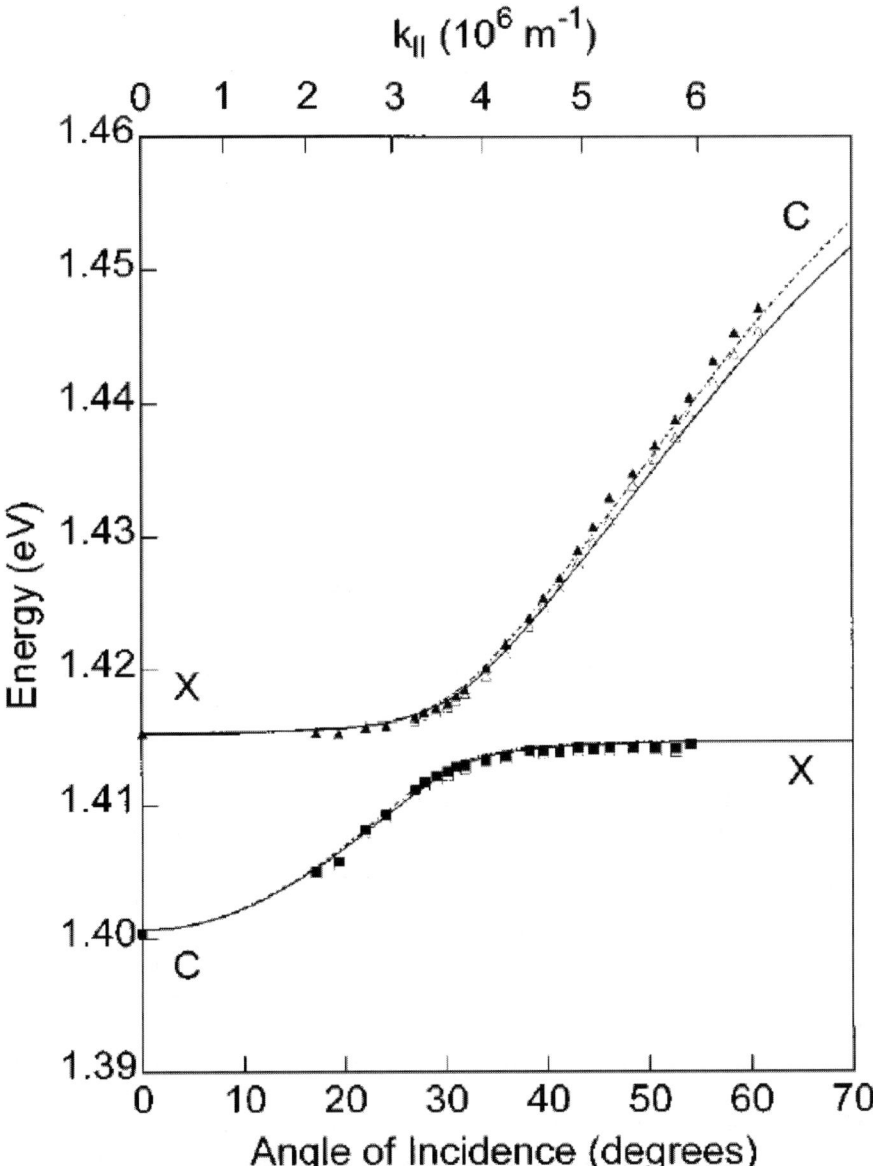

Fig. 30. Energy positions of reflectivity dips for a quantum well embedded microcavity as a function of the angle of incidence (or of $k_\parallel$) (from Ref. [59]). The symbols denote experimental points.

## 9.2. BULK MICROCAVITY POLARITONS

The problem of calculating exciton-polaritons in bulk microcavities can be better handled in linear response theory by the transfer matrix model [60]. At the border of each layer of thickness $\ell_j$ the transfer matrix is simply given by the classical optics expression:

$$\begin{pmatrix} \cos(\frac{\omega}{c} n_j \ell_j) & -\frac{i}{n_j} \sin(\frac{\omega}{c} n_j \ell_j) \\ -i n_j \sin(\frac{\omega}{c} n_j \ell_j) & \cos(\frac{\omega}{c} n_j \ell_j) \end{pmatrix}. \quad (72)$$

Inside the microcavity, close to the exciton resonance, we use $n = \sqrt{\varepsilon}$ and the dielectric function in $k$-space

$$\varepsilon(\omega, k) = \varepsilon_\infty + \frac{4\pi \beta \omega_0^2}{\omega_0^2 + \frac{\hbar k^2}{M_{ex}^*} - \omega^2 - i\gamma \omega}, \quad (73)$$

where $M_{ex}^*$ is the total exciton mass, to obtain the electric fields $E(z)$ and the polarization fields

$$\vec{P}(z) = \sum_{\alpha=1}^{2} \tilde{n}_\alpha^2 \left( E_\alpha^+ e^{i k_\alpha z} + E_\alpha^- e^{-i k_\alpha z} \right), \quad (74)$$

where $\tilde{n}_\alpha^2 = \frac{c^2 k_\alpha^2}{\omega^2} - \varepsilon_\infty$, and $\alpha$ denotes the two different $\vec{k}$ vectors for each frequency. Using the Pekar A.B.C. condition ($P = 0$ at $|z| = L_c/2$), one finds relations between the four field amplitude variables $E_\alpha^\pm$ and obtains the transfer matrix for the microcavity. The total transfer matrix $M$ can be obtained as the product of the individual matrices of the microcavity spacer and of the layers of the D.B.R. mirrors [60]. The total reflectivity and the related optical properties can be calculated in the usual way [53] to obtain:

$$r = \frac{M_{21} + n_{sub} M_{22} - M_{11} - n_{sub} M_{12}}{M_{21} + n_{sub} M_{22} + M_{11} + n_{sub} M_{12}}, \quad (75)$$

$n_{sub}$ being the refractive index of the substrate which supports the microstructure on one side.

The confinement by D.B.R. can be on one side only (open cavity) or on both sides (closed cavity). The D.B.R. can produce a null phase shift on reflection ($r^+$) or a $\pi$ phase shift ($r^-$) according to the relative magnitude of the first and second layer of the D.B.R. This can be used to control the exciton–photon interaction, which is more relevant for the $r^+$ type M.C. [60], as can be intuitively understood by the greater average field intensities in this case.

The observation of bulk exciton polaritons in microcavities also reveals the existence of a satellite structure due to the quantization of the center of mass

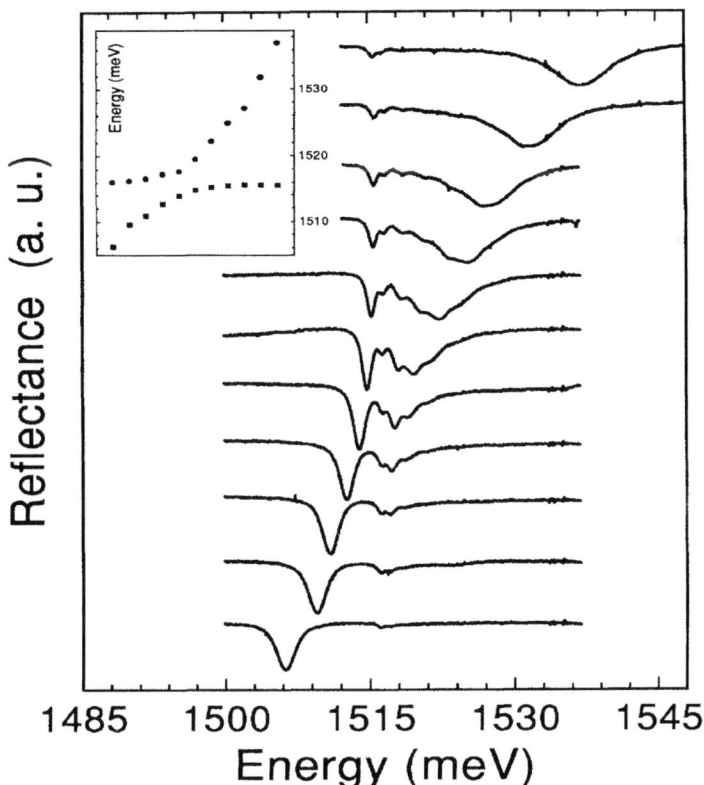

Fig. 31. Reflectivity of M.C.s in the frequency region near the bulk exciton. The thickness $L_c$ decreases from top to bottom (the seventh curve corresponds to degeneracy). The satellite structure due to exciton confinement is clearly visible when one approaches resonance (from Ref. [62]).

motion of the exciton in the growth direction:

$$E_n = \hbar\omega_0 + \frac{n^2\hbar^2\pi^2}{2 M_{\text{ex}}^* L_c^2}. \tag{76}$$

This quantization has been theoretically predicted by D'Andrea and Del Sole [61] and experimentally observed in thin layers of GaAs [60]. In the case of fully closed microcavities the strong coupling between bulk excitons and photons produces the two cavity polaritons described above with a very large Rabi splitting, as in the Q.W. implanted M.C., but with a satellite structure due to the exciton confined motion (76). We show in Figure 31 the experimental evidence of the above described effects in the reflectivity of a $\lambda/2$ GaAs/AlGaAs M.C., with the satellite structure clearly observed when the two polariton modes are strongly mixed [62]. Also the increase of the interaction with $r^+$ reflecting mirrors has been observed.

Another observation to be made concerns the larger value of the polariton splitting at the cross point in microcavities, compared to the usual longitudinal-transverse splitting in bulk material, due to the increased electron-photon coupling. A comparison of the Rabi polariton splittings of Q.W. implanted and bulk microcavities can also be made. They depend on the exciton oscillator strengths, which favors the former, but also on the active layer thickness, which favors the latter.

The analysis described above has been extended by Vladimirova et al. [56] to study the in-plane spatial dispersion of the two-dimensional exciton-polaritons, as in the case of Q.W. implanted cavities. Polariton waves having electric field components normal to the interface ($Z$ or $L$ polaritons described in the case of Q.W.) can be excited only with oblique incidence geometry and polarization in the plane of incidence. This also allows the determination of spatial dependence by varying the angle of incidence, as shown for surface and Q.W. polaritons.

### 9.3. Quantum Theory of Polaritons in M.C.

As shown in detail for the case of bulk polaritons, also in the case of M.C. polaritons one can develop a full quantum theory by considering the electromagnetic radiation in the cavity and the exciton states as field particles with interaction only between states with the same $|\vec{k}_\parallel|$. The interacting Hamiltonian can be made diagonal recovering the results of the semiclassical approach and obtaining the appropriate wave functions. Such a program has been carried out by Savona et al. [63,64]. It has the advantage, as in the general case of the quantum theory of light, of allowing calculations of the statistical properties of the radiation and of the luminescence spectrum. For a general analysis we recommend the lectures by A. Quattropani [65].

As an example we show the simple case of the interaction of an exciton with the cavity mode, with the Hamiltonian expressed in terms of creation and destruction operators for the exciton ($b^\dagger$, $b$) and for the cavity mode ($a^\dagger$, $a$) as

$$H = \hbar\omega_{ex} b^+ b + \hbar\omega_c a^+ a + \hbar C(b^+ - b)(a^+ + a) + \hbar D(a^+ + a)^2, \quad (77)$$

with $C$ and $D$ coupling constants analogous to (9b) and (9c) for the case of the bulk. Introducing the polariton operators:

$$\alpha_i = W_i a + X_i b + Y_i a^+ + Z_i b^+, \quad (78)$$

with $i = 1, 2$ for lower and upper polariton, and requiring

$$[\alpha_i, \alpha_j^+] = \delta_{i,j}, \quad (79)$$

we can reduce the Hamiltonian to the diagonal form:

$$H_{\text{diag.}} = \hbar\Omega_1 \alpha_1^+ \alpha_1 + \hbar\Omega_2 \alpha_2^+ \alpha_2, \tag{80}$$

where the eigenfrequencies $\Omega_1$ and $\Omega_2$ can be determined from the condition

$$[\alpha_i, H] = [\alpha_i, H_{\text{diag.}}] = \hbar\Omega_i \alpha_i, \tag{81}$$

which gives for $\Omega$ the compatibility equation

$$(\omega_{\text{ex}}^2 - \Omega^2)(\omega_c^2 - \Omega^2) + 4D\omega(\omega_{\text{ex}}^2 - \Omega^2) - 4C^2\omega_{\text{ex}}\omega_c = 0. \tag{82}$$

This, for $\Omega$ close to $\omega_c$ and $\omega_{\text{ex}}$, and with $D = C^2/\omega_{\text{ex}}$ as in the bulk case, gives for the upper and lower polariton the approximate solutions

$$\Omega^\pm = \frac{\omega_c + \omega_{\text{ex}}}{2} \pm \frac{1}{2}\sqrt{(\omega_c - \omega_{\text{ex}})^2 + C^2}, \tag{83}$$

in agreement with expression (70) of the semiclassical approach.

This does not include the linewidth effects, which can be considered by using the Stahl–Baslev approach or the Cho approach described in Section 4.

One can also consider the quantum mechanical treatment of the coupling of the polariton cavity modes with the continuum of the radiation modes outside the cavity, thus obtaining the radiative broadening, as in the case of Q.W., and the microcavity polariton luminescence. For this and other related problems we recommend Ref. [16] and more specifically [64].

## 9.4. APPLICATIONS

The field of semiconductor microcavities is of interest also for new photonic applications related to the specific nature of the exciton polaritons described above. In this case, particularly appropriate are three-dimensional microcavities, where lateral confinement is obtained by cutting the M.C. along the plane to form a cylinder (pillar microcavities). For this development we refer to a recent review paper by L.C. Andreani [65]. One must also consider the relevance of spatial disorder in localizing the microcavity polaritons, for which the reader can consult Ref. [66].

As examples of possible applications one can consider the optical bistability, the enhancement of photoluminescence, and polariton M.C. lasers.

It has been demonstrated that the large increase in the radiation intensities in microcavity produces saturation effects which decrease the oscillator strength, with a behavior of the type

$$f = \frac{f_0}{1 + n/n_S}, \tag{84}$$

where $n_S$ is a saturation density which reduces the oscillator strength to half its original value. This consequently bleaches the Rabi splitting, and this is expected

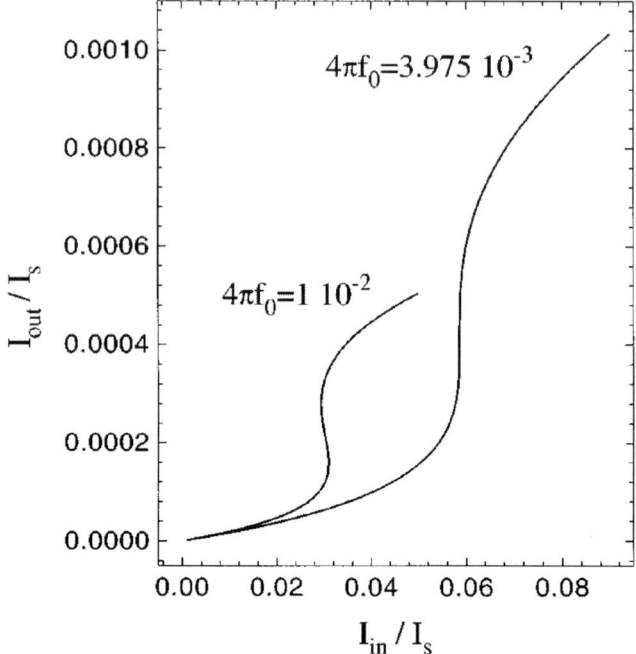

Fig. 32. Computed transmitted power of a M.C. with a GaAs Q.W. as a function of the radiation intensity (referred to saturation) for two values of the polariton linewidth (from Ref. [67]).

to produce an optical bistability when the optical constants of the microcavity are computed as functions of the light intensity. We report in Figure 32 the results of detailed calculations of the transmitted power as a function of the incoming power, performed on the basis of rate equations related to the oscillator strengths (84). We can see a clear evidence of a bistability effect in the transmitted power with a relatively small light intensity [67].

Another interesting possibility is to greatly enhance the photoluminescence by inserting in bulk or in Q.W. implanted M.C. some organic material to which the excitation can be transferred by Forster effect (dipole–dipole interaction at close distance) or by radiation coupling. The organic material can be in the form of ordered bulk material or of a Q.W., in which case cavity polaritons result as shown above, or in the form of disordered organic semiconductors, for which the lower polariton branch only exists for small values of the $k_\parallel$-vector ($\lambda_c \gg$ mean distance between molecules) [68]. In both cases a very large Rabi splitting occurs in M.C. containing organic material as a result of the large oscillator strength of Frenkel excitons. The energy of Q.W. or bulk cavity polaritons can then be transferred to the upper polariton branch of the organic polariton and from this decay by fast

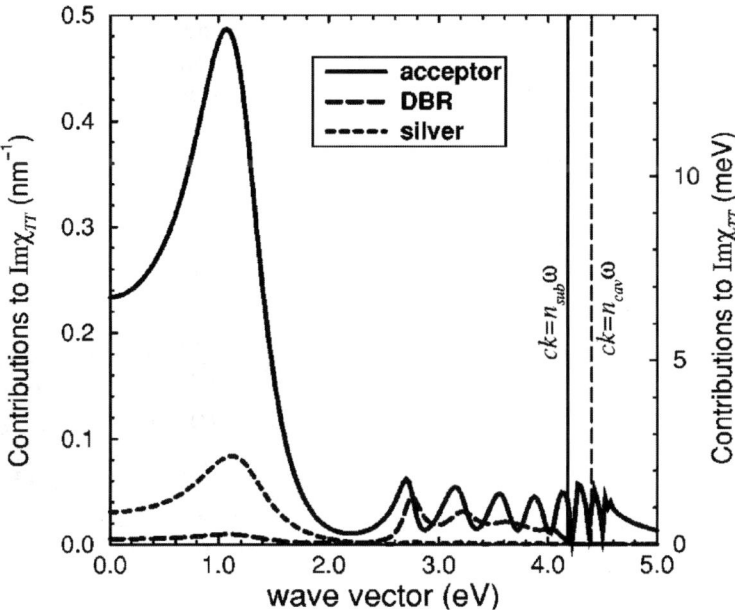

Fig. 33. Energy loss of a M.C. containing an organic material as acceptor and a Q.W. as donor (from Ref. [69]). The large decay is due to energy transfer. Comparison of DBR and silver reflectors is also shown.

phonon emission to the lower incoherent branch and produce a strong photoluminescence typical of the organic material. To show that the presence of organic material increases the optical energy loss of the cavity we report in Figure 33 the result of a specific calculation on an organic implanted microcavity [69].

Another feature of microcavity polaritons which is attracting great interest is the possibility of obtaining a very large density of polaritons at $\vec{k}_\parallel = 0$, the minimum energy state, due to their boson-like nature. This has suggested the enticing possibility of Bose–Einstein condensation of polaritons in microcavities. A very promising feature related to large density polaritons in microcavities is the possibility of polariton lasers. In this case the amplification of light is not related to population inversion, but to "stimulated scattering" by which pairs of polaritons collide with one another and produce a minimum energy and momentum polariton and a higher energy and momentum polariton, as shown in Figure 34. Since the scattering rate is proportional to the number of initial and final states the existence of a reservoir of final states at $k_\parallel = 0$ exponentially increases their number and consequently the photoluminescence normal to the planes. This is a microdevice which converts pump photons into the output beams of different frequency as an optical parametric oscillator [70]. The efficiency of the process is of course dependent on the density at which polaritons are destroyed, and promising results

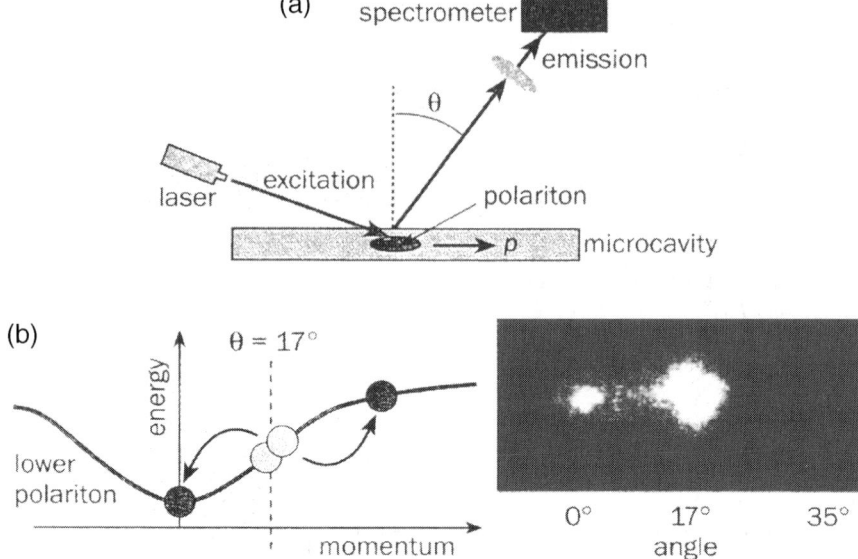

Fig. 34. (a) Experimental apparatus for stimulated photoemission from a microcavity. (b) Polariton collision which produces an excited polariton at large $k_\parallel$ and a polariton at $k_\parallel = 0$. Gain in luminescence occurs when the pump laser enters the sample at an angle of 17°. From Ref. [70].

have been obtained [70]. Very recently, possible evidence for cavity polariton Bose–Einstein condensation has been inferred from the second-order coherence of the emitted light [71].

## Acknowledgements

The author is deeply indebted to Vladimir Agranovich and Giuseppe La Rocca for their suggestions, and to Leonardo Silvestri for his help in preparing the typescript.

## References

1. G. La Rocca, in: V.M. Agranovich, G.F. Bassani (Eds.), Electronic Excitations in Organic Based Nanostructures, in: Thin Films and Nanostructures, Vol. 31, Elsevier, San Diego, 2003, Chapter 2; F. Bassani, G. Pastori Parravicini, Electronic States and Optical Transitions in Solids, Pergamon Press, Oxford, 1975;
Y. Yu, M. Cardona, Fundamentals of Semiconductors, Springer, Berlin, 1991.
2. L.C. Andreani, F. Bassani, A. Quattropani, *Nuovo Cimento D* 10 (1988) 1473;

V.K. Sribnaya, K.B. Tolpygo, E.P. Troitskaya, *Sov. Phys.-Solid State* 20 (1978) 977;
L. Pirozzi, F. Bassani, E. Tosatti, M. Tosi, G. Harbeke, *Solid State Comm.* 24 (1977) 15.
3. R. Peierls, "Momentum and pseudomomentum of light and sound", in: F. Bassani, F. Fumi, M. Tosi (Eds.), Highlights of Condensed Matter Physics, Course LXXXIX of the E. Fermi School, North-Holland, Amsterdam, 1985, p. 137.
4. K. Huang, *Proc. Roy. Soc. London Ser. A* 208 (1951) 352;
K.B. Tolpygo, *Zh. Eksper. Teoret. Fiz.* 20 (1950) 497;
M. Born, K. Huang, Dynamical Theory of Crystal Lattices, Oxford University Press, Oxford, 1952.
5. R. Loudon, "Nonlinear optics with polaritons", in: N. Bloembergen (Ed.), Nonlinear Spectroscopy, Course LXIV of the Enrico Fermi School, SIF, Bologna, 1977, p. 296.
6. U. Fano, *Phys. Rev.* 103 (1956) 1202.
7. J.J. Hopfield, *Phys. Rev.* 112 (1958) 1555.
8. V.M. Agranovich, *Zh. Exper. Teoret. Fiz.* 37 (1959) 430; *Sov. Phys. JETP* 10 (1960) 307.
9. F. Bassani, F. Ruggiero, A. Quattropani, *Nuovo Cimento D* 7 (1986) 700;
M. Combescot, O. Betbeder-Matibet, *Solid State Comm.* 80 (1991) 1011.
10. A. Quattropani, L.C. Andreani, F. Bassani, *Nuovo Cimento D* 7 (1986) 55.
11. F. Bassani, L.C. Andreani, "Exciton-polariton states in insulators and semiconductors", in: U. Grassano, N. Terzi (Eds.), Excited State Spectroscopy of Solids, Course XCVI of the E. Fermi School, SIF, Bologna, 1987, pp. 1–36.
12. S.I. Pekar, Crystal Optics and Additional Light Waves, Benjamin-Cummings, Menlo Park, CA, 1983.
13. V.M. Agranovich, D.L. Mills (Eds.), Surface Polaritons, Modern Problems in Condensed Matter Sciences, Vol. 1, North-Holland, Amsterdam, 1982.
14. F. Bassani, F. Tassone, L.C. Andreani, "Excitons and polaritons in quantum wells", in: A. Stella, L. Miglio (Eds.), Superlattices and Interfaces, Course CXVII of the E. Fermi School, SIF, Bologna, 1993, pp. 187–216.
15. L.C. Andreani, "Optical transitions, excitons, and polaritons in bulk and low-dimensional semiconductor structures", in: E. Burstein, C. Weissbuch (Eds.), Confined Electrons and Photons, Plenum Press, New York, 1995, pp. 57–111.
16. V. Savona, C. Piermarocchi, A. Quattropani, P. Schwendimann, F. Tassone, "Optical properties of microcavity polaritons", in: Phase Transitions, Vol. 68, Gordon and Breach, OPA Overseas Publisher Association, New York, 1999, pp. 169–279.
17. S.I. Pekar, *Sov. Phys. JETP* 6 (1958) 785.
18. W.R. Heller, A. Marcus, *Phys. Rev.* 84 (1951) 809.
19. V.M. Agranovich, V.L. Ginzburg, Crystal Optics with Spatial Dispersion and Excitons, Springer, Berlin, 1984;
K.B. Tolpigo, *Phys. Status Solidi B* 72 (1975) K43.
20. M. Artoni, J.L. Birman, *Phys. Rev. B* 44 (1986) 3736;
J.L. Birman, M. Artoni, B.-S. Wang, *Phys. Rep.* 194 (1990) 367–377.
21. A. Stahl, I. Baslev, Electrodynamics of the Semiconductor Band Edge, Springer-Verlag, Berlin, 1987;
G. Czajkowski, F. Bassani, L. Silvestri, Rivista del Nuovo Cimento, submitted for publication;
K. Cho, Optical Response of Nanostructures, Microscopic Nonlocal Theory, Springer-Verlag, Berlin, 2003.
22. F. Bassani, G. Czajkowski, A. Tredicucci, P. Schillak, *Nuovo Cimento D* 15 (1993) 337.
23. D. Fröhlich, E. Mahler, P. Wiesner, *Phys. Rev. Lett.* 26 (1971) 554;
B. Hönerlage, A. Bivas, Vu Duy Phach, *Phys. Rev. Lett.* 41 (1978) 49;
J.K. Pribram, G.L. Koos, F. Bassani, J.P. Wolfe, *Phys. Rev. B* 28 (1983) 1048.

24. Y. Chen, F. Bassani, J. Massies, C. Deparis, G. Neu, *Europhys. Lett.* 14 (1991) 483;
    A. Tredicucci, Y. Chen, F. Bassani, J. Massies, C. Deparis, G. Neu, *Phys. Rev. B* 47 (1993) 10348.
25. F. Bassani, Y. Chen, G. Czajkowski, A. Tredicucci, *Phys. Status Solidi B* 180 (1993) 115.
26. M.V. Lebedev, M.I. Strashnikova, V.B. Timofeev, V.V. Chernyi, *JETP Lett.* 39 (1984) 440.
27. J. Lagois, B. Fisher, "Surface exciton polaritons from an experimental viewpoint", in: V.M. Agranovich, D.L. Mills (Eds.), Surface Polaritons, in: Modern Problems in Condensed Matter Sciences, Vol. 1, North-Holland, Amsterdam, 1982, p. 69.
28. V.M. Agranovich, "Effects of the transition layer and spatial dispersion in the spectra of surface polaritons", in: V.M. Agranovich, D.L. Mills (Eds.), Surface Polaritons, in: Modern Problems in Condensed Matter Sciences, Vol. 1, North-Holland, Amsterdam, 1982, p. 187.
29. A.A. Maradudin, D.L. Mills, *Phys. Rev. B* 7 (1973) 2782.
30. I.T. Hirabayashi, Y. Koda, Y. Tokura, J. Murata, Y. Kancko, *J. Phys. Soc. Japan* 40 (1977) 1215.
31. G. La Rocca, in: V.M. Agranovich, G.F. Bassani (Eds.), Electronic Excitations in Organic Based Nanostructures, in: Thin Films and Nanostructures, Vol. 31, Elsevier, San Diego, 2003, Chapter 2; G. Bastard, Wave Mechanics Applied to Semiconductor Nanostructures, Les Editions de Physique, Les Ulis, Paris, 1989.
32. S.R. Forrest, *Chem. Rev.* 97 (1997) 1793;
    V.M. Agranovich, G.C. La Rocca (Eds.), Organic Nanostructures: Science and Applications, Course CIL of the Varenna E. Fermi School, SIF, Bologna, 2002.
33. L.C. Andreani, F. Bassani, A. Pasquarello, Symmetry Properties and Selection Rules of Excitons in Quantum Wells, in: Symmetry in Nature, Scuola Normale Superiore, Pisa, 1989.
34. V.M. Agranovich, O. Dubovsky, *Sov. Phys. JETP Lett.* 3 (1966) 345.
35. Ya. Aaaviksoo, Ya. Lippmaa, T. Renoit, *Opt. Spectrosc. (USSR)* 62 (1987) 419.
36. L.C. Andreani, F. Bassani, *Phys. Rev. B* 41 (1990) 7536.
37. K. Cho, *Solid State Commun.* 33 (1980) 911.
38. D. Fröhlich, P. Köhler, E. Meneses-Pacheco, G. Khitrova, G. Weimann, in: A. D'Andrea, R. Del Sole, R. Girlanda, A. Quattropani (Eds.), Proceedings of the International Meeting on the Optics of Excitons and Confined Systems, in: The Institute of Physics Conference Series, Vol. 123, Bristol, 1992;
    D. Fröhlich, R. Wille, W. Schlapp, G. Weimann, *Phys. Rev. Lett.* 59 (1987) 1748.
39. F. Tassone, F. Bassani, L.C. Andreani, *Il Nuovo Cimento D* 12 (1990) 1673.
40. L.C. Andreani, "Excitons and polaritons in confined systems", in: B. Deveaud, A. Quattropani, P. Schwendimann (Eds.), Course CL of the Varenna E. Fermi School, SIF, Bologna, 2003, p. 105.
41. L.C. Andreani, F. Tassone, F. Bassani, *Solid State Comm.* 77 (1991) 641.
42. J. Martinez-Pastor, A. Vinattieri, L. Carraresi, M. Colocci, P. Rossignol, G. Weimann, *Phys. Rev. B* 47 (1993) 10456;
    M. Gurioli, A. Vinattieri, M. Colocci, C. Deparis, J. Massies, G. Neu, A. Bosacchi, S. Franchi, *Phys. Rev. B* 44 (1991) 3115.
43. B. Deveaud, F. Clérot, N. Ray, K. Satzke, B. Sermage, D.S. Katzer, *Phys. Rev. Lett.* 67 (1991) 2355.
44. K. Cho, *J. Phys. Soc. Japan* 55 (1986) 4113.
45. V.M. Agranovich, in: Surface Polaritons, in: Modern Problems in Condensed Matter Sciences, Vol. 1, North-Holland, Amsterdam, 1982;
    V.M. Agranovich, V. Ginzburg, Crystal Optics with Spatial Dispersion and Excitons, Springer-Verlag, Berlin, 1984.
46. R. Atanasov, F. Bassani, V.M. Agranovich, *Phys. Rev. B* 49 (1994) 2658.
47. M.W. Berz, L.C. Andreani, E.F. Steigmeier, F.K. Reinhart, *Solid State Commun.* 80 (1991) 553;
    R. Atanasov, F. Bassani, V.M. Agranovich, *Jour. de Physique II* 3 (C5) (1993) 63.
48. F. Tassone, F. Bassani, L.C. Andreani, *Phys. Rev. B* 45 (1992) 6023.

49. L. Banyai, I. Galbraith, C. Ell, H. Haug, *Phys. Rev. B* 36 (1987) 6099;
    G. La Rocca, in: V.M. Agranovich, G.F. Bassani (Eds.), Electronic Excitations in Organic Based Nanostructures, in: Thin Films and Nanostructures, Vol. 31, Elsevier, San Diego, 2003, Chapter 2.
50. F. Tassone, F. Bassani, *Il Nuovo Cimento D* 14D (1992) 1241; *Il Nuovo Cimento D* 17D (1995) 1743.
51. D.S. Citrin, *Phys. Rev. Lett.* 69 (1992) 3393; *Solid State Comm.* 84 (1992) 281; *Phys. Rev. B* 47 (1993) 3832;
    D.S. Citrin, in: E. Burstein, C. Weissbuch (Eds.), Confined Electrons and Photons, Plenum Press, NY, 1995.
52. V. Savona, "Linear optical properties of semiconductor microcavities with embedded quantum wells", in: QED Phenomena and Applications of Microcavities and Photonic Crystals, Springer-Verlag, Berlin, submitted for publication.
53. M. Born, E. Wolf, Principles of Optics, Pergamon Press, Oxford, 1993.
54. V. Savona, L.C. Andreani, P. Schwendimann, A. Quattropani, *Solid State Comm.* 93 (1995) 733.
55. M.S. Skolnick, T.A. Fisher, D.M. Whittaker, "Strong coupling phenomena in quantum microcavity structures", *Sem. Sci. Technol.* 13 (1998) 645–669.
56. M.R. Vladimirova, A.V. Kavokin, M. Kaliteevski, *Phys. Rev. B* 54 (1996) 14566.
57. D.S. Citrin, *Phys. Rev. B* 50 (1994) 5497; *Phys. Rev. B* 54 (1996) 16425.
58. G. Panzarini, L.C. Andreani, A. Armitage, D. Baxter, M.S. Skolnick, V.N. Astratov, J.S. Roberts, A.V. Kavokin, M.A. Kaliteevski, M.R. Vladimirova, *Phys. Rev. B* 59 (1999) 5082.
59. D. Baxter, M. Skolnick, A. Armitage, V.N. Astratov, D.M. Whittaker, T.A. Fisher, J.S. Roberts, D.J. Mowbray, M.A. Kaliteevski, *Phys. Rev. B* 56 (1997) R10032.
60. Y. Chen, A. Tredicucci, F. Bassani, *Phys. Rev. B* 52 (1995) 1800.
61. A. D'Andrea, R. Del Sole, *Phys. Rev. B* 41 (1990) 413.
62. A. Tredicucci, Y. Chen, V. Pellegrini, M. Börger, M. Sorba, F. Beltram, F. Bassani, *Phys. Rev. Lett.* 75 (1995) 3907;
    A. Tredicucci, C. Yong, V. Pellegrini, C. Deparis, *Appl. Phys. Lett.* 66 (1995) 2388.
63. V. Savona, F. Tassone, *Solid State Comm.* 95 (1995) 673.
64. V. Savona, F. Tassone, C. Piermarocchi, A. Quattropani, P. Schwendimann, *Phys. Rev. B* 53 (1996) 13051;
    V. Savona, C. Weissbuch, *Phys. Rev. B* 54 (1996) 10835.
65. A. Quattropani, Quantum Theory of Quantum Well Polaritons in Semiconductor Microcavities, Lectures given at Scuola Normale Superiore, 2002.
66. V. Savona, "Properties of semiconductor nanostructures in the presence of disorder", in: B. Deveaud, A. Quattropani, P. Schwendimann (Eds.), Course CL of the International School of Physics E. Fermi, IOS Press, Amsterdam, 2003, p. 219.
67. A. Tredicucci, Y. Chen, V. Pellegrini, M. Börger, F. Bassani, *Phys. Rev. A* 54 (1996) 3493.
68. V.M. Agranovich, M. Litinskaia, D.G. Lidzey, submitted for publication.
69. D.M. Basko, F. Bassani, G. La Rocca, V.M. Agranovich, *Phys. Rev. B* 62 (2000) 15692.
70. J.J. Baumberg, *Physics World* 15 (2002) 37.
71. H. Deng, G. Weihs, C. Santori, J. Bloch, Y. Yamamoto, *Science* 298 (2002) 199.

Chapter 4

# Optics and Nonlinearities of Excitons in Organic Multilayered Nanostructures and Superlattices

V.M. AGRANOVICH AND A.M. KAMCHATNOV

*Institute of Spectroscopy, Russian Academy of Sciences, Troitsk, Moscow Reg., 142190, Russia*

1. Introduction . . . . . . . . . . . . . . . . . . . . . . . . . . . . . . . . . . . . . . . . . 185
2. Dielectric Constant Tensor of Long Period Organic Superlattices with Isotropic Layers . . 186
3. Dielectric Constant Tensor of Long Period Organic Superlattices with Anisotropic Layers . 190
4. Optical Nonlinearities in Organic Multilayers . . . . . . . . . . . . . . . . . . . . . . . 194
5. Dielectric Tensor for Short Period Organic Superlattices . . . . . . . . . . . . . . . . . 197
6. Gas-Condensed Matter Shift and Possibility to Govern Spectra of Frenkel Excitons . . . . 201
7. Fermi Resonance Interface Modes in Organic Superlattices . . . . . . . . . . . . . . . . 207
   References . . . . . . . . . . . . . . . . . . . . . . . . . . . . . . . . . . . . . . . . 219

## 1. Introduction

Recent advances in molecular beam deposition methods allowed one to prepare molecular multilayered structures analogous to inorganic superlattices and quantum well structure. These molecular structures are held together by weak van der Waals forces rather than the valence or ionic forces, giving greater freedom and flexibility in preparing structures of high optical quality. In contrast to bulk materials, one can tailor these nanoscale structures in order to change physical parameters of interest. These novel engineered materials open up a new field of research, which is very promising from the technological as well as scientific point of view.

In particular, superlattices are of great interest for the study of phenomena which arise at interfaces between different media. These artificial layered crystals may be considered as systems with "condensed" interfaces since the total area of their interfaces is proportional to the volume of the system. Under these conditions some specific surface and quasi-two-dimensional effects must make an important contribution into the bulk crystal optics. For example, their macroscopic electrodynamics corresponds to uniaxial rather than isotropic crystal optics.

Different interactions taking place at the interfaces are of great importance and can be responsible for the appearance of new linear and nonlinear optical effects. Thus, it can be shown that in such crystalline layered molecular structures excitations can be concentrated near interfaces between different layers. Various types of Fermi resonances can be used as a universal mean for achieving optical bi-stability and multi-stability. New states, Fermi resonance interface modes and Fermi resonance interface solitons were also predicted.

In this chapter we shall present elementary treatment of some effects mentioned above.

## 2. Dielectric Constant Tensor of Long Period Organic Superlattices with Isotropic Layers

First we shall consider the simplest case of superlattice formed by two types of alternating layers with thicknesses $l_1$ and $l_2$ whose electromagnetic properties are characterized by isotropic dielectric constants $\varepsilon_1$ and $\varepsilon_2$, respectively. This means that $l_1$ and $l_2$ are greater than the crystal lattice constants so that macroscopic electrodynamics can be applied to such layers. If the superlattice interacts with electromagnetic wave with wavelength $\lambda = 2\pi/k$ much greater than $l_1$ and $l_2$, then its electromagnetic properties can also be described by some averaged dielectric constant tensor which components can be expressed in terms of $\varepsilon_1$, $\varepsilon_2$ and $l_1$, $l_2$ [1]. To this end, we use the relation

$$\mathbf{D}^{(1)} = \varepsilon_1(\omega)\mathbf{E}^{(1)}, \qquad \mathbf{D}^{(2)} = \varepsilon_2(\omega)\mathbf{E}^{(2)}, \qquad (1)$$

and the boundary conditions which read that tangential component $\mathbf{E}_t$ of $\mathbf{E}$ and normal component $\mathbf{D}_n$ of $\mathbf{D}$ are continuous at the interfaces [2]. Let us take tangential components of (1) and average them over period of the superlattice,

$$\mathbf{D}_t \equiv \frac{1}{l_1 + l_2}\left(l_1 \mathbf{D}_t^{(1)} + l_2 \mathbf{D}_t^{(2)}\right) = \frac{1}{l_1 + l_2}\left(l_1 \varepsilon_1(\omega)\mathbf{E}_t^{(1)} + l_2 \varepsilon_2(\omega)\mathbf{E}_t^{(2)}\right).$$

Since $\mathbf{E}_t$ is continuous at the interfaces and variation of $\mathbf{E}_t$ inside each layer at $l_1, l_2 \ll \lambda$ can be neglected, then $\mathbf{E}_t^{(1)} \cong \mathbf{E}_t^{(2)} \cong \overline{\mathbf{E}}_t$, where $\overline{\mathbf{E}}_t$ is the value of the electric field strength averaged over the superlattice. Hence we obtain

$$\mathbf{D}_t = \varepsilon_\perp(\omega)\overline{\mathbf{E}}_t, \qquad (2)$$

where

$$\varepsilon_\perp(\omega) = \frac{l_1 \varepsilon_1(\omega) + l_2 \varepsilon_2(\omega)}{l_1 + l_2}. \qquad (3)$$

In a similar way, averaging of normal components of Eqs. (1) gives

$$\mathbf{E}_n \equiv \frac{1}{l_1 + l_2}\left(l_1 \mathbf{E}_n^{(1)} + l_2 \mathbf{E}_n^{(2)}\right) = \frac{1}{l_1 + l_2}\left(\frac{l_1}{\varepsilon_1(\omega)}\mathbf{D}_n^{(1)} + \frac{l_2}{\varepsilon_2(\omega)}\mathbf{D}_n^{(2)}\right),$$

and since $\mathbf{D}_n$ is continuous at the interfaces and change little inside the layers, we obtain

$$\mathbf{E}_n = \frac{1}{\varepsilon_\parallel(\omega)}\mathbf{D}_n, \qquad (4)$$

where

$$\varepsilon_\|(\omega) = \frac{1}{l_1+l_2}\left(\frac{l_1}{\varepsilon_1(\omega)} + \frac{l_2}{\varepsilon_2(\omega)}\right). \tag{5}$$

This means that the dielectric constant tensor of the superlattice consisting of optically isotropic layers has the form characteristic for uniaxial crystal. If the layers are parallel to the $(x, y)$ plane and normal to the axis $z$, then the dielectric tensor of the superlattice has the components

$$\varepsilon_{ij} = \varepsilon_i \delta_{ij}, \tag{6}$$

where

$$\varepsilon_x = \varepsilon_y = \varepsilon_\perp(\omega), \qquad \varepsilon_z = \varepsilon_\|(\omega). \tag{7}$$

Let us consider several simple problems typical for optics of superlattices.

### 2.1. Plane Wave in the Bulk

First let us find plane wave solution in the bulk of the superlattice. For monochromatic wave with frequency $\omega$ the Maxwell equations take the form [2]

$$i\omega \mathbf{H} = c(\nabla \times \mathbf{E}), \qquad i\omega \mathbf{D} = -c(\nabla \times \mathbf{H}), \tag{8}$$

and since in the plane wave all quantities are proportional to $\exp(i\mathbf{kr})$, these equations reduce to

$$\omega \mathbf{H} = c(\mathbf{k} \times \mathbf{E}), \qquad \omega \mathbf{D} = -c(\mathbf{k} \times \mathbf{H}). \tag{9}$$

Exclusion of $\mathbf{H}$ from these equations gives

$$(\omega^2/c^2)\mathbf{D} = k^2(\mathbf{E} - \mathbf{k}(\mathbf{kE})/k^2). \tag{10}$$

In our uniaxial case we may assume that the wavevector $\mathbf{k}$ lies in the $(x, z)$ plane,

$$\mathbf{k} = (k\sin\theta, 0, k\cos\theta), \tag{11}$$

where $\theta$ is the angle between $\mathbf{k}$ and the $z$ axis. By definition of $\varepsilon_{ij}$ (Eqs. (6), (7)) we have

$$D_x = \varepsilon_\perp E_x, \qquad D_y = \varepsilon_\perp E_y, \qquad D_z = \varepsilon_\| E_z. \tag{12}$$

Hence, the $y$-component of Eq. (10) gives the dispersion relation

$$k^2 = (\omega^2/c^2)\varepsilon_\perp \tag{13}$$

for ordinary wave with polarization $\mathbf{E} = (0, E_y, 0)$, and $x$- and $z$-components give the system for $E_x$ and $E_z$ which has a nontrivial solution, if

$$\frac{\cos^2\theta}{\varepsilon_\perp} + \frac{\sin^2\theta}{\varepsilon_\|} = \frac{\omega^2}{c^2 k^2} = \frac{1}{n^2}, \tag{14}$$

where $n$ is the refraction index for the extraordinary wave with vector **E** lying in the same plane as the wavevector **k** and the optical axis of the superlattice "crystal".

## 2.2. SURFACE WAVE PROPAGATING ALONG SUPERLATTICE LAYERS

Now let semi-space $z > 0$ be occupied by a superlattice with effective optical axis directed along $z$, and semi-space $z < 0$ is occupied by isotropic medium with dielectric constant $\varepsilon = -|\varepsilon| < 0$. We shall look for the solution for electromagnetic wave localized near the interface between the two media and propagating along the $x$ axis with magnetic field $\mathbf{H} = (0, H, 0)$ directed along the $y$ axis. Then from the second equation (8) we have

$$D_x = -\frac{ic}{\omega}\frac{\partial H}{\partial z}, \qquad D_y = 0, \qquad D_z = \frac{ic}{\omega}\frac{\partial H}{\partial x},$$

and hence at $z > 0$ (medium $I$) the electric field components are equal to

$$E_x^I = -\frac{ic}{\omega\varepsilon_\perp}\frac{\partial H^I}{\partial z}, \qquad E_y^I = 0, \qquad E_z^I = \frac{ic}{\omega\varepsilon_\parallel}\frac{\partial H^I}{\partial x}. \tag{15}$$

Substitution of these formulas into the first equation (8) with taking into account that all field variables do not depend on $y$ yields the equation for $H^I$:

$$\frac{1}{\varepsilon_\parallel}\frac{\partial^2 H^I}{\partial x^2} + \frac{1}{\varepsilon_\perp}\frac{\partial^2 H^I}{\partial z^2} = -\frac{\omega^2}{c^2}H^I. \tag{16}$$

In a similar way we obtain for the medium $II$ ($z < 0$)

$$\frac{1}{\varepsilon}\left(\frac{\partial^2 H^{II}}{\partial x^2} + \frac{\partial^2 H^{II}}{\partial z^2}\right) = -\frac{\omega^2}{c^2}H^{II}. \tag{17}$$

We look for the solution of these equations in the localized near interface form

$$H(x, z) = \begin{cases} H_0 \exp(ikx - \kappa_1 z), & z > 0, \\ H_0 \exp(ikx + \kappa_2 z), & z < 0, \end{cases} \tag{18}$$

where continuity of $H_y$ at the interface is already taken into account. Substitution of Eq. (18) into (16) gives expressions for $\kappa_1$ and $\kappa_2$:

$$\kappa_1 = \sqrt{\frac{\varepsilon_\perp}{\varepsilon_\parallel}k^2 - \varepsilon_\perp\frac{\omega^2}{c^2}}, \qquad \kappa_2 = \sqrt{k^2 + |\varepsilon|\frac{\omega^2}{c^2}}. \tag{19}$$

One more boundary condition of continuity of $E_x$ at the interface between media $I$ and $II$,

$$\frac{1}{\varepsilon_\perp}\frac{\partial H^I}{\partial z} = \frac{1}{\varepsilon}\frac{\partial H^{II}}{\partial z}, \tag{20}$$

yields the equation

$$-\frac{\kappa_1}{\varepsilon_\perp} = \frac{\kappa_2}{\varepsilon}. \quad (21)$$

It is clear that for existence of such surface mode either $\varepsilon_\perp$ or $\varepsilon$ must be negative and we supposed above that $\varepsilon = -|\varepsilon| < 0$. Now substitution of (19) into (21) and simple transformations give the dispersion relation for this surface wave:

$$k^2 = \frac{\omega^2}{c^2} \cdot \frac{|\varepsilon||\varepsilon_\|(|\varepsilon| + \varepsilon_\perp)}{|\varepsilon|^2 - \varepsilon_\perp \varepsilon_\|}. \quad (22)$$

If $\varepsilon_\perp = \varepsilon_\| = \varepsilon_1$, $\varepsilon = \varepsilon_2$, then we return to the well-known formula

$$k^2 = \frac{\omega^2}{c^2} \cdot \frac{|\varepsilon_2|\varepsilon_1}{|\varepsilon_2| - \varepsilon_1} \quad (23)$$

for surface wave at the interface between two media with homogeneous dielectric properties.

Now let lower media have a dielectric constant equal to that of one of the layers, say, $\varepsilon = \varepsilon_2$. Then with the use of Eqs. (3) and (5), Eq. (22) transforms again to the form (23), but the value of the parameter $\kappa_1$ which determines decay of the surface wave in the upper superlattice (see (19)) differs from that for the homogeneous medium.

## 2.3. Surface Wave along Interface Perpendicular to Layers

Now let the interface between the superlattice and homogeneous medium with dielectric constant $\varepsilon$ be situated at the $(y, z)$ plane and $z$ axis be directed perpendicular to the layers. We suppose that the surface wave propagates along $z$ axis and **H** is directed along $y$ axis:

$$\mathbf{H} = (0, H(x, z), 0); \quad (24)$$

then from the Maxwell equations we find that in the superlattice $x < 0$ the electric field has components

$$\mathbf{E}^I = \left( -\frac{ic}{\omega \varepsilon_\perp} \frac{\partial H^I}{\partial z}, 0, \frac{ic}{\omega \varepsilon_\|} \frac{\partial H^I}{\partial x} \right), \quad (25)$$

and in the homogeneous medium $x > 0$ it has components

$$\mathbf{E}^{II} = \left( -\frac{ic}{\omega \varepsilon} \frac{\partial H^{II}}{\partial z}, 0, \frac{ic}{\omega \varepsilon} \frac{\partial H^{II}}{\partial x} \right). \quad (26)$$

Looking for the solution in the form

$$H(x, z) = \begin{cases} H_0 \exp(ikz - \kappa_1 x), & x > 0, \\ H_0 \exp(ikz + \kappa_2 x), & x < 0, \end{cases} \quad (27)$$

and satisfying the condition of continuity of $E_z$ at $x = 0$,

$$\frac{1}{\varepsilon_\|}\frac{\partial H^I}{\partial x}\bigg|_{x=0} = \frac{1}{\varepsilon}\frac{\partial H^{II}}{\partial x}\bigg|_{x=0}, \tag{28}$$

we arrive at the dispersion relation

$$k^2 = \frac{\omega^2}{c^2} \cdot \frac{|\varepsilon|\varepsilon_\perp(|\varepsilon| + \varepsilon_\perp)}{|\varepsilon|^2 - \varepsilon_\perp\varepsilon_\|}. \tag{29}$$

If $\varepsilon = \varepsilon_2$, then this formula reduces to

$$k^2 = \frac{\omega^2}{c^2} \cdot \frac{|\varepsilon_2|\varepsilon_1}{|\varepsilon_2| - \varepsilon_1} \cdot \frac{\varepsilon_\perp}{\varepsilon_\|}. \tag{30}$$

## 3. Dielectric Constant Tensor of Long Period Organic Superlattices with Anisotropic Layers

### 3.1. Dielectric Tensor of a Superlattice

The formulas (3) and (5) can be generalized to the case of anisotropic dielectric constants in the layers [3].

Let us assume that dielectric properties of the layers are determined by tensors $\varepsilon_{ij}^\mu(\omega)$, $\mu$ is the layer number, $\mu = 1, 2, \ldots, \sigma$, $\sigma$ being the number of the layers in a unit cell; the lattice period is $L = \sum_\mu l_\mu$ is the thickness of the $\mu$th layer. Again we assume that the layers are parallel to the $(x, y)$ plane and perpendicular to the $z$ axis. As before, at interfaces of the layers the field components $E_1^\mu$, $E_2^\mu$, $D_3^\mu$ are continuous, and therefore it is convenient to express the field components $D_1^\mu$, $D_2^\mu$, $E_3^\mu$ in terms of $E_1^\mu$, $E_2^\mu$, $D_3^\mu$. To this end we write anisotropic generalization of Eqs. (1):

$$D_1^\mu = \varepsilon_{11}^\mu E_1^\mu + \varepsilon_{12}^\mu E_2^\mu + \varepsilon_{13}^\mu E_3^\mu,$$
$$D_2^\mu = \varepsilon_{21}^\mu E_1^\mu + \varepsilon_{22}^\mu E_2^\mu + \varepsilon_{23}^\mu E_3^\mu, \tag{31}$$
$$D_3^\mu = \varepsilon_{31}^\mu E_1^\mu + \varepsilon_{32}^\mu E_2^\mu + \varepsilon_{33}^\mu E_3^\mu,$$

and from the last equation obtain

$$E_3^\mu = \frac{1}{\varepsilon_{33}^\mu}D_3^\mu - \frac{\varepsilon_{31}^\mu}{\varepsilon_{33}^\mu}E_1^\mu - \frac{\varepsilon_{32}^\mu}{\varepsilon_{33}^\mu}E_2^\mu. \tag{32}$$

Then substitution of this equation into the first two equations (31) gives

$$D_1^\mu = \left(\varepsilon_{11}^\mu - \frac{\varepsilon_{13}^\mu\varepsilon_{31}^\mu}{\varepsilon_{33}^\mu}\right)E_1^\mu + \left(\varepsilon_{12}^\mu - \frac{\varepsilon_{13}^\mu\varepsilon_{32}^\mu}{\varepsilon_{33}^\mu}\right)E_2^\mu + \frac{\varepsilon_{13}^\mu}{\varepsilon_{33}^\mu}D_3^\mu,$$

$$D_2^\mu = \left(\varepsilon_{21}^\mu - \frac{\varepsilon_{23}^\mu \varepsilon_{31}^\mu}{\varepsilon_{33}^\mu}\right) E_1^\mu + \left(\varepsilon_{22}^\mu - \frac{\varepsilon_{23}^\mu \varepsilon_{32}^\mu}{\varepsilon_{33}^\mu}\right) E_2^\mu + \frac{\varepsilon_{23}^\mu}{\varepsilon_{33}^\mu} D_3^\mu. \tag{33}$$

Eqs. (32) and (33) provide a starting point for averaging procedure similar to one used above. With the accuracy of order $\sim L/\lambda \ll 1$ the fields $\mathbf{E}^\mu$ and $\mathbf{D}^\mu$ inside the layers can be considered as constant and due to continuity of $E_1^\mu$, $E_2^\mu$, $D_3^\mu$ we have

$$E_1 = E_1^\mu, \quad E_2 = E_2^\mu, \quad D_3 = D_3^\mu, \quad \mu = 1, 2, \ldots, \sigma. \tag{34}$$

Then averaging over the superlattice period according to the rules

$$E_i = \frac{1}{L} \sum_\mu E_i^\mu l_\mu, \qquad D_i = \frac{1}{L} \sum_\mu D_i^\mu l_\mu \tag{35}$$

yields at once

$$\begin{aligned} D_1 &= a_{11} E_1 + a_{12} E_2 + a_{13} D_3, \\ D_2 &= a_{21} E_1 + a_{22} E_2 + a_{23} D_3, \\ E_3 &= b_{31} E_1 + b_{32} E_2 + b_{33} D_3, \end{aligned} \tag{36}$$

where

$$a_{11} = \frac{1}{L} \sum_\mu \left(\varepsilon_{11}^\mu - \frac{\varepsilon_{13}^\mu \varepsilon_{31}^\mu}{\varepsilon_{33}^\mu}\right) l_\mu \equiv \langle \varepsilon_{11} - \varepsilon_{13}\varepsilon_{31}/\varepsilon_{33}\rangle,$$

$$a_{12} = \langle \varepsilon_{12} - \varepsilon_{13}\varepsilon_{32}/\varepsilon_{33}\rangle, \quad a_{13} = \langle \varepsilon_{13}/\varepsilon_{33}\rangle, \tag{37}$$

$$a_{21} = \langle \varepsilon_{21} - \varepsilon_{23}\varepsilon_{32}/\varepsilon_{33}\rangle, \quad a_{22} = \langle \varepsilon_{22} - \varepsilon_{23}\varepsilon_{32}/\varepsilon_{33}\rangle, \quad a_{23} = \langle \varepsilon_{23}/\varepsilon_{33}\rangle,$$

$$b_{31} = -\langle \varepsilon_{31}/\varepsilon_{33}\rangle, \quad b_{32} = -\langle \varepsilon_{32}/\varepsilon_{33}\rangle, \quad b_{33} = \langle 1/\varepsilon_{33}\rangle.$$

The angle brackets denote an arithmetic average over the superlattice period. It is clear that this average is equivalent to averaging of the fields over the thickness $d \ll \lambda$, where $d \gg L$.

Thus we find that

$$D_i = \varepsilon_{ij}^{SL} E_j, \tag{38}$$

where the superlattice dielectric tensor is determined as follows:

$$\varepsilon_{11}^{SL} = \langle \varepsilon_{11} - \varepsilon_{13}\varepsilon_{31}/\varepsilon_{33}\rangle + \langle \varepsilon_{13}/\varepsilon_{33}\rangle \langle \varepsilon_{31}/\varepsilon_{33}\rangle / \langle 1/\varepsilon_{33}\rangle,$$

$$\varepsilon_{12}^{SL} = \langle \varepsilon_{12} - \varepsilon_{13}\varepsilon_{32}/\varepsilon_{33}\rangle + \langle \varepsilon_{23}/\varepsilon_{33}\rangle \langle \varepsilon_{31}/\varepsilon_{33}\rangle / \langle 1/\varepsilon_{33}\rangle,$$

$$\varepsilon_{13}^{SL} = \langle \varepsilon_{13}/\varepsilon_{33}\rangle / \langle 1/\varepsilon_{33}\rangle,$$

$$\varepsilon_{21}^{SL} = \langle\varepsilon_{11} - \varepsilon_{23}\varepsilon_{31}/\varepsilon_{33}\rangle + \langle\varepsilon_{23}/\varepsilon_{33}\rangle\langle\varepsilon_{31}/\varepsilon_{33}\rangle/\langle 1/\varepsilon_{33}\rangle,$$

$$\varepsilon_{22}^{SL} = \langle\varepsilon_{22} - \varepsilon_{23}\varepsilon_{32}/\varepsilon_{33}\rangle + \langle\varepsilon_{23}/\varepsilon_{33}\rangle\langle\varepsilon_{32}/\varepsilon_{33}\rangle/\langle 1/\varepsilon_{33}\rangle,$$

$$\varepsilon_{23}^{SL} = \langle\varepsilon_{23}/\varepsilon_{33}\rangle/\langle 1/\varepsilon_{33}\rangle,$$

$$\varepsilon_{13}^{SL} = \langle\varepsilon_{31}/\varepsilon_{33}\rangle/\langle 1/\varepsilon_{33}\rangle, \quad \varepsilon_{32}^{SL} = \langle\varepsilon_{32}/\varepsilon_{33}\rangle/\langle 1/\varepsilon_{33}\rangle,$$

$$\varepsilon_{33}^{SL} = \langle 1/\varepsilon_{33}\rangle^{-1}. \tag{39}$$

It follows from the above relations that the tensor $\varepsilon_{ij}^{SL}$ is symmetric with respect to interchange of the indices $i$ and $j$ only if the tensors $\varepsilon_{ij}^{\mu}$ are symmetric (i.e., in the absence of the external magnetic field). Moreover, these relations allow one to investigate the decrease of symmetry of the tensor $\varepsilon_{ij}^{SL}$ that can occur in some cases.

In a previously studied case with $\varepsilon_{ij}^{\mu} = \varepsilon^{\mu}\delta_{ij}$, the tensor $\varepsilon_{ij}^{SL}$ acquires the symmetry of a uniaxial crystal (see relations (6), (7)). However, no decrease in symmetry occurs if the layers have a uniaxial crystal symmetry with the optical axis perpendicular to the interfaces of the layers. If in at least one of the layers the optical axis is directed in a different way, the superlattice symmetry decreases. Thus if this axis in one of the layers is parallel to the interface of the layers, the superlattice has an orthorhombic symmetry. Eqs. (39) can also be used for treatment of other cases.

### 3.2. MAGNETOOPTICAL EFFECTS IN SUPERLATTICES

Let us discuss the influence of a static magnetic field assuming that the dielectric constant in the layers is a scalar $\varepsilon_0^{\mu}$. In the presence of a static magnetic field and up to terms linear in this field the dielectric constant in the layers is

$$\varepsilon_{ij}^{\mu} = \varepsilon_0^{\mu}\delta_{ij} + i\gamma_{ijl}^{\mu}H_l^0, \tag{40}$$

where $H^0$ is the static magnetic field, and the pseudotensor $\gamma_{ijl}^{\mu}$ can be written as

$$\gamma_{ijl}^{\mu} = \gamma^{\mu}e_{ijl}, \tag{41}$$

$e_{ijl}$ being the totally antisymmetric tensor of the third rank. It is clear that the dielectric tensor of a superlattice in the approximation linear in the field $\mathbf{H}^0$ must have the form:

$$\varepsilon_{ij}^{SL} = \varepsilon_{ii}^{SL}\delta_{ij} + ie_{ijm}\gamma_{ml}^{SL}H_l^0, \tag{42}$$

where the gyration tensor $\gamma_{ml}^{SL}$ is yet unknown. In a uniaxial crystal we have $\gamma_{ml} = \gamma_{mm}\delta_{ml}$, $\gamma_{11} = \gamma_{22} \equiv \gamma^{\perp}$, $\gamma_{33} \equiv \gamma^{\parallel}$, so that the problem is that of finding the quantities $\gamma^{\perp}$ and $\gamma^{\parallel}$. In order to find these quantities it is sufficient to consider two particular cases of orientation of the magnetic field $\mathbf{H}^0$:

(1) Assume that the magnetic field $\mathbf{H}^0$ is directed along the $z$ axis. Then the nondiagonal nonzero elements of the tensor $\varepsilon_{ij}^\mu$ are

$$\varepsilon_{12}^\mu = -\varepsilon_{21}^\mu = i\gamma^\mu H^0,$$

so that according to (39) we obtain

$$\varepsilon_{ij}^{SL} = \varepsilon_{ii}^{SL}\delta_{ii} + i\gamma^\| e_{ij3} H^0,$$

where

$$\varepsilon_{11}^{SL} = \varepsilon_{22}^{SL} \equiv \varepsilon^\perp = \langle\varepsilon\rangle, \qquad \varepsilon_{33}^{SL} \equiv \varepsilon^\| = \langle 1/\varepsilon\rangle^{-1}, \tag{43}$$

and

$$\gamma^\| = \frac{1}{L}\sum_\mu l_\mu \gamma^\mu \equiv \langle\gamma\rangle. \tag{44}$$

(2) If the magnetic field is parallel to the interfaces of layers and is directed, for example, along the $y$ axis, then the only nondiagonal nonzero components of the tensor are:

$$\varepsilon_{31}^\mu = -\varepsilon_{13}^\mu = i\gamma^\mu H^0.$$

In this case, according to (39), the tensor $\varepsilon_{ij}^{SL}$ is determined by the relation

$$\varepsilon_{ij}^{SL} = \varepsilon_{ii}^{SL}\delta_{ij} + i\gamma^\perp e_{ijl} H^0, \tag{45}$$

where

$$\gamma^\perp = \langle\gamma/\varepsilon\rangle/\langle 1/\varepsilon\rangle. \tag{46}$$

It is clear that at arbitrary orientation

$$\varepsilon_{ij}^{SL} = \varepsilon_{ii}^{SL}\delta_{ij} + i e_{ijl} g_l, \tag{47}$$

where the gyration vector $\mathbf{g}$ is determined by the relation

$$g_l = \gamma_{ll} H_l^0. \tag{48}$$

### 3.3. Influence of Static Electric Field

In derivation of (39) we assumed that the medium was nonmagnetic. Only in this case one can consider the field $\mathbf{H}^0$ to be independent of $\mu$ and thus having the same value in all layers of the superlattice. In the case of applied electric field such approximation cannot be justified. Therefore, in the presence of a static electric field instead of (40) we have to use the following relation:

$$\varepsilon_{ij}^\mu = \varepsilon_0^\mu \delta_{ij} + \chi_{ijl}^\mu E_l^{0\mu}, \tag{49}$$

where $\chi_{ijl}$ is the tensor of the third rank symmetric with respect to interchange of indices $i$ and $j$. Let us assume that the superlattice is formed by layers with cubic crystal symmetry, so that the components of the tensor $\chi_{ijl}$ can be written in the form

$$\chi_{ijl}^{\mu} = \chi^{\mu}|e_{ijl}|, \tag{50}$$

where $\chi^{\mu}$ is some constant and $|e_{ijl}|$ is the absolute value of the component of the totally antisymmetric tensor $e_{ijl}$. Since nondiagonal components of the tensor $\varepsilon_{ij}^{\mu}$ are linear in the field $\mathbf{E}^{0\mu}$, according to (39) all diagonal tensor components acquire only quadratic in the field $E^0$ corrections which in our approximation have to be neglected. For nondiagonal components of the tensor $\varepsilon_{ij}^{SL}$ in linear in the field $E^0$ approximation we find from (39)

$$\begin{aligned}\varepsilon_{12}^{SL} &= \varepsilon_{21}^{SL} = \langle \chi E_3^0 \rangle, \\ \varepsilon_{13}^{SL} &= \varepsilon_{31}^{SL} = \langle \chi E_2^0/\varepsilon \rangle/\langle 1/\varepsilon \rangle, \\ \varepsilon_{23}^{SL} &= \varepsilon_{32}^{SL} = \langle \chi E_1^0/\varepsilon \rangle/\langle 1/\varepsilon \rangle. \end{aligned} \tag{51}$$

Since for a static electric field we may take for any $\mu$ that $E_1^{0\mu} = E_1^0$, $E_1^{0\mu} = E_2^0$, $E_3^{0\mu} = (1/\varepsilon^{\mu})\varepsilon_{33}^{SL} E_3^0$, where $\mathbf{E}^0$ is the static field averaged over the superlattice period and $\varepsilon^{\mu} \equiv \varepsilon_0^{\mu}(\omega = 0)$, we find that the tensor $\varepsilon_{ij}^{SL}$ can be written as

$$\varepsilon_{ij}^{SL} = \varepsilon_{ii}^{SL}\delta_{ij} + \chi_{ijl}^{SL} E_l^0, \tag{52}$$

where the tensor $\chi_{ijl}^{SL}$ is determined as follows:

$$\begin{aligned}\chi_{ijl}^{SL} &= |e_{ijl}|\chi_{ll}^{SL}, \\ \chi_{11}^{SL} &= \chi_{22}^{SL} = \langle \chi/\varepsilon \rangle/\langle 1/\varepsilon \rangle, \qquad \chi_{33}^{SL} = \langle \chi/\varepsilon \rangle/\langle 1/\varepsilon \rangle. \end{aligned} \tag{53}$$

The above relations can turn useful in all situations when the influence of an external electric field static field on the dielectric tensor has to be taken into account, e.g., in the theory of Raman scattering by polaritons.

## 4. Optical Nonlinearities in Organic Multilayers

Let us turn now to nonlinear optical properties of superlattices. We shall use the method of the preceding sections which reproduces very simply the results of Ref. [4] under natural assumption of smallness of nonlinear effects.

Now each layer is described by dielectric tensor $\varepsilon_{\mu}$ (which is taken for simplicity scalar) and by the nonlinear susceptibility. For example, in case of the third order (Kerr) nonlinearity we have for the layer $\mu$

$$\mathbf{D}_\mu = \varepsilon_\mu \mathbf{E}_\mu + \chi_\mu^{(3)} |\mathbf{E}_\mu|^2 \mathbf{E}_\mu, \tag{54}$$

where again we suppose that the bulk nonlinear susceptibility $\chi_\mu^{(3)}$ corresponds to isotropic medium.

It is easy to see that the nonlinear properties of the superlattice depend crucially on the direction of polarization of the light wave. Since the nonlinear expression (54) cannot be split into tangential and normal components, we shall consider these two cases separately.

The problem is very simple for the case of tangential polarization, when the electric field $\mathbf{E}_\mu$ is polarized in the plane of the layers and the wave vector can be directed as perpendicular to the layers, so along them. Due to continuity of tangential components of $\mathbf{E}_\mu$ at interfaces we have $\mathbf{E}_\mu = \mathbf{E}$ and averaging of (54) yields at once

$$\mathbf{D} = \varepsilon_\perp \mathbf{E}_\mu + \chi_\perp^{(3)} |\mathbf{E}|^2 \mathbf{E}, \tag{55}$$

where $\varepsilon_\perp$ is given by the usual expression

$$\varepsilon_\perp = \frac{1}{L} \sum_\mu l_\mu \varepsilon_\mu \tag{56}$$

and

$$\chi_\perp^{(3)} = \frac{1}{L} \sum_\mu l_\mu \chi_\mu^{(3)}. \tag{57}$$

Thus, for this polarization the nonlinear susceptibility is averaged in the same way as linear one.

The situation changes drastically for the case of normal polarization of $\mathbf{E}_\mu$ when the electric field vector is directed perpendicular to the layers and wave vector is directed along them. In this case $\mathbf{D}_\mu$ is continuous at interfaces so that $\mathbf{D}_\mu = \mathbf{D}$. Now averaging of Eq. (54) yields

$$\mathbf{E} \equiv \frac{1}{L} \sum_\mu l_\mu \mathbf{E}_\mu = \frac{1}{\varepsilon_\parallel} \mathbf{D} - \frac{1}{L} \sum_\mu l_\mu \frac{\chi_\mu^{(3)}}{\varepsilon_\mu} |\mathbf{E}_\mu|^2 \mathbf{E}_\mu, \tag{58}$$

where

$$\frac{1}{\varepsilon_\parallel} = \frac{1}{L} \sum_\mu \frac{l_\mu}{\varepsilon_\mu} \tag{59}$$

is the usual linear expression for this component of the averaged dielectric tensor. Since nonlinear effects are small compared with linear ones, we can consider the second term in (58) as a small correction and replace here $\mathbf{E}_\mu$ by $\mathbf{D}_\mu/\varepsilon_\mu =$

$\mathbf{D}/\varepsilon_\mu = (\varepsilon_\|/\varepsilon_\mu)\mathbf{E}$ so that (58) takes the form

$$\mathbf{D} = \varepsilon_\|\mathbf{E} + \chi_\|^{(3)}|\mathbf{E}|^2\mathbf{E}, \tag{60}$$

where

$$\chi_\|^{(3)} = \frac{1}{L}\sum_\mu l_\mu \left(\frac{\varepsilon_\|}{\varepsilon_\mu}\right)^4 \chi_\mu^{(3)}. \tag{61}$$

Note that effective nonlinear susceptibility $\chi_\|^{(3)}$ can be enhanced in the multilayered structure compared with its bulk value [4]. To show this, let us consider a simple example of superlattice with two kinds of layers ($\mu = 1, 2$) in elementary cell where only one layer ($\mu = 1$) is nonlinear. Then Eq. (61) takes the form

$$\chi_\|^{(3)} = \frac{l_1}{L}\left(\frac{\varepsilon_\|}{\varepsilon_1}\right)^4 \chi_1^{(3)} = \frac{L^3 l_1}{[l_1 + (L-l_1)(\varepsilon_1/\varepsilon_2)]^4}\chi_1^{(3)}. \tag{62}$$

The depending on $l_1$ factor has maximum at

$$l_1 = \frac{L}{3}\frac{\varepsilon_1}{\varepsilon_2 - \varepsilon_1} < L$$

for $\varepsilon_2 > 4\varepsilon_1/3$ equal to

$$\left(\frac{\chi_\|^{(3)}}{\chi_1^{(3)}}\right)_{max} = \frac{1}{3}\frac{\varepsilon_1}{\varepsilon_2 - \varepsilon_1}\left(\frac{3\varepsilon_2}{4\varepsilon_1}\right)^4 \tag{63}$$

which is greater than unity and grows fast with increase of the ratio $(3\varepsilon_2/4\varepsilon_1) > 1$. This enhancement of nonlinear susceptibility in layered structures was confirmed experimentally [5].

In the case of quadratic nonlinearity we shall consider simple example of interaction of two waves with frequencies $\omega$ and $2\omega$, so that in each layer there are two material relations

$$\begin{aligned}D_\mu(2\omega) &= \varepsilon_\mu(2\omega)E_\mu(2\omega) + \chi_\mu^{(2)}\left(E_\mu(\omega)\right)^2, \\ D_\mu(\omega) &= \varepsilon_\mu(\omega)E_\mu(\omega) + \chi_\mu^{(2)}E_\mu(2\omega)E_\mu^*(\omega),\end{aligned} \tag{64}$$

where all field variables have the same direction. Again for the case of tangential polarization we obtain simple average of the nonlinear susceptibility for both frequencies,

$$\chi_\|^{(2)} = \frac{1}{L}\sum_\mu l_\mu \chi_\mu^{(2)}. \tag{65}$$

For normal polarization of fields a calculation similar to the considered above for the third order nonlinearity case yields

$$D(2\omega) = \varepsilon_\|(2\omega)E(2\omega) + \chi_\|^{(2)}\left(E(\omega)\right)^2,$$

$$D(\omega) = \varepsilon_\|(\omega)E(\omega) + \chi_\|^{(2)}E(2\omega)E^*(\omega), \qquad (66)$$

where

$$\chi_\|^{(2)} = \frac{1}{L}\sum_\mu l_\mu \frac{\varepsilon_\|(2\omega)}{\varepsilon_\mu(2\omega)}\left(\frac{\varepsilon_\|(\omega)}{\varepsilon_\mu(\omega)}\right)^2 \chi_\mu^{(2)}. \qquad (67)$$

This formula was obtained in [4] by a different method.

## 5. Dielectric Tensor for Short Period Organic Superlattices

The above treatment assumes that electromagnetic properties of each layer can be described by macroscopic dielectric tensor $\varepsilon_{ij}$. Comparison with experiment shows that this approach is accurate enough even for layers thickness about a few lattice constants. However, for such layers the microscopic approach is also of considerable interest and it becomes inevitable for superlattices consisting of monolayers. In this approach molecules are assumed to be characterized by an effective polarizability appropriate to the crystal environment which could in principle be calculated quantum mechanically but in practice is deduced from experimental quantities and used to interrelate them.

From microscopic point of view, it is necessary to treat the variation of electric field on an atomic scale and to recognize that the field responsible for polarizing a molecule is neither the applied field nor the macroscopic field but rather the local field given by the sum of the applied field and the field due to the surrounding polarized molecules. Since the polarization of the molecules depends on the local field, we have to calculate the local field in a self-consistent manner. In our treatment we shall follow [6].

Let the superlattice consist of molecular multilayers and each multilayer consist of monolayers labelled $\mu$, where $\mu$ ranges from 1 to the number of layers $N$ in the repeating unit of the superlattice. The layers are assumed to be subjected to a uniform applied field $\mathbf{E}^0$, which could be of an electromagnetic wave provided its wavelength greatly exceeds the thickness of the repeating unit. Then the local polarizing electric field at a molecule in monolayer $\mu$ is

$$\mathbf{E}_\mu^{local} = \mathbf{E}^0 + \sum_{\mu'}\mathbf{t}_{\mu\mu'}\cdot\mathbf{P}_{\mu'}. \qquad (68)$$

Here $t_{\mu\mu'}$ is the sum of dipole tensors

$$t_{\mu\mu'} = \sum_{n,n'} \nabla\nabla \frac{1}{r}\bigg|_{\mathbf{r}=\mathbf{r}_{\mu n}-\mathbf{r}_{\mu'n'}} \qquad (69)$$

between a molecule in layer $\mu$ and all molecules in layer $\mu'$, which is defined to be zero for $\mu, n = \mu', n'$, where $n, n'$ label locations of molecules in layers $\mu$ and $\mu'$, respectively. Here $\mathbf{p}_{\mu'}$ is the induced dipole moment of molecules in layers $\mu'$. If it is assumed that interactions between adjacent layers are negligible, then

$$t_{\mu\mu'} = t_\mu \delta_{\mu\mu'} \qquad (70)$$

with $t_\mu$ the dipole sum within layer $\mu$.

Experimental observables are related to the macroscopic electric field $\mathbf{E}$ which in layer $\mu$ in case of plane geometry of superlattice structure may be defined by [6]

$$\mathbf{E}_\mu = \mathbf{E}^0 - 4\pi \mathbf{n}(\mathbf{n} \cdot \mathbf{P}_\mu), \qquad (71)$$

where $\mathbf{P}_\mu$ is the layer polarization $\mathbf{p}_\mu/v_\mu$, with $v_\mu$ the volume per molecule in layer $\mu$, and $\mathbf{n}$ is the unit vector normal to the layers. This definition agrees with the usual condition of continuity of tangential component $\mathbf{E}_\mu^t$ of $\mathbf{E}_\mu$ at an interface. Indeed, $\mathbf{E}_\mu^t$ is the difference between $\mathbf{E}_\mu$ and the normal component $\mathbf{n}(\mathbf{n} \cdot \mathbf{E}_\mu)$, i.e.,

$$\mathbf{E}_\mu^t = \mathbf{E}_\mu - \mathbf{n}(\mathbf{n} \cdot \mathbf{E}_\mu). \qquad (72)$$

Then substitution for $\mathbf{E}_\mu$ from Eq. (71) yields

$$\mathbf{E}_\mu^t = \mathbf{E}^0 - 4\pi \mathbf{n}(\mathbf{n} \cdot \mathbf{E}^0) \qquad (73)$$

which is the tangential component of $\mathbf{E}^0$ and is a constant, independent of $\mu$. The electric displacement is defined by

$$\mathbf{D}_\mu = \mathbf{E}_\mu + 4\pi \mathbf{P}_\mu. \qquad (74)$$

Its normal component $\mathbf{D}_\mu^n$ is $\mathbf{n}(\mathbf{n} \cdot \mathbf{D}_\mu)$, which on substitution from Eq. (71) yields

$$\mathbf{D}_\mu^n = \mathbf{n}(\mathbf{n} \cdot \mathbf{E}^0). \qquad (75)$$

This is the normal component of $\mathbf{E}^0$, which is the normal component of the electric displacement outside the superlattice and is constant independent of $\mu$. Thus Eq. (71) agrees with the condition that the normal component of the electric displacement is constant across an interface.

In what follows we shall confine ourselves to the case (70), that is assume that interactions between adjacent layers are negligible. Then from Eqs. (68) and (71) we find that the local field is related to the macroscopic field by

$$\mathbf{E}_\mu^{\text{local}} = \mathbf{E}_\mu + \frac{1}{v_\mu}\mathbf{L}_\mu \cdot \mathbf{p}_\mu. \tag{76}$$

Here $\mathbf{L}_\mu$ is the layer Lorentz-factor tensor

$$\mathbf{L}_\mu = v_\mu \mathbf{t}_\mu + 4\pi\, \mathbf{nn}. \tag{77}$$

The induced dipole moment is related to the local field by

$$\mathbf{p}_\mu = \alpha_\mu \cdot \mathbf{E}_\mu^{\text{local}}, \tag{78}$$

where $\alpha_\mu$ is the polarizability tensor for molecules in layer $\mu$. Substitution in Eq. (76) yields

$$\mathbf{E}_\mu^{\text{local}} = \mathbf{E}_\mu + \frac{1}{v_\mu}\mathbf{L}_\mu \cdot \alpha_\mu \cdot \mathbf{E}_\mu^{\text{local}}, \tag{79}$$

and this equation at once gives

$$\mathbf{E}_\mu^{\text{local}} = \left(1 - \frac{1}{v_\mu}\mathbf{L}_\mu \cdot \alpha_\mu\right)^{-1} \mathbf{E}_\mu \equiv \mathbf{d}_\mu \cdot \mathbf{E}_\mu, \tag{80}$$

where $\mathbf{d}_\mu$ is the local-field tensor. From Eq. (78) we have

$$\mathbf{P}_\mu = \alpha_\mu \cdot \mathbf{d}_\mu \mathbf{E}_\mu, \tag{81}$$

so that the linear susceptibility $\chi_\mu^{(1)}$ defined via

$$\mathbf{P}_\mu = \chi_\mu^{(1)} \cdot \mathbf{E}_\mu \tag{82}$$

is obtained as

$$\chi_\mu^{(1)} = \alpha_\mu \cdot \mathbf{d}_\mu. \tag{83}$$

Now we can relate the macroscopic fields to the applied fields. From Eqs. (82) and (71) we obtain

$$\mathbf{E}_\mu = \mathbf{E}^0 - 4\pi\left(\mathbf{n}\cdot\chi_\mu^{(1)}\mathbf{E}_\mu\right). \tag{84}$$

This relation can be inverted to give

$$\mathbf{E}_\mu = \left[1 - \frac{4\pi\,\mathbf{n}\cdot(\mathbf{n}\cdot\chi_\mu^{(1)})}{1 + 4\pi\,\mathbf{n}\cdot\chi_\mu^{(1)}\cdot\mathbf{n}}\right]\cdot\mathbf{E}^0 \equiv \mathbf{S}_\mu \cdot \mathbf{E}^0. \tag{85}$$

To obtain the total polarization, we have to sum over monolayers $\mu$,

$$\mathbf{P} = \frac{1}{L}\sum_\mu l_\mu \mathbf{P}_\mu, \tag{86}$$

where $l_\mu$ is the thickness of the layer $\mu$ and $L = \sum_\mu l_\mu$ is the thickness of the repeating assembly of monolayers. Substitution for $\mathbf{P}_\mu$ from (82) yields

$$\mathbf{P} = \frac{1}{L} \sum_\mu l_\mu \chi_\mu^{(1)} \cdot \mathbf{E}_\mu, \tag{87}$$

and hence from Eq. (85) the total polarization is related to the applied field by

$$\mathbf{P} = \frac{1}{L} \sum_\mu l_\mu \chi_\mu^{(1)} \cdot \mathbf{S}_\mu \mathbf{E}^0. \tag{88}$$

To obtain the multilayer susceptibility, we need the average macroscopic field over $N$ layers defined by

$$\mathbf{E} = \frac{1}{L} \sum_\mu l_\mu \mathbf{E}_\mu, \tag{89}$$

whence

$$\mathbf{E} = \frac{1}{L} \sum_\mu l_\mu \mathbf{S}_\mu \cdot \mathbf{E}_\mu \equiv \mathbf{S} \cdot \mathbf{E}^0. \tag{90}$$

The multilayer susceptibility $\chi_\mu^{(1)}$ relates $\mathbf{P}$ to $\mathbf{E}$ according to

$$\mathbf{P} = \chi^{(1)} \cdot \mathbf{E}. \tag{91}$$

Comparison of Eqs. (88) and (90) then yields

$$\chi^{(1)} = \sum_\mu \frac{l_\mu}{L} \chi_\mu^{(1)} \cdot \mathbf{S}_\mu \cdot \mathbf{S}^{-1}. \tag{92}$$

Thus the multilayers susceptibility is a sum of the individual layers susceptibilities weighted by two factors. The first $l_\mu/L$ gives the layer contribution to the polarization, and the second $\mathbf{S}_\mu \cdot \mathbf{S}^{-1}$ gives the layer contribution to the macroscopic field. When susceptibility $\chi^{(1)}$ is known, the dielectric tensor can be found from the relation

$$\varepsilon = 1 + 4\pi \chi. \tag{93}$$

Let us show that for macroscopic layers these formulas reduce to those derived in the preceding section. To return to index notation, we assume that $\mathbf{n}$ has components $(0, 0, 1)$ in a Cartesian system, where axes 1 and 2 lie in the layers and 3 is the direction of the normal to them. Then from Eqs. (85) and (93) we find

$$\mathbf{S}_\mu = \begin{pmatrix} 1 & 0 & 0 \\ 0 & 1 & 0 \\ -\varepsilon_{13}^{(\mu)}/\varepsilon_{33}^{(\mu)} & -\varepsilon_{23}^{(\mu)}/\varepsilon_{33}^{(\mu)} & 1/\varepsilon_{33}^{(\mu)} \end{pmatrix}, \tag{94}$$

where $\varepsilon_{ij}^{(\mu)}$ is the $ij$ component of the dielectric tensor for layer $\mu$. Averaging of $\mathbf{S}_\mu$ according to Eq. (85) and inversion of the result gives

$$\mathbf{S}^{-1} = \begin{pmatrix} 1 & 0 & 0 \\ 0 & 1 & 0 \\ \frac{\langle \varepsilon_{13}/\varepsilon_{33}\rangle}{\langle 1/\varepsilon_{33}\rangle} & \frac{\langle \varepsilon_{23}/\varepsilon_{33}\rangle}{\langle 1/\varepsilon_{33}\rangle} & \frac{1}{\langle 1/\varepsilon_{33}\rangle} \end{pmatrix}. \quad (95)$$

At last, Eq. (92) yields for averaged dielectric tensor the equation

$$\varepsilon = \sum_\mu \frac{l_\mu}{L} \varepsilon^\mu \cdot \mathbf{S}_\mu \cdot \mathbf{S}^{-1} = \langle \varepsilon \cdot \mathbf{S} \rangle \cdot \mathbf{S}^{-1}. \quad (96)$$

Performing the averages and matrix multiplication yields expressions for the components of $\varepsilon$ equivalent to Eqs. (39).

The present treatment is capable in principle of treating variations in the dielectric response in the interface region. However, the assumption that interactions between layers are negligible (see Eq. (70)) precludes any discrimination between the dielectric response in the bulk and interface regions, and moreover gives only an approximation to the bulk dielectric tensor. Only in this way can the macroscopic and microscopic treatments be compared. Microscopic approach admits also extension to interacting layers [6,7].

## 6. Gas-Condensed Matter Shift and Possibility to Govern Spectra of Frenkel Excitons

Now we turn to discussion of properties of excitons in layered structures. Superlattices are systems with "condensed" interfaces, since in these artificial layered crystals the total area of interfaces is proportional to the volume. In these conditions the specific surface and quasi-two-dimensional effects must make an important contribution to the bulk crystal properties [8].

First we consider the properties and the role of interfaces with the boundary of anthracene crystal with vacuum. Of course, it is a particular case of boundary. However, this case has been investigated in many experiments and therefore can be considered as some kind of experimental foundation of the approach we will use.

It can be considered now as well-established that 2D exciton state—the lowest electronic excitation of external monolayer of anthracene crystal—is blue shifted by 204 cm$^{-1}$ with respect to the bottom of exciton band in the bulk. The electronic transition of the first monolayer lies between the bulk value and the isolated molecule value which is blue shifted by 2000 cm$^{-1}$ with respect to the bulk. The excited electronic state of the surface molecular monolayer is clearly seen in emission at low temperature. The monolayer next to the surface is blue shifted by 10 cm$^{-1}$ and the following one by 2 cm$^{-1}$. The nature of these blue shifts is

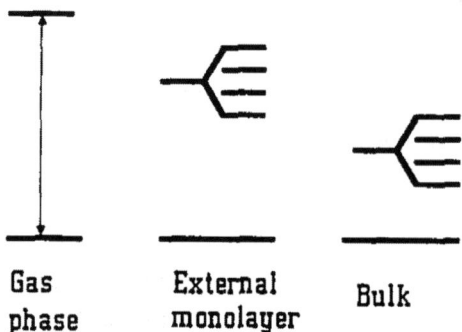

Fig. 1. The levels near the boundary of a molecular crystal.

now well understood and is related with the absence of neighbors for molecules in the external monolayer from vacuum side. Therefore, for these molecules the gas-condensed matter (G-CM) shift of electronic transition frequency is smaller than the G-CM shift in the bulk (see Figure 1; we assume that the surface corresponds to $(a, b)$ plane of anthracene crystal). For temperatures low compared with the blue shift, the surface layer acts as an isolated monolayer and is an ideal system for investigation of two-dimensional excitons. Such excitons at weak dephasing should exhibit a superradiant radiative decay [9]. This ultrafast decay of anthracene films was first observed in picosecond measurements conducted by Aaviksoo et al. [10]. Relative quantum yield measurements of the bulk and the surface emission indicate that the decay of the monolayer is purely radiative with a very small contribution of relaxation to the bulk. The picosecond timescales observed in these experiments were a first, and a beautiful example of superradiance in two-dimensional excitons.

After these short remarks let us return again to the first monolayers of the anthracene crystal. As the width of the exciton band in this crystal for wave vectors directed along axis $C'$ (i.e., along the normal to the $(a, b)$ plane) is very small ($\sim 5$ cm$^{-1}$), we can state that on the exciton's approaching the surface its energy increases (Figure 2), interaction of exciton with surface is repulsive and we have here some type of dead layer for bulk exciton. To go ahead it is necessary to recollect how the gas-condensed matter shift can be calculated. It is known from the theory of molecular (Frenkel) excitons1 [11] that this shift appears due to the difference between the energies of interaction of the excited molecule (molecular state $f$) and the unexcited molecule (molecular ground state 0) with all other molecules of the crystals in ground state:

$$D^{AA} = \sum_m D^{AA}_{nm}, \qquad (97)$$

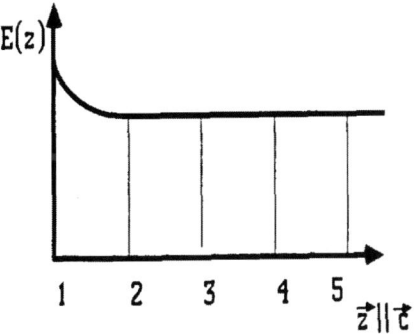

Fig. 2. The dependence $E(z)$ at the boundary with vacuum.

where

$$D_{nm}^{AA} = \langle \phi_n^{fA} \phi_m^{0A} | V_{nm} | \phi_n^{fA} \phi_m^{0A} \rangle - \langle \phi_n^{0A} \phi_m^{0A} | V_{nm} | \phi_n^{0A} \phi_m^{0A} \rangle,$$

$\phi_n^r$ ($r = 0A, fA$) denotes the wave functions of molecule A in ground ($0A$) and in excited ($fA$) state, $V_{nm}$ is the operator of the Coulomb interaction between the molecules $n$ and $m$ determined by the coordinates of their electrons and nuclei $m \equiv \mathbf{m} \equiv (m_1, m_2, m_3)$; usually the values $D_{nm} < 0$ are negative for molecules with inversion symmetry (anthracene, naphthalene, etc.). For molecules $n$ lying on the boundary with vacuum (to which corresponds, for example, space $m_3 < 0$) the summation over molecules $m$ with $m_3 < 0$ is excluded. Therefore, the respective value $D \equiv D_S^{0A}$ ($S$ denotes a surface, $A$ a molecule, $0$ a vacuum) can be written as

$$D^{0A} = \sum_{m_3 > 0} D_{nm}^{AA}, \tag{98}$$

where $n = (0, 0, 1)$ (here, for simplicity, we consider crystals with one molecule per unit cell), $|D_S| < |D|$ and we obtain the case represented in Figures 1, 2. In this particular case the blue shift value is equal to

$$\Delta D \equiv D_S^{0A} - D = -\sum_{m_3 \leqslant 0} D_{nm}^{AA} > 0. \tag{99}$$

Now let us consider what will occur if the crystal under consideration has a boundary with another molecular crystal $B$. Obviously, the value of G-CM shift will change. The shift in this case can be written as

$$D_S \equiv D_S^{BA} = \sum_{m_3 > 0} D_{nm}^{AA} + \sum_{b_3 < 0} D_{nm}^{AB}, \tag{100}$$

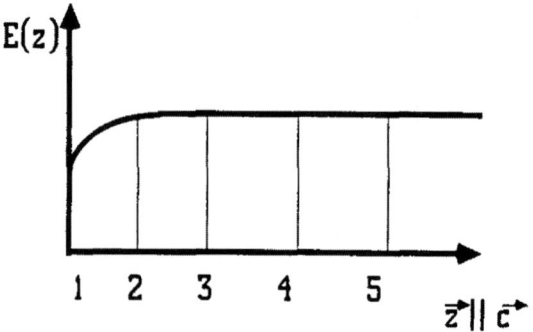

Fig. 3. The dependence $E(z)$ in the case $S_B > S_A$.

where

$$D_{nm}^{AB} = \langle \phi_n^{fA} \phi_m^{0B} | V_{nm} | \phi_n^{fA} \phi_m^{0B} \rangle - \langle \phi_n^{0A} \phi_m^{0B} | V_{nm} | \phi_n^{0A} \phi_m^{0B} \rangle.$$

Therefore, the shift of level for molecules in the first monolayer in comparison to the bulk value is equal to

$$\Delta D = D_S^{BA} - D^{AA} = \sum_{m_3 < 0} \left( D_{nm}^{AB} - D_{nm}^{AA} \right). \tag{101}$$

Although each of the values $D_{nm}^{AA}$ and $D_{nm}^{AB}$ for the lowest electronic molecular excitations is negative, as a rule, we no longer have the possibility to do the definite statement with respect to the sign of the molecular level shift $\Delta D$: this shift can be for different pairs of molecules $A$ and $B$ either positive or negative. Thus, if the molecules $B$ possess in ground state relatively small static multipoles (let us denote their value conditionally by $S_B$) the shift $\Delta D$ should be positive, as happens for anthracene at the boundary with vacuum (by definition $S_B \gg S_0$) (Figures 1, 2). If we have the opposite case and respective multipoles of type $B$ molecules are large enough ($S_B \gg S_A$) the shift $\Delta D$ can become negative. In this case instead of the situation expressed in Figure 2 we obtain attraction of excitons to the surface (Figure 3). Let us recall now that we are interested in organic superlattices and assume that we have under discussion the superlattice of type $B\,A\,B\,A\,B$. Then we have the case shown in Figure 4 for the exciton energy $E(z)$ at $S_A > S_B$ and for the case $S_A < S_B$ the dependence $E(z)$ is different (see Figure 5; for definiteness Figures 4 and 5 correspond to five lattice constants in $A$-layer).

Let us stress an important circumstance now. The molecule and crystal of anthracene which we have used above as a well-investigated example, possess an inversion center. The dipole moment operator for such molecules can have only nondiagonal nonzero matrix elements so that the quantities $D_{nm}$ depend only on

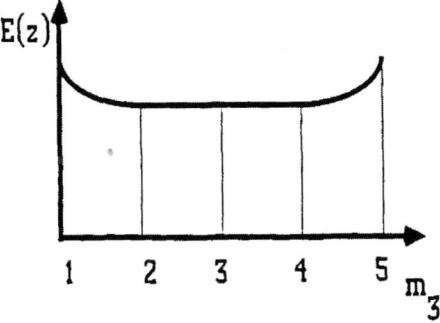

Fig. 4. The dependence $E(z)$ for $A$ layer in the case $S_A > S_B$.

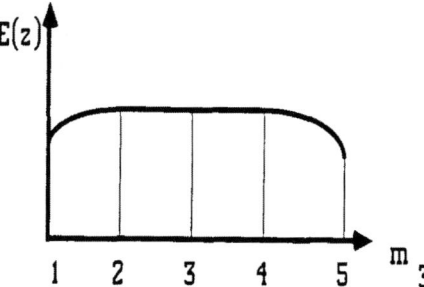

Fig. 5. The dependence $E(z)$ for $A$ layer in the case $S_B > S_A$.

the quadrupole moment and higher moments. Therefore, the quantities $D_{nm}$ in this case containing only diagonal matrix elements of the operator of intermolecular interaction decrease rapidly with increasing $n - m$ and it proves sufficient to take into account the interaction between nearest neighbors in order to calculate their contribution to the exciton energy. That is why the shift of molecular levels in comparison to the bulk value for anthracene molecules is important only for the first and probably for the second external monolayers in anthracene crystal.

For the crystals without an inversion center the long-range dipole–dipole interaction becomes very important, while the special role of external monolayers (together with the dependence of exciton spectra on the layer thickness) becomes weaker.

Let us consider now the conditions under which the dependencies of the type expressed in Figures 4, 5 have physical sense. It is useful here to return again to the case of anthracene crystal in the vacuum-crystal contact region. The homogeneous width $\gamma$ of the lowest exciton transition in anthracene crystal at low temperature is, as was mentioned, of the order of $10-20 \text{ cm}^{-1}$, i.e., of the order of the width of exciton band $\Delta$ (for wave vectors $k$ which are parallel to the axis $C'$). Therefore, in layers of anthracene type in which the inequalities

$$|\Delta D| \gg \gamma, \Delta,$$

are fulfilled, the curvature of the bottom of exciton zone like shown in Figures 4, 5 can be very important and it is necessary to take it into account in the discussion of optical or electro-optical processes in organic multilayer structures. In this case we have a system with large inhomogeneous broadening and this effect can be important even in the cases where the homogeneous width is of the order of the inhomogeneous width. Clearly, in the crystals to which Figure 4 corresponds the excitons created as a result of absorption of the high energy photons ($\hbar\omega > E(z)$) should be concentrated after energy relaxation in the region of the minimum of the function $E(z)$, that is in one of the middle monolayers (the monolayer $m_3 = 3$ in Figure 4). In the structures of another type to which Figure 5 corresponds the excitons should be concentrated in the interface region.

Let us now consider organic multilayer systems in which molecules possess in their stationary states the static dipole moments. If in this case the dipole–dipole interaction dominates, the relations for G-CM shift take a more definite form. In this case the quantity $D_{nm}^{AB}$ can be written in the following way:

$$D_{nm}^{AB} = \left(p_i^{fA} - p_i^{0A}\right) t_{ij}(n,m) \left(p_j^{0B}\right),$$

where $p^{fA}$ and $p^{0A}$ are static dipole moments of molecule $A$ in the ground (0) and excited ($f$) state, $p^{0B}$ the static dipole moment of molecule $B$ in the ground state, $t_{ij}(n,m)$ the tensor determining the interaction of dipoles situated at points $n$ and $m$. Therefore, according to Eq. (101) we obtain

$$\Delta D = \sum_{m, m_3 < 0} \mu_i^{A(0,f)} t_{ij}(n,m) \mu_j^{B(0)}, \tag{102}$$

where

$$\mu_i^{A(0,f)} = \left(p_i^{fA} - p_i^{0A}\right), \quad \mu_j^{B(0)} = \left(p_j^{0B} - p_j^{0A}\right).$$

Thus, in this case the shift $\Delta D$ is equal to the energy of interaction of the difference dipole moment $\mu^{A(0,f)}$ at the site $n$ with the set of dipoles $\mu^{B(0)}$ located on the lattice sites $m$ with negative component $m_3$. Note again that in dependence on the directions of $\mu^{A(0,f)}$ and $\mu^{B(0)}$ the shift $\Delta D$ can be either positive or negative. Let $\mu^{A(0,f)} \simeq \mu^{B(0)} \simeq 5$ Debye, lattice constant $a \simeq 5$ Å, then equation for $\Delta D$ gives an order-of-magnitude estimate $\Delta D \simeq 1000$ cm$^{-1}$. This very crude estimation indicates nevertheless that for molecules without inverse symmetry the shifts $\Delta D$ will be determined by dipole–dipole interaction.

## 7. Fermi Resonance Interface Modes in Organic Superlattices

### 7.1. FERMI RESONANCE IN MOLECULES

Fermi resonance is a phenomenon which takes place in vibrational or electronic spectra of molecules. For example, let a molecule have two vibrational modes with frequencies $\omega_a$ and $\omega_b$. If the second order resonance condition $2\omega_a \simeq \omega_b$ is fulfilled, then the $\hbar\omega_b$ transition in infrared spectrum can be split into two lines of comparable intensity and the second line cannot be explained as a result of interaction of light with the vibrational $a$ mode because the transitions with excitation of two $\hbar\omega_a$ quanta are forbidden due to well-known $n \to n \pm 1$ selection rule for harmonic oscillator. E. Fermi explained [12] this experimental observation as a result of nonlinear resonance interaction of two vibrational modes with each other. Since that time the notion of Fermi resonance has been generalized to the processes with participation of different types of quanta (e.g., $\omega_1 + \omega_2 \simeq \omega_3$, $\omega_1 + \omega_2 \simeq \omega_3 - \omega_4$, and so on) and to electronic types of excitations as well. Further generalizations were suggested for Fermi resonance interactions of collective modes in molecular crystals and other macroscopic systems, so that Fermi resonance phenomenon became part of not only molecular physics but solid state physics also. In case of multilayer crystalline organic structures their spectrum is created by the "overlapping" of the spectra of different crystalline compounds and new Fermi resonances arise due to the anharmonicity across the interface, which may be quite interesting from various point of view. For example, one may suppose that Fermi resonance interaction between different molecules composing the layered structure can play an important role in its optical properties.

At first, we shall consider the Fermi resonance theory in molecules from classical mechanics point of view. Let $q_a$ and $q_b$ be two normal coordinates corresponding to the vibrational degrees of freedom of the molecule with eigenfrequencies $\omega_a$ and $\omega_b$ (we neglect all other degrees of freedom of the molecule). Then the Lagrangian of the vibrational motion can be written as follows,

$$L = \frac{M_a \dot{q}_a^2}{2} + \frac{M_b \dot{q}_b^2}{2} - U(q_a, q_b), \qquad (103)$$

where the overdot stands for the derivative with respect to time $t$, $M_a$ and $M_b$ are "mass" coefficients, and the potential energy $U(q_a, q_b)$ can be expanded into series with respect to powers of variables $q_a$ and $q_b$,

$$U(q_a, q_b) = U_2(q_a, q_b) + U_3(q_a, q_b) + \cdots. \qquad (104)$$

Since $q_a$ and $q_b$ are normal coordinates, in harmonic approximation we have a diagonal form of the potential energy:

$$U_2(q_a, q_b) = \frac{M_a \omega_a^2}{2} q_a^2 + \frac{M_b \omega_b^2}{2} q_b^2. \qquad (105)$$

The next nonlinear term in (104) corresponds to the third degrees of $q$'s:

$$U_3(q_a, q_b) = \sum_{m+n=3} \alpha_{mn} q_a^m q_b^n \qquad (106)$$

(the sum over cubic terms $\alpha_{03} q_b^3 + \alpha_{12} q_a q_b^2 + \cdots$). It is clear that not all terms in the sum (106) are equally important under the Fermi resonance condition

$$2\omega_a \simeq \omega_b. \qquad (107)$$

Indeed, in the equations of motion

$$M_a(\ddot{q}_a + \omega_a^2 q_a) = -\partial U_3/\partial q_a = -3\alpha_{30} q_a^2 - 2\alpha_{21} q_a q_b - \alpha_{12} q_b^2,$$
$$M_b(\ddot{q}_b + \omega_b^2 q_b) = -\partial U_3/\partial q_b = -\alpha_{21} q_a^2 - 2\alpha_{12} q_a q_b - 3\alpha_{03} q_b^2, \qquad (108)$$

the different terms on the right-hand side have different physical sense. For example, the terms

$$-3\alpha_{30} q_a^2 \quad \text{and} \quad -3\alpha_{03} q_b^2$$

describe a weak nonlinearity of separate eigenmodes and are not responsible for their interaction at all; hence we can omit them from the sum (106). The other terms differ from each other by their time dependence. In harmonic approximation we have

$$q_a = q_a^0 \cos(\omega_a t + \phi_a), \qquad q_b = q_b^0 \cos(\omega_b t + \phi_b), \qquad (109)$$

and the potential (106) is a small perturbation leading to slow variation of $q_a^0$ and $q_b^0$. Substitution of (109) into the right-hand sides of (108) leads to resonant and non-resonant terms. For example, in the first equation (108) the left-hand side oscillates with approximately harmonic frequency $\omega_a$ whereas the term

$$q_b^2 \propto \cos^2 \omega_b t = (1 + \cos 2\omega_b t)/2$$

does not contain such a harmonic and describes a non-resonant interaction of modes $q_a$ and $q_b$. Hence we must hold in this equation only the term

$$q_a q_b \propto \cos \omega_a \cos \omega_b = \left[\cos(\omega_b + \omega_a)t + \cos(\omega_b - \omega_a)t\right]/2$$

which according to the condition (107) contains the resonant "force" oscillating with frequency $\omega_b - \omega_a \simeq \omega_a$. Analogously, in the second equation (108) we must hold only the term $\alpha_{21} q_a^2$ which contains the resonant "force" oscillating approximately with the frequency $2\omega_a \simeq \omega_b$ of the mode $q_b$. Both these terms arise from the potential energy $\alpha_{21} q_a^2 q_b$, which includes in addition some non-resonant terms which must be removed in the so-called *rotating wave approximation* we use here. To this end, it is convenient to introduce into the classical equations of motion

$$M_a(\ddot{q}_a + \omega_a^2 q_a) + 2\alpha_{21} q_a q_b = 0, \qquad M_b(\ddot{q}_b + \omega_b^2 q_b) + \alpha_{21} q_a^2 = 0, \quad (110)$$

where we took into account the relevant $\alpha_{21} q_a^2 q_b$ interaction term only, the complex variables

$$A = \sqrt{\frac{M_a}{2\omega_a}} (\omega_a q_a + i\dot{q}_a), \qquad B = \sqrt{\frac{M_b}{2\omega_b}} (\omega_b q_b + i\dot{q}_b), \quad (111)$$

and their complex conjugate $A^*$ and $B^*$, so that

$$q_a = \frac{1}{\sqrt{2M_a\omega_a}}(A + A^*), \qquad q_b = \frac{1}{\sqrt{2M_b\omega_b}}(B + B^*), \quad (112)$$

$$\dot{q}_a = i\sqrt{\frac{\omega_a}{2M_a}}(A^* - A), \qquad \dot{q}_b = i\sqrt{\frac{\omega_b}{2M_b}}(B^* - B). \quad (113)$$

Simple transformation yields the following equations for complex amplitudes $A$ and $B$:

$$\dot{A} = -i\omega_a A - 2 \cdot \frac{\alpha_{21} i}{2M_a\omega_a\sqrt{2M_b\omega_b}}(A + A^*)(B + B^*), \quad (114)$$

$$\dot{B} = -i\omega_b B - \frac{\alpha_{21} i}{2M_a\omega_a\sqrt{2M_b\omega_b}}(A + A^*)^2. \quad (115)$$

In harmonic approximation the last terms in these equations can be omitted, so in this case we get

$$A(t) = A(0)\exp(-i\omega_a t), \qquad B(t) = B(0)\exp(-i\omega_b t). \quad (116)$$

As we stressed above, the rotating wave approximation consists of taking into account only the terms varying with time with approximately the same frequencies, that is, according to the resonance condition (107) we neglect in Eq. (114) all interaction terms except $A^*B \propto \exp[-i(\omega_b - \omega_a)t]$ and in Eq. (115) all interaction terms except $A^2 \propto \exp(-2i\omega_a t)$. As a result, we arrive at the following simple system of equations

$$i\dot{A} = \omega_a A + 2\Gamma A^* B, \qquad i\dot{B} = \omega_b B + \Gamma A^2, \quad (117)$$

where the notation for the interaction constant

$$\Gamma = \frac{\alpha_{21}}{2M_a\omega_a\sqrt{2M_b\omega_b}} \quad (118)$$

is introduced.

Eqs. (117) arise in various physical contexts. In particular, they describe the process of second harmonic generation in nonlinear optics (see, e.g., [13]). In fact, they can be applied to any process in which two classical oscillating modes (waves) transform one into the other under the resonance condition $2\omega_a \simeq \omega_b$.

It is important that Eqs. (117) can be derived from the Lagrangian

$$L = \frac{i}{2}(\dot{A}^*A - A^*\dot{A} + \dot{B}^*B - B^*\dot{B}) + \omega_a A^*A + \omega_b B^*B$$
$$+ \Gamma(A^{*2}B + A^2B^*), \qquad (119)$$

where $A$, $A^*$, $B$, $B^*$ are considered as independent variables. Then the Lagrange equations

$$\frac{d}{dt}\frac{\partial L}{\partial \dot{A}^*} - \frac{\partial L}{\partial A^*} = 0, \qquad \frac{d}{dt}\frac{\partial L}{\partial \dot{B}^*} - \frac{\partial L}{\partial B^*} = 0$$

reproduce Eq. (117). The classical Hamiltonian has the form

$$H = \omega_a A^*A + \omega_b B^*B + \Gamma(A^2 B^* + A^{*2}B), \qquad (120)$$

where pairs of canonically conjugate variables are $A$, $A^*$ and $B$, $B^*$, respectively. Then the Hamiltonian form of the equations of motion is as follows:

$$i\dot{A} = \partial H/\partial A^*, \qquad i\dot{A}^* = -\partial H/\partial A,$$
$$i\dot{B} = \partial H/\partial B^*, \qquad i\dot{B}^* = -\partial H/\partial B. \qquad (121)$$

Having the Hamiltonian treatment, it is easy to proceed to quantum mechanical formulation of the Fermi resonance problem. The classical variables (111) and their complex conjugates correspond to the "annihilation" and "creation" operators of quantum oscillator:

$$A \to \hbar^{1/2}\hat{a}, \quad A^* \to \hbar^{1/2}\hat{a}^\dagger, \quad B \to \hbar^{1/2}\hat{b}, \quad B^* \to \hbar^{1/2}\hat{b}^\dagger, \qquad (122)$$

where "dagger" denotes the Hermitian conjugation and the operators $\hat{a}$, $\hat{a}^\dagger$, and $\hat{b}$, $\hat{b}^\dagger$ obey the usual commutation relations

$$[\hat{a}, \hat{a}^\dagger] = 1, \qquad [\hat{b}, \hat{b}^\dagger] = 1, \qquad [\hat{a}, \hat{b}] = [\hat{a}^\dagger, \hat{b}^\dagger] = 0. \qquad (123)$$

The classical Hamiltonian (120) converts into quantum mechanical one in the following way

$$\widehat{H} = \hbar\omega_a \hat{a}^\dagger\hat{a} + \hbar\omega_b \hat{b}^\dagger\hat{b} + \hbar^{3/2}\Gamma(\hat{a}^2\hat{b}^\dagger + \hat{a}^{\dagger 2}\hat{b}), \qquad (124)$$

where the product of non-commuting variables is replaced by their mean value:

$$A^*A \to \frac{\hbar}{2}(\hat{a}^\dagger\hat{a} + \hat{a}\hat{a}^\dagger) = \hbar(\hat{a}^\dagger\hat{a} + \tfrac{1}{2}),$$

and zero oscillation terms are dropped.

Now we are ready to discuss the energy splitting of molecular Fermi resonance states. From classical point of view, the stationary states correspond to a purely periodic dependence of the amplitudes $A$ and $B$ on time:

$$A = A_0 \exp\left(-\frac{i\Omega}{2}t\right), \qquad B = B_0 \exp(-i\Omega t). \tag{125}$$

So we seek the solution of Eqs. (117) in the form (125) which leads to the system

$$\begin{aligned}(\omega_a - \Omega/2)A_0 + 2\Gamma A_0^* B_0 &= 0, \\ (\omega_b - \Omega)B_0 + \Gamma A_0^2 &= 0.\end{aligned} \tag{126}$$

We multiply the first equation by $A_0$, introduce the "intensity" of vibration

$$I = |A_0|^2, \tag{127}$$

analogous to the number of quanta, and eliminate $A_0^2$ and $B_0$ from this system, which gives the equation

$$(\omega_a - \Omega/2)(\omega_b - \Omega) = 2\Gamma^2 I \tag{128}$$

for calculation of the frequency $\Omega$. This equation has two roots

$$\Omega_{1,2} = \omega_a + \omega_b/2 \pm \sqrt{(\omega_a - \omega_b/2)^2 + 4\Gamma^2 I}. \tag{129}$$

We see that our classical system has two "eigenfrequencies" depending on intensity $I$ of $A$-mode. In quantum terms, the intensity $I$ is proportional to the number of $a$ quanta, $n_a = \hat{a}^\dagger \hat{a} = I/\hbar$, $n_b = \hat{b}^\dagger \hat{b} = I/(2\hbar)$, which in the classical limit are much greater than unity, $n_a = 2n_b \gg 1$. However in usual infrared experiments the lines observed correspond to the excitation of the state with $n_a = 2$, $n_b = 1$. Let us try to apply our classical formula (129) to this quantum region, i.e., we substitute in it $I = A^*A = 2\hbar$ (two $a$-quanta). As a result we obtain that the energies of these two states are equal to

$$E_{1,2} = \hbar\Omega_{1,2} = \hbar\omega_a + \hbar\omega_b/2 \pm \sqrt{(\hbar\omega_a - \hbar\omega_b/2)^2 + 8\hbar^3\Gamma^2}. \tag{130}$$

It is interesting that this semiclassical formula almost reproduces the exact quantum result. Indeed, the interaction term in quantum Hamiltonian (124) couples the states

$$|\psi_1\rangle = |2_a 0_b\rangle \quad \text{and} \quad |\psi_2\rangle = |0_a 1_b\rangle, \tag{131}$$

where $|n_a m_b\rangle = |n\rangle_a |m\rangle_b$ is the oscillators' state with $n$ $a$-quanta and $m$ $b$-quanta. Using the well-known relations

$$\begin{aligned}\hat{a}|n\rangle_a &= \sqrt{n}|n-1\rangle_a, & \hat{a}^\dagger|n\rangle_a &= \sqrt{n+1}|n+1\rangle_a, \\ \hat{b}|m\rangle_b &= \sqrt{m}|m-1\rangle_b, & \hat{b}^\dagger|m\rangle_b &= \sqrt{m+1}|m+1\rangle_b,\end{aligned} \tag{132}$$

we obtain the matrix elements of the Hamiltonian in the basis (131):

$$H = \begin{pmatrix} 2\hbar\omega_a & \sqrt{2}\hbar^{3/2}\Gamma \\ \sqrt{2}\hbar^{3/2}\Gamma & \hbar\omega_a \end{pmatrix}. \qquad (133)$$

The eigenstates' energies are determined by the secular equation

$$(2\hbar\omega_a - E)(\hbar\omega_b - E) = 2\hbar^3\Gamma^2$$

and differs by factor 2 from the semiclassical result (128) with $I = 2\hbar$. Thus, we see that such a molecule has two energy levels $E_{1,2}$ each connected with the ground state by one-quantum transition due to large $b$-component in both eigenfunctions. This leads to the observed splitting of the infrared and Raman spectra lines. Just such splitting of the Raman spectra was explained by E. Fermi in his pioneering article [12].

## 7.2. Fermi-Resonance Wave in a Two-Layer System

The next natural step is discussion of Fermi resonance effects in molecular crystals. Let molecules having Fermi resonance between intramolecular vibrations form a molecular crystal due to weak (van der Waals) forces. Then the individual molecular vibrational excitations discussed above become coupled to each other and form collective Fermi resonance bands. We shall consider here a simple two-layer 1D model with intermolecular interaction between only nearest neighbors (see Figure 6).

The quantum mechanical Hamiltonian can be written in the form

$$\hat{H} = \sum_m \left[\hbar\omega_a \hat{a}_m^\dagger \hat{a}_m + \hbar\omega_b \hat{b}_m^\dagger \hat{b}_m + \hbar^{3/2}\Gamma\left(\hat{a}_m^2 \hat{b}_m^\dagger + \hat{a}_m^{\dagger 2}\hat{b}_m\right)\right]$$
$$+ \sum_m \left[\hbar V_a\left(\hat{a}_{m+1}^\dagger \hat{a}_m + \hat{a}_m^\dagger \hat{a}_{m+1}\right) + \hbar V_b\left(\hat{b}_{m+1}^\dagger \hat{b}_m + \hat{b}_m^\dagger \hat{b}_{m+1}\right)\right], \quad (134)$$

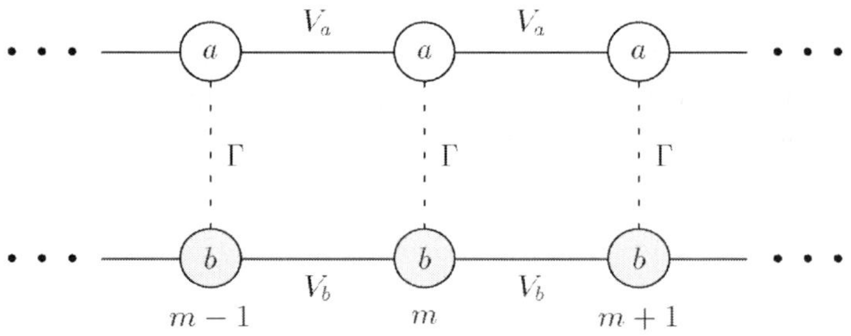

Fig. 6. Two-layer one-dimensional model of a crystal with one molecule in elementary cell of each layer. Vibrational modes in $a$ and $b$ molecules satisfy the Fermi resonance condition $2\omega_a \simeq \omega_b$ and interact with excitations in neighboring cells with constants $V_a$ and $V_b$, respectively. Constant $\Gamma$ describes the nonlinear interaction between molecules in adjacent layers.

where $m$ is the index of the lattice site, the first sum represents molecular Hamiltonians in each site (see Eq. (124)), and the second sum corresponds to the intermolecular interaction of vibrations—the term $\hbar V_a \hat{a}_{m+1}^\dagger \hat{a}_m$ describes a transition of one $a$-quantum from the site $m$ to the site $m+1$, and analogous interpretation have the other terms. In classical approximation we obtain the Hamiltonian

$$H = \sum_m \left[ \omega_a A_m^* A_m + \omega_b B_m^* B_m + \Gamma \left( A_m^2 B_m^* + A_m^{*2} B_m \right) \right]$$

$$+ \sum_m \left[ V_a \left( A_{m+1}^* A_m + A_m^* A_{m+1} \right) \right.$$

$$\left. + V_b \left( B_{m+1}^* B_m + B_m^* B_{m+1} \right) \right], \quad (135)$$

and the following equations of motion for complex amplitudes $A_m$ and $B_m$:

$$i \partial A_m / \partial t = \partial H / \partial A_m^* = \omega_a A_m + V_a (A_{m-1} + A_{m+1}) + 2 \Gamma A_m^* B_m,$$

$$i \partial B_m / \partial t = \partial H / \partial B_m^* = \omega_b B_m + V_b (B_{m-1} + B_{m+1}) + \Gamma A_m^2. \quad (136)$$

Let us look for the solution in the form of plane wave

$$A_m = A \exp[-i(\Omega t - Km)/2], \qquad B_m = B \exp[-i(\Omega t - Km)] \quad (137)$$

(we assume that the lattice constant is equal to unity). Then the infinite system reduces to a simple system of two algebraic equations

$$(\omega_a - \Omega/2 + 2V_a \cos(K/2))A + 2\Gamma A^* B = 0,$$

$$(\omega_b - \Omega + 2V_b \cos K)B + \Gamma A^2 = 0, \quad (138)$$

which actually coincides with the system (126). If there were no Fermi resonance interaction ($\Gamma = 0$), then we would have two linear modes with well-known dispersion laws

$$\Omega_1(K) = 2\omega_a + 4V_a \cos(K/2), \qquad \Omega_2(K) = \omega_b + 2V_b \cos K. \quad (139)$$

Fermi resonance coupling between molecular vibrations leads to interaction of these linear modes with each other which gives rise to mixed waves. To obtain their dispersion law, we again introduce the intensity

$$I = |A|^2$$

and reduce the system to the equation

$$(\omega_a - \Omega/2 + 2V_a \cos(K/2))(\omega_b - \Omega + 2V_b \cos K) = 2\Gamma^2 I \quad (140)$$

with the solutions

$$\Omega_{1,2}(K) = \omega_a + \omega_b/2 + 2V_a \cos(K/2) + V_b \cos K$$

$$\pm \left[ (\omega_a - \omega_b/2 + 2V_a \cos(K/2) - V_b \cos K)^2 + 8\Gamma^2 I \right]^{1/2}.$$

These expressions define the dispersion laws of normal modes arising from linear plane waves due to nonlinear Fermi resonance interaction. It is important that the nonlinearity leads to the dependence of the dispersion laws on the intensity $I$ of vibrations. Such a dependence gives rise to the soliton solutions discussed in the following sections.

The relations (140) can be considered as a good enough approximation only in the limit of large intensity

$$I \gg \hbar. \qquad (141)$$

Nevertheless, as we saw in the preceding section, semiclassical formulas give exact enough results even in the quantum region $I \geqslant \hbar$. More exact relations can be easily derived by means of quantum mechanical treatment [14–16].

### 7.3. Fermi Resonance Interface Waves

Now we shall consider the case when one interface separates two simple cubic 2D crystals composed of different molecules of $a$ and $b$ types interacting across the interface. In two dimensions with line interface (3D generalization is straightforward) the equations of motion for the amplitudes $A_{mn}$ and $B_{mn}$ read (compare with Eqs. (136))

$$\vdots \qquad \vdots$$

$$n = 2: i\dot{A}_{2,m} = \omega_a A_{2,m} + V_a(A_{2,m-1} + A_{2,m+1} + A_{1,m} + A_{3,m}),$$

$$n = 1: i\dot{A}_{1,m} = \omega_a A_{1,m} + V_a(A_{1,m-1} + A_{1,m+1} + A_{2,m}) + 2\Gamma A^*_{1,m} B_{0,m},$$

$$n = 0: i\dot{B}_{0,m} = \omega_b B_{0,m} + V_b(B_{0,m-1} + B_{0,m+1} + B_{-1,m}) + \Gamma A^2_{1,m},$$

$$n = -1: i\dot{B}_{-1,m} = \omega_b B_{-1,m} + V_b(B_{-1,m-1} + B_{-1,m+1} + B_{-2,m} + B_{0,m}),$$

$$\vdots \qquad \vdots \qquad (142)$$

where the equations with $n > 2$ and $n < -1$ have the same structure as at $n = 2$ and $n = -1$, respectively. They describe the propagation of excitations in bulk $a$ and $b$ crystals above and below the interface. New interface ("surface") modes arise due to Fermi resonance interaction across the interface [17–19]. We look for the solution in the form

$$A_{n,m} = A e^{-\kappa_a(n-1)} \exp[-i(\Omega t - Km)/2], \quad \text{for } n \geqslant 1,$$
$$B_{n,m} = B e^{\kappa_b n} \exp[-i(\Omega t - Km)], \quad \text{for } n \leqslant 0, \qquad (143)$$

where the amplitudes of vibrations exponentially decay as we go away from the interface. Then Eqs. (142) with $n \geqslant 2$ and $n \leqslant -1$ are solved by these functions provided

$$\Omega/2 = \omega_a + 2V_a \cos(K/2) + 2V_a \cosh \kappa_a,$$
$$\Omega = \omega_b + 2V_b \cos K + 2V_b \cosh \kappa_b. \tag{144}$$

These equations give us the values of $\kappa_a$ and $\kappa_b$ as functions of $K$ and $\Omega$. Then Eqs. (142) reduce to

$$V_a \exp(\kappa_a) A = 2\Gamma A^* B, \tag{145}$$

$$V_b \exp(\kappa_b) B = \Gamma A^2. \tag{146}$$

We multiply (145) by $A$ and introduce the intensity of vibrations

$$I = |A|^2, \tag{147}$$

so that Eqs. (145), (146) give

$$V_a V_b \exp(\kappa_a + \kappa_b) = 2\Gamma^2 I. \tag{148}$$

Elimination of $V_a e^{\kappa_a}$ and $V_b e^{\kappa_b}$ from this equation can be done with the use of Eqs. (144), and yields the relation defining implicitly the dependence of $\Omega$ on $K$, i.e., the dispersion laws of the Fermi resonance modes

$$\left\{ \Omega/2 - \omega_a - 2V_a \cos(K/2) + \text{sgn}\bigl(\Omega/2 - \omega_a - 2V_a \cos(K/2)\bigr) \right.$$
$$\left. \times \sqrt{(\Omega/2 - \omega_a - 2V_a \cos(K/2))^2 - 4V_a^2} \right\}$$
$$\times \left\{ \Omega - \omega_b - 2V_b \cos K + \text{sgn}(\Omega - \omega_b - 2V_b \cos K) \right.$$
$$\left. \times \sqrt{(\Omega - \omega_b - 2V_b \cos K)^2 - 4V_b^2} \right\} = 8\Gamma^2 I. \tag{149}$$

We see that they are modified compared to the case of Fermi resonance waves (140), (141) in infinite 1D cubic crystal. At $I = 0$ (or $\Gamma = 0$), when interaction across the interface disappears, Eq. (149) reproduces two dispersion relations for surface waves in half-infinite crystals.

Note that addition of a new dimension to 1D cubic crystal with "point" interface leads to the replacements

$$\omega_a \to \omega_a + 2V_a \cos \frac{K}{2}, \qquad \omega_b \to \omega_b + 2V_b \cos K, \tag{150}$$

which are a result of translational invariance along the interface following from the interaction terms in the Hamiltonian (we omit index $n$ numbering sites along the axis perpendicular to the interface)

$$\sum_m \left[ V_a \bigl(a_m^\dagger a_{m-1} + a_{m-1}^\dagger a_m\bigr) + V_b \bigl(b_m^\dagger b_{m-1} + b_{m-1}^\dagger b_m\bigr) \right].$$

If we add the third dimension (labelled by index $l$) along the interface, we can take it into account by means of the following replacements in the formulas for 1D case

$$\omega_a \to \omega_a + 2V_a\left(\cos\frac{K_m}{2} + \cos\frac{K_l}{2}\right),$$

$$\omega_b \to \omega_b + 2V_b(\cos K_m + \cos K_l),$$

(151)

$K_m$ and $K_l$ being the wave vectors along $m$ and $l$ axes, respectively. Thus, we conclude that in calculations of dispersion laws of waves propagating in superlattices with plane interfaces, it is sufficient to consider first the one-dimensional models with the axis directed perpendicular to the interfaces. Then the general formulas for dispersion laws can be obtained by means of replacements (150) or (151). So we reduce 2D or 3D problems to 1D problem with only one coordinate directed perpendicular to interfaces.

### 7.4. BISTABLE ENERGY TRANSMISSION THROUGH THE INTERFACE WITH FERMI RESONANCE INTERACTION

Interesting phenomena can take place in systems with Fermi resonance under influence of external electromagnetic field. Here we consider one example of such behavior—bistable energy transmission through the interface with Fermi resonance interaction [18]. To make calculations easier, we shall consider the following simplified model. Let monomolecular layer of $a$ molecules be deposited on the plane surface of a crystal made of $b$ molecules. For 1D case such a system is shown in Figure 7.

Let $a$ molecules interact with electromagnetic field

$$\mathcal{E}(t) = E + E^*, \qquad E = E_0 \exp(-i\omega_L t).$$

(152)

in resonance with $\omega_a$ vibrations, $\omega_L \approx \omega_a \simeq \omega_b/2$, so that we can neglect the direct pumping of $\omega_b$ excitations. We take the dipole moment of $a$ molecules to be linear in their coordinates $q_a$ so that the electromagnetic interaction term in Lagrangian is proportional to (see Eqs. (112))

$$q_a\mathcal{E}(t) = \frac{1}{\sqrt{2M_a\omega_a}}(A + A^*)(E + E^*) \simeq \frac{1}{\sqrt{2M_a\omega_a}}(AE^* + A^*E),$$

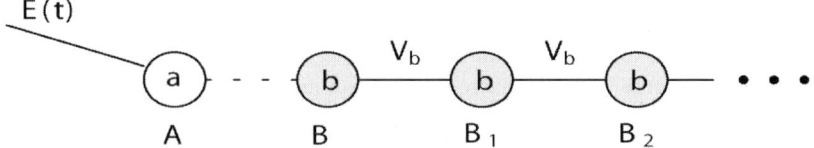

Fig. 7. The sketch of 1D interface structure under influence of electromagnetic field.

where we have omitted the fast oscillating terms $AE \propto \exp(-2i\omega_L t)$ and $A^*E^* \propto \exp(2i\omega_L t)$ according to our rotating wave approximation. Thus we have

$$L_{\text{int}} = \mu(AE^* + A^*E), \qquad (153)$$

where $\mu$ up to a constant factor is the dipole moment of $a$ molecules. Correspondingly, the equation of motion of $a$ molecule in our 1D model reads

$$i\partial A/\partial t = \omega_a A + 2\Gamma A^* B + \mu E(t). \qquad (154)$$

The $b$ molecules do not interact with electromagnetic field and their equations of motion have usual form

$$i\partial B/\partial t = \omega_b B + \Gamma A^2 + V_b B_1,$$

$$i\partial B_1/\partial t = \omega_b B_1 + V_b(B + B_2), \ldots. \qquad (155)$$

Now we have driving force $E(t) = E_0 \exp(-i\omega_L t)$ so that the $a$ molecule oscillates with this laser field frequency $\omega_L$ and due to Fermi resonance interaction across the interface this leads to oscillations of $b$ molecules with frequency $2\omega_L$. As a result, we obtain an algebraic system for amplitudes $A, B, B_1, \ldots$:

$$(\omega_a - \omega_L)A + 2\Gamma A^* B + \mu E_0 = 0,$$

$$(\omega_b - 2\omega_L)B + \Gamma A^2 + V_b B_1 = 0,$$

$$(\omega_b - 2\omega_L)B_1 + V_b(B + B_2) = 0, \ldots. \qquad (156)$$

Equations for amplitudes $B_1, B_2, \ldots$ are linear and solved by

$$B_n = \exp(ipn)B_{n-1}, \quad B_1 = \exp(ip)B, \qquad (157)$$

where the wave vector $p$ is determined by the equation

$$2\omega_L = \omega_b + 2V_b \cos p. \qquad (158)$$

Then we get from the first two equations of the system (156)

$$B = \frac{\exp(ip)\Gamma A^2}{V_b}$$

$$= \frac{2\Gamma A^2}{2\omega_L - \omega_b + \text{sgn}(2\omega_L - \omega_b)\sqrt{(2\omega_L - \omega_b)^2 - 4V_b^2}}, \qquad (159)$$

and

$$A\left[\omega_L - \omega_a - \frac{4\Gamma^2|A|^2}{2\omega_L - \omega_b + \text{sgn}(2\omega_L - \omega_b)\sqrt{(2\omega_L - \omega_b)^2 - 4V_b^2}}\right]$$

$$= \mu E. \qquad (160)$$

Now we introduce the "pumping" intensity

$$I_{\text{pump}} = |\mu E|^2/(\omega_a - \omega_L)^2 \qquad (161)$$

and arrive at the following equation

$$I(1 - DI)^2 = I_{\text{pump}} \qquad (162)$$

which determines implicitly the intensity of vibrations

$$I = |A|^2 \qquad (163)$$

as a function of the pumping intensity $I_{\text{pump}}$, where

$$D = \frac{4\Gamma^2}{(\omega_L - \omega_a)\left[2\omega_L - \omega_b + \text{sgn}(2\omega_L - \omega_b)\sqrt{(2\omega_L - \omega_b)^2 - 4V_b^2}\right]}. \qquad (164)$$

The cubic equation (162) can have three roots which indicates on bistability—two values of $I$ correspond to one pumping intensity (third root corresponds to an unstable state). The plot of the function $I(I_{\text{pump}})$ is shown in Figure 8, where one can see the bistability region $0 \leqslant I_{\text{pump}} \leqslant 0.15 I_c$ (with $I_c = 1/D = 1$).

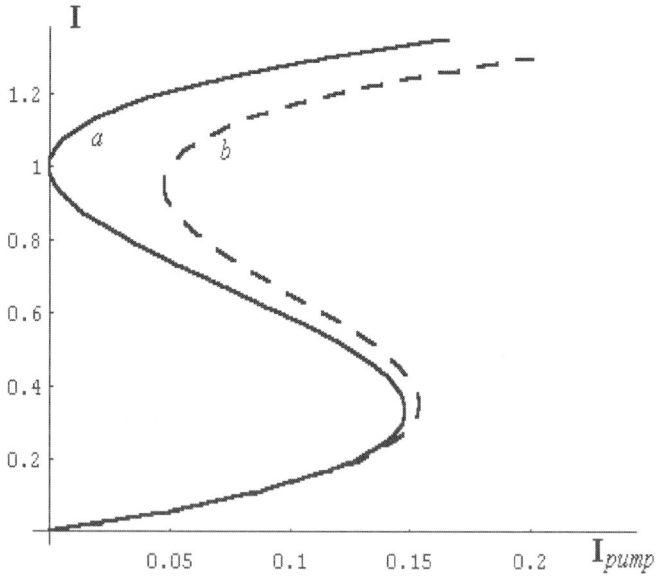

Fig. 8. Dependence of intensity $I$ of $a$ vibrations on pumping intensity: (a) without damping; (b) with damping.

More details can be found in [18]. Here we note only that there exists nonzero solution of (162) even at vanishing pumping:

$$I = 1/D, \tag{165}$$

i.e., when the oscillation frequency of $b$ molecules $\Omega$ satisfies the following equation

$$(\Omega/2 - \omega_a)\left[\Omega - \omega_b + \text{sgn}(\Omega - \omega_b)\sqrt{(\Omega - \omega_b)^2 - 4V_b^2}\right] = 4\Gamma^2 I. \tag{166}$$

This vibrational state existing without pumping is just the Fermi resonance interface state.

In conclusion we note that interfaces introduce many new phenomena in physics of multilayer structures which can find numerous applications to science and technology.

## References

1. V.M. Agranovich, V.E. Kravtsov, "Notes on crystal optics of superlattices", *Solid State Commun.* 55 (1985) 85;
   M. Born, E. Wolf, Principles of Optics, Pergamon, Oxford, 1968, Section 14.5.2.
2. L.D. Landau, E.M. Lifshitz, Electrodynamics of Continuous Media, Pergamon, Oxford, 1960.
3. V.M. Agranovich, "Dielectric permeability and influence of external fields on optical properties of superlattices", *Solid State Commun.* 78 (1991) 747.
4. R.W. Boyd, J.E. Sipe, "Nonlinear optical susceptibilities of layered composite materials", *J. Opt. Soc. Am. B* 11 (1994) 297.
5. G.L. Fisher, R.W. Boyd, R.J. Gehr, et al., "Enhanced nonlinear optical response of composite materials", *Phys. Rev. Lett.* 74 (1975) 1871.
6. R.W. Munn, "Microscopic theory of dielectric response for molecular multilayers", *J. Chem. Phys.* 99 (1993) 1404.
7. N.S. Tyu, "Local field effects and tensors of dielectric permeability in organic superlattices", *Solid State Commun.* 90 (1994) 667.
8. V.M. Agranovich, "Excitons in organic superlattices", *Mol. Cryst. Liq. Cryst.* 230 (1993) 13.
9. V.M. Agranovich, O.A. Dubovsky, "Effect of retarded interaction on exciton spectrum", *JETP Letters* 3 (1966) 223.
10. Ya. Aaviksoo, Ya. Lippmaa, T. Reinot, "Measurement of a picosecond time decay for anthracene surface", *Opt. Spectrosc. USSR* 62 (1987) 419.
11. A.S. Davydov, Theory of Molecular Excitons, Plenum Press, New York, 1971.
12. E. Fermi, "Über den Ramaneffekt des Kohlendioxyds", *Zs. f. Phys.* 71 (1931) 250.
13. A.C. Newell, J.V. Moloney, Nonlinear Optics, Addison-Wesley, Redwood City, 1992.
14. V.M. Agranovich, "Biphonons and Fermi resonance in vibrational spectra of crystals", in: V.M. Agranovich, R.M. Hochstrasser (Eds.), Spectroscopy and Excitation Dynamics of Condensed Molecular Systems, North-Holland, Amsterdam, 1983, p. 83.
15. V.M. Agranovich, I.I. Lalov, "Effects of strong anharmonicity in spectra of optical phonons and polaritons", *Uspekhi Fiz. Nauk* 146 (1985) 267; *Sov. Phys. Uspekhi* 28 (1985) 484.
16. V.M. Agranovich, O.A. Dubovsky, "Phonon multimode spectra: biphonons and triphonons in crystals with defects", in: R.J. Elliott, I.P. Ipatova (Eds.), Optical Properties of Mixed Crystals, North-Holland, Amsterdam, 1988, p. 297.

17. V.M. Agranovich, O.A. Dubovsky, "Fermi resonance interface modes in organic multilayer structures", *Chem. Phys. Lett.* 210 (1993) 458.
18. V.M. Agranovich, J.B. Page, "Fermi resonance interface modes and bistable energy transmission through the interface", *Phys. Lett. A* 183 (1993) 395.
19. V.M. Agranovich, A.M. Kamchatnov, "Fermi resonance interface solitons", *Pis'ma Zh. Exp. Teor. Fiz.* 59 (1994) 397; *JETP Lett.* 59 (1994) 425.

## Chapter 5

# Mixing of Frenkel and Charge-Transfer Excitons and Their Quantum Confinement in Thin Films

### MICHAEL HOFFMANN

*Institut für Angewandte Photophysik, Technische Universität Dresden, 01062 Dresden, Germany*

1. Introduction . . . . . . . . . . . . . . . . . . . . . . . . . . . . . . . . . . . . . . . 221
2. Electronic Frenkel and Charge-Transfer Excitons in Rigid One-Dimensional Crystals . . . 226
3. Strong Coupling of the Electronic Excitations with Internal Phonon Modes . . . . . . . . . 245
4. Applications and Consequences for Quantum Confinement . . . . . . . . . . . . . . . . 276
5. Conclusion . . . . . . . . . . . . . . . . . . . . . . . . . . . . . . . . . . . . . . . 288
   Acknowledgements . . . . . . . . . . . . . . . . . . . . . . . . . . . . . . . . . . . 289
   References . . . . . . . . . . . . . . . . . . . . . . . . . . . . . . . . . . . . . . . 290

## 1. Introduction

For opto-electronic applications of organic semiconductors, a microscopic understanding of the lowest excited states is essential for further developments. This chapter reviews a minimum set of phenomena that are relevant for one particular class of organic semiconductors: quasi-one-dimensional molecular crystals. It is based on the concepts of exciton theory for molecular crystals and it therefore excludes polymers and amorphous molecular solids.

Within a one-dimensional theory, the major effects of Frenkel exciton transfer, mixing of Frenkel and charge-transfer (CT) excitons and coupling to one intramolecular vibrational mode are described within a common framework. This framework was developed to model the exciton states of PTCDA (3,4,9,10-perylenetetracarboxylic dianhydride, see Figure 1) and related perylene pigment dyes.

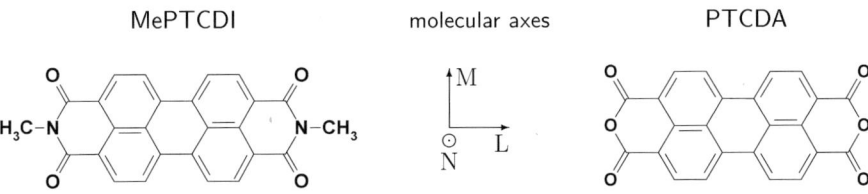

Fig. 1. Chemical structures of the investigated molecules and definition of the molecular axes. MePTCDI = N-N′-dimethylperylene-3,4,9,10-dicarboximide, PTCDA = 3,4,9,10-perylenetetracarboxylic dianhydride.

To understand the optical properties of a semiconductor crystal, one can use two different approaches: One approach is to start from the viewpoint of one-particle states for charge carriers (valence and conduction bands) and then to refine this picture by introducing correlation effect. The correlated many-particle states (e.g., Wannier–Mott excitons) are called large-radius excitons. This approach is useful if the exciton formation is a small energetic effect compared to the formation of the one-particle bands. For example, in GaAs, the one-particle bandwidths are in the order of 2 eV, whereas the binding energy of the Wannier–Mott exciton is only 4.2 meV (cf., e.g., [1]).

The alternative approach starts from the correlated many-particle states of isolated molecules and then introduces intermolecular interactions (small radius exciton theories). This approach is useful if the exciton binding energy is large compared to the one-particle bandwidths and the optical spectra are more closely related to the molecular states than to one-particle crystal states. Such a situation is typical for molecular crystals, which are characterized by small Van der Waals interactions between the molecules. Even for crystals with coplanar stacking and relatively strong interactions, e.g., PTCDA, the binding energy of the lowest excitons is on the order of 1 eV,[1] compared to estimated one-particle bandwidths on the order of 0.2 eV (see Section 4.1).

The most basic small radius exciton theory considers only Frenkel excitons, i.e., crystal states are described as superpositions of neutral molecular states. That means, all states in which electrons would be transferred from one molecule to another one are excluded, and the exciton radius defined as the mean separation between the excited electron and the remaining hole is necessarily smaller than one lattice constant. Frenkel exciton theory is now a standard tool described in many reviews and monographs (e.g., [3–5]). It was extensively applied in the third quarter of the 20th century to describe optical properties of naphthalene and anthracene crystals. In its mature versions, it tries to include as many interactions as possible to obtain quantitative predictions, in particular for the well defined phenomenon of Davydov splitting (see, e.g., [6]).

Current interest in molecular crystals concentrates on materials with promising properties for opto-electronics. Especially the demand for reasonably high charge-carrier mobilities leads to materials with strong intermolecular interactions. In connection with the demand for exciton energies in the visible range, this draws attention to aromatic dyes in which the molecules form stacks with a close coplanar arrangement of the aromatic system. Prominent examples are derivatives of the perylene tetracarboxylic acid or phthalocyanines. We will use two perylene derivatives as model compounds for this situation: MePTCDI (N-N'-

---

[1] Binding energy of the lowest CT state in PTCDA from Ref. [2]. The lowest Frenkel exciton lies even 0.1 eV below, see Section 4.1.

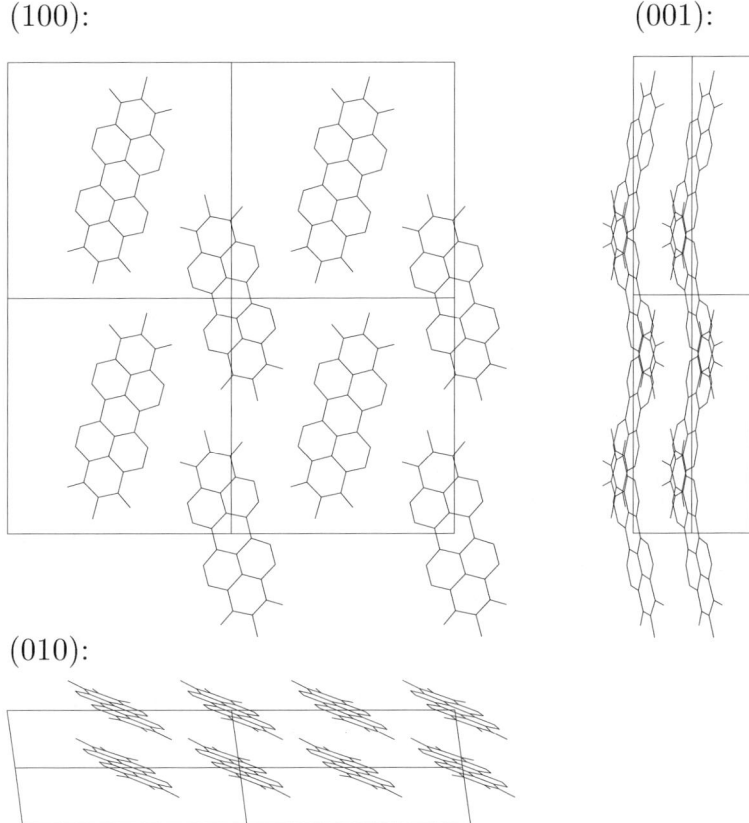

Fig. 2. Crystal structure of MePTCDI from data of Ref. [85]. We show the projections of $2 \times 2$ unit cells onto the $b$–$c$-plane (100), the $a$–$b$-plane (001) and onto the $a$–$c$-plane (010). The crystal structure is monoclinic, space group $P2_1/c$, $Z = 2$ molecules per unit cell, $a = 3.874$ Å, $b = 15.580$ Å, $c = 14.597$ Å, $\beta = 97.65°$.

dimethylperylene-3,4,9,10-dicarboximide) and PTCDA (3,4,9,10-perylenetetracarboxylic dianhydride, see Figure 1).

PTCDA has become a paradigm because it readily forms highly ordered films [7,8], while related perylene derivatives have solar cells applications [9–11]. Several works have sought to understand the PTCDA absorption spectrum and related properties of its electronic excitations [12–18]. A whole class of PTCDA-derivatives is provided by various substituents in place of the methyl group in MePTCDI. The crystal structures of such PTCDA-derivatives are always characterized by molecular stacks, as shown for MePTCDI in Figure 2. The geometrical arrangement suggests that the interactions within the stacks will be much stronger than inter-stack interactions. Therefore, we call such materials quasi-one-dimensional. An experimental evidence of the quasi-one-dimensional nature is given by the value of the Davydov splitting, which characterizes the inter-stack interaction between the translationally non-equivalent molecules of the unit cell. In MePTCDI, this Davydov splitting was observed, and it is indeed much smaller than the effects related to interactions within the stacks [16].

A major advantage of PTCDA-derivatives is simple and accessible molecular behavior. The lowest electronic excitation is a dipole allowed $\pi-\pi^*$ transition with a strong transition dipole along the long molecular axis (e.g., [19,20]). It couples predominantly to one effective vibrational mode of the carbon-backbone (cf. Section 3.1), which causes the vibronic progression seen in the absorption spectrum in Figure 3(a) between 2.2 and 3.3 eV. The next higher allowed electronic transition is the small feature at 3.4 eV in Figure 3(a), which is related to an $M$-axis polarized transition [19,20].

Since the optical spectra of the isolated molecule are determined by the conjugated system they are almost independent of the outer substituents [21]. However, the changes upon crystal formation depend sensitively on the actual crystal structure. Klebe et al. [21] discuss empirically the relation between the crystal absorption spectra and the crystal geometry for a large set of PTCDA-derivatives. Because of the quasi-one-dimensional nature of the crystals, and because of the nearly constant distance between the molecular planes in the stacks, the crucial parameter is the lateral displacement of neighboring molecules. We illustrate this dependence for our two model compounds MePTCDI and PTCDA in Figures 3(b) and 3(c).

On a very rough scale, the crystal spectra in Figures 3(b) and 3(c) still show similarities with the monomer spectrum, which corresponds to the nature of a molecular crystal and motivates the approach by small radius exciton theories. However, on the scale of the vibronic progression (0.17 eV), the differences between the monomer and the crystal spectrum are pronounced. This effect of the intermolecular interactions is much stronger than for the lowest transition in the early model compound anthracene. These strong interactions require the inclusion of charge-transfer states in the small radius exciton description.

This chapter will present the basic models that are needed to describe the lowest exciton states in such quasi-one-dimensional systems. In Section 2, we describe the purely electronic problem for a small radius exciton theory in a one-dimensional molecular crystal. We emphasize the relation between the general Merrifield Hamiltonian and minimum models involving just Frenkel and nearest-neighbor charge-transfer states. In Section 3, the tools for the inclusion of internal phonons are discussed. Section 4 demonstrates the path to quantitative applications of these models. In Section 4.1, it is shown that the electronic Hamiltonian for Frenkel and CT excitons, together with coupling to internal phonons,

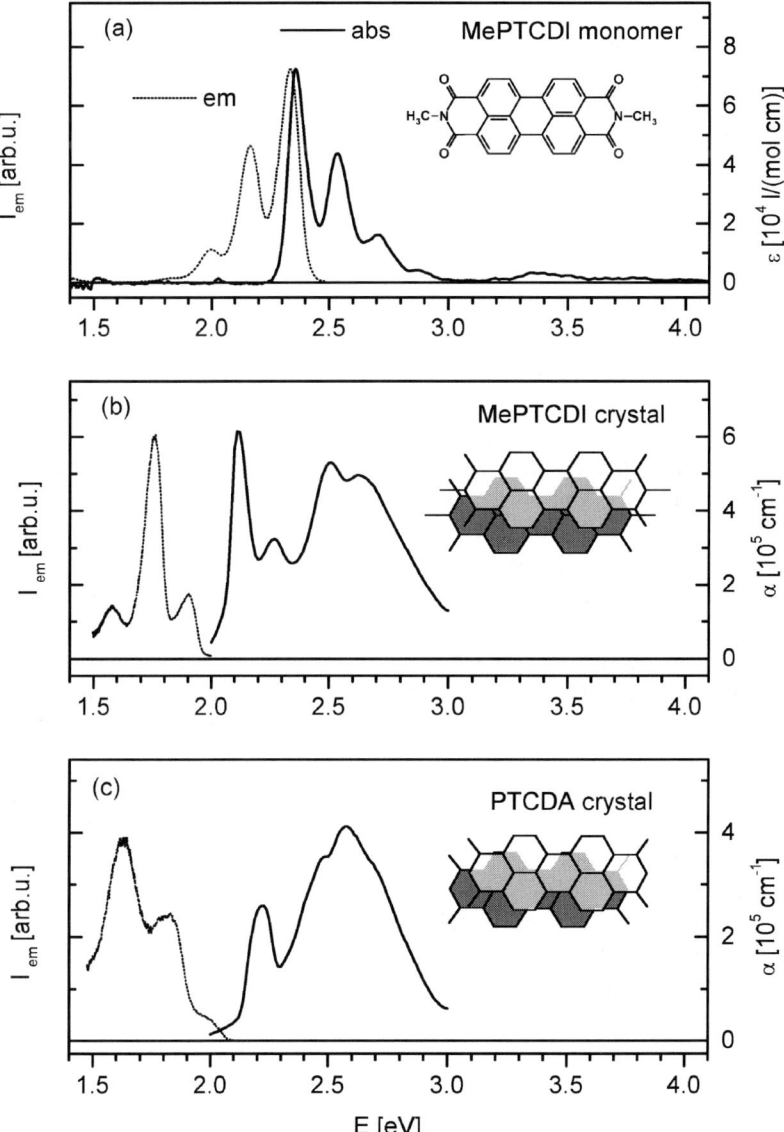

Fig. 3. Absorption and emission spectra of the monomer unit compared to crystal spectra. (a) cw spectra of MePTCDI dissolved in chloroform at room temperature, absorption spectrum (at concentration 0.5 µM) from [16], emission spectrum at concentration 0.3 µM. (b) and (c): absorption spectra of thin polycrystalline films at 10 K, from [16], emission spectra of small single crystals, MePTCDI from [106], PTCDA from [88]. The insets in (b) and (c) show the arrangement of nearest stack neighbors projected onto the molecular planes.

can give a realistic description of absorption spectra of the model compounds. In Section 4.2, the implications of such models for finite systems are discussed.

## 2. Electronic Frenkel and Charge-Transfer Excitons in Rigid One-Dimensional Crystals

### 2.1. Localized Basis States in Real and Momentum Space

Let us first consider the purely electronic states of a one-dimensional molecular crystal. That means, we presume that the positions of all nuclei are fixed in space. In a molecular crystal the constituting molecules retain their individuality to a high degree. This molecular structure provides the basis for small radius exciton theories, in which the *changes* of the electronic structure relative to a reference system of non-interacting molecules are considered.

In many cases, the problem can be conveniently posed in form of empirical model Hamiltonians. That means, one defines the system by introducing a basis set of localized excited states and by choosing the matrix elements between them. Then, the properties of the model can be studied and compared to experiments. On this level, an explicit microscopic definition of the basis states and matrix elements is not necessary.

Let $|\varphi_n^f\rangle$ denote such an excited basis state localized at the lattice position $n$ in the one-dimensional crystal. In simple cases, this might be a localized neutral excitation (Frenkel exciton) or an ionized molecule (charge carrier). Generally, it might be a state with an arbitrary internal structure, which has to be specified by further variables (e.g., the radius of a charge-transfer state). This set of internal variables shall be denoted by $f$. These states are assumed to form an orthonormal basis:

$$\langle \varphi_m^g | \varphi_n^f \rangle = \delta_{mn} \delta_{gf}. \tag{1}$$

We denote the electronic ground state of the crystal by $|o\rangle$ and introduce creation and annihilation operators as

$$X_{nf}^\dagger |o\rangle = |\varphi_n^f\rangle, \tag{2}$$

$$X_{nf} |\varphi_n^f\rangle = |o\rangle. \tag{3}$$

The Hamiltonian in the representation of these localized basis states consists of on-site energies

$$\varepsilon_f = \langle \varphi_n^f | H | \varphi_n^f \rangle \tag{4}$$

and transfer integrals

$$J_{nf,mg} = \langle \varphi_m^g | H | \varphi_n^f \rangle \quad (n \neq m). \tag{5}$$

Because of the translational symmetry in the crystal, the on-site energies do not depend on the site-index $n$ and the transfer integrals only depend on the distance between the site indices:

$$J_{nf,mg} = J_{(n-m)f,0g}. \tag{6}$$

In second quantization, the Hamiltonian can be written as

$$H = \varepsilon_f \sum_{nf} X_{nf}^\dagger X_{nf} + \sum_{\substack{nf \\ mg}}{}' J_{nf,mg} X_{nf}^\dagger X_{mg}. \tag{7}$$

The prime at the summation symbol indicates that the terms for $n = m$ are to be omitted. In this notation, the transfer terms can be visualized as a transfer of the exciton of type $g$ at site $m$ into an exciton of type $f$ at site $n$ ("$mg \Rightarrow nf$").

The translational symmetry of the crystal can be readily used to split the complete Hilbert space of basis states $|\varphi_{nf}\rangle$ into nonmixing subspaces of definite total momentum. We assume a periodicity length of $N$ and transform all operators into their momentum space representation:

$$X_{kf} = \frac{1}{\sqrt{N}} \sum_n e^{-ikn} X_{nf}. \tag{8}$$

Here, the coordinate $k$ of the quasi-momentum takes the values

$$k = \frac{2\pi}{N} l \quad \text{with } l = \ldots, -1, 0, +1, \ldots \text{ and } -\frac{N}{2} < l \leqslant +\frac{N}{2}. \tag{9}$$

The back-transformation is given by

$$X_{nf} = \frac{1}{\sqrt{N}} \sum_k e^{ikn} X_{kf}. \tag{10}$$

In this momentum space representation, operators with different $k$ do not mix anymore:

$$X_{kf}^\dagger X_{k'g} = \delta_{kk'} X_{kf}^\dagger X_{kg}. \tag{11}$$

This can be proven by inserting the momentum space representation (8) into (11)

$$X_{kf}^\dagger X_{k'g} = \frac{1}{N} \sum_{nn'} e^{i(kn - k'n')} X_{nf}^\dagger X_{n'g}$$

and introducing a new summation index $m = n' - n$. Because of the translational symmetry and the periodic boundary conditions, we can substitute $X_{nf}^\dagger X_{(n+m)g} \Rightarrow X_{0f}^\dagger X_{mg}$ and obtain

$$X_{kf}^\dagger X_{k'g} = \frac{1}{N} \sum_n e^{-i(k-k')n} \sum_m e^{-ikm} X_{0f}^\dagger X_{mg}.$$

Using the identity

$$\sum_n e^{-i(k-k')n} = N\delta_{kk'} \quad \text{for all } k, k' \text{ from Eq. (9)}, \tag{12}$$

Eq. (11) is obtained.

Now, the back-transformation (10) can be inserted into the real space representation of $H$ from Eq. (7). If the relation (11) as well as the arguments for its proof are used, the Hamiltonian takes the form:[2]

$$H = \sum_k H_k, \tag{13}$$

$$H_k = \sum_f \varepsilon_f X_{kf}^\dagger X_{kf} + \sum_{fg} L_k^{fg} X_{kf}^\dagger X_{kg}. \tag{14}$$

Here, the symbol $L_k^{fg}$ is used to abbreviate the lattice sum

$$L_k^{fg} = \sum_{m \neq 0} e^{ikm} J_{0f,mg}. \tag{15}$$

Eq. (14) describes the mixing of the various momentum space basis states (8) within the subspace of a given total momentum $k$. The dimension of this $k$-subspaces is determined by the number $F$ of localized basis states at a fixed position. The aim of an empirical small-radius exciton model is the identification of a small number of relevant basis states.

## 2.2. Model Hamiltonians for Frenkel and Charge-Transfer States

One of the simplest examples is the following Frenkel exciton Hamiltonian. As localized basis states one considers states $|n\rangle$ in which molecule number $n$ is in the first excited state whereas all other molecules are in the electronic ground state. A nearest-neighbor hopping model for Frenkel excitons can now be written as

$$H_{NN}^{FE} = \varepsilon_{FE} \sum_n a_n^\dagger a_n + J \sum_n (a_n^\dagger a_{n+1} + a_{n+1}^\dagger a_n). \tag{16}$$

Here, $\varepsilon_{FE}$ denotes the on-site energy of a localized Frenkel exciton and $J$ the nearest-neighbor exciton transfer integral (hopping integral). If introduced as an empirical model, Hamiltonian (16) reflects no more than heuristic assumptions

---

[2] Cf. Ref. [4, p. 123] or [22] for the case of Frenkel excitons.

about the physical situation. One assumes that only one excited state of the molecule is important, one assumes that localized basis states can be introduced corresponding to this excited molecular state and one assumes that only the nearest-neighbor matrix elements are important. A strict microscopic justification and a derivation of the precise meaning of the model parameters is another task on which we will not focus here.

In the case of the Frenkel exciton Hamiltonian (16), the connection to microscopic theory is well established (a widely available review is given in Ref. [4]). In this case, the localized functions $|n\rangle$ are strictly identified with the eigenfunctions of the noninteracting case, and their orthogonality has to be introduced as an approximation. From the analysis of the Schrödinger equation with the complete interaction Hamiltonian, the model parameters can be related to exact expressions using molecular wave functions. For example, it is seen that the on-site energy $\varepsilon_{FE}$ deviates from the excitation energy of the noninteracting molecule by a solvent shift term. Furthermore, the nature of the involved approximations becomes clear. In this case, one is working in the Heitler–London approximation, which is valid only for $|J| \ll \varepsilon_{FE}$.

Using the representation in momentum space, $k$-states according to Eq. (8) can be introduced:

$$a_k = \frac{1}{\sqrt{N}} \sum_n e^{-ikn} a_n. \tag{17}$$

Then, the Frenkel exciton Hamiltonian (16) takes the diagonal form

$$H_{NN}^{FE} = \sum_k H_k, \tag{18}$$

$$H_k = (\varepsilon_f + 2J \cos k) a_k^\dagger a_k. \tag{19}$$

Thus, the momentum space Frenkel excitons $a_k^\dagger |o\rangle$ are already eigenstates of the Hamiltonian $H_{NN}^{FE}$. They form an energy band

$$E(k) = \varepsilon_f + 2J \cos k, \quad k \in [0, \pi]. \tag{20}$$

The Frenkel exciton model from Eq. (16) can be directly extended to the case of a three-dimensional crystal with inclusion of arbitrary exciton transfer integrals $J_{\mathbf{n},\mathbf{m}}$. Then, it reads

$$H_{3D}^{FE} = \varepsilon_{FE} \sum_{\mathbf{n}} a_{\mathbf{n}}^\dagger a_{\mathbf{n}} + {\sum_{\mathbf{n},\mathbf{m}}}' J_{\mathbf{n},\mathbf{m}} a_{\mathbf{n}}^\dagger a_{\mathbf{m}}. \tag{21}$$

Here, the site indices $\mathbf{n}$, $\mathbf{m}$ are three-dimensional vectors. The second sum runs over all pairs $\mathbf{n}$, $\mathbf{m}$ with $\mathbf{n} \neq \mathbf{m}$, which is symbolized by the prime at the summation index. Such a type of exciton model was already suggested by Frenkel in 1931 [23,24] and first applied to molecular crystals by Davydov in 1948 [25].

One way to extend the two-level model (21) is to include higher excited states of the molecule. For the one-dimensional crystal, this leads to model Hamiltonians of the form

$$H = \sum_{nf} \varepsilon_f a_{nf}^\dagger a_{nf} + \sum_{\substack{nf \\ mg}}{}' J_{nf,mg} a_{nf}^\dagger a_{mg}, \qquad (22)$$

where $a_{nf}^\dagger$ now creates a localized Frenkel exciton in level $f$ at molecule $n$ with on-site energy $\varepsilon_f$ and $J_{nf,mg}$ denotes the hopping integrals between the various localized states. The prime at the summation excludes the terms with $n = m$.

The interaction of various molecular excited states (mixing of molecular configurations) as in Eq. (22) was extensively investigated. It was first considered by Craig [26] to explain the experimentally observed polarization ratios of the Davydov components in anthracene crystals. A general treatment in second quantization was given by Agranovich [27]. Reviews are available in Refs. [4,5,28].

Another extension of the two-level model (21) is the inclusion of charge-transfer (CT) states. In the one-dimensional crystal, a localized CT state can be written as

$$|n, f\rangle = c_{n,f}^\dagger |o\rangle, \qquad (23)$$

meaning that an electron is transferred from molecule $n$, where it leaves a hole, to molecule $n + f$. If arbitrary electron–hole distances $f$ are allowed, the complete set of the CT basis states $|n, f\rangle$ can describe not only bound CT states but also unbound states corresponding to free charge carriers.

CT states are the lowest electronic excitations in so-called charge-transfer crystals, which are mixed crystals containing donor and acceptor type molecules. Prominent example are anthracene-PMDA (anthracene as donor and pyromellitic dianhydride as acceptor) or TTF-TCNQ (tetrathiafulvalene as donor and tetracyano-p-quinodimethane as acceptor). The fundamental optical properties of such crystals are reviewed, e.g., in Ref. [29], and electronic model Hamiltonians in second quantization are reviewed, e.g., in Ref. [30]. In such charge-transfer crystals, the lowest electronic excitations are pure CT states and Frenkel excitons can be neglected. We are interested, however, in one-component molecular crystals. In this case, CT excitons are expected to lie energetically above—but maybe close—to the lowest Frenkel excitons. Then, both types of states have to be included in model Hamiltonians, and the true eigenstates can be of mixed character. The possibility of such mixed states was first pointed out in 1957 by Lyons [31]. A detailed theoretical investigation of this general case was presented 1961 in the work of Merrifield [32]. This model was extended for a description of absorption by Hernandez and Choi [33] and for the inclusion of several Frenkel excitons by Pollans and Choi [34]. In anthracene crystals, CT states are essential to describe electro-absorption spectra [35–39] and charge-generation mechanisms [40]. However, their effect on the low energy optical absorption and emission spectra is still very small. In quasi-one-dimensional crystals with close coplanar arrangement of the molecules, Frenkel and CT states are considered to be strongly mixed even in the lowest energy region [15–17,41] and thus this mixing is an essential feature even for the description of the linear absorption spectra.

The general one-dimensional Merrifield model for Frenkel–CT mixing [32] considers one electron (position $n_e$) and one hole (position $n_h$), which can perform nearest neighbor hops independently of each other. The corresponding transfer integrals are $t_e$ and $t_h$, respectively. Electron–hole correlation is introduced by a Coulombic attraction potential $V(n_e - n_h)$. The special situation $n_e = n_h$ is considered as a Frenkel exciton with an on-site energy $\varepsilon_{FE}$. This Frenkel exciton can perform arbitrary hops with hopping integrals $J_{n,m}$ as in the Frenkel exciton Hamiltonian (22).

In Merrifield's work, the localized basis states are expressed by electron and hole positions and the Frenkel exciton occurs just as a special position. Now, we will write down his Hamiltonian in our notation of Frenkel and CT excitons. For Merrifield's general case, our notation becomes much less convenient but it reveals the path towards simplified small radius exciton Hamiltonians. First, we explicitly split the Merrifield Hamiltonian into a Frenkel part $H^{FE}$, a CT part $H^{CT}$ and a term that mixes Frenkel and CT states $H^{FE-CT}$:

$$H_{Merri} = H^{FE}_{Merri} + H^{CT}_{Merri} + H^{FE-CT}_{Merri}. \qquad (24)$$

The Frenkel part corresponds to the Frenkel exciton Hamiltonian (22) with just one molecular configuration:

$$H^{FE}_{Merri} = \varepsilon_{FE} \sum_n a_n^\dagger a_n + {\sum_{nm}}' J_{n,m} a_n^\dagger a_m. \qquad (25)$$

The CT part treats the nearest-neighbor hopping of electrons and holes, which always means the transfer of a CT state with separation $f$ into a CT state with separation $f \pm 1$. Since $f$ can be arbitrarily large, these CT basis states can also describe the motion of an unbound (free) electron–hole pair. However, all Frenkel states (which correspond to an electron–hole separation $f = 0$) have to be excluded now, which we denote by primes at the summation symbols:

$$H^{CT}_{Merri} = {\sum_{nf}}' V(f) c_{n,f}^\dagger c_{n,f}$$

$$+ {\sum_{nf}}' \left[ t_h \underbrace{c_{n+1,f-1}^\dagger c_{n,f}}_{n \xrightarrow{h} n+1} + t_h \underbrace{c_{n-1,f+1}^\dagger c_{n,f}}_{n \xrightarrow{h} n-1} + \text{h.c.} \right]$$

$$+ \sum_{nf}{}' \left[ t_e \underbrace{c^\dagger_{n,f+1} c_{n,f}}_{n+f \xrightarrow{e} n+f+1} + t_e \underbrace{c^\dagger_{n,f-1} c_{n,f}}_{n+f \xrightarrow{e} n+f-1} + \text{h.c.} \right]. \tag{26}$$

The first sum in this notation covers the on-site energies of the CT states, which only depend on the separation $f$. The second sum includes all nearest-neighbor hops of a hole ("$n \xrightarrow{h} n \pm 1$"), and the third sum the corresponding hops of an electron ("$n+f \xrightarrow{e} n+f \pm 1$").

Finally, the Frenkel–CT mixing part in Merrifield's Hamiltonian can be expressed as:

$$H^{\text{FE-CT}}_{\text{Merri}} = \sum_n \left[ t_e \underbrace{a^\dagger_n c_{n,+1}}_{n+1 \xrightarrow{e} n} + t_e \underbrace{a^\dagger_n c_{n,-1}}_{n-1 \xrightarrow{e} n} + \text{h.c.} \right]$$

$$+ \sum_n \left[ t_h \underbrace{a^\dagger_n c_{n+1,-1}}_{n+1 \xrightarrow{h} n} + t_h \underbrace{a^\dagger_n c_{n-1,+1}}_{n-1 \xrightarrow{h} n} + \text{h.c.} \right]. \tag{27}$$

Here, the first sum describes the processes in which a nearest neighbor CT state (separation $|f|=1$) is transferred into a Frenkel exciton by a hop of the electron to the position of the hole ("$n \pm 1 \xrightarrow{e} n$"). The reverse transfer of a Frenkel into a CT state is included in the Hermitian conjugated part. The second sum describes the analogous process for the jump of the hole ("$n \pm 1 \xrightarrow{h} n$").

At this stage, the separation of the charge hopping processes into $H^{\text{CT}}_{\text{Merri}}$ and $H^{\text{FE-CT}}_{\text{Merri}}$ seems like an unnecessary complication. However, this separation makes an important point obvious: The charge hops in $H^{\text{CT}}_{\text{Merri}}$ and in $H^{\text{FE-CT}}_{\text{Merri}}$ are not identical, since they connect different kinds of states. Only in an uncorrelated one-particle picture, it would not make a difference for the hopping integrals whether the final position is neutral or occupied by an opposite charge. Strictly speaking, even the hopping integrals that connect the various CT states in $H^{\text{CT}}_{\text{Merri}}$ might be different from each other. Using one value for all one-particle hopping integrals is a clearly definable approximation in the model. An explicit distinction of the various hops becomes more important with the rising availability of computational methods that allow microscopic calculations of both dissociation integrals and one-particle hopping integrals on highly correlated levels.

As a second important point, the separation into $H^{\text{CT}}_{\text{Merri}}$ and $H^{\text{FE-CT}}_{\text{Merri}}$ allows the approximate separation of low-lying small radius excitons. Let us first consider $H^{\text{CT}}_{\text{Merri}}$. The basis states that are mixed by this CT Hamiltonian have on-site energies $V(f)$. In the simplest approximation, which was also used by Merrifield, the on-site energies follow Coulomb's law $V(f) = V_1/|f|$. Then, the separation between the lowest CT states ($|f|=1$) and the second lowest ones ($|f|=2$) is $V_1/2$. In our model compound PTCDA, the lowest CT states lie at approximately

2.3 eV (cf. Section 4.1) giving a separation of 1.2 eV. Detailed microscopic models predict a separation of 0.5 eV [2].

The mixing of this states with the higher states is determined by the charge carrier hopping integrals $t$. In molecular crystals, these hopping integrals are small and even for quasi-one-dimensional materials such as PTCDA, values on the order of 0.05 eV are typically discussed (cf. Section 4.1). These values are much smaller than the separation from higher CT states. Thus, the lowest CT states will mix only weakly with the higher CT states. In contrast, the separation of the higher CT states becomes smaller and tends to zero for $f \to \infty$. Thus, these higher states will strongly mix with each other.

From this discussion we can derive the qualitative structure of the eigenstates of $H_{\text{Merri}}^{\text{CT}}$. The lowest eigenstates will be dominated by the two lowest CT basis states ($f = -1, +1$). In the momentum space representation, the CT Hamiltonian for the lowest states becomes

$$H_{\text{Merri}}^{\text{CT}}(k) = V(1)\left(c_{k,-1}^{\dagger} c_{k,-1} + c_{k,+1}^{\dagger} c_{k,+1}\right). \tag{28}$$

Thus, the two lowest CT basis states $|k, +1\rangle$ and $|k, -1\rangle$ are two degenerate eigenstates. However, these states do not yet represent the symmetry of the Hamiltonian for inversion about any site $n$. Therefore, one can introduce symmetry adapted basis states

$$\frac{|k, +1\rangle \pm |k, -1\rangle}{\sqrt{2}}.$$

These symmetry adapted states are the simplest version of small radius CT states in a crystal with translational and inversion symmetry. The mixing of such states with Frenkel excitons will be the main topic of the following sections.

For the higher CT states, the matrix elements between CT states with different separation $f$ will be in the same order or larger than their decreasing energetic separation. Therefore, it is not possible anymore to isolate weakly interacting subspaces in the Hamiltonian that correspond to CT states of a certain radius $f$. In Merrifield's analytical solution [32], this situation is treated exactly. Qualitatively, the situation now becomes similar to a Wannier–Mott exciton since for large exciton radii the discrete lattice structure of the molecular crystal can be replaced by a continuum model with effective masses for the charge carriers.[3]

Now, we will consider the situation that only the lowest CT states $|k, \pm 1\rangle$ can mix with the Frenkel states $|k, 0\rangle$. If we neglect all higher CT states in the Merrifield Hamiltonian (24), we arrive at a much simpler Hamiltonian that includes

---

[3]The equivalence between both models can be directly seen in the difference equations (18) of Ref. [32] for the expansion coefficients $\alpha(n)$. For large radii $n$, $(\alpha(n-1)+\alpha(n+1))/2$ can be approximated by the second derivative $\partial^2 \alpha/\partial n^2$ and the difference equation transforms into the Schrödinger equation for the relative coordinate of an electron–hole pair in a Coulomb potential.

only nearest-neighbor CT states and mixing of these states with Frenkel excitons at the site of either the electron or the hole. If, as a further simplification, we also reduce the Frenkel exciton hopping to nearest neighbors, we arrive at a nearest-neighbor Hamiltonian of the form

$$H_{NN} = H_{NN}^{FE} + H_{NN}^{CT} + H_{NN}^{FE-CT}, \tag{29}$$

$$H_{NN}^{FE} = \varepsilon_{FE} \sum_n a_n^\dagger a_n + J \sum_n (a_n^\dagger a_{n+1} + a_{n+1}^\dagger a_n),$$

$$H_{NN}^{CT} = \varepsilon_{CT} \sum_{nf} c_{nf}^\dagger c_{nf},$$

$$H_{NN}^{FE-CT} = \sum_n \left\{ t_e \big( \underbrace{a_n^\dagger c_{n,+1}}_{n+1 \xrightarrow{e} n} + \underbrace{a_n^\dagger c_{n,-1}}_{n-1 \xrightarrow{e} n} \big) + t_h \big( \underbrace{a_n^\dagger c_{n+1,-1}}_{n+1 \xrightarrow{h} n} + \underbrace{a_n^\dagger c_{n-1,+1}}_{n-1 \xrightarrow{h} n} \big) + \text{h.c.} \right\}.$$

In this nearest-neighbor version of the Merrifield Hamiltonian, the FE–CT mixing term is identical with the original equation (27).

After Fourier transformation $a_n \to a_k$ and $c_n \to c_k$ according to Eq. (8), the Hamiltonian becomes:

$$H_{NN} = \sum_k \big( H_{NN}^{FE}(k) + H_{NN}^{CT}(k) + H_{NN}^{FE-CT}(k) \big), \tag{30}$$

$$H_{NN}^{FE}(k) = (\varepsilon_{FE} + 2J \cos k) a_k^\dagger a_k,$$

$$H_{NN}^{CT}(k) = \varepsilon_{CT} \big( c_{k,+1}^\dagger c_{k,+1} + c_{k,-1}^\dagger c_{k,-1} \big),$$

$$H_{NN}^{FE-CT}(k) = a_k^\dagger \big[ (t_e + t_h e^{-ik}) c_{k,+1} + (t_e + t_h e^{ik}) c_{k,-1} \big] + \text{h.c.}$$

Now, the Frenkel–CT mixing part in $H_{NN}^{FE-CT}(k)$ can be further simplified by introducing symmetry adapted CT operators with even or odd symmetry with respect to change of the direction of the charge-transfer:[4]

$$\tilde{c}_{k\pm} \equiv \frac{1}{\sqrt{2} t_k} \big[ (t_e + t_h e^{-ik}) c_{k,+1} \pm (t_e + t_h e^{+ik}) c_{k,-1} \big], \tag{31}$$

where

$$t_k \equiv \sqrt{t_+^2 \cos^2 \frac{k}{2} + t_-^2 \sin^2 \frac{k}{2}} \tag{32}$$

---

[4] $\tilde{c}_{k+}$ is directly chosen as the term [...] multiplied with an—at first unknown—normalization constant. The normalization constant $1/(\sqrt{2} t_k)$ is then obtained by demanding the correct expression of $H_{NN}^{CT}(k)$ in Eq. (34).

with

$$t_\pm \equiv t_e \pm t_h. \tag{33}$$

The Hamiltonian for the CT states then simplifies to:

$$H_{NN}^{CT}(k) = \varepsilon_{CT}\{\tilde{c}_{k+}^\dagger \tilde{c}_{k+} + \tilde{c}_{k-}^\dagger \tilde{c}_{k-}\}, \tag{34}$$

$$H_{NN}^{FE-CT}(k) = \sqrt{2}\, t_k a_k^\dagger \tilde{c}_{k+} + \text{h.c.} \tag{35}$$

Now, the odd operator $\tilde{c}_{k-}$ does not mix with the Frenkel operators $a_k$ anymore. The odd CT state $\tilde{c}_{k-}^\dagger|o\rangle = |k,1\rangle_-$ is a nonmixing eigenstate of $H_{NN}^{CT}(k)$ and thereby of the complete Frenkel–CT Hamiltonian (30). It lies at the energy $\varepsilon_{CT}$ of the localized CT states.

The remaining even part of the $k$-dependent Hamiltonian can be represented by the two even basis states $|k,0\rangle = a_k^\dagger|o\rangle$ and $|k,1\rangle_+ = \tilde{c}_{k+}^\dagger|o\rangle$. In this representation, it has the form of a $2 \times 2$ matrix:

$$\underline{H}_{NN+}(k) = \begin{pmatrix} \varepsilon_{FE} + 2J\cos k & \sqrt{2}\, t_k \\ \sqrt{2}\, t_k & \varepsilon_{CT} \end{pmatrix}. \tag{36}$$

The two eigenstates $|\Psi_j(k)\rangle$ ($j = 1, 2$) of this problem are linear combinations

$$|\Psi_j(k)\rangle = u_{FEj}(k)|k,0\rangle + u_{CTj}(k)|k,1\rangle_+. \tag{37}$$

The composition of the states is characterized by the squared coefficients $|u|^2$ in terms of a Frenkel exciton character

$$F_{FEj}(k) \equiv |u_{FEj}(k)|^2 = |\langle\Psi_j(k)|a_k^\dagger|o\rangle|^2 \tag{38}$$

and a CT character

$$F_{CTj}(k) \equiv |u_{CTj}(k)|^2 = |\langle\Psi_j(k)|\tilde{c}_{k+}^\dagger|o\rangle|^2. \tag{39}$$

From the matrix representation (36), one can directly see that the off-diagonal term between Frenkel and CT states is entirely given by $t_k$. This term depends in a characteristically $k$-dependent way on the combination of the electron and hole transfer integrals. In particular, one has at the boundaries of the Brillouin zone:

$$t_{k=0} = t_+ = t_e + t_h,$$

$$t_{k=\pi} = t_- = t_e - t_h.$$

If one looks, e.g., only at absorption spectra ($k = 0$), the two parameters $t_e$ and $t_h$ are reduced to one effective parameter $t_+$.

For illustration, we show several concrete exciton band structures of this Hamiltonian in Figures 4 and 5. Since an additive constant in the on-site energies has no physical meaning in this model, we always take $\varepsilon_{FE} = 0$. An example without FE–CT mixing is shown in Figure 4(a) ($t_e = t_h = 0$). All other pictures demonstrate

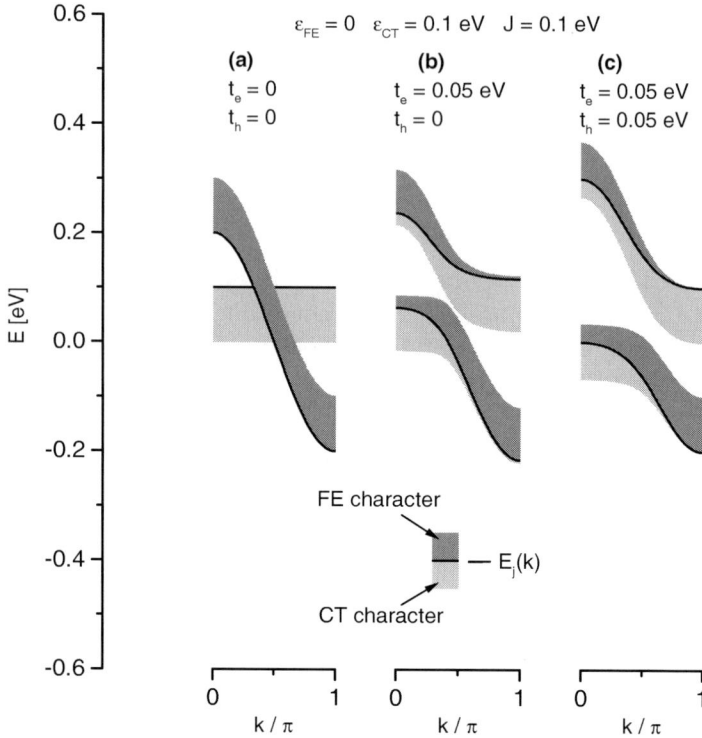

Fig. 4. Exciton bands for increasing charge-transfer parameters $t_e$, $t_h$ and constant $\varepsilon_{FE}$, $\varepsilon_{CT}$, $J$. The FE character $F_{FEj}$ of each band is indicated by the upper stripe, the CT character $F_{CTj}$ by the lower one (cf. Eqs. (38), (39)). (a) With $t_e = 0$ and $t_h = 0$, all interactions between the FE and CT states are turned off. The pure CT excitons form a dispersionless band at $\varepsilon_{CT}$, the Frenkel excitons form a pure FE band with bandwidth $4J$. (b) With one charge-transfer integral ($t_e$) nonzero, FE and CT excitons can mix. Both bands have a mixed character and their crossing is avoided. The composition of the bands varies with $k$. For $t_h > 0$ and $t_e = 0$, the picture would be the same. (c) Now, both CT integrals are nonzero, which increases the overall mixing and repulsion of the bands. In the shown special case $t_e = t_h$, the bands are not mixed at $k = \pi$.

the situation for various mixing situations, as explained in the figure captions. The characters of the bands (FE or CT) are always indicated by the upper and lower stripes. The vertical height of each stripe is always proportional to the character $F(k)$.

The visualization scheme of Figures 4 and 5 easily allows to realize the two main features that determine the exciton band structure. The first important feature is the overall dispersion of the contributing Frenkel exciton. The pure Frenkel exciton band is a simple cosine function as seen in Figure 4(a). If this band is mixed with other bands, also the FE character is distributed over all mixed bands.

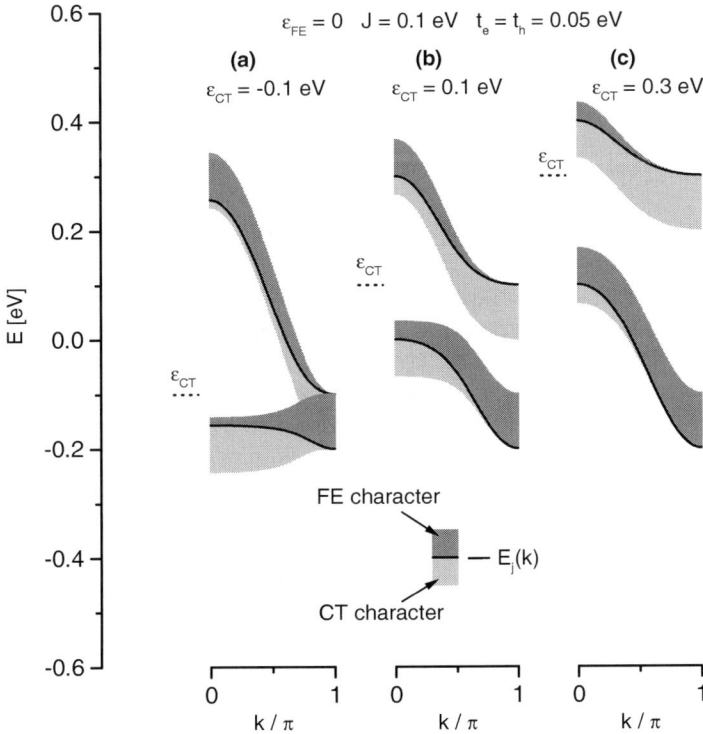

Fig. 5. Exciton bands for varying CT on-site energy $\varepsilon_{CT}$ and constant $\varepsilon_{FE}$, $J$, $t_e = t_h$. The FE character $F_{FEj}$ of each band is indicated by the upper stripe, the CT character $F_{CTj}$ by the lower one (cf. Eqs. (38), (39)). Because of $t_e = t_h$, the FE band and the CT band mix strongly at $k = 0$ but do not mix at $k = \pi$ (cf. Figure 4). (a) With $\varepsilon_{CT} = -0.1$ eV, the CT state is energetically well below the FE band at $k = 0$ (cf. Figure 4(a)) and strong mixing occurs only for intermediate $k$. (b) With $\varepsilon_{CT} = 0.1$ eV, both bands are strongly mixed at $k = 0$. (c) With $\varepsilon_{CT} = 0.3$ eV, the CT band lies above the pure FE band and both states are weakly mixed.

However, the center of mass of the FE character

$$\overline{E}_{FE}(k) \equiv \sum_j F_{FEj}(k) E_j(k) \tag{40}$$

exactly follows the dispersion of the pure FE band:

$$\overline{E}_{FE}(k) = \varepsilon_{FE} + 2J \cos k. \tag{41}$$

The same holds for the center of mass of the CT character:

$$\overline{E}_{CT}(k) \equiv \sum_j F_{CTj}(k) E_j(k), \tag{42}$$

and

$$\overline{E}_{CT}(k) = \varepsilon_{CT}. \tag{43}$$

Both identities (41) and (43) follow directly from the theory of unitary transformations: If $\underline{U} = (u_{ij})$ is the transformation matrix that diagonalizes $\underline{H}$ by diag$(E_j) = \underline{U}^\dagger \underline{H} \underline{U}$, then

$$\sum_j u_{ij}^2 E_j = H_{ii}. \tag{44}$$

The physical meaning of Eqs. (41) and (43) is very intuitive: No matter in which way the FE and CT states mix, the original dispersion of these states remains as a center of mass dispersion of the corresponding characters. Therefore, the upper stripes of the FE character will always disperse to lower energies and thus keep on average their original dispersion. This average dispersion can easily be tracked by the eye. By the same token, the average position of the lower stripes represents the dispersion-less band of the pure CT state.

### 2.3. Characters and Transition Dipoles of the Eigenstates

With knowledge of the eigenstates (37) of our considered Frenkel–CT Hamiltonian (29), we can investigate the transition dipoles of these states. Let $\hat{\vec{P}}$ be the total transition dipole operator, which is a one-electron operator acting on all electrons in the system:

$$\hat{\vec{P}} = \sum_i e\vec{r}_i. \tag{45}$$

First, we decompose the transition dipole into a Frenkel and CT component by means of Eq. (37):

$$\vec{P}_j = \langle \Psi_j(k) | \hat{\vec{P}} | o \rangle$$

$$= \underbrace{u_{FEj}(k)\langle k, 0 | \hat{\vec{P}} | o \rangle}_{\equiv \vec{P}_{FEj}} + \underbrace{u_{CTj}(k)_+ \langle k, 1 | \hat{\vec{P}} | o \rangle}_{\equiv \vec{P}_{CTj}}. \tag{46}$$

The FE component $\vec{P}_{FEj}$ becomes

$$\vec{P}_{FEj} = u_{FEj}(k)\frac{1}{\sqrt{N}}\sum_n e^{ikn}\underbrace{\langle n, 0 | \hat{\vec{P}} | o \rangle}_{=\vec{p}_{FE}}$$

$$= u_{FEj}(k)\vec{p}_{FE}\frac{1}{\sqrt{N}}\underbrace{\sum_n e^{ikn}}_{\stackrel{(12)}{=} N\delta_{k,0}} = u_{FEj}(k)\delta_{k,0}\sqrt{N}\vec{p}_{FE}. \tag{47}$$

Here, we have introduced the transition dipole $\vec{p}_{FE}$ of a localized Frenkel exciton basis state:

$$\vec{p}_{FE} \equiv \langle n, 0 | \hat{\vec{P}} | o \rangle. \tag{48}$$

This local Frenkel exciton transition dipole represents the transition from the ground state of the crystal to a localized Frenkel excitation at any position $n$. It corresponds to the transition dipole of an isolated molecule, although it is not strictly identical with it due to the formal difference between the localized crystal functions and the wave functions of an isolated molecule.

The CT component in Eq. (46) becomes by means of Eq. (31)

$$\vec{P}_{CTj} = u_{CTj}(k) \frac{1}{\sqrt{2}t_k} \left\{ \left(t_e + t_h e^{ik}\right) \langle k, +1 | \hat{\vec{P}} | o \rangle + \left(t_e + t_h e^{-ik}\right) \langle k, -1 | \hat{\vec{P}} | o \rangle \right\}$$

$$= u_{CTj}(k) \frac{1}{\sqrt{2}t_k} \left\{ \left(t_e + t_h e^{ik}\right) \frac{1}{\sqrt{N}} \sum_n e^{ikn} \underbrace{\langle n, +1 | \hat{\vec{P}} | o \rangle}_{= \vec{p}_{CT}/\sqrt{2}} \right.$$

$$\left. + \left(t_e + t_h e^{-ik}\right) \frac{1}{\sqrt{N}} \sum_n e^{ikn} \underbrace{\langle n, -1 | \hat{\vec{P}} | o \rangle}_{= \vec{p}_{CT}/\sqrt{2}} \right\} \tag{49}$$

$$= u_{CTj}(k) \delta_{k,0} \sqrt{N} \frac{\vec{p}_{CT}}{\sqrt{2}} \underbrace{\frac{1}{\sqrt{2}t_k} \left\{ \left(t_e + t_h e^{ik}\right) + \left(t_e + t_h e^{-ik}\right) \right\}}_{= \sqrt{2} \text{ for } k=0}$$

$$= u_{CTj}(k) \delta_{k,0} \sqrt{N} \vec{p}_{CT}. \tag{50}$$

In analogy to Eq. (48), we have used the transition dipole of a localized CT basis state, which will be discussed again in Eq. (71):

$$\vec{p}_{CT}/\sqrt{2} \equiv \langle n, +1 | \hat{\vec{P}} | o \rangle = \langle n, -1 | \hat{\vec{P}} | o \rangle. \tag{51}$$

Both in Eq. (47) for the Frenkel transition dipole and in Eq. (50) for the CT transition dipole, the $k$ selection rule follows mathematically from identity (12). Combination of Eqs. (47) and (50) in Eq. (46) gives the final expression

$$\vec{P}_j = \delta_{k,0} \sqrt{N} \left[ u_{FEj}(k) \vec{p}_{FE} + u_{CTj}(k) \vec{p}_{CT} \right]. \tag{52}$$

The oscillator strength $f_j$ of a transition $|o\rangle \to |\Psi_j\rangle$ with transition dipole $\vec{P}_j$ is defined as

$$f_j = \frac{2m E_j}{e^2 \hbar^2} \vec{P}_j^2, \tag{53}$$

with $m$ being the free electron mass and $E_j$ the transition energy. In simple cases, e.g., for molecules in vacuum, $f_j$ directly determines the peak area in an absorption spectrum.

Typically, the CT transition dipole is very small compared to the Frenkel transition dipole. If we neglect the CT transition dipole, the FE part of the oscillator strength becomes:

$$f_{\text{FE}j} = \frac{2mE_j}{e^2\hbar^2} N \vec{p}_{\text{FE}}^2 \times F_{\text{FE}j}(0). \tag{54}$$

In our case, we are always interested in a group of transitions that span an energy region given by the order of magnitude of the interaction parameters and by the variation in the on-site energies. This relevant energy region is typically small compared to the total on-site energies $\varepsilon_{\text{FE}}$ and $\varepsilon_{\text{CT}}$. Furthermore, an arbitrary offset in the on-site energies would only appear as an additive constant in our model Hamiltonians. Therefore, the factor $E_j$ in Eq. (54) can be seen as a mere proportionality constant and the relation simplifies to

$$f_{\text{FE}j} \propto F_{\text{FE}j}(0). \tag{55}$$

In this way, the Frenkel character introduced in Eq. (38) as the FE contents in the eigenstate obtains an additional meaning: The FE character at $k = 0$ directly gives the FE part of the oscillator strength of the corresponding state. This statement is very obvious for the case of purely electronic states, but it remains valid and becomes more valuable for vibronic states (exciton–phonon mixtures) in Section 3.

If Frenkel and CT basis-states are strongly mixed, both resulting eigenstates can have a significant Frenkel character (see, e.g., Figures 4 and 5). This redistribution of the oscillator strength from the Frenkel state to all eigenstates is depicted by the notion that the CT states borrow oscillator strength from the Frenkel states.

If the CT transition dipole has a significant size, it can contribute to the total transition dipole (52) of the Frenkel–CT mixed eigenstates. Since the coefficients $u_{\text{CT}j}(k)$ vary for the different eigenstates, not only the size but even the direction of the total transition dipole can be different for the different eigenstates. The pure CT contribution to the oscillator strength is proportional to the CT character:

$$f_{\text{CT}j} = \frac{2mE_j}{e^2\hbar^2} N \vec{p}_{\text{CT}}^2 \times F_{\text{CT}j}(0). \tag{56}$$

Typically, this CT contribution cannot be directly observed since it is masked by the much stronger Frenkel contribution. But if $\vec{p}_{\text{CT}} \not\parallel \vec{p}_{\text{FE}}$, this CT oscillator strength could be directly probed by light with polarization perpendicular to $\vec{p}_{\text{FE}}$.

As for the Frenkel part, the CT part of the oscillator strength is essentially given by the CT character at $k = 0$:

$$f_{\text{CT}j} \propto F_{\text{CT}j}(0). \tag{57}$$

Because of the proportionality of the oscillator strengths to the corresponding character, the characters are also named spectral weights.

## 2.4. Direction of Charge-Transfer Transition Dipoles

Let us now discuss the small CT transition dipole $\vec{p}_{CT}$ from Eq. (51) in more detail. For this, we consider as the simplest model system a dimer of two identical closed-shell molecules A and B. Let us assume that the molecules themselves have inversion symmetry and that there is an additional center of inversion between the molecules, which always holds if the dimer corresponds to translationally equivalent nearest-neighbor molecules from a crystal with inversion symmetry. The inversion symmetry is fulfilled in many typical examples of one-component molecular crystals, and it simplifies the situation considerably.

Let us further assume that the electronic structure of the isolated molecules can be represented by just two molecular orbitals, the HOMO (highest occupied molecular orbital) and the LUMO (lowest unoccupied molecular orbital). Thus, we have a one-particle basis set of four monomer orbitals ($H_A$, $H_B$, $L_A$, $L_B$) and we can construct many-particle configurations by distributing the four electrons on these orbitals. Note that this monomer orbital set is not completely orthogonal since $\langle H_A | H_B \rangle$ and $\langle L_A | L_B \rangle$ can be nonzero. We only have

$$\langle H_A | L_A \rangle = \langle H_B | L_B \rangle = 0 \tag{58}$$

since the orbitals from each molecule are orthogonal by definition. Furthermore, from the inversion symmetry and the orthogonality (58), one can show that

$$\langle H_A | L_B \rangle = \langle H_B | L_A \rangle = 0. \tag{59}$$

From the monomer orbitals $H_{A/B}$, $L_{A/B}$, we can directly construct the following localized basis states:[5]

$$|ME_A\rangle = |H_A \to L_A\rangle$$
$$= \frac{1}{\sqrt{2}} \{ |L_A \bar{H}_A H_B \bar{H}_B\rangle^{(-)} + |H_A \bar{L}_A H_B \bar{H}_B\rangle^{(-)} \}, \tag{60}$$

$$|ME_B\rangle = |H_B \to L_B\rangle$$
$$= \frac{1}{\sqrt{2}} \{ |H_A \bar{H}_A L_B \bar{H}_B\rangle^{(-)} + |H_A \bar{H}_A H_B \bar{L}_B\rangle^{(-)} \}, \tag{61}$$

$$|CT_{A \to B}\rangle = |H_A \to L_B\rangle$$
$$= \frac{1}{\sqrt{2}} \{ |L_B \bar{H}_A H_B \bar{H}_B\rangle^{(-)} + |H_A \bar{L}_B H_B \bar{H}_B\rangle^{(-)} \}, \tag{62}$$

---

[5] We use a quantum chemical notation for spin orbitals as, e.g., in Ref. [42].

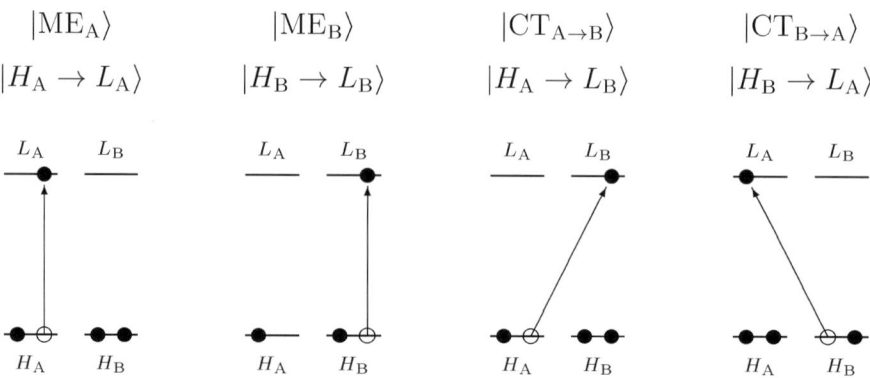

Fig. 6. Scheme of the localized configurations for an idealized dimer with four monomer orbitals. All configurations represent excited singlet states and therefore no spin is indicated for the electrons.

$$|CT_{B \to A}\rangle = |H_B \to L_A\rangle$$
$$= \frac{1}{\sqrt{2}} \{|H_A \bar{H}_A L_A \bar{H}_B\rangle^{(-)} + |H_A \bar{H}_A H_B \bar{L}_A\rangle^{(-)}\}. \qquad (63)$$

The two molecular excitations $|ME_{A/B}\rangle$ represent excited states of the isolated molecules. Furthermore, there are two charge-transfer excitations $|CT_{A \to B}\rangle$ and $|CT_{B \to A}\rangle$, which represent the transfer of an electron to the other molecule. These configurations are depicted in Figure 6. They are all strictly orthogonal to the ground state, which is obvious for the molecular excitations and which also follows from Eq. (59) for the CT excitations.[6]

However, there is no strict orthogonality between the molecular and the CT excitations, since the scalar products $\langle ME_A | CT_{A \to B}\rangle$ and $\langle ME_B | CT_{B \to A}\rangle$ include the overlap $\langle L_A | L_B\rangle$. This means, the CT states $|CT_{A \to B}\rangle$ and $|CT_{B \to A}\rangle$ constructed from orbitals of isolated molecules are only approximations for the CT basis states needed in the crystal Hamiltonian $H_{NN}$ in Eqs. (29). For a strict microscopic description of the CT crystal basis states, a further orthogonalization of either the molecular orbitals (leading to Wannier orbitals) or of the localized many-electron configurations would be necessary.

For a qualitative discussion of the CT transition dipole, we will now ignore the difference between the true orthogonal crystal basis and the localized basis states derived from monomer orbitals. Then, the transition dipole of the CT state

---

[6] In order to calculate the overlap $^{(-)}\langle H_A \bar{H}_A H_B \bar{H}_B | CT_{A \to B}\rangle$ of the localized CT state with the ground state, one has to represent the CT state in an orthogonal orbital basis, since for the nonorthogonal basis $H_{A,B}$, $L_{A,B}$ the standard rules for Slater determinants cannot be applied. A suitable orthogonalized orbital basis is given by symmetric and antisymmetric linear combinations of the monomer orbitals.

$\vec{p}_{B \to A} = \langle CT_{B \to A} | \hat{\vec{P}} | o \rangle$ can be related to the molecular monomer orbitals. This problem was already discussed in detail by Mulliken [43], and we can use his expression[7]

$$\vec{p}_{B \to A} = -e \langle L_A | \vec{r} | H_B \rangle. \tag{64}$$

Thus far, this result is very intuitive: The CT transition dipole is given as the dipole moment of the transition density

$$\varrho_{AB}(\vec{r}) = L_A(\vec{r}) H_B(\vec{r}) \tag{65}$$

by

$$\vec{p}_{B \to A} = -e \int d^3 r \, \vec{r} \times \varrho_{AB}(\vec{r}). \tag{66}$$

The structure of the transition density cloud $\varrho_{AB}(\vec{r})$, however, cannot be easily predicted for complicated molecules.

Mulliken and Person [44, pp. 23ff] give a qualitative discussion for the case of a $\sigma$-type overlap between atomic orbitals. In this case, the transition density forms a small cloud without nodes and is localized between the two molecules. Then, $\int d^3 r \, \vec{r} \varrho_{AB}$ can be approximated by

$$\int d^3 r \, \vec{r} \varrho_{AB} \approx \vec{R}_{AB} \int d^3 r \, \varrho_{AB} = \vec{R}_{AB} S_{AB}, \tag{67}$$

where $S_{AB}$ is the overlap $\langle L_A | H_B \rangle$ and $\vec{R}_{AB}$ is the average position of the transition density cloud. Based on this approximation, the CT transition dipole in a donor–acceptor complex is derived as[8]

$$\vec{p}_{B \to A} = -e \sqrt{2} \frac{S_{AB}}{\sqrt{1 + S_{AB}^2}} (\vec{R}_{AB} - \vec{R}_B). \tag{68}$$

This result is very intuitive: The CT transition dipole can be visualized as the transfer of an electron from the donor (position $\vec{R}_B$) to the overlap region between donor and acceptor (position $\vec{R}_{AB}$). The direction of $\vec{p}_{B \to A}$ is along the connection line between donor and acceptor.

We want to emphasize that this intuitive picture completely breaks down in our case of a dimer with inversion symmetry. All effects included in Eq. (68)

---

[7] Mulliken actually considered a dimer of an electron donor B and a chemically different acceptor A, without inversion symmetry. Then, Eq. (59) does not hold and Mulliken had to consider the mixing with the ground state as well as terms arising from the overlap $\langle L_A | H_B \rangle$. Eq. (64) is accordingly a simplified version for his expression for $\mu_{01}$.

[8] See [43], Eqs. (18) and (19). We give only the leading term for strong CT character of the excited state.

result from the nonzero overlap $S_{AB}$ between the donor HOMO and the acceptor LUMO, and they vanish for the symmetric dimer. The next nonzero term (Eq. (66)) in the symmetric dimer results not from the average position $\vec{R}_{AB}$ of the transition density cloud $\varrho_{AB}(\vec{r})$ but from its internal structure.

As another consequence of the inversion symmetry, the true eigenstates within the CT manifold will also obey the symmetry. These symmetry adapted CT states are:

$$|CT_\pm\rangle \equiv \frac{|CT_{A\to B}\rangle \pm |CT_{B\to A}\rangle}{\sqrt{2}}. \tag{69}$$

The transition dipoles of these symmetry-adapted CT states are

$$\vec{p}_{CT\pm} = \frac{\vec{p}_{B\to A} \pm \vec{p}_{A\to B}}{\sqrt{2}}. \tag{70}$$

The two CT-transition dipoles $\vec{p}_{B\to A}$ and $\vec{p}_{A\to B}$ in Eq. (70) are equal because of the inversion symmetry, which can also be seen from Eq. (64):

$$\vec{p}_{B\to A} = \vec{p}_{A\to B} \equiv \vec{p}_{CT}/\sqrt{2}. \tag{71}$$

Thus, Eq. (70) becomes:

$$\vec{p}_{CT+} = \vec{p}_{CT}, \tag{72}$$

$$\vec{p}_{CT-} = 0. \tag{73}$$

In the case of coplanar stacked aromatic molecules, the structure of the transition density cloud cannot be easily approximated. $\varrho_{AB}$ is formed by two complicated $\pi$ orbitals, which overlap in the region between the molecules. There it forms a flat, quasi-two-dimensional cloud with dimensions of the molecular size. Furthermore, like the contributing monomer orbitals, it has a complicated nodal structure with lobes of alternating sign. Therefore, it can be expected that the transition dipole $\vec{p}_{B\to A}$ will lie approximately in the plane of this quasi-two-dimensional transition density cloud, i.e., parallel to the molecular planes. If there are no additional symmetries, the actual direction cannot be estimated. As an illustration, we show the monomer orbitals $L_A$ and $H_B$ and the resulting transition density for a MePTCDI dimer in Figure 7. There, it is clearly visible how the transition density is formed in the plane between the molecules. In conclusion, the direction the CT transition dipole and the underlying leading terms are very different for a symmetric dimer compared to the classical donor–acceptor complexes studied by Mulliken [43].

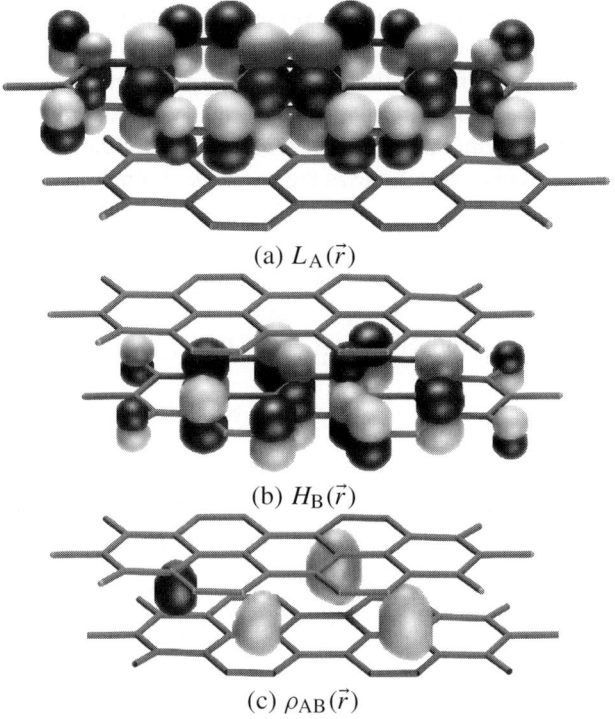

Fig. 7. Formation of the transition density cloud in a dimer of MePTCDI molecules. (a) LUMO $L_A$. (b) HOMO $H_B$. (c) Illustration of the transition density $\varrho_{AB} = H_B(\vec{r}) \times L_A(\vec{r})$. The molecular structure corresponds to nearest-neighbors along the stacking direction ($a$-axis) from the experimental crystal structure [85], see Figure 2. The molecular orbitals were calculated with the ZINDO/S method as implemented by HyperChem [107], the transition density was calculated from these orbitals. Its precise structure is not very accurate since the orthogonality relations are not exact in the ZINDO/S scheme, but the location between the molecules can be illustrated. Calculations and pictures by K. Schmidt.

## 3. Strong Coupling of the Electronic Excitations with Internal Phonon Modes

### 3.1. Exciton–Phonon Coupling in the Isolated Molecule

In Section 2, we have considered exciton states in a rigid lattice, which is a purely electronic problem. Now, we want to include exciton–phonon coupling, i.e., coupling of the electronic states to lattice vibrations. In this work, we consider only one type of exciton–phonon coupling, namely coupling to internal phonon modes (vibrations along intramolecular nuclear coordinates). These vibrations occur already in the isolated molecule, and their effect is usually demonstrated by a

Franck–Condon picture as in Figure 8. If the exciton–phonon coupling constant (see below) is on the order of one, the internal exciton–phonon coupling leads to a pronounced "vibronic progression" in the absorption and emission spectra of the isolated molecules. Such a progression is very conspicuous in the monomer spectra of our model compounds MePTCDI and PTCDA, as can be seen in Figure 3(a). In this section, we summarize the notation for exciton–phonon coupling in the isolated molecule, and in Sections 3.2–3.8 we deal with exciton–phonon coupling in one-dimensional systems.

Let us for simplicity assume that the isolated molecule has just one nuclear coordinate $q$. In Born–Oppenheimer approximation, the molecular wave functions can be split into an electronic and a vibrational part:

$$\psi_{fv}(\vec{r}, q) = \varphi_f(\vec{r}, q) \times \chi_{fv}(q), \tag{74}$$

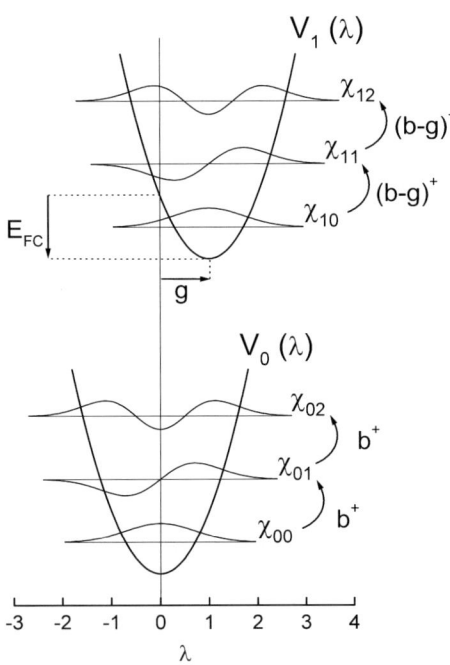

Fig. 8. Schematic energy diagram for the vibrational potentials of ground ($V_0$) and excited ($V_1$) state along the dimensionless coordinate $\lambda$. The excited state potential is displaced on the $\lambda$-axis by the exciton–phonon coupling constant $g$, which corresponds to a vibrational reorganization energy $E_{FC} = \hbar\omega g^2$. The vibrational wave functions $\chi_{fv}(\lambda)$ are shown for the lowest three levels. The operators $b^\dagger$ create phonons in the ground state potential, the displaced operators $(b-g)^\dagger$ create phonons in the excited state potential.

where $f$ numbers the electronic and $v$ the vibrational levels; $\varphi_f(\vec{r}, q)$ is the electronic wave function, which depends explicitly on all electron coordinates $\vec{r}$ and parametrically on the nuclear coordinate $q$; $\chi_{fv}(q)$ are the vibrational wave functions in the electronic state $f$ and vibrational level $v$. They depend only on the nuclear coordinate $q$. In Born–Oppenheimer approximation, the Schrödinger equation for the molecule separates into an electronic problem at a fixed nuclear coordinate $q$ and a vibrational problem for the nuclear coordinate $q$.

From the electronic problem, the total energy of the molecule in electronic state $f$ is given as a function $V_f(q)$ of the nuclear coordinate. Therefore, the vibrational Hamiltonian in state $f$ becomes:

$$H_f^{\text{vib}} = -\frac{\hbar^2}{2M_{\text{eff}}}\nabla^2 + V_f(q). \tag{75}$$

For small elongations from the equilibrium position, the vibrational potential becomes a harmonic potential

$$V_f(q) = \frac{1}{2}M_{\text{eff}}\omega_f^2(q - q_{0f})^2 + v_f, \tag{76}$$

with effective mass $M_{\text{eff}}$, angular frequency $\omega_f$ and a total energy offset $v_f$.

Since the potential $V_1(q)$ in the excited state depends on all electrons but only a few electrons take part in the excitation, the curvature of $V_1(q)$ differs not strongly from that of the potential $V_0(q)$ in the ground-state. Hence, the vibronic spacing is very similar. For example, in the case of MePTCDI in chloroform, the difference of $\hbar\omega$ derived from the absorption and emission spectra amounts to 3 meV compared to $\hbar\omega = 170$ meV (values from Figure 3). We therefore make the further approximation that the vibrational spacings $\hbar\omega_f$ are identical for the ground and excited state.

As an abbreviation, we introduce the dimensionless nuclear coordinate[9]

$$\lambda \equiv \sqrt{\frac{M_{\text{eff}}\omega}{2\hbar}}(q - q_{00}), \tag{77}$$

which is centered around the equilibrium position $q_{00}$ of the electronic ground state $f = 0$. Now, the vibrational potentials are in the electronic ground state

$$V_0(\lambda) = \hbar\omega\lambda^2 + v_0 \tag{78}$$

and in the excited state

$$V_1(\lambda) = \hbar\omega(\lambda - g)^2 + v_1. \tag{79}$$

---

[9] In the literature, several ways are common to introduce dimensionless coordinates and exciton–phonon coupling constants.

The exciton–phonon coupling constant $g$ introduced here is the displacement of the exited state potential $V_1$ with respect to the ground state potential $V_0$ along the dimensionless coordinate $\lambda$. Both potentials are illustrated in Figure 8. If one introduces the vertical excitation energy

$$E_{\text{vert}} \equiv V_1(0) - V_0(0) \tag{80}$$

for the excitation energy in a rigid molecule ($\lambda \stackrel{!}{=} 0$), the excited state potential can be written as

$$V_1(\lambda) = \hbar\omega(\lambda - g)^2 + v_0 + E_{\text{vert}} - E_{\text{FC}}. \tag{81}$$

Here, the vibrational reorganization energy (Franck–Condon energy) $E_{\text{FC}}$ gives the energy gain for the geometry relaxation of the electronically excited molecule after a "vertical excitation" into the new equilibrium geometry. The reorganization energy is trivially connected with the exciton–phonon coupling constant by

$$E_{\text{FC}} = \hbar\omega g^2. \tag{82}$$

Optical absorption corresponds to the transitions $|\psi_{00}\rangle \to |\psi_{1\nu}\rangle$, since for $\hbar\omega = 0.17$ eV $\gg kT$ only the lowest vibrational level of the electronic ground state is occupied. The transition dipole moment is given by the matrix element with the dipole moment operator $\hat{\vec{P}}$:

$$\vec{p}^\nu = \langle \psi_{1\nu} | \hat{\vec{P}} | \psi_{00} \rangle. \tag{83}$$

This matrix element refers to integration over all electron and nucleus coordinates. In the product representation (74), the vibrational part $\chi_{f\nu}$ only depends on the nucleus coordinates, whereas the electronic part $\varphi_f$ contains these nucleus coordinates as a variable that varies slowly compared to the electron coordinates. Therefore, the integration can be split into an electronic and a vibrational matrix element (e.g., [45, p. 305]):

$$\vec{p}^\nu = \underbrace{\langle \varphi_1 | \hat{\vec{P}} | \varphi_0 \rangle}_{\vec{p}_{\text{FE}}} \times \underbrace{\langle \chi_{1\nu} | \chi_{00} \rangle}_{S\binom{\nu}{0}}. \tag{84}$$

Here, we have introduced the electronic transition dipole moment of the lowest molecular excitation $\vec{p}_{\text{FE}}$ and the vibronic overlap factors $S\binom{\nu}{0}$. The oscillator strength $f$ for a transition with energy $E$ is defined as $f = 2mE|\vec{p}|^2/(e^2\hbar^2)$ (cf. Eq. (53)), where $\vec{p}$ is the transition dipole moment. Thus, the oscillator strength (corresponding to the absorption peak area) of the $\nu$th transition is

$$f_\nu = \frac{2m}{e^2\hbar^2} \cdot E^\nu |\vec{p}_{\text{FE}}|^2 \times S^2\binom{\nu}{0}. \tag{85}$$

The squared values $S^2\binom{v}{0}$ are called the Franck–Condon factors. They determine the intensity distribution in the vibronic progression. In the considered case of a single vibrational coordinate $\lambda$ and two harmonic potentials displaced by $g$, the Franck–Condon factors can be analytically expressed as (e.g., [46]):

$$S^2\binom{v}{0} = |\langle \chi_{1v}|\chi_{00}\rangle|^2 = \frac{g^{2v}}{v!}e^{-g^2}. \tag{86}$$

For large coupling constants $g$, the maximum intensity occurs at high vibrational levels, whereas for $g = 0$ exclusively the lowest vibrational level is excited ($f_v = \delta_{0v}$). For a coupling constant of $g = 1$, which is visualized in Figure 8 and represents the order of magnitude in PTCDA-derivatives, the lowest and second lowest vibrational level obtain approximately the same oscillator strengths ($f_1 \approx f_0$). This leads to the characteristic spectrum in Figure 3(a), where the areas of the 0–0 peak and of the 0–1 peak are on the same order of magnitude.

In order to extend the molecular vibrational Hamiltonian (75) to an aggregate Hamiltonian, the introduction of phonon creation and annihilation operators is very helpful:[10]

$$b^\dagger \equiv \lambda - \frac{1}{2}\frac{\partial}{\partial \lambda}, \tag{87}$$

$$b \equiv \lambda + \frac{1}{2}\frac{\partial}{\partial \lambda}. \tag{88}$$

Then, the vibrational Hamiltonian in the electronic ground state becomes:

$$H_0^{\text{vib}} = \hbar\omega\left(b^\dagger b + \frac{1}{2}\right) + v_0. \tag{89}$$

For the vibrational Hamiltonian in the excited state, it is more convenient to introduce displaced operators because of the potential displacement in Eq. (81):

$$\tilde{b}^\dagger \equiv b^\dagger - g, \tag{90}$$

$$\tilde{b} \equiv b - g. \tag{91}$$

Then, from comparison with Eq. (89), the Hamiltonian in the excited state can be written as

$$H_1^{\text{vib}} = \hbar\omega\left(\tilde{b}^\dagger \tilde{b} + \frac{1}{2}\right) + v_0 + E_{\text{vert}} - E_{\text{FC}}. \tag{92}$$

---

[10] A good summary can be found in Ref. [47, pp. 248–252]; a comprehensive introduction is given, e.g., in Ref. [48].

After inserting relations (90), (91) and (82) for $E_{FC}$, we can express the excited state Hamiltonian with the undisplaced operators $b^\dagger$, $b$:

$$H_1^{\text{vib}} = \hbar\omega\left(b^\dagger b - g(b^\dagger + b) + \frac{1}{2}\right) + v_0 + E_{\text{vert}}. \quad (93)$$

Now, it is possible to combine the vibrational Hamiltonians for the ground and excited state by introducing electronic operators. For this, we assume that the electronic problem is a strict two-level problem with the electronic ground state $|\phi_0\rangle$ and the exited state $|\phi_1\rangle$. We define electronic operators $a^\dagger$ and $a$ by

$$a^\dagger|\phi_0\rangle \equiv |\phi_1\rangle, \quad a|\phi_0\rangle \equiv 0,$$
$$a^\dagger|\phi_1\rangle \equiv 0, \quad a|\phi_1\rangle \equiv |\phi_0\rangle. \quad (94)$$

That means, $a^\dagger$ creates an exciton at the molecule, $a$ destroys it and $a^\dagger a$ is the exciton number operator.

Now, both operators $H_0^{\text{vib}}$ and $H_1^{\text{vib}}$ can be combined by multiplying the additional terms in $H_1^{\text{vib}}$ with $a^\dagger a$. We obtain the monomer Hamiltonian

$$H_{\text{mono}} = \hbar\omega\left(b^\dagger b + \frac{1}{2}\right) + v_0 + a^\dagger a \times \left(-\hbar\omega g(b^\dagger + b) + E_{\text{vert}}\right). \quad (95)$$

Here, the first term is the familiar Hamiltonian of the harmonic oscillator, which describes the internal phonons in the molecule. The constant offset $v_0$ for the total ground state energy of the molecule is not relevant here. $a^\dagger a$ is zero for the electronic ground state but one for the excited state. In the excited state, the linear exciton–phonon coupling $-\hbar\omega g(b^\dagger + b)$ and the vertical excitation energy $E_{\text{vert}}$ are added. The exciton–phonon coupling, which is the most interesting term, entirely results from the displacement of the exited state vibrational potential.

### 3.2. THE HOLSTEIN HAMILTONIAN FOR EXCITON–PHONON COUPLING

Now, we can extend the monomer Hamiltonian (95) to a one-dimensional chain. In the first step, we neglect intermolecular interactions and formally add the monomer Hamiltonians $H_{\text{mono}}(n)$ of the molecules, which are numbered by the index $n = 1, \ldots, N$:

$$\sum_n H_{\text{mono}}(n) = \sum_n \left[\hbar\omega\left(b_n^\dagger b_n + \frac{1}{2}\right) + v_0 \right.$$
$$\left. + a_n^\dagger a_n \left(-\hbar\omega g(b_n^\dagger + b_n) + E_{\text{vert}}\right)\right]. \quad (96)$$

We are exclusively interested in the one-exciton subspace of this problem ($\sum_n a_n^\dagger a_n = 1$). The energy of the lowest state in this one-exciton subspace is given by the lowest vibrational level of $N-1$ molecules in the electronic ground state and the lowest vibrational level of one molecule in the excited state:

$$(N-1) \times \left[\hbar\omega\left(0+\frac{1}{2}\right)+v_0\right] + \left[\hbar\omega\left(0+\frac{1}{2}\right)+v_0+E_{\text{vert}}-E_{\text{FC}}\right]$$

$$= N \times \left[\frac{1}{2}\hbar\omega+v_0\right]+E_{\text{vert}}-E_{\text{FC}}.$$

We want to use this state as the reference state and set the zero of the energy axis to its energy. Then, the Hamiltonian (96) in the non-interacting case becomes (with using $E_{\text{FC}} = g^2\hbar\omega$)

$$H_{\text{non-inter}} = H^{\text{ph}} + H^{\text{FE-ph}}, \tag{97}$$

with the phonon part

$$H^{\text{ph}} = \hbar\omega \sum_n b_n^\dagger b_n \tag{98}$$

and the (Frenkel) exciton–phonon coupling part

$$H^{\text{FE-ph}} = \hbar\omega \sum_n a_n^\dagger a_n \left(-g(b_n^\dagger + b_n) + g^2\right). \tag{99}$$

We call this non-interacting case the molecular limit.

Now, we add to the non-interaction Hamiltonian (97) one of the simplest electronic interaction terms, namely a nearest-neighbor Frenkel exciton hopping $H^{\text{FE}}_{\text{elec}}$ as introduced in Eq. (16):

$$H^{\text{FE}}_{\text{elec}} = J \sum_n \left(a_n^\dagger a_{n+1} + a_{n+1}^\dagger a_n\right), \tag{100}$$

where $J$ is the Frenkel exciton transfer integral. With this term, we obtain the classical Holstein Hamiltonian

$$H^{\text{FE}}_{\text{Hol}} = H^{\text{FE}}_{\text{elec}} + H^{\text{ph}} + H^{\text{FE-ph}}. \tag{101}$$

This Hamiltonian is one of the simplest model systems for exciton–phonon coupling. $H^{\text{FE}}_{\text{elec}}$ is a purely electronic operator for Frenkel-exciton hopping, $H^{\text{ph}}$ is a purely vibrational operator for internal phonons and $H^{\text{FE-ph}}$ is a linear coupling term, which couples excitons and phonons locally (i.e., if they occupy the same molecule).

The Hamiltonian (101) is commonly called Holstein Hamiltonian because of two fundamental works [49,50] by Holstein on transport properties in this model system. The ground state of this Hamiltonian describes the states of free charge carriers or excitons that are responsible for electric conduction or exciton diffusion. Therefore, the ground state has been extensively studied by approximate analytical methods (reviews, e.g., in Refs. [5,51–54]). Currently, much effort is also spent in obtaining numerical solutions for the full parameter range by variational approaches [55–61], direct diagonalization [62–66], quantum Monte-Carlo calculations [67–70], and density-matrix renormalization-group techniques [71]. Compared to this, the properties of higher states have been much less investigated. These excited vibronic states, however, are essential for an understanding of optical absorption spectra. The relevant issues were identified in the initial studies of molecular crystals and limiting cases were analyzed (see, e.g., [3,72]). For intermediate cases, however, only a few quantitative studies have been published. These include direct diagonalization studies of dimers [15,73], variational and direct-diagonalization studies of aggregates [74–78], and a discussion of the second lowest vibronic state in an infinite chain [66]. Here, we will present the numerical approach described in Ref. [41], which was specifically developed in the context of our work on PTCDA-related quasi-one-dimensional crystals.

### 3.3. Basis Functions for Numerical Diagonalization

Our aim is to find the low energy eigenstates of the Holstein Hamiltonian (101) within the one-exciton manifold. For this, we use a numerical approach based on direct diagonalization. At first, we define a set of basis functions for the Hilbert space on which the Hamiltonian acts. The basis functions should be close to the final solution. Then, already a small and finite subset of functions can reasonably represent the solution.

In our case, we obtain the basis functions from the reference system given by the non-interacting case ($J = 0$, molecular limit). In a localized picture, the eigenstates in the molecular limit are simple product functions of molecular vibronic states. One exciton is localized at site $n$ is and the vibrational wave functions at this site are given by oscillator functions in the displaced potential $V_1$. At all other sites, which we count relative to the position of the exciton, the vibrational wave functions are oscillator functions in the ground state potential $V_0$. These localized eigenstates can be used as basis functions for representing the solution of the complete Hamiltonian.

Thus, the basis functions can be written as

$$|n\underline{v}\rangle \equiv |n\rangle \times |\ldots v_{-1} \tilde{v}_0 v_1 \ldots\rangle \equiv a_n^\dagger |o_{\text{el}}\rangle \times \tilde{B}_{n\underline{v}}^\dagger |o_{\text{vib}}\rangle. \qquad (102)$$

Here, the first factor describes the electronic part of a localized Frenkel exciton at site $n$. The second factor describes the vibrational wave function of the chain. It is created by the action of the vibrational operator $\tilde{B}_{n\underline{v}}^\dagger$ on the vibrational ground state:

$$\tilde{B}_{n\underline{v}}^\dagger = \underbrace{\frac{1}{\sqrt{v_0!}} (\tilde{b}_n^\dagger)^{v_0} \cdot e^{-\frac{g^2}{2}} e^{g b_n^\dagger}}_{\text{displaced on } n} \times \underbrace{\prod_{m \neq 0} \frac{1}{\sqrt{v_m!}} (b_{n+m}^\dagger)^{v_m}}_{\text{undisplaced otherwise}}. \qquad (103)$$

Here, the first factor ("displaced") describes internal phonons in the displaced potential at the site $n$ of the exciton, where the factor $e^{-g^2/2} e^{gb_n^\dagger}$ transforms the undisplaced vibrational ground state $|o_n\rangle$ into the displaced ground state $|\tilde{o}_n\rangle$ (e.g., [47, p. 249]). The second factor ("undisplaced") describes internal phonons at all sites different from $n$ in the undisplaced potential.

The phonon-cloud state $|\underline{\nu}\rangle$ contains the phonon occupation numbers $\nu_m$ around the exciton for all lattice sites. In the long notation $|\ldots\nu_{-1}\tilde{\nu}_0\nu_1\ldots\rangle$, the special position of the exciton ($m=0$) is denoted by the tilde. A complete phonon-cloud basis for a chain of $N$ molecules consists of $N$-boson states and leads to huge basis sets even for small occupation numbers. But a far smaller basis is sufficient to calculate the absorption spectrum.

In the molecular limit, optical absorption from the electronic and vibrational ground state only creates phonons at the site of the electronic excitation, i.e., only phonon clouds of the form $|\ldots 00\tilde{\nu}_0 00\ldots\rangle$. We call such clouds *joint configurations*. In contrast, excited states with any $\nu_m \neq 0$ for $m \neq 0$ cannot be reached optically. We call these clouds *separated configurations*, since there is at least one phonon excitation separated from the exciton position. An example of a joint and of a separated configuration is illustrated in Figures 9 and 10, respectively. In the

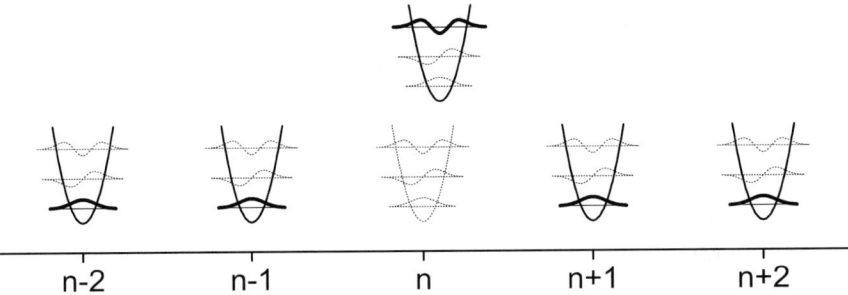

Fig. 9. Illustration of a joint configuration (here: $|\nu\rangle = |\ldots 00\tilde{2}00\ldots\rangle$).

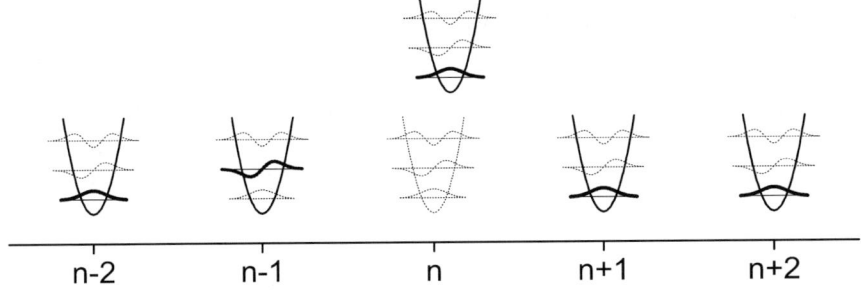

Fig. 10. Illustration of a separated configuration (here: $|\nu\rangle = |\ldots 01\tilde{0}00\ldots\rangle$).

molecular limit, the absorption spectrum can be explained by considering exclusively the joint configurations since the separated configurations would not mix with the joint ones and they have no transition dipole moment with the ground state on their own.

For $|J| > 0$, the separated configurations can mix with the joint configurations. That means, optical absorption creates a state in which phonons are excited at arbitrary distance from the exciton site. However, the contribution of separated configurations decreases with increasing exciton–phonon separation. Thus, the photo-excited exciton will be surrounded by a *localized* phonon cloud. The localized nature of phonon clouds is the motivation for our choice of basis functions. Instead of $N$-dimensional cloud states $|\underline{v}\rangle$, a finite range $|v_{-M} \ldots \tilde{v}_0 \ldots v_M\rangle$, with $M$ denoting the extension of the phonon cloud, will be sufficient. Numerically, $M$ can be increased until convergence is reached. Important qualitative insight can already be obtained from the inclusion of just nearest-neighbor phonon clouds ($M = 1$).

The choice of the displaced basis functions in Eq. (102) corresponds to applying the polaron canonical transformation (Lang–Firsov transformation) to a set of basis functions, in which all vibrational functions (including the site $n$ of the exciton) are oscillator functions in the ground state potential ([79] or see also, e.g., [5, p. 98], [53, p. 25]).

With the restriction to local phonon clouds around the exciton, we Fourier transform the basis states (102):

$$|k\underline{v}\rangle \equiv \frac{1}{\sqrt{N}} \sum_n e^{ikn} |n\underline{v}\rangle. \quad (104)$$

These states represent an exciton "dressed" with a local phonon cloud. The index $k$ gives the quasi-momentum of the whole object, i.e., the dressed exciton, and $k$ is a good quantum number due to translational symmetry. Thus, for any given $k$ the basis set consists only of a set of phonon cloud configurations. We emphasize that in contrast to the real-space basis (102), the momentum-space basis functions (104) are not Born–Oppenheimer separable into a product of a purely electronic and a purely vibrational part.

Having specified the basis states, the Hamiltonian can be represented as a matrix. Application of $H_{\text{Hol}}^{\text{FE}}$ to the real space states from (102) yields the matrix elements:

$$\langle m\underline{\mu} | H_{\text{Hol}}^{\text{FE}} | n\underline{v} \rangle = \delta_{m,n} \langle \underline{\mu} | \underline{v} \rangle \sum_i v_i$$
$$+ J \left[ \delta_{m,n-1} \mathcal{F}_{-1}\left(\frac{\underline{\mu}}{\underline{v}}\right) + \delta_{m,n+1} \mathcal{F}_{+1}\left(\frac{\underline{\mu}}{\underline{v}}\right) \right]. \quad (105)$$

The first term in this compact notation results from the operators $H^{\text{ph}}$ and $H^{\text{FE-ph}}$. They contain no interactions between different sites and thus simply count the phonons in the Lang–Firsov basis. The overlap factor $\langle \underline{\mu} | \underline{v} \rangle$ stands for the total overlap of two phonon clouds centered at the same lattice site. It is nonzero only for identical clouds due to the orthogonality of the oscillator functions:

$$\langle \underline{\mu} | \underline{v} \rangle = \prod_i \delta_{\mu_i, v_i}. \tag{106}$$

The second term in Eq. (105) results from the purely electronic Frenkel transfer process $H^{\text{FE}}_{\text{elec}}$. The vibrational part of the basis functions factors out and leads to the Franck–Condon overlaps $\mathcal{F}_{\pm 1}$ for the total vibronic overlap of the phonon cloud $\underline{v}$ centered at $n$ and the phonon cloud $\underline{\mu}$ centered at $m = n \pm 1$:

$$\mathcal{F}_{-1} = S\binom{\mu_0}{v_{-1}} \cdot S\binom{v_0}{\mu_{+1}} \cdot \prod_{i \neq 0, 1} \langle \mu_i | v_{i-1} \rangle, \tag{107}$$

$$\mathcal{F}_{+1} = S\binom{v_0}{\mu_{-1}} \cdot S\binom{\mu_0}{v_{+1}} \cdot \prod_{i \neq -1, 0} \langle \mu_i | v_{i+1} \rangle. \tag{108}$$

Here, $S\binom{v}{\mu}$ is the overlap between a displaced oscillator function with quantum number $v$ and an undisplaced function with quantum number $\mu$ [80]

$$S\binom{v}{\mu} \equiv \left\langle \frac{1}{\sqrt{\mu!}} (b^\dagger)^\mu o \, \bigg| \, \frac{1}{\sqrt{v!}} (\tilde{b}^\dagger)^v \tilde{o} \right\rangle$$

$$= \frac{e^{-\frac{g^2}{2}}}{\sqrt{\mu! v!}} \sum_{i=0}^{\min(\mu,v)} \frac{(-1)^{v-i} g^{\mu+v-2i} \mu! v!}{i!(\mu-i)!(v-i)!}. \tag{109}$$

It is obvious that in the Lang–Firsov basis the strength $g$ of the exciton–phonon coupling enters only through the magnitude of the factors $\mathcal{F}_{\pm 1}$ in the inter-site hopping term.

In the momentum space representation (104), the Hamiltonian matrix becomes

$$\langle k\underline{\mu} | H^{\text{FE}}_{\text{Hol}} | k\underline{v} \rangle = \langle \underline{\mu} | \underline{v} \rangle \hbar \omega \sum_i v_i$$

$$+ J \left[ e^{-ik} \mathcal{F}_{-1} \binom{\underline{\mu}}{\underline{v}} + e^{+ik} \mathcal{F}_{+1} \binom{\underline{\mu}}{\underline{v}} \right]. \tag{110}$$

For general momenta $k$, these matrix elements are complex numbers. For our intended application to spectroscopy, the values at the Brillouin-zone edges ($k = 0, \pi$) are of interest, and there the matrix elements are real. Representing the final eigenstates as

$$|\Psi_j(k)\rangle = \sum_{\underline{v}} u_{\underline{v}j}(k)|k\underline{v}\rangle, \quad (111)$$

we obtain the eigenvalue problem

$$\sum_{\underline{\mu}} \langle k\underline{\mu}|H_{\text{Hol}}^{\text{FE}}|k\underline{v}\rangle \cdot u_{\underline{\mu}j} = E_j \cdot u_{\underline{v}j} \quad (112)$$

for the real matrix $\langle k\underline{\mu}|H_{\text{Hol}}^{\text{FE}}|k\underline{v}\rangle$. Its eigenvalues $E_j$ and eigenstates $|\Psi_j(k)\rangle$ are the stationary solutions of the Holstein Hamiltonian (101).

### 3.4. Transition Dipoles and Phonon Clouds of the Eigenstates

The properties of the eigenstates (111) are easily computed. We start with the Frenkel exciton (FE) character defined as in Eq. (38) by the projection onto a purely electronic Frenkel exciton state $a_k^\dagger|o\rangle$:

$$F_{\text{FE}j}(k) \equiv |\langle \Psi_j(k)|a_k^\dagger|o\rangle|^2. \quad (113)$$

By use of the explicit expression (111) and the inverse Fourier transformations, one obtains:

$$F_{\text{FE}j}(k) = \left|\sum_{\underline{v}} u_{\underline{v}j}^*(k)\langle n\underline{v}|a_n^\dagger|o\rangle\right|^2. \quad (114)$$

The matrix element $\langle n\underline{v}|a_n^\dagger|o\rangle$ can be split into its electronic and vibrational components according to Eq. (102):

$$\langle n\underline{v}|a_n^\dagger|o\rangle = \langle a_n^\dagger \tilde{B}_{n\underline{v}}^\dagger o|a_n^\dagger|o\rangle$$

$$= \underbrace{\langle a_n^\dagger o_{\text{el}}|a_n^\dagger|o_{\text{el}}\rangle}_{=1} \times \langle \tilde{B}_{n\underline{v}}^\dagger o_{\text{vib}}|o_{\text{vib}}\rangle.$$

The vibrational overlap factor in this equation can be formally evaluated by use of Eq. (103). Very obviously, it gives a factor $S$ from position $n$, where the displaced vibrational wave function of level $v_0$ overlaps with the undisplaced vibrational ground state. At all other lattice positions $n+r$, we have the Kronecker symbol for the orthogonality of the wave functions in the same undisplaced potential:

$$\langle n\underline{v}|a_n^\dagger|o\rangle = \langle \tilde{B}_{n\underline{v}}^\dagger o_{\text{vib}}|o_{\text{vib}}\rangle = S\binom{v_0}{0}\prod_{r\neq 0}\delta_{v_r,0}. \quad (115)$$

Thus, the Frenkel exciton character becomes:

$$F_{\text{FE}j}(k) = \left| \sum_{\underline{v}} u^*_{\underline{v}j}(k) S\binom{v_0}{0} \prod_{r \neq 0} \delta_{v_r,0} \right|^2. \tag{116}$$

The characters obey the sum rule

$$\sum_j F^2_{\text{FE}j}(k) = 1. \tag{117}$$

since

$$\sum_j F^2_{\text{FE}j}(k) = \sum_j |\langle \Psi_j(k)|a^\dagger_k|o\rangle|^2$$

$$= \sum_j \langle o|a_k|\Psi_j(k)\rangle\langle \Psi_j(k)|a^\dagger_k|o\rangle$$

$$= \langle o|a_k \underbrace{\sum_j |\Psi_j(k)\rangle\langle \Psi_j(k)|}_{=1} a^\dagger_k|o\rangle$$

$$= 1.$$

As in the electronic problem, the exciton character determines the oscillator strength of a $k=0$ state. The transition dipole of state $|\Psi_j(k)\rangle$ is

$$\vec{P}_j = \langle \Psi_j(k)|\hat{\vec{P}}|o\rangle$$

$$= \sum_{\underline{v}} u^*_{\underline{v}j}(k)\langle k\underline{v}|\hat{\vec{P}}|o\rangle$$

$$= \sum_{\underline{v}} u^*_{\underline{v}j}(k) \frac{1}{\sqrt{N}} \sum_n e^{ikn} \langle n\underline{v}|\hat{\vec{P}}|o\rangle. \tag{118}$$

For the transition dipole of the basis state $|n\underline{v}\rangle$, we have to use the approximate separability of the transition dipole moment into an electronic part and a vibrational overlap factor as in Eq. (84) for the isolated molecule. Then, the transition dipole of each basis state $|n\underline{v}\rangle$ becomes

$$\langle n\underline{v}|\hat{\vec{P}}|o\rangle = \langle a^\dagger_n \tilde{B}^\dagger_{n\underline{v}}o|\hat{\vec{P}}|o\rangle \tag{119}$$

$$= \underbrace{\langle a^\dagger_n o_{\text{el}}|\hat{\vec{P}}|o_{\text{el}}\rangle}_{\vec{P}_{\text{FE}}} \times \underbrace{\langle \tilde{B}^\dagger_{n\underline{v}}o_{\text{vib}}|o_{\text{vib}}\rangle}_{\text{Eq. (115)}} \tag{120}$$

$$= \vec{p}_{\text{FE}} \times S\begin{pmatrix} v_0 \\ 0 \end{pmatrix} \prod_{r \neq 0} \delta_{v_r,0}. \qquad (121)$$

Thus, with identity (12) the transition dipole becomes

$$\vec{P}_j = \delta_{k,0}\sqrt{N}\vec{p}_{\text{FE}} \times \sum_{\underline{v}} u^*_{\underline{v}j}(k) S\begin{pmatrix} v_0 \\ 0 \end{pmatrix} \prod_{r \neq 0} \delta_{v_r,0}.$$

Comparison of this equation with the expression (116) for the Frenkel exciton character shows that the squared transition dipole is given by the character

$$|\vec{P}_j|^2 = \delta_{k,0} N \vec{p}_{\text{FE}}^{\,2} F_{\text{FE}j}(k), \qquad (122)$$

and the oscillator strength (see Eq. (53)) of a $k = 0$ state is proportional to the Frenkel character:

$$f_j = \frac{2mE_j}{e^2\hbar^2} N \vec{p}_{\text{FE}}^{\,2} \times F_{\text{FE}j}(0). \qquad (123)$$

This equation is identical to the corresponding expression (54) for the FE oscillator strength in the electronic problem, since the transition dipole is a purely electronic operator and its action on a vibronic state is only determined by the electronic character of this state.

As an illustration, we show in Figure 11 the results of such a calculation for $k = 0$ and the parameters $J = 0.5\hbar\omega$ and $g = 1$. The energy levels $E_j$ of the eigenstates are arranged at a vertical energy axis in the left part. Their FE character $F_{\text{FE}j}$ is indicated by the horizontal length of each stick. The lowest state appears as a solitary stick at $E_1 = 0.0074\hbar\omega$. At higher energies, the spectrum consists of many densely packed lines resulting from the mixture of the various phonon cloud configurations in the basis set. The numerical spectrum remains discrete only since the basis is finite. To illustrate the dense vibronic manifold, we always convolve stick spectra with a Gaussian of constant standard derivation ($\sigma = 0.15\hbar\omega$) and show the broadened spectrum using a convenient scaling factor.

Another important property of a vibronic state $|\Psi_j(k)\rangle$ is the internal structure of its phonon cloud. One measure to characterize it is the set of expectation values $\langle \widehat{N}_m \rangle$ for the occupation number operators:

$$\langle \widehat{N}_m \rangle \equiv \left\langle \sum_n a^\dagger_n a_n b^\dagger_{n+m} b_{n+m} \right\rangle. \qquad (124)$$

These occupation numbers show how many phonons are excited at the oscillator that is $m$ lattice spacings apart from the exciton. Note that they depend on the displacement chosen for the oscillator functions in the basis set. Thus, they are no observable quantities. They are mainly important for choosing a reasonable basis set: Since numerically for each relative site $m$, only states up to a predefined number

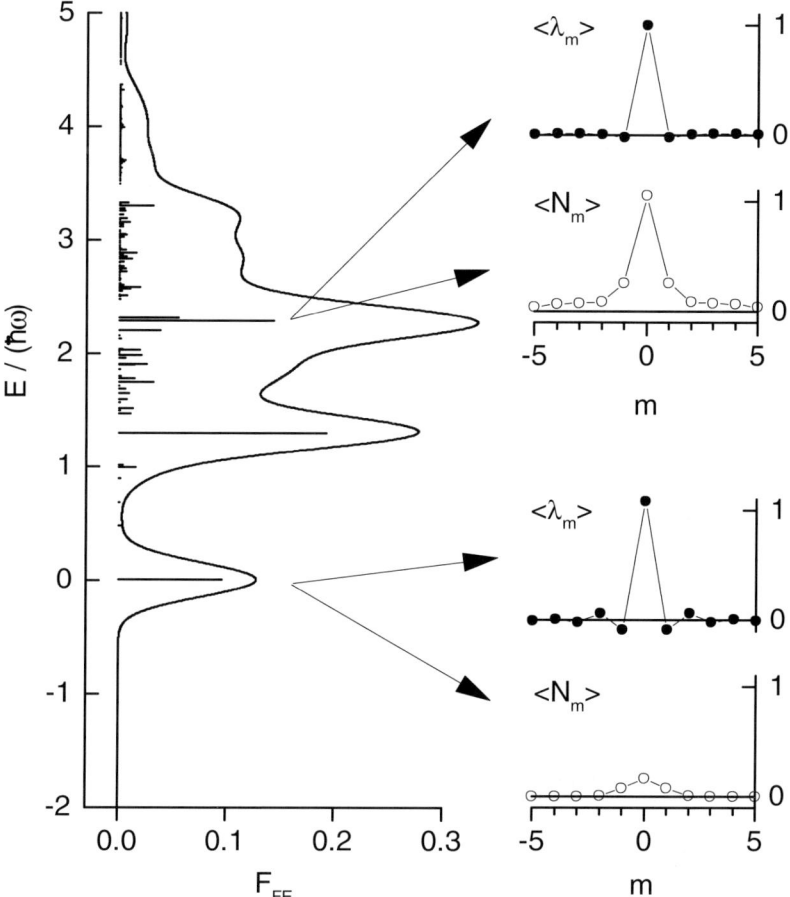

Fig. 11. Illustration of the eigenstates and their properties for a numerical solution of the Holstein model (101) with parameters $J = 0.5\hbar\omega$ and $g = 1$ at total momentum $k = 0$. In the left panel, the optically active eigenstates are shown at a vertical energy axis. The sticks indicate the FE character $F_{\text{FE}j}$ of each state according to Eq. (116). For a visualization of the resulting spectrum, the stick spectrum is convolved with a Gaussian (standard derivation $\sigma = 0.15\hbar\omega$) and the broadened spectrum is scaled for easy superposition (here: area $\int f(E)\,dE = 0.5\hbar\omega$). In the right panels, the occupation number clouds $\langle \hat{N}_m \rangle$ and displacement clouds $\langle \hat{\lambda}_m \rangle$ are shown for two particular eigenstates (cf. comments to Eqs. (124) and (125)).

$\nu_m^{\max}$ can be included in the basis set, it must be assured that $\langle \hat{N}_m \rangle \ll \nu_m^{\max}$. These phonon occupation numbers are again illustrated in Figure 11 for two representative eigenstates of high spectral weight. For the lowest state at $E_1 = 0.0074\hbar\omega$, there are 0.16 phonons at the exciton site ($m = 0$), and the total phonon number is $\sum_m \langle N_m \rangle = 0.34$. In the molecular limit, this state would be the zero-phonon state, but the hopping term $J$ leads to a nonzero phonon occupation number. At a higher state $E_{41} = 2.28\hbar\omega$, the total phonon number is 2.12 with a peak value of $\langle \tilde{N}_0 \rangle = 1.05$. This state originates from the 2-phonon state in the molecular limit. Electronic delocalization leads to broad phonon clouds.

A description of the phonon cloud that is independent of the basis set can be provided by the expectation values of the displacement operators:

$$\langle \lambda_m \rangle \equiv \left\langle \sum_n a_n^\dagger a_n \frac{b_{n+m}^\dagger + b_{n+m}}{2} \right\rangle. \tag{125}$$

This displacement cloud $\langle \lambda_m \rangle$ gives the average distortion from equilibrium (along the dimensionless normal coordinate $\lambda$) at a molecule which is $m$ sites from the exciton. Note that the exciton itself is completely delocalized in real space and so is its displacement cloud. This delocalization follows directly from the assumed perfect translational symmetry. The values $\langle \lambda_m \rangle$ as a function of the distance $m$ only show the spatial correlation between the electronic excitation and the lattice distortion.

With respect to the basis representation (102), the displacement cloud of a state $|\Psi_j(k)\rangle$ (111) is obtained as:

$$\langle \lambda_m \rangle = \sum_{\underline{\mu}\,\underline{\nu}} u_{\underline{\mu}j}^* u_{\underline{\nu}j} \times \left( \prod_{r \neq m} \delta_{\mu_r, \nu_r} \right)$$

$$\times \left( \frac{\sqrt{\nu_m + 1}}{2} \delta_{\mu_m, \nu_m+1} + \frac{\sqrt{\nu_m}}{2} \delta_{\mu_m, \nu_m-1} + g \delta_{m,0} \delta_{\mu_0, \nu_0} \right). \tag{126}$$

Again, Figure 11 may serve as an illustration. There, the displacement clouds are shown for the same representative states that were analyzed in terms of occupation number clouds. The narrow clouds show that the actual lattice distortion is much more localized around the exciton than the broad occupation number clouds might suggest. This difference results from the fact that the vibronic wavefunction in the actual eigenstates cannot be accurately represented by single oscillator functions of the special Lang–Firsov basis.

### 3.5. Truncated Phonon Basis and Symmetry Adaptation

By now, the formal tools for calculating and analyzing the eigenstates of the Holstein Hamiltonian (101) have been collected. The only remaining issue is how to truncate the infinite phonon-cloud basis to a number that allows numerical diagonalization. For this, we first restrict the basis to cloud states of the form

$$|\underline{\nu}_M\rangle = |\nu_{-M} \ldots \tilde{\nu}_0 \ldots \nu_{+M}\rangle, \tag{127}$$

as motivated below Eq. (103). That means, only phonon clouds localized at the $2M+1$ molecules around the exciton are included. Strongly delocalized or even free phonons can only be approximated using large $M$.

Second, for each position in the phonon-cloud we restrict the maximum occupation number:

$$v_m \leqslant v_m^{\max}. \tag{128}$$

In this way, the localized nature of the phonon cloud can better be taken into account by considering only small occupation numbers $v_m^{\max}$ at sites far away from the exciton. A typical cut-off vector as used for the calculation in Figure 11 has $M = 5$ and looks like $|12345\widetilde{6}54321\rangle$.

Third, among these states we include only those for which the total number of phonons does not exceed a given maximum:

$$\sum_m v_m \leqslant v_{\text{tot}}^{\max}. \tag{129}$$

In this way, high energy basis states are excluded. Since the overlap factors for states with high vibrational excitation decrease rapidly, these states do not appear in the absorption spectrum. Condition (129) is only effective for $v_{\text{tot}}^{\max} < \sum_m v_m^{\max}$, but typically it can be used as a strong restriction (e.g., $v_{\text{tot}}^{\max} = 6$ in Figure 11).

Now, we have arrived at a fairly complex description for the cut-off conditions of the basis set, given by the numbers $M$, $\underline{v}^{\max}$, $v_{\text{tot}}^{\max}$. However, this complex scheme allows to choose a basis just large enough to represent the optically active eigenstates of the Hamiltonian.

The minimum radius, $M = 0$, is an important special case of the phonon basis in which electronic and vibrational excitations are always at the same site, just as in the $J = 0$ limit. These joint exciton–phonon configurations (see illustration in Figure 9) can be considered as distinct molecular excited states and treated within the standard framework of Frenkel exciton theory. Following Broude, Rashba and Sheka [47, p. 185], we call this the *molecular vibron model*:

$$M = 0. \tag{130}$$

The molecular vibron model follows naturally from the exciton concept and was successfully applied to early interpretations of crystal spectra [26]. The approximation is additionally justified if—beyond the simplest Holstein Hamiltonian (101)—the phonon energy differs between the electronic ground and excited state of the molecule (cf. Ref. [3, p. 87ff] or Ref. [47, p. 198f]).

To find a suitable phonon basis for concrete calculations, we start with the molecular vibron model and gradually increase the phonon basis until the obtained absorption spectrum converges. This procedure is demonstrated in Ref. [41].

In addition to the general truncation scheme, in some cases the dimension of the phonon basis can be reduced by symmetry. For the Frenkel exciton problem in this section, we have inversion symmetry about the exciton's site. So we can introduce symmetry adapted basis states $|k\underline{v}\rangle_\pm$ in which the phonon cloud is either symmetric (+) or antisymmetric (−) with respect to inversion about its center. Inversion of the phonon cloud in the non-adapted basis (104) shall be denoted by a bar:

$$|\underline{\bar{v}}\rangle: \bar{v}_n = v_{-n}. \tag{131}$$

Even the non-adapted basis contains some symmetric phonon-clouds ($\underline{\bar{v}} = \underline{v}$). For all other states, a symmetry adaption has to be chosen. Thus, the symmetry adapted states can be obtained as:

$$|k\underline{v}\rangle_+ = \begin{cases} |k\underline{v}\rangle & \text{for } \underline{\bar{v}} = \underline{v}, \\ \frac{1}{\sqrt{2}}(|k\underline{v}\rangle + |k\underline{\bar{v}}\rangle) & \text{for } \underline{\bar{v}} \neq \underline{v}, \end{cases}$$

$$|k\underline{v}\rangle_- = \frac{1}{\sqrt{2}}(|k\underline{v}\rangle - |k\underline{\bar{v}}\rangle) \quad \text{for } \underline{\bar{v}} \neq \underline{v}. \tag{132}$$

Now, the symmetric subspace spanned by the $|k\underline{v}\rangle_+$ states does not mix with the antisymmetric subspace spanned by the $|k\underline{v}\rangle_-$ states and the diagonalization can be done separately for both subspaces. For a large cut-off radius of the phonon cloud, the dimension of the two subspaces is roughly one half of the original basis. Furthermore, the transition dipoles of all antisymmetric states vanish exactly and only the symmetric space is needed for the absorption spectrum.

### 3.6. THE LIMIT FOR WEAK INTERMOLECULAR ELECTRONIC COUPLING

In order to illustrate the qualitative effects that arise from exciton transfer, we will now apply perturbation theory for the limit of weak electronic coupling ($J \ll g\hbar\omega$). The reference system is given by the molecular limit ($J = 0$). Then, the molecular vibron model (130) gives an exact description of the optically active states, which form an equally spaced vibronic progression (cf. Figure 11(a)). We consider the lowest (zero-phonon) and second lowest (one-phonon) molecular vibronic states.

The lowest state of the unperturbed system is $|k\underline{v}\rangle$ with $|\underline{v}\rangle = |\ldots 0\tilde{0}0 \ldots\rangle$. This state at $E_1^{(0)} = 0$ is non-degenerate, and application of first order perturbation theory gives immediately

$$E_1^{(1)} = \langle k\underline{v}|H_{\text{Hol}}^{\text{FE}}|k\underline{v}\rangle = 2J\cos(k) \times S^2\binom{0}{0} = 2J\cos(k) \times e^{-g^2}. \tag{133}$$

This result is well known from small polaron theory for zero temperature. The width $4J$ of the purely electronic band is renormalized by the overlap factor $e^{-g^2}$ since the exciton moves together with its displacement cloud.

The second lowest state of the unperturbed system is, in the molecular vibron model, $|k\underline{v}\rangle$ with $|\underline{v}\rangle = |\ldots 0\tilde{1}0 \ldots\rangle$. This is the molecular one-phonon state. Considering a complete phonon basis, the molecular one-phonon state is degenerate with all other dark basis states that contain one phonon excitation at an arbitrary exciton–phonon separation $n$. A perturbation $J > 0$ will mix all these states and lift their degeneracy.

This can be analyzed by writing down the matrix elements (110) for the states of the one-phonon manifold. We define the state $|k\underline{v}(n)\rangle$ by a phonon cloud with the structure $v_i = \delta_{i,n}$ and analogously for $|k\underline{\mu}(m)\rangle$: $\mu_i = \delta_{i,m}$. The matrix representation (110) then becomes

$$H_{mn} = \langle k\underline{\mu}(m)|H_{\text{Hol}}^{\text{FE}}|k\underline{v}(n)\rangle$$
$$= \delta_{m,n}\hbar\omega + Je^{-g^2}\left(W_{mn} + g^2 V_{mn}\right), \quad (134)$$

where

$$W_{mn} = \delta_{m,n+1} \times e^{-ik} + \delta_{m,n-1} \times e^{+ik} \quad (135)$$

$$= \begin{pmatrix} \ddots & & & & & & \\ \cdots & 0 & e^{ik} & 0 & 0 & 0 & \cdots \\ \cdots & e^{-ik} & 0 & e^{ik} & 0 & 0 & \cdots \\ \cdots & 0 & e^{-ik} & 0 & e^{ik} & 0 & \cdots \\ \cdots & 0 & 0 & e^{-ik} & 0 & e^{ik} & \cdots \\ \cdots & 0 & 0 & 0 & e^{-ik} & 0 & \cdots \\ & & & & & & \ddots \end{pmatrix},$$

and $V_{mn}$ is a matrix that has nonzero-elements only for $|m|, |n| \leqslant 1$:

$$V_{mn} = \begin{pmatrix} \ddots & & & & & & \\ \cdots & 0 & 0 & 0 & 0 & 0 & \cdots \\ \cdots & 0 & 0 & -e^{ik} & e^{ik} & 0 & \cdots \\ \cdots & 0 & -e^{-ik} & 2\cos k & -e^{ik} & 0 & \cdots \\ \cdots & 0 & e^{-ik} & -e^{-ik} & 0 & 0 & \cdots \\ \cdots & 0 & 0 & 0 & 0 & 0 & \cdots \\ & & & & & & \ddots \end{pmatrix}. \quad (136)$$

The non-diagonal contributions in $W_{mn}$ and $V_{mn}$ mix the joint configuration from the molecular vibron model with separated configurations. However, $V_{mn}$ only mixes the states where the phonon is located either at the exciton site or at its nearest neighbor. Therefore, $W_{mn}$ and $V_{mn}$ act in completely different ways.

Let us first discuss the case of $g \ll 1$ and neglect $V_{mn}$ in Eq. (134). For $k = 0$ or $k = \pi$, $W_{mn}$ is the Hamiltonian of a nearest-neighbor hopping particle on an infinite chain with open boundary. This gives a wave-like solution. In contrast to the ordinary hopping problem, the exact consideration of the specific boundary conditions is essential now. Only then, the correct amplitude at the special site $n = 0$ can be obtained; and this amplitude alone determines the exciton character. Thus, one obtains the eigenstates

$$|\Psi_j\rangle = \frac{1}{\sqrt{M+1}} \sum_{n=-M}^{M} \sin\left(\frac{n j \pi}{2M+2}\right) |k\underline{\nu}(n)\rangle \tag{137}$$

with

$$j = 1, 2, \ldots, 2M+1. \tag{138}$$

Their energies are

$$E_j^{(1\text{ph}, g \ll 1)} = \hbar\omega \pm 2J e^{-g^2} \cos\left(\frac{j\pi}{M+1}\right), \tag{139}$$

where $\pm$ refers to $k = 0$ and $k = \pi$, respectively. The FE character of state $j$ at $k = 0$ follows from Eq. (116). It has only two values depending on the index $j$:

$$F_{\text{FE}j} = \begin{cases} \frac{1}{M+1} \cdot g^2 e^{-g^2} & \text{for odd } j, \\ 0 & \text{for even } j. \end{cases} \tag{140}$$

The $M$ states with even $j$ and zero FE character belong to the subspace of the antisymmetric states in the symmetry adapted basis (132). The $M+1$ optically active states with odd $j$ are the symmetric states. These active states form a band of equally absorbing states with a total width of $4J e^{-g^2}$. The total FE character of these active states sums up to $g^2 e^{-g^2}$ representing the value of the molecular limit. In all these states, the phonon cloud is not localized around the exciton but consists of a standing phonon wave. We emphasize that this behavior is the limit for small $g$. In this limit, the total FE character of the considered one-phonon band gives only a small feature in the overall absorption spectrum since the major part of the FE character is concentrated in the zero-phonon state.

Complementary, the $V_{mn}$ part in the perturbation expression (134) mixes only the cloud states with phonon excitations at or next to the exciton site. Therefore, in the limit of large $g$, the basis set can be reduced to include only local phonon cloud configurations up the nearest neighbor ($M = 1$). Using the symmetry adapted basis functions (132), the symmetric one-phonon subspace consists only of two phonon configurations: $|\Phi_1(k)\rangle = |k\rangle|0\tilde{1}0\rangle_+$ and $|\Phi_2(k)\rangle = |k\rangle|1\tilde{0}0\rangle_+$. The Hamiltonian in the representation of these two states takes the form

$$H_{mn} = \delta_{mn} \hbar\omega + 2J e^{-g^2} \cos(k) \times \begin{pmatrix} g^2 & \frac{1-g^2}{\sqrt{2}} \\ \frac{1-g^2}{\sqrt{2}} & \frac{1}{2}g^2 \end{pmatrix} \tag{141}$$

with eigenvalues

$$E_{\pm}^{(1\text{ph}, g \gg 1)} = \hbar\omega + 2J\cos(k) \cdot g^2 e^{-g^2} \cdot \frac{3}{4}\left(1 \pm \sqrt{1 - \frac{16}{9g^2} + \frac{8}{9g^4}}\right). \quad (142)$$

Thus, the zero-order energy $E = \hbar\omega$ splits into two bands $E_\pm(k)$. Similarly to the perturbation-in-$J$ treatment of the lowest state (133), the electronic bandwidth $4J$ is multiplied by an overlap factor $g^2 e^{-g^2}$ which corresponds to the interaction of the transition-dipole moments of the molecular one-phonon state. However, there are *two* states now. In the limit $g \to \infty$, their energies tend to:

$$E_{+}^{(1\text{ph}, g \to \infty)} \to \hbar\omega + 2J\cos k \cdot g^2 e^{-g^2} \cdot \frac{3}{4}, \quad (143)$$

$$E_{-}^{(1\text{ph}, g \to \infty)} \to \hbar\omega. \quad (144)$$

In this limit, both states still have an FE character of $F_{\text{FE}+} \to \frac{2}{3}$ and $F_{\text{FE}-} \to \frac{1}{3}$.

This splitting into two states which both carry spectral weight is entirely caused by the delocalization of the phonon cloud. Such a delocalization is neglected in the simplest approach of the molecular vibron model (130), which would mean the neglect of state $|\Phi_2(k)\rangle$ in Hamiltonian (141). Looking at the non-diagonal term in Hamiltonian (141) suggests, and closer inspection of the full one-phonon subspace Hamiltonian (134) confirms: For the special value $g = 1$, the molecular vibron state $|k\underline{v}(n = 0)\rangle$ decouples from all other phonon cloud configurations. Only in this case, the molecular vibron model becomes exact (in the one-phonon subspace) and yields one energy band at

$$E^{(1\text{ph}, g=1)} = \hbar\omega + 2J\cos k \cdot g^2 e^{-g^2} \quad (145)$$

which includes the complete FE character of the one-phonon state ($g^2 e^{-g^2}$).

To give an illustration of the phenomena in the one-phonon subspace and to show the relevance of the described limiting cases, we show a numerical solution in Figure 12. For this, we solved the Hamiltonian (134) numerically for a phonon cloud of radius $M = 20$ at the total momentum $k = 0$. For $k = \pi$, the spectra only have to be mirrored with respect to $E = \hbar\omega$.

In Figure 12(a), the "exact" numerical results (graph 1) are shown for a relatively small $g = 0.5$. The tendency of a broad band with constant FE character is clearly visible. This bandwidth is compared to the width the free phonon part $W_{mn}$ from Eq. (139) in graph 2. Both agree very well. The molecular vibron model ($M = 0$) would give a single active sate at $(E - \hbar\omega)/J = 0.389$ (position indicated by graph 3). This state would represent the weighted center of the exact band but it would veil the large splitting ($\Delta E/J \approx 1.55$).

In Figure 12(c), the numerical solution is shown for a rather large $g = 1.5$ (graph 1). It clearly approaches the two active states from the nearest neighbor

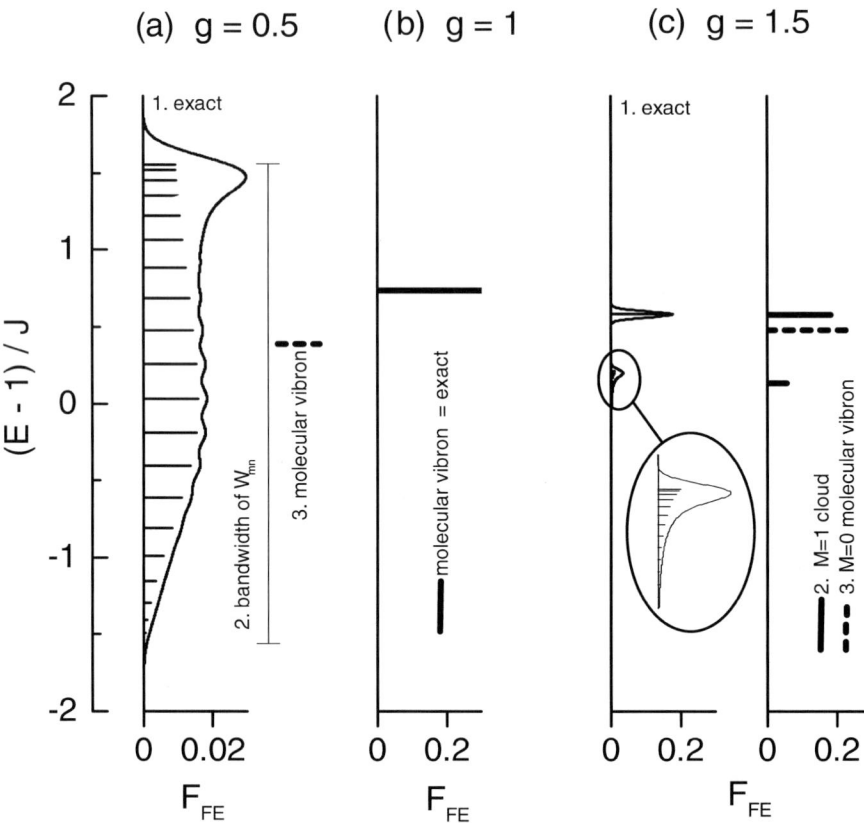

Fig. 12. Perturbative treatment $J \to 0$ of the one-phonon subspace for three coupling parameters $g$. The "exact" stick spectra are numerical solutions of the one-phonon Hamiltonian (134) for a phonon-cloud radius of $M = 20$. The envelopes are convolutions of the stick spectra with Gaussians of appropriate width. Figure 12(a) represents the small-$g$ case, where a broad one-phonon sideband is formed. The "exact" solution in graph 1 is compared to the bandwidth of the free-phonon part ($W_{mn}$ from Eq. (139)) in graph 2 and to the position of the single active state from the molecular vibron model (130) in graph 3. Figure 12(b) represents the $g = 1$ case, where the molecular vibron model becomes exact. Figure 12(c) represents the large-$g$ case, in which the exciton interacts mainly with a nearest neighbor phonon cloud. The "exact" numerical solution in graph 1 resembles the approximate solution (141) for a nearest neighbor cloud ($M = 1$) in graph 2. The single state from the molecular vibron model ($M = 0$) is shown in graph 3.

cloud (radius $M = 1$) given by Eq. (142), which is shown in graph 2. For comparison, the result of the molecular vibron model ($M = 0$) is also shown in graph 3. As for $g < 1$, the molecular vibron model can only represent the weighted center of the one-phonon states but not their qualitative splitting. Note that for both cases

$g < 1$ and $g > 1$ the correct splittings of the one-phonon states are on the same order as the perturbation parameter $J$.

The situation for energies above the one-phonon subspace becomes more complex and will not be considered here. Already in the two-phonon subspace, which is spanned by all zero-order basis states with a total phonon number 2, there occurs a high degeneracy of various cloud configurations. The numerical calculations in Ref. [41] confirm that for not too strong electronic coupling ($J \stackrel{<}{\approx} 0.5\hbar\omega$) and $g$ in the order of 1, the approximation of highly localized phonon clouds or even the molecular vibron model yields a good description of the full absorption spectrum.

### 3.7. NUMERICAL SOLUTIONS FOR VARIOUS ELECTRONIC COUPLING STRENGTHS

In this section, we want to give an impression of absorption spectra for various conditions. We always consider an electron–phonon coupling constant of $g = 1$, which is a typical order of magnitude for the strongly coupled modes in $\pi$-conjugated systems. The numerical spectra were calculated with phonon cloud radius $M = 5$, a maximum total phonon number $v_{\text{tot}}^{\max} = 6$ (cf. Eq. (129)) and a phonon cloud cut-off vector of $|\underline{v}^{\max}\rangle = |12345\widetilde{6}54321\rangle$ (cf. Eq. (128)) corresponding to 4485 symmetric basis states in the symmetry adapted basis (132). This ensures a sufficient accuracy for all shown spectra (on the order of the graphic resolution) except for the numerically demanding case $J = 1\hbar\omega$, $k = 0$. In this case, deviations on the order of about 10% might occur on the high energy side ($E > 3\hbar\omega$).

As discussed in the previous section, for $g = 1$ the molecular vibron model (cf. Eq. (130)) is a good approximation for weak electronic coupling ($J \ll g\hbar\omega$). At first, we illustrate the quality of this approximation for intermediate positive $J$ at $k = 0$ (top of the band). In Figure 13, we compare the discrete vibronic states resulting from the molecular vibron model with the complete numerical solutions. For $J = 0.5\hbar\omega$, the molecular vibron model still gives a qualitatively reasonable description of the spectrum. The main effect of the delocalized cloud basis in the high energy region is a broadening of the spectra. The lowest state, however, moves considerably from $E_1 = 0.229\hbar\omega$ in the molecular vibron model to $E_1 = 0.0074\hbar\omega$ in the largest basis set. For $J = 1\hbar\omega$, the deviations between the molecular vibron model and the exact solution are already on the same order as all structures in the spectrum.

To give an overview about general trends, we show characters and electronic bands for $g = 1$ and various $J$ in Figure 14. The electronic bands $E_{\text{elec}}(k)$ in this figure are the solutions of the electronic Hamiltonian $H_{\text{elec}}^{\text{FE}} + g^2\hbar\omega$ (cf. Eq. (100)). The molecular Franck–Condon relaxation energy $g^2\hbar\omega$ has to be added to comply with our energy axis definition, in which the relaxed molecular zero-phonon

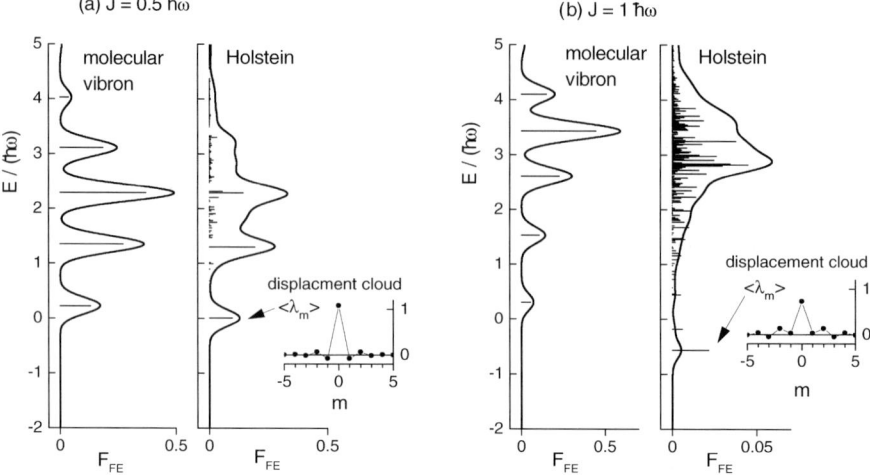

Fig. 13. Comparison of the molecular vibron model (left panels) and the numerical solution for the complete Holstein problem (right panels) for two exciton transfer integrals: (a) $J = 0.5\hbar\omega$, (b) $J = 1\hbar\omega$. Exciton–phonon coupling is $g = 1$ and quasi-momentum is $k = 0$.

state lies at $E = 0$. The states and spectra resulting from the complete Holstein Hamiltonian $H_{\text{Hol}}^{\text{FE}}$ (Eq. (101)) are shown at $k = 0$ and $k = \pi$.

In the noninteracting case $J = 0$ (Figure 14(a)), the electronic dispersion is zero and the spectra are the vibronic states of the isolated molecule. The molecular zero phonon state lies below the electronic band by the amount of the Franck–Condon relaxation energy $g^2\hbar\omega$.

For a moderately small interaction $J = 0.5\hbar\omega$ (Figure 14(b)), the spectrum still shows distinct peaks reminding of the molecular vibronic states. However, the shape of the spectrum is changed. In particular, the spectrum loses one fingerprint of a vibronic progression in the molecular case, namely the constant energy spacing between the peaks. The centers of mass of the spectra are shifted upwards at the top of the band ($k = 0$) and downwards at the bottom of the band ($k = \pi$), corresponding to the dispersion of the electronic band. The bandwidth of the lowest dressed exciton state is $\Delta E = 0.62\hbar\omega$, which is still similar to the weak-electronic coupling limit of $4Je^{-g^2} = 0.74\hbar\omega$.

For stronger interaction, $J = 1\hbar\omega$ (Figure 14(c)), the spectra already show the tendency towards the opposite limiting case. Instead of a number of vibronic peaks, just one major peak close to the position of the electronic bands starts to emerge. Furthermore, there remains a small one-phonon sideband approximately one vibrational quantum above the bottom of the band structure (at $k = \pi$). The one-phonon sideband at $k = 0$ corresponds to a mixture of the lowest $k = \pi$

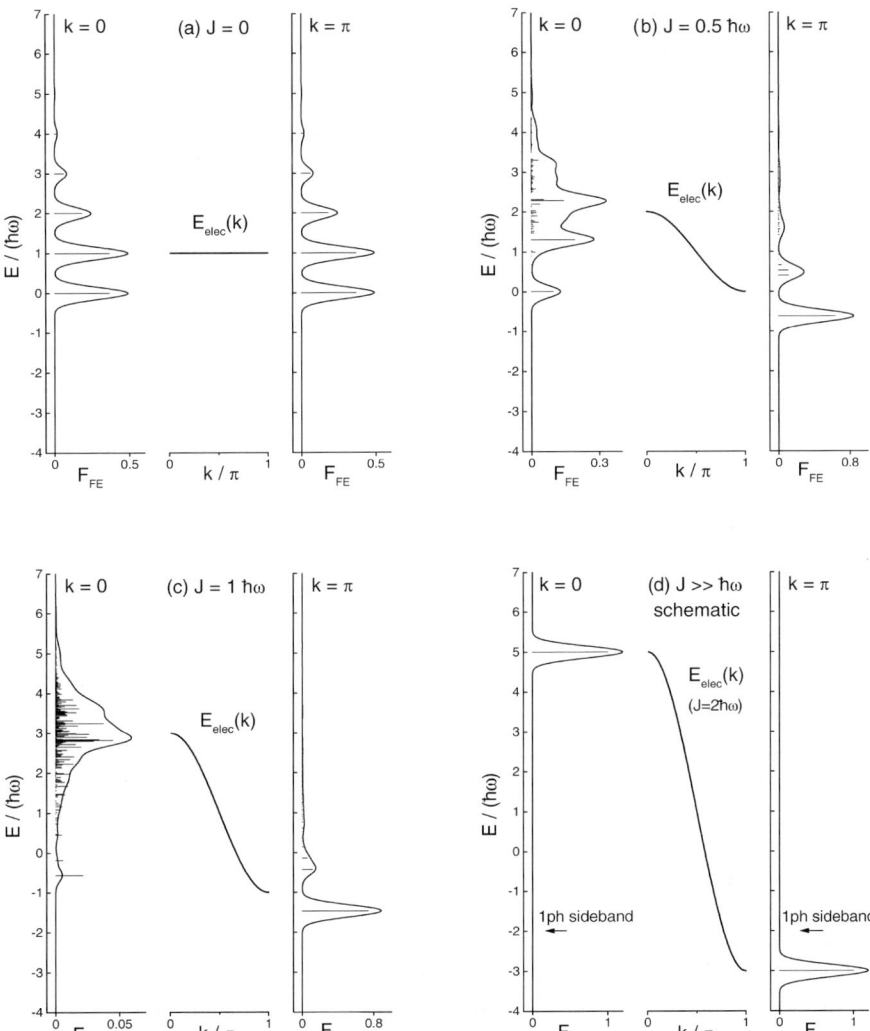

Fig. 14. Overview of band structures in the Holstein model for fixed exciton phonon coupling constant $g = 1$ and various exciton hopping integrals $J$. The vibronic spectra at $k = 0$ and $k = \pi$ were calculated and represented as for Figure 11. For the limiting case $J \gg \hbar\omega$, we show only schematic spectra at energy positions corresponding to the electronic bands for $J = 2\hbar\omega$.

dressed exciton and a $k = -\pi$ phonon. For further increasing $J$, this one-phonon sideband will more and more lose its spectral weight. Nevertheless, there remains a low-lying $k = 0$ state which determines the overall dispersion of the lowest state to approximately one-vibrational quantum, independently of the stronger and stronger electronic dispersion $4J$.[11]

The limit of strong electronic coupling is schematically depicted in Figure 14(d). The electronic band is shown for the example of $J = 2\hbar\omega$. The given absorption spectra, however, are not calculated but only serve as a schematic illustration. In the strong coupling limit, the electronic bandwidth $4J$ is large compared to the molecular Franck–Condon relaxation energy $g^2\hbar\omega$, the exciton hopping is "fast" compared to the exciton–phonon coupling and the Born–Oppenheimer approximation should be applied to the whole crystal as one object (cf. Ref. [81]). The total lattice displacement $\sum \langle \lambda_m \rangle = g$ is now equally distributed over the $N \to \infty$ molecules. Therefore, the total relaxation energy $E_{FC} = g^2\hbar\omega \sum \langle \lambda_m \rangle^2$ tends to zero. Figuratively speaking, the very fast exciton loses its phonon cloud. Compared to the molecular limit (lowest state at $E = 0$), the lowest state will now be given by the purely electronic band at $E = 2J \cos k + g^2\hbar\omega$. Because of the vanishing relaxation energy, higher vibronic states have no spectral weight and the absorption spectrum consists of a narrow line at the electronic energy. The position of the one-phonon side-bands is also indicated, but these side-bands have vanishing spectral weight in the limit $J \to \infty$.

### 3.8. THE HOLSTEIN HAMILTONIAN WITH CHARGE-TRANSFER STATES

The Holstein Hamiltonian for Frenkel excitons (101) can be very naturally extended to include charge-transfer (CT) states. Let $c_{n,f}^\dagger$ be the creation operator for a nearest-neighbor CT state in which an electron is transferred from lattice site $n$ to site $n + f$ ($f = \pm 1$). The molecular limit is again defined as the case where no transfer interactions (neither Frenkel exciton transfer nor charge transfer nor Frenkel–CT interactions) are considered. Then, the electronic CT Hamiltonian is

$$H^{CT} = D \sum_{n,f} c_{n,f}^\dagger c_{n,f}, \qquad (146)$$

with $D$ being the on-site energy of a CT state in the molecular limit (relative to the Frenkel exciton on-site energy at zero in our energy units).

The electron or hole excitation of the CT state are assumed to couple to the same effective vibrational coordinate $\lambda$ as the Frenkel exciton. With the electron–phonon coupling constant $g_e$ and the hole–phonon coupling constant $g_h$, the linear

---

[11] This behavior of the dispersion depends on the exciton–phonon coupling constant $g$. If the exciton–phonon coupling energy $g^2\hbar\omega$ is large compared to the electronic bandwidth $4J$, the bandwidth of the lowest dressed sate takes the value of the weak-electronic coupling limit $4Je^{-g^2}$, which can be smaller than $\hbar\omega$. For a detailed discussion, see, e.g., Ref. [60].

coupling between CT states and phonons is described by the Hamiltonian

$$H^{\text{CT-ph}} = \hbar\omega \sum_{n,f} c^\dagger_{n,f} c_{n,f}$$

$$\times \left[ -g_h(b^\dagger_n + b_n) - g_e(b^\dagger_{n+f} + b_{n+f}) + g_h^2 + g_e^2 \right]. \quad (147)$$

These expressions are analogous to the Frenkel-exciton–phonon coupling in Eq. (99). The term $(g_h^2 + g_e^2)\hbar\omega$ is the vibrational relaxation energy of a CT state in the molecular limit. As in Eq. (99), this term is added to align the on-site energy $D$ of the CT states to its value in the molecular limit.

For the electronic mixing between Frenkel and CT excitons, we use the corresponding part from the Merrifield Hamiltonian (27):

$$H^{\text{FE-CT}} = \sum_n \left[ t_e \left( a^\dagger_n c_{n,+1} + a^\dagger_n c_{n,-1} \right) \right.$$

$$\left. + t_h \left( a^\dagger_n c_{n+1,-1} + a^\dagger_n c_{n-1,+1} \right) + \text{h.c.} \right]. \quad (148)$$

Thus, the extended Holstein Hamiltonian for Frenkel and CT excitons becomes

$$H^{\text{FCT}}_{\text{Hol}} = H^{\text{FE}}_{\text{Hol}} + H^{\text{CT}} + H^{\text{CT-ph}} + H^{\text{FE-CT}}. \quad (149)$$

This Hamiltonian corresponds to the dimer Hamiltonian used in Ref. [15].

A natural extension of the basis states $|n\underline{v}\rangle$ from Eq. (102) is obtained by including the new electronic degree of freedom $f$. The value $f = 0$ shall denote the former Frenkel exciton basis states:

$$\left[ |nf\underline{v}\rangle \right]_{f=0} \equiv |n\underline{v}\rangle. \quad (150)$$

A Lang–Firsov-type basis for CT states ($f = \pm 1$) follows in complete analogy to Eqs. (102) and (103) for the Frenkel excitons:

$$\left[ |nf\underline{v}\rangle \right]_{f=\pm 1} \equiv c^\dagger_{n,f}|o_{\text{el}}\rangle \times B^\dagger_{nf\underline{v}}|o_{\text{vib}}\rangle, \quad (151)$$

with

$$B^\dagger_{nf\underline{v}} \equiv \underbrace{\frac{1}{\sqrt{v_0!}} (b^\dagger_n - g_h)^{v_0} \cdot e^{-\frac{g_h^2}{2}} e^{g_h b^\dagger_n}}_{g_h\text{-disp. on } n} \times \underbrace{\frac{1}{\sqrt{v_f!}} (b^\dagger_{n+f} - g_e)^{v_f} \cdot e^{-\frac{g_e^2}{2}} e^{g_e b^\dagger_{n+f}}}_{g_e\text{-disp. on } n+f}$$

$$\times \underbrace{\prod_{m\neq 0,f} \frac{1}{\sqrt{v_m!}} (b^\dagger_{n+m})^{v_m}}_{\text{undisp. otherwise}}. \quad (152)$$

Here, $b_n^\dagger - g_h$ creates phonons in the $g_h$ displaced potential at the hole position $n$ and $b_n^\dagger - g_e$ creates phonons in the $g_e$-displaced potential at the electron position $n + f$. At all other sites, the vibrational potential is not displaced.

The real-space basis states from Eqs. (150), (151) can again be Fourier-transformed to momentum-space basis states with total momentum $k$:

$$|kf\underline{v}\rangle \equiv \frac{1}{\sqrt{N}} \sum_n e^{ikn} |nf\underline{v}\rangle. \tag{153}$$

As for the Frenkel problem, the matrix elements of the Frenkel–CT Holstein Hamiltonian (149) can be derived in a straightforward way. The final expressions become lengthy due to various overlap factors and we omit them here. The basis can be reduced to a manageable size by a truncation scheme as for the Frenkel problem. Then, the eigenstates $|\Psi_j(k)\rangle$ at $k = 0$ or $k = \pi$ can again be obtained by standard diagonalization methods for real matrices in the form

$$|\Psi_j(k)\rangle = \sum_{f\underline{v}} u_{f\underline{v}j}(k) |kf\underline{v}\rangle. \tag{154}$$

As for the Frenkel exciton problem, the most important property for characterizing an eigenstate $|\Psi_j(k)\rangle$ is its electronic character. Since the states are now constructed from different electronic states (Frenkel and CT), we have to distinguish different electronic characters.

The Frenkel exciton character is given as in definition (113) by the projection onto a Frenkel exciton:

$$F_{\text{FE}j}(k) = \left|\langle \Psi_j(k) | a_k^\dagger | o \rangle\right|^2. \tag{155}$$

Using the decomposition (154) into basis states, the FE character becomes:

$$F_{\text{FE}j}(k) = \left|\sum_{f\underline{v}} u^*_{f\underline{v}j} \langle kf\underline{v} | a_k^\dagger | o \rangle\right|^2. \tag{156}$$

The matrix element $\langle kf\underline{v} | a_k^\dagger | o \rangle$ is nonzero only for Frenkel-type basis states ($f = 0$). Then, it reduces to the same expression as for the Frenkel-only problem (see Eq. (116)) and we get:

$$F_{\text{FE}j}(k) = \left|\sum_{\underline{v}} u^*_{0\underline{v}j} S\binom{v_0}{0} \prod_{r \neq 0} \delta_{v_r,0}\right|^2. \tag{157}$$

For describing the CT character of the states, we use a projection onto the symmetric CT states

$$\tilde{c}^\dagger_{k+} |o\rangle = \frac{1}{\sqrt{2t_k}} \left\{ (t_e + t_h e^{+ik}) c^\dagger_{k,+1} + (t_e + t_h e^{-ik}) c^\dagger_{k,-1} \right\} |o\rangle,$$

which were introduced in Eq. (31) for the electronic problem. Then, the CT character becomes

$$F_{\mathrm{CT}j}(k) \equiv \left| \langle \Psi_j(k) | \tilde{c}_{k+}^\dagger | o \rangle \right|^2 \quad (158)$$

$$= \left| \sum_{f\underline{v}} u^*_{f\underline{v}j} \langle kf\underline{v} | \tilde{c}_{k+}^\dagger | o \rangle \right|^2. \quad (159)$$

The matrix element can be decomposed into an electronic part and a vibrational overlap factor:

$$\langle kf\underline{v} | \tilde{c}_{k+}^\dagger | o \rangle = \frac{1}{\sqrt{2} t_k} \left\{ \left( t_e + t_h e^{ik} \right) \delta_{f,+1} + \left( t_e + t_h e^{-ik} \right) \delta_{f,-1} \right\}$$

$$\times \langle \tilde{B}^\dagger_{nf\underline{v}} o_{\mathrm{vib}} | o_{\mathrm{vib}} \rangle.$$

Naturally, the electronic part of the projection onto the symmetric CT state becomes zero for a pure FE basis state ($f = 0$). The vibrational overlap factor contains two overlap terms for the electron and hole displaced lattice site:

$$\langle \tilde{B}^\dagger_{nf\underline{v}} o_{\mathrm{vib}} | o_{\mathrm{vib}} \rangle = S_{gh} \binom{v_0}{0} \times S_{ge} \binom{v_f}{0} \times \prod_{r \neq 0, f} \delta_{v_r, 0}.$$

At the important special points $k = 0$ and $k = \pi$, the electronic factor simplifies greatly and the projection of the basis state becomes for both $k$:

$$\langle kf\underline{v} | \tilde{c}_{k+}^\dagger | o \rangle = \frac{1}{\sqrt{2}} \times S_{gh} \binom{v_0}{0} \times S_{ge} \binom{v_f}{0} \times \prod_{r \neq 0, f} \delta_{v_r, 0}. \quad (160)$$

Inserting this matrix element into Eq. (159) gives the formula for the CT character of a numerically obtained eigenstate.

The transition dipole can be discussed similarly to the Frenkel-only case (see Eq. (118)), and one obtains a Frenkel and a CT transition dipole component as in the electronic problem (see Eq. (46)). For the experimental interpretation, we will later assume that the CT component is a negligible contribution ($p_{\mathrm{CT}} \ll p_{\mathrm{FE}}$), and then the squared transition dipole becomes proportional to the Frenkel character at $k = 0$ as in Eq. (118):

$$|\vec{P}_j|^2 \approx |\vec{P}_{\mathrm{FE}j}|^2 = \delta_{k,0} N \vec{p}_{\mathrm{FE}}^2 F_{\mathrm{FE}j}(k). \quad (161)$$

A representative calculation is shown in Figures 15 and 16 for the parameters $J = 0.5\hbar\omega$, $g = 1$, $D = 0$, $t_e = t_h = 0.5\hbar\omega$. The Frenkel part of this parameter set corresponds to the calculation in Figure 11. The basis cut-off vector for the phonon-space was $|\underline{v}\rangle^{\max} = |123\tilde{4}5432 1\rangle$ with $v_{\mathrm{tot}}^{\max} = 5$, resulting in 4332 basis states. An additional CT state is assumed at resonance with the Frenkel state

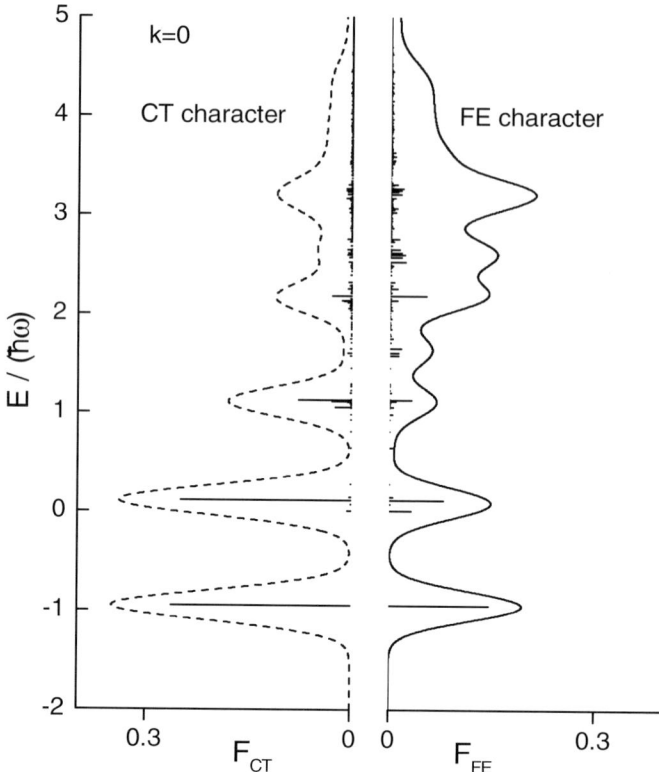

Fig. 15. Eigenstates of the extended Holstein model for Frenkel–CT mixing (149) at total momentum $k = 0$. Parameters: $J = 0.5\hbar\omega$, $g = 1$, $D = 0$, $t_e = t_h = 0.5\hbar\omega$, $g_e = g_h = 1/\sqrt{2}$. The Frenkel parameters and the illustration correspond to Figure 11. $F_{FE}$ shows the spectral weights (Frenkel character) of the Frenkel-part, $F_{CT}$ shows the spectral weights of the symmetric CT part. The broadened spectra are both normalized to an area of $0.5\hbar\omega$.

($D = 0$). The charge-transfer integrals $t_e$ and $t_h$ are chosen equal to the Frenkel hopping integral to give an illustration for strong Frenkel–CT mixing.

For the electron and hole coupling parameters, we used $g_e = g_h = g/\sqrt{2}$, which corresponds to equal relaxation energy for the CT state and the Frenkel exciton. In contrast to the Frenkel exciton–phonon coupling constant $g$, $g_e$ and $g_h$ are not easily accessible since absorption spectra of the ions would be needed. Alternatively, one might use quantum chemical calculations or at least qualitative arguments: Perylene's $\pi$-system is alternant. Simple Hückel theory then gives equal and opposite charges in the cation and anion, with half-filled HOMO and LUMO, respectively, while both are half-filled in the excited state. We have $g_e = g_h = g/2$ for noninteracting electrons. The Pariser–Parr–Pople model of in-

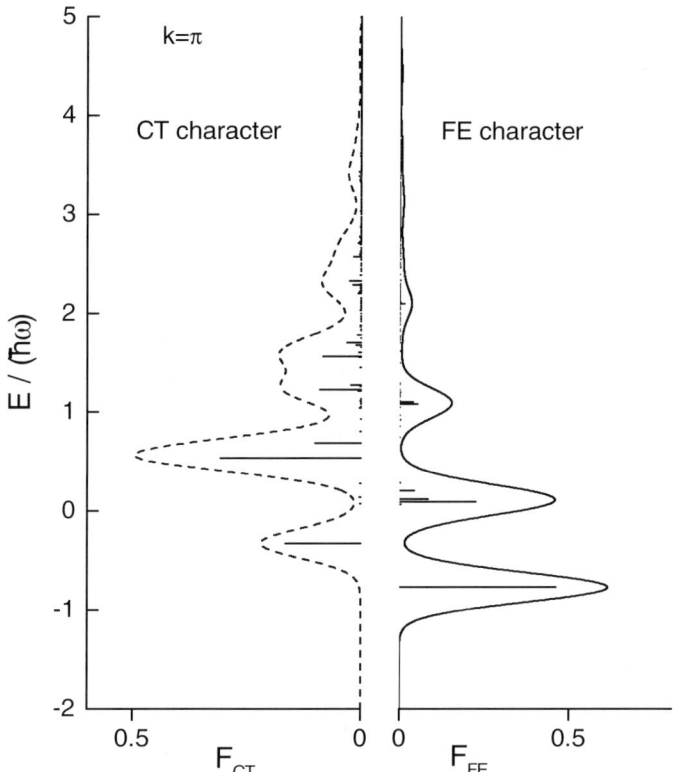

Fig. 16. Eigenstates of the extended Holstein model for Frenkel–CT mixing (149) at total momentum $k = \pi$. Parameters as in Figure 15. Because of $t_e = t_h$, the electronic FE and CT states do not mix and all eigenstates have either pure FE or pure CT character.

teracting $\pi$-electrons yields $g_e = g_h$ for systems with electron–hole symmetry. The bond order changes and relaxation energy of the singlet excitation in anthracene or trans-stilbene are now approximately half that of the triplet, which in turn is comparable to the relaxation energy of dication or dianion [82,83]. Our initial choice of equal relaxation energy for the Frenkel and CT excitation follows the correlated case, although this is a guess and PTCDA does not have e–h symmetry.

At the top of the band ($k = 0$, Figure 15), the energetic degeneracy and the large charge-transfer integrals lead to a strong mixing of Frenkel and CT states throughout the whole spectrum. The overall distribution of the spectral weights gives more Frenkel character to the higher states as a result of the positive $J$. The FE character in Figure 15 should be compared to the Frenkel-only problem from Figure 11. In the Frenkel-only problem, the lowest state gave rise to a single peak in the broadened spectrum at $E \approx 0$. This peak is now split into two well separated peaks at $E \approx -1\hbar\omega$ and $E \approx 0$. In such a way, strong mixing with CT states can add new features to the absorption spectrum even if their intrinsic transition dipoles are zero ($\vec{p}_{CT} = 0$) as it was discussed for the electronic problem in Section 2.

At the bottom of the band ($k = \pi$, Figure 16), the symmetry of the CT integrals ($t_e = t_h$) in this special case decouples the electronic Frenkel and CT states (cf. discussion below Eq. (39)). Therefore, the electronic character of all states is either purely Frenkel or purely CT. Only some indirect mixing is introduced by the phonon part of the Hamiltonian, which mainly affects the vibronic structure of the CT-character states.

## 4. Applications and Consequences for Quantum Confinement

### 4.1. Description of PTCDA-Derivatives

In Section 3.8, the energies $E_j$ and transition dipoles $P_j$ (Eq. (161)) of the eigenstates of the one-dimensional Holstein problem for mixed Frenkel–CT states were obtained. These quantities are essential but not yet sufficient for the description of a real absorption spectrum of a quasi-one-dimensional molecular crystal.

We still assume that *all* inter-stack interactions are on a much smaller energy scale than the in-stack interaction $J$. That means, the energy spectrum of the one-dimensional model is in first approximation not affected. However, the direction of the transition dipoles is determined by the complete three-dimensional crystal structure. In PTCDA and MePTCDI and many other organic crystals, the unit cell contains two non-equivalent molecules. Then, the transition dipoles of the non-equivalent molecules A and B couple and form two Davydov components ($\beta = p, s$) with orthogonal transition dipoles:

$$\vec{P}_{j\beta} = \frac{\vec{P}_j(A) \pm \vec{P}_j(B)}{\sqrt{2}}. \tag{162}$$

For the crystal structure of PTCDA and MePTCDI, the $p$-direction is given as the crystallographic $b$ axis. The $s$ direction lies approximately in the (102) plane since the molecular planes of both inequivalent molecules are roughly parallel to the (102) plane (within 5° [84] for PTCDA and within 10° for MePTCDI, derived from [85]).

Knowing the transition dipoles per unit cell, the transverse dielectric constant for perturbation by an external light wave polarized along the $\beta = p, s$ directions can be expressed as a sum over the excited states (cf., e.g., [4,86]):

$$\varepsilon_\beta^0(E) = 1 + \frac{8\pi}{v} \sum_j \frac{\vec{P}_{j\beta}^2 E_j}{E_j^2 - E^2 - i\hbar\Gamma E}. \tag{163}$$

Here, $v$ is the volume of the unit cell and $\Gamma^{-1}$ the life time of the excited states. Furthermore, the energies now have to be taken as the absolute energies with respect to the total ground state. Thus, the excitation energy $E_{00}$ of our reference state (zero-phonon state of the molecular limit) has to be included.

Eq. (163) is rigorous for any quantum system if *all* excited states are included. However, we are considering only the lowest electronic excitation. Therefore, we include the contribution of the higher states (mixing of molecular configurations) by using a phenomenologically modified formula for the dielectric function:

$$\varepsilon_\beta(E) = \varepsilon_\beta^{\text{bg}} + \frac{8\pi}{v} \sum_j \frac{(f_\beta^{\text{bg}} \vec{P}_{j\beta})^2 E_j}{E_j^2 - E^2 - i\hbar\Gamma E}. \quad (164)$$

Here, $\varepsilon_\beta^{\text{bg}}$ is a background dielectric constant that represents the value of $\varepsilon_\beta(0)$ corresponding to a crystal in which the considered lowest electronic excitation would not exist. $f_\beta^{\text{bg}}$ is a screening factor describing the modification of the acting field by the higher transitions. Furthermore, the higher transitions will modify the Frenkel exciton hopping integral $J$ and thereby all the eigenstates of the system. Since we treat $J$ as an effective fitting parameter anyway, the effect of the higher transitions onto $J$ is not important here but should be remembered in any microscopic interpretation of $J$. Such a background modification of the dielectric function was discussed for a simple model system of one purely electronic Frenkel exciton in a cubic crystal in Ref. [5]. In our general case, the effect of the higher transitions represented in the background parameters is also anisotropic in nature.

The dielectric function (164) includes a Lorentzian broadening of the individual eigenstates due to a finite lifetime $\Gamma^{-1}$. In a typical situation, however, there are several other sources of a much larger broadening: (i) coupling to further low energy vibrations, (ii) splitting of the single effective vibrational mode, which actually consists of several nearly degenerate modes, and (iii) inhomogeneous broadening. To account for all these effects empirically, we replace each eigenstate of the Holstein model $|\Psi_j\rangle$ by a Gaussian distribution of states with standard deviation $\sigma_j$ as, e.g., done in Ref. [87]. The individual broadenings $\sigma_j$ have no microscopic meaning and should be seen as no more than a convenient tool to compare the spectrum from the eigenstates of the Holstein model to an experimental spectrum. Practically, we assigned constant values of $\sigma_j$ for 4 separate regions of the spectrum in order to have only 4 different broadening parameters. The individual Lorentzian linewidth is assumed to be much smaller than the $\sigma_j$ and does not contribute anymore.

From the complex dielectric function (164), the complex refractive index $(n + i\kappa)^2 = \varepsilon$ and the absorption coefficient $\alpha = 2E/(\hbar c)\kappa$ can be calculated for the special light waves that propagate perpendicular to the $p$–$s$ plane and are polarized along the $p$ or $s$ direction. For general directions, the complex rules of crystal optics would have to be considered. We note that the consideration of the *absolute* absorption coefficient is essential for describing the shape of solid state spectra. The microscopic models provide predictions only about the *relative* spectral distribution of the transition dipoles, which determines the shape of the imaginary part $\varepsilon_2(E)$ of the dielectric function. The shape of the absorption spectrum $\alpha(E)$ however, is strongly influenced by the variation of the refractive index in the absorption region ($\alpha = E\varepsilon_2/(\hbar c n)$) and the variation of $n$ is again determined by the absolute absorption coefficient. Only if $\alpha$ is very small, as typically for spectroscopy of solutions ("dilute limit"), $n$ does not vary and the shape of the absorption spectrum is directly given by the distribution of the spectral weights (exciton characters).

For PTCDA and MePTCDI, it is possible to create vapor-deposited polycrystalline films with a high preferential orientation such that the (102) crystal planes and thus approximately the $p$ and $s$ directions always lie parallel to the substrate. Only the azimuthal orientation, i.e., the orientation of the $p$ and $s$ directions within the substrate plane, is difficult to control.

In Ref. [41], low-temperature absorption spectra of such vapor deposited films were used to obtain model parameters for the Holstein Hamiltonian by fitting. The values for the monomer were taken from the solution spectra (see Figure 3): $\hbar\omega = 0.17$ eV and $g = 0.88$ (for both PTCDA and MePTCDI). For the electron and hole coupling parameters, $g_e = g_h = g/\sqrt{2}$ was used as in Section 3.8. Furthermore, only one value was used for the charge-transfer integrals: $t = t_e = t_h$. This simplification is motivated, since for absorption only the value $t_+ = t_e + t_h$ enters the electronic problem (cf. the discussion below Eq. (39)) and since comparable values are suggested by quantum chemical calculations (cf. Ref. [15]). Then, there remain four essential parameters in the model Hamiltonian: Frenkel exciton transfer integral $J$, CT separation $D$, charge-transfer integral $t$, and the zero-point reference energy $E_{00}$ of the molecular limit. It has to be noted that the electronic Frenkel–CT mixing at a given quasi-momentum $k$ is only determined by the absolute value of the transfer integrals (cf. Eqs. (32), (36)) and thus only $|t|$ can be derived. The key-parameters obtained in Ref. [41] are for PTCDA: $J = 42$ meV, $D = 97$ meV, $|t| = 42$ meV, $E_{00} = 2.23$ eV, and for MePTCDI: $J = 46$ meV, $D = 240$ meV, $|t| = 57$ meV, and $E_{00} = 2.13$ eV.[12]

The structure of the corresponding eigenstate spectrum at $k = 0$ is best visualized by the Frenkel and CT characters of the states. We illustrate these characters in Figures 17(b) and 18(b) in the same scheme as in Figures 15 and 16. In these fits, the composition of the optically active states at $k = 0$ shows a strong mixing of Frenkel and CT excitons. The Frenkel character determines the absorption coefficient $\alpha$. The comparison of the experimental absorption coefficient and the

---

[12] In Ref. [41], the values of $t_+$ were given.

## Exciton band structure scheme PTCDA

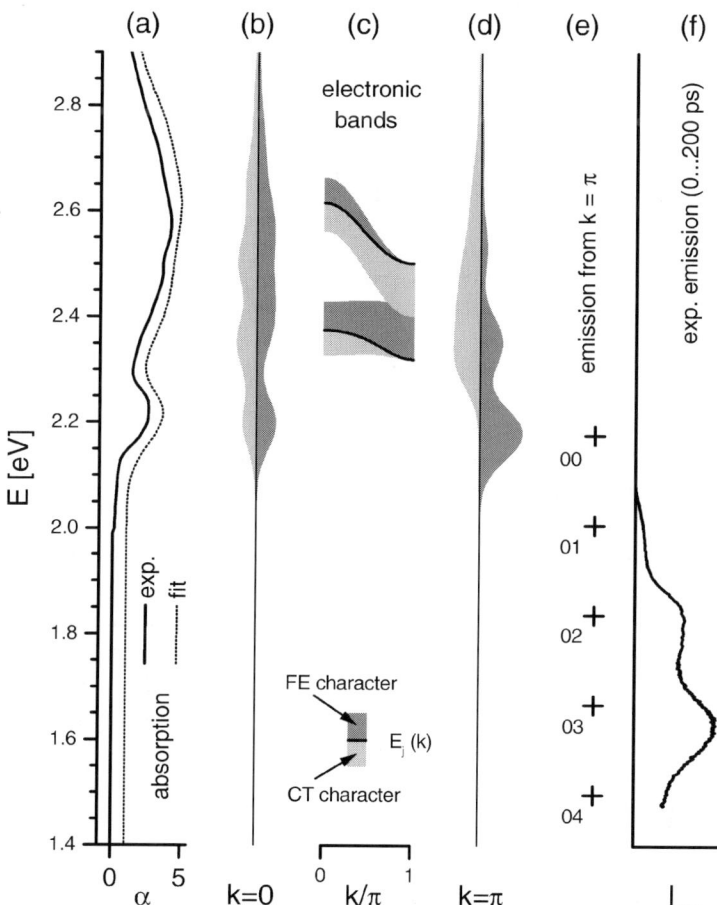

Fig. 17. Suggested exciton band structure in PTCDA and experimental spectra. (a) solid line: experimental low temperature absorption spectrum of thin poly-crystalline film as reported in Ref. [41], $\alpha$ in $10^5$ cm$^{-1}$; dotted line: $\alpha$ from model fit (plotted with offset $+1 \times 10^5$ cm$^{-1}$). (b) Vibronic model states at $k = 0$. The right side (dark shading) gives the Frenkel character $F_{\text{FE}\,j}$ of the states $j$ from Eq. (156), the left side the CT character $F_{\text{CT}\,j}$ from Eq. (158). Instead of the closely lying individual states, we show a broadened spectrum that summarizes the net contribution. The Frenkel character at $k = 0$ determines the oscillator strength and corresponds to the model absorption spectrum in (a). (c) Electronic bands $E_j(k)$ corresponding to Eq. (165). These bands show the overall dispersion and the $k$-dependent Frenkel–CT mixing, which result from the electronic interaction parameters. (d) Vibronic model states for $k = \pi$. (e) Emission energies for transitions from the lowest $k = \pi$ state to the vibrational levels of the electronic ground state. The highest transition (00) is strictly dipole-forbidden. (f) transient emission spectrum of a PTCDA single crystal at 10 K (time window 0...200 ps), from Ref. [88].

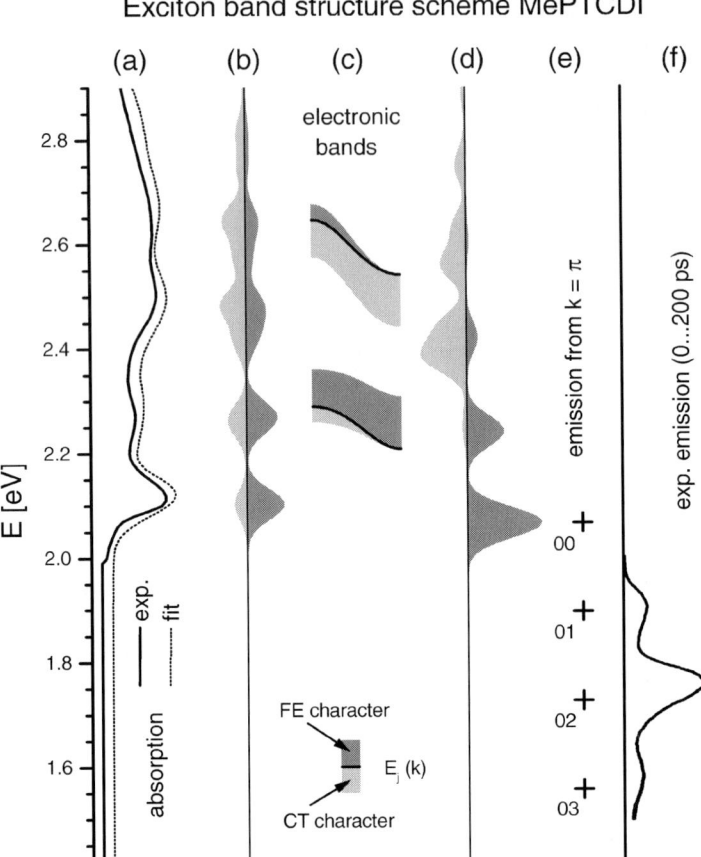

Fig. 18. Suggested exciton band structure in MePTCDI and experimental spectra. For detailed explanations see Figure 17. The experimental absorption spectrum in (a) is measured at 10 K at a highly oriented poly-crystalline film with polarization parallel to the strong Davydov component (crystallographic $b$-axis), cf. Ref. [16]. The transient emission spectrum in (f) is measured at 4 K at a single crystal, time window 0...200 ps, from Ref. [88].

model fit is given in the panels (a). The characteristic difference between the absorption coefficient spectrum in (a) and the distribution of the Frenkel characters in (b) is entirely caused by the spectral shape of the refractive index $n$, which becomes small at energies above the major absorption region.

It has to be emphasized that the fitting parameters contain many uncertainties. In particular for PTCDA, the absorption spectrum alone is not specific enough to determine the situation uniquely. Similarly good fits of the experimental spectra can be obtained for different parameter sets with varying degree of CT mixing. Even total neglect of CT states would give a satisfactory fit with a Frenkel transfer integral of 70 meV. Such a value corresponds to the three-dimensional Frenkel exciton model for PTCDA in Ref. [18] with a nearest-neighbor hopping of 82 meV. For MePTCDI, the absorption spectrum has a more characteristic shape with four main peaks. This spectrum can only be fitted within this framework by assuming a strong Frenkel–CT mixing.

The strongest support for the assumption of Frenkel–CT mixing is provided by the interpretation of electro-absorption rather than by linear absorption spectra [15,17], since a pure Frenkel exciton model would not explain the strong response to electrical fields perpendicular to the molecular planes. In Ref. [17], a three-dimensional version of a Frenkel–CT Hamiltonian (in the molecular vibron approximation) was used to model electro-absorption spectra of PTCDA. The strongest effects were confirmed to arise from the one-dimensional stacks. Hamiltonian parameters could be obtained from fits and also confirmed by microscopic calculations. The essential transfer parameters corresponding to our one-dimensional version were obtained as $J = 180$ meV, $D = 130$ meV, $t = -55$ meV [17]. These parameters from the electro-absorption model essentially agree with our suggestion for the linear absorption spectra. The largest discrepancy is in the Frenkel exciton transfer integral $J$, where our smaller value mainly results from the use of a dielectric function.

Apart from the remaining uncertainty about the Frenkel–CT mixing, the obtained charge-transfer integrals would imply single-particle bandwidths $4t_e$, $4t_h$ on the order of no more than 0.2 eV, which is still smaller than the total exciton binding energy. Thus, the qualitative picture is consistent with the approach of a small radius exciton theory (cf. discussion in Section 1).

We will now discuss following Ref. [88] what the proposed model for the absorbing states ($k = 0$) means for the complete exciton band structure. In order to rationalize the $k$ dependencies, we first concentrate on the purely electronic bands shown in Figures 17(c) and 18(c). As in Section 3.7, the electronic bands are given by the purely electronic parts of the Holstein Hamiltonian (149):

$$H_{\text{elec}} = H_{\text{elec}}^{\text{FE}} + H_{\text{elec}}^{\text{CT}} + H_{\text{elec}}^{\text{FE-CT}} + g^2 \hbar \omega. \tag{165}$$

This electronic Hamiltonian corresponds to the electronic problem in Eq. (29), were the resulting bands and characters were visualized in Figures 4 and 5. As for the Frenkel exciton case in Section 3.7, the Franck–Condon relaxation energy of the molecular zero-phonon state is added because this state defines the reference energy. The electronic bands in the Holstein model proposed for PTCDA and MePTCDI show that the electronic Frenkel character disperses to lower energies as a result of the positive Frenkel exciton hopping integral $J$. The center of mass of the CT character remains at a constant position since a CT dispersion is not considered in the model. Furthermore, for this special choice of charge-transfer integrals $t_e = t_h$, the Frenkel and CT states do not mix at $k = \pi$ (see discussion below Eq. (39)).

The complete vibronic eigenstate structure at $k = \pi$ is shown in Figures 17(d) (PTCDA) and 18(d) (MePTCDI). As at $k = 0$, the inclusion of exciton–phonon coupling at $k = \pi$ transforms the two electronic states into a broad vibronic spectrum. With the model parameters from the absorption fit, the lowest $k = \pi$ state lies at $E = 2.18$ eV for PTCDA and at $E = 2.06$ eV for MePTCDI.

The band bottom of the exciton band structure is a starting point for the discussion of emission spectra since all photo-excited states will rapidly relax to these lowest states. The transition from these $k = \pi$ states to the total ground state is strictly dipole-forbidden, at least in perfect crystals. However, transitions to $k = \pi$ phonons in the electronic ground state are allowed. We indicate the resulting transition energies (including the forbidden 00-transition) in the panels (e). For a qualitative comparison, these transition energies are compared to transient low-temperature photoluminescence spectra of single crystals in the panels (f) of both figures. The comparison shows that the energetic positions of the 01 and 02 transition do approximately agree with the peaks in the emission spectra. This supports the order of magnitude of the model parameters, especially the relatively small value of the Frenkel hopping $J$. A larger $J$, as could be expected from quantum chemical arguments and from models without dielectric function [15,17], would give a larger separation between the lowest absorption peak and the 01-emission peak.

The assignment of the emission spectra still is very tentative, since the spectra do not show an exact vibronic progression and the decay times of the peaks are slightly different, which becomes much more pronounced at higher temperatures. Furthermore, the emission spectra sensitively depend on the concrete sample configurations, on the considered time scale (up to cw) and on the temperature. This leads to widely varying emission spectra and assignments. It is not clear at this stage to what extent extrinsic defects or further intrinsic effects determine the emission behavior. In particular, a strong coupling to external phonon modes, corresponding to excimer emission, can be expected as an additional effect (cf. the explicit treatment for perylene in Ref. [89]). Such effects might be reasons for the discrepancies, which are particularly pronounced in the lowest PTCDA emission peak. A detailed discussion of time-resolved PTCDA emission spectra was recently given in Refs. [90,91]. These works are based on a pure Frenkel exciton band structure model within the molecular vibron approximation [18]. The various spectral features and their temperature dependencies are explained by including the small interactions between the non-equivalent molecules, which leads to Davydov splitting of the exciton bands, and by considering vibrational relaxation along several different coordinates.

Since the fine details of the experimental emission spectra are very complex there is a large number of conceivable models for their interpretation. At present, there are not enough tests available that would allow independent confirmation of many model aspects. In particular, the complexity of model assumptions seems to be comparable to the complexity the explained observations. Therefore, it is very desirable to identify basic and unique features in the excited state structure of the considered class of materials. The exciton band structure within a Frenkel–CT–Holstein model provides such a unified framework for the description of absorption spectra and it is a starting point for a discussion of emission spectra and relaxation processes after photo-excitation.

### 4.2. Inclusion of Finite Size and Quantum Confinement Effects

Up to here, the treatment of molecular crystals was presented for the case of an infinite chain. Real systems are always finite, and many systems of current interest cover the complete range of one-dimensional chain lengths starting from a single molecule. For example, thin films consisting of one or just a few monolayers of PTCDA can be prepared by organic molecular beam epitaxy (OMBE) and they can be studied with spectroscopic methods (e.g., [7,92]). Thus, it is important to know down to which chain length $N$ the picture of an infinite chain is still correct. It is even more important to investigate which qualitatively new effects can occur in finite systems. We will discuss these questions in the framework of the one-dimensional exciton models presented in the previous sections.

The existence of a finite system size $N$ affects all model descriptions at the very beginning. The starting point still is a model Hamiltonian expressed by on-site energies and interaction terms for localized excitations. However, the site index is restricted now ($n = 1, \ldots, N$) and the boundary conditions play a major role. Furthermore, the sites are not equivalent anymore and therefore the on-site energies $\varepsilon_{FE}$ and exciton transfer integrals $J$ might have site-dependent values.

Let us illustrate the effects for the simplest example of the nearest-neighbor Frenkel exciton model from Eq. (16). In a finite chain, it can be written as:

$$H_{NN}^{FE}(N) = \sum_{n=1}^{N} \varepsilon_{FE}(n) a_n^\dagger a_n + \sum_{n=1}^{N-1} J(n) \left( a_n^\dagger a_{n+1} + a_{n+1}^\dagger a_n \right). \qquad (166)$$

In this notation, the boundary conditions are expressed by the index range for the hopping term: Transfer from site $n = 1$ is only possible towards higher indices and transfer from site $n = N$ only towards lower indices. Even if the site-dependence of $\varepsilon_{FE}(n)$ and $J(n)$ is neglected, the boundary conditions in Hamiltonian (166) do not allow the separation into nonmixing subspaces with different quasi-momentum (Eq. (11)). Thus, a general diagonalization scheme has to work with basis states in real space. Even for a Frenkel–CT–Holstein model as in Eq. (149), a numerical diagonalization can be done in real-space along similar lines as presented for the momentum-space basis states, and only the number of basis states increases by a factor $N$.

The solutions of our example in Eq. (166) for constant $\varepsilon_{FE}(n) = \varepsilon_{FE}$ and $J(n) = J$ are well known (e.g., [93]). The $N$ eigenstates have energies

$$E_j(N) = \varepsilon_{FE} + 2J \cos \frac{\pi j}{N+1}, \quad j = 1, 2, \ldots, N. \tag{167}$$

The size dependence of the lowest and highest state energy is shown in Figure 19. The difference between the highest and lowest state corresponds to the exciton bandwidth in the infinite system and it approaches its value $\Delta E = 4J$ for the limit $N \to \infty$. For the dimer ($N = 2$), the bandwidth has already half of its maximum value. This size dependence results entirely from the effect that the exciton is confined in its hopping motion by the system boundaries. Therefore, the effect can be approximately described by the picture of a "particle in a box".

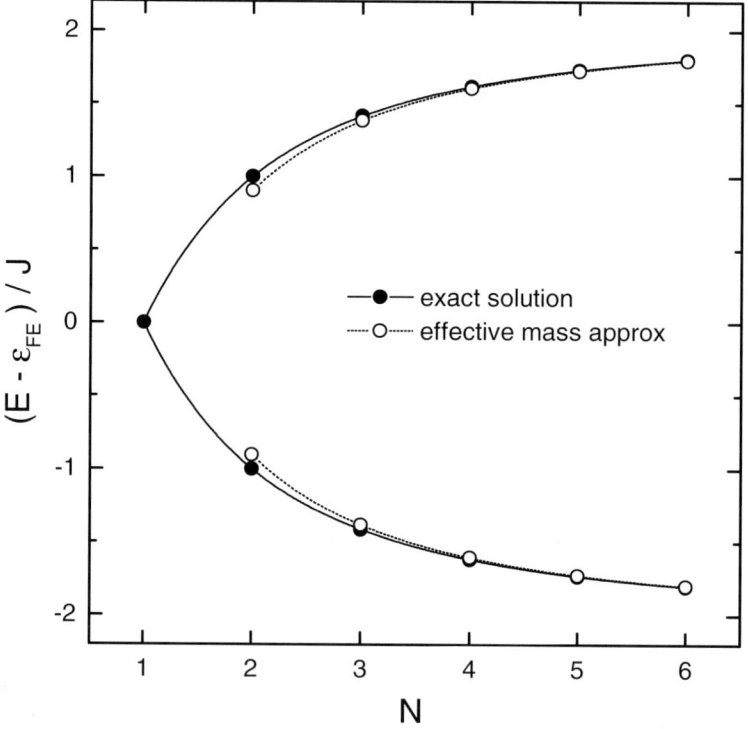

Fig. 19. Size dependence of the highest and lowest Frenkel exciton state in a linear nearest-neighbor hopping model (Eq. (166)) for constant $\varepsilon_{FE}$ and $J$. The "exact solution" shows the state energies from Eq. (167), the "effective mass approximation" shows the band edges from Eq. (171).

We will illustrate this particle-in-a-box behavior by an alternative approach based on the effective mass approximation (cf. also [94]): In the infinite system, the exciton states for Hamiltonian (166) are entirely characterized by their dimensionless quasi-momentum $k$ and their dispersion relation is (cf. Eq. (20))

$$E_k^{\text{eff}} = \varepsilon_{\text{FE}} + 2J \cos k. \qquad (168)$$

The effective mass of the exciton follows as

$$m_{\text{eff}}^{-1} = \frac{a^2}{\hbar^2} \frac{\partial^2 E_k^{\text{eff}}}{\partial k^2} = -\frac{a^2}{\hbar^2} \times 2J \cos k, \qquad (169)$$

where $a$ is the lattice constant. We now treat this exciton as a real particle with mass $m_{\text{eff}}$, which is confined within the chain. Thus, the wave function should be zero at the non-available lattice positions $n = 0$ and $n = N + 1$. This corresponds to a potential well of size $L = (N + 1)a$ with infinitely high boundaries. The ground state energy of the particle in such a well is $\pi^2 \hbar^2 / (2 m_{\text{eff}} L^2)$. This ground state energy corresponds to the energy shift $\Delta E^{\text{eff}}$ that is induced by the finite system size:

$$\Delta E^{\text{eff}} = \frac{\pi^2 \hbar^2}{2 m_{\text{eff}} L^2}. \qquad (170)$$

From Eqs. (168) and (170), the energy $E_k^{\text{eff}}(N)$ of the confined exciton follows as

$$E_k^{\text{eff}}(N) = E_k^{\text{eff}} + \Delta E^{\text{eff}}$$

$$= \varepsilon_{\text{FE}} + 2J \cos(k) \times \left(1 - \frac{1}{2} \frac{\pi^2}{(N+1)^2}\right). \qquad (171)$$

We show this energy from the effective mass approximation for $k = 0$ and $k = \pi$ in comparison with the exact energies in Figure 19. The effective mass approximation works very good even down to a chain length of $N = 2$, where the deviation is still only $J(\pi^2 - 9)/18 = 0.048 J$.

The comparison in Figure 19 illustrates for the case of the one-dimensional Frenkel exciton: The effect of the finite system size can approximately be described by treating the exciton as a particle in a box, where the effective mass of the particle is given by the dispersion relation. This approach can be extended for the case that the exciton has a more complicated structure, i.e., if the localized basis states have additional internal degrees of freedom (cf. Section 2.1). Even in this case, the translational symmetry of the infinite system assures that the eigenstates can be classified by their total momentum $k$ and further quantum numbers for their internal structure. The dispersion $E(k)$ of a given state determines its effective mass and can thereby cause a quantum size effect as given by Eq. (170). However, this simple picture obviously breaks down if the internal structure of the exciton state is affected by changes of the system size. Such a change must occur if the internal structure is related to a characteristic size (quantum length) of the exciton and the system size becomes comparable to or smaller than this intrinsic quantum length.

For the simple Frenkel exciton in Figure 19, there is no internal degree of freedom and therefore no intrinsic quantum length. For the Merrifield model (see Eq. (24)) of one electron and one hole on a one-dimensional chain, the eigenstates do have an internal structure and a characteristic length is given by the mean separation between electrons and holes. For the Holstein model (see Eq. (101)), an internal structure is introduced by the structure of the phonon cloud and a characteristic quantum length follows from the extension of the lattice distortion $\langle \lambda_m \rangle$ around the exciton (cf. Figure 11). In all such cases with internal structure, one can therefore distinguish two finite size effects: (I) A *quantum size effect* according to Eq. (170) that results from the confinement of the center-of-mass motion and (II) a *quantum confinement effect* that results from a confinement of the internal exciton structure.

A mathematically explicit version of the heuristic distinction between quantum size and quantum confinement effects is shown by Kayanuma for three-dimensional Wannier–Mott excitons [95]. In this case, the intrinsic quantum length of the exciton is the electron–hole separation $R$. Kayanuma shows that for a system size $L \gg R$ ("weak confinement" in the classification of Kayanuma), the size dependence of the eigenstate energies is given by Eq. (170). For $L \stackrel{<}{\approx} R$ ("strong confinement"), the internal structure of the exciton is drastically changed, and the eigenstates are rather characterized by the confinement of the single particles.

From this discussion we can conclude that the internal structure, more specifically the intrinsic quantum length $R$, of exciton states is crucial for estimating the effects of a finite system size. In the framework of the Frenkel-charge-transfer Holstein model (Eq. (149)) from this chapter, there are three effects that give rise to an intrinsic quantum length:

(I) Because of the CT states involved, one has a non-trivial electron–hole separation. However, since only nearest-neighbor CT states are included, this electron–hole separation has an upper limit of one lattice constant and quantum-confinement is of very limited interest.

(II) In a finite system, the subtle boundary conditions for CT states can lead to the appearance of Tamm-like surface states [96]. These surface states have a decay length which is entirely given by the parameters in the Hamiltonian and which can be in the order of many lattice constants. Thus, this decay length has the nature of an intrinsic quantum length. If the finite chain becomes smaller than the quantum length of the surface states, their internal structure is changed, i.e., quantum-confinement occurs [97,98].

(III) Exciton–phonon coupling leads to the appearance of phonon clouds, i.e., the exciton is surrounded by a lattice displacement $\langle \lambda_m \rangle$ (see Eq. (125)), which is illustrated for specific examples in Figures 11 and 13. For the scenario presented in Section 4.1, i.e., for the strongly coupled internal vibrations in PTCDA derivatives, the phonon clouds have extensions not much larger than one lattice constant. However, one might also consider phonon modes which are not that strongly coupled. With respect to such modes, one has the situation of strong electronic coupling, and the approach of localized phonon clouds from Section 3 is not appropriate. Now, one can use the continuum model for large phonon clouds described by Rashba [99]. This situation was studied for Frenkel excitons by Agranovich et al. in Ref. [100]. The intrinsic quantum length (extension of the phonon cloud) in this case is $R = Ja/(2g^2\hbar\omega)$,[13] and $R$ can easily become large for small coupling constants or small phonon energies. This results in quantum confinement effects, which are quantitatively discussed in Ref. [100].

In a realistic finite system, not only the quantum-size effect according to Eq. (170) and the quantum confinement effects (I)–(III) from above may occur, but also the possibility of site-dependent on-site and interaction energies should be considered as in Eq. (166). However, it is very difficult to predict trends for the dependencies $\varepsilon_{FE}(n)$, $\varepsilon_{CT}(n)$, $J(n)$, $t_{e/h}(n)$.

A strong effect is expected for the variation of the on-site energy of the Frenkel exciton. For this case, at least some knowledge is available for the lowest transition in anthracene. The total gas-to-crystal shift is on the order of 2500 cm$^{-1}$ (0.3 eV),[14] whereas surface states were experimentally identified with energy differences to the bulk states of about 200 cm$^{-1}$ (20 meV) [101]. In Ref. [101], the energetic separation was interpreted as being mainly due to the gas-to-crystal shift of the molecules from the outermost surface layer. The shift for the next (sub-surface) layer was already considerably smaller. Thus, one can assume that $\varepsilon_{FE}(n)$ is almost constant for $n > 1$, and even that the change at $n = 1$ is by far not as large as the upper limit given by the gas-to-crystal shift.

The on-site energy $\varepsilon_{CT}$ of the CT states is strongly affected by the electronic polarization of the surrounding molecules. Thus, close to the surface, strong changes of $\varepsilon_{CT}(n)$ can be expected. Modern microscopic calculations of polarization energies seem to allow estimations for this effect using a methodology that has already been applied for the polarization energy of ions in finite systems [102].

The Frenkel transfer integral $J$ is also strongly influenced by dielectric screening from the surrounding molecules and therefore can be expected to change significantly close to the surface. Thus, one has to cope with a variety of different effects and unknown parameters and it is not surprising that even for the experimentally well investigated case of PTCDA a coherent picture has not been presented yet. Shifts in the absorptions spectra of PTCDA/NTCDA multilayer structures

---

[13]This follows from Eq. (26) of Ref. [100]. Note that there the meaning of $g$ is different.
[14]The shift is given as the difference of the vapor-phase transition (27688 cm$^{-1}$ taken from [101]) and the center of the $\parallel b$ and $\perp b$ polarized crystal transitions (values taken from [4]).

[103] were initially interpreted as quantum-confinement (in the strict sense discussed above) of Wannier–Mott excitons. However, a pure Wannier–Mott exciton picture seems not to be adequate for PTCDA (see discussion in Section 4.1). An alternative interpretation for the shifts is given in terms of a varying contribution from the gas-to-crystal shift at the surface by Agranovich et al. in Ref. [104]. The situation becomes even more complicated due the possible mixture of different crystal phases in thin layers [105]. Thus, a clear identification of quantum size and quantum confinement effects in PTCDA-related quasi-one-dimensional crystals requires considerably more information than currently available. In particular, one has to find ways in which the various contributing effects can be separately estimated from independent theoretical or experimental methods.

## 5. Conclusion

In this chapter, we illustrated the application of basic exciton models for describing optical spectra of quasi-one-dimensional molecular crystals. We concentrated on two essential effects: Mixing of electronic Frenkel and charge-transfer excitations and strong coupling to one internal molecular vibration. For these models, we gave a comprehensive and self-contained discussion with emphasis on illustrational examples.

In Section 2, the electronic states were described as one-dimensional collective excitations consisting of molecular excitations (Frenkel excitons) and nearest-neighbor charge-transfer excitations. This special case of the more general Merrifield model [32] seems to be suitable for molecular crystals that form stacks with close coplanar arrangement of the molecular planes. The essential parameters are given by the nearest-neighbor Frenkel transfer integral $J$, the separation of the CT states from the molecular excitation $D$ and the electron and hole transfer integrals $t_e$ and $t_h$. The interplay of these parameters is illustrated in terms of band structure plots showing the two resulting bands and their mixed character. The mixing of two electronic states can lead to two peaks in the absorption spectra—even if the CT states have a vanishing intrinsic transition dipole moment. This effect is one probable candidate for the explanation of broadening or peak-splitting in absorption spectra of quasi-one-dimensional crystals. An alternative explanation would be Davydov splitting due to Frenkel-type interactions of non-equivalent molecules. This Davydov splitting, however, is expected to be significantly smaller [16]. We also discussed the qualitative nature of the intrinsic CT transition dipole.

Section 3 considers strong coupling of the electronic excitations to internal molecular vibrations, which is a typical feature in many molecules of current interest. This coupling is most obvious in the vibronic progression in absorption spectra of isolated molecules. In a one-dimensional crystal, even the simplest model description for such exciton–phonon coupling (Holstein Hamiltonian) leads to a complicated many-particle problem that cannot be generally solved. Here, we describe a numerical approach that is conceptually simple and allows the calculation of optical spectra by standard numerical diagonalization tools. This approach is appropriate for weak electronic coupling and exciton–phonon coupling constants in the order of one. This means in other words: The isolated molecules show a pronounced vibronic progression (0–0 and 0–1 peak of comparable height) and the spectral changes in the crystal concern only an energy range comparable to the range of the vibronic progression.

In Section 4, we outlined the application of the basic models to experimental spectra of thin organic films. The first step is a realistic description of bulk optical constants. The capabilities and limitations of the minimum models are illustrated for spectra of two archetypal perylene derivatives. For very thin layers, a number of additional finite size and quantum confinement effects can occur. We qualitatively discussed the nature of such effects in the framework of our minimum model. A general distinction can be made on the basis of the internal exciton structure. If, as in the pure electronic Frenkel case, the exciton has no internal structure, the finite system size leads to a simple particle-in-a-box behavior. This quantum size effect is still the leading effect if the system size is larger than any intrinsic length scale of a more complicated exciton. As soon as the system size affects the internal structure, we speak of quantum confinement. Intrinsic quantum lengths responsible for quantum confinement can occur in the idealized problem in two ways: (I) The electronic problem for Frenkel and CT states can lead to surface states with a finite decay length. (II) Exciton–phonon coupling leads to phonon clouds with finite extension around the electronic excitation. Further effects are expected to arise from a site dependent modification of the Hamiltonian parameters.

At the present stage, there is still no general agreement about the specific interpretation of experimental spectra in terms of microscopic exciton models. Even in the case of bulk spectra, the available information is typically not sufficient for a unique parameterization. Furthermore, a comparison between different interpretations is often difficult since typically different aspects are explained by different models. More certainty can be expected if not just single materials are studied but general trends are investigated and described by a common framework.

## Acknowledgements

Most of the material reviewed here is a result of fruitful collaborations that I enjoyed during the past years. The aspects of mixed Frenkel–CT excitons in Section 2 were developed together with Vladimir M. Agranovich, Karin Schmidt and Karl Leo, and the exciton–phonon problem in Section 3 was investigated together with Zoltan G. Soos. I am very thankful to these collaborators for their help and inspiration. To K. Schmidt I am also indebted for discussions on quantum confinement and for providing Figure 7. I want to thank all members of the Institut für Angewandte Photophysik at Technische Universität Dresden for their support. Financial support by Deutsche Forschungsgemeinschaft is gratefully acknowledged.

## References

1. C. Kittel, Introduction to Solid State Physics, John Wiley & Sons, New York, 1996.
2. E.V. Tsiper, Z.G. Soos, *Phys. Rev. B* 64 (2001) 195124.
3. D.P. Craig, S.H. Walmsley, Excitons in Molecular Crystals, W.A. Benjamin, Inc., New York, 1968.
4. A.S. Davydov, Theory of Molecular Excitons, Plenum Press, New York, 1971.
5. V.M. Agranovich, M.D. Galanin, Electronic Excitation Energy Transfer in Condensed Matter, North-Holland, Amsterdam, 1982.
6. D.W. Schlosser, M.R. Philpott, *J. Chem. Phys.* 77 (1982) 1969.
7. S.R. Forrest, *Chem. Rev.* 97 (1997) 1793.
8. E. Umbach, K. Glöckler, M. Sokolowski, *Surface Science* 402–404 (1998) 20.
9. C.W. Tang, *Appl. Phys. Lett.* 48 (1986) 183.
10. P. Peumans, V. Bulović, S.R. Forrest, *Appl. Phys. Lett.* 76 (2000) 2650.
11. D. Meissner, J. Rostalski, *Synth. Met.* 121 (2001) 1551.
12. P.M. Kazmaier, R. Hoffmann, *J. Am. Chem. Soc.* 116 (1994) 9684.
13. E.I. Haskal, Z. Shen, P.E. Burrows, S.R. Forrest, *Phys. Rev. B* 51 (1995) 4449.
14. V. Bulović, P.E. Burrows, S.R. Forrest, J.A. Cronin, M.E. Thompson, *Chem. Phys.* 210 (1996) 1.
15. M.H. Hennessy, Z.G. Soos, R.A. Pascal Jr., A. Girlando, *Chem. Phys.* 245 (1999) 199.
16. M. Hoffmann, K. Schmidt, T. Fritz, T. Hasche, V.M. Agranovich, K. Leo, *Chem. Phys.* 258 (2000) 73.
17. G. Mazur, P. Petelenz, M. Slawik, *J. Chem. Phys.* 118 (2003) 1423.
18. I. Vragović, R. Scholz, M. Schreiber, *Europhys. Lett.* 57 (2002) 288.
19. M. Sadrai, L. Hadel, R.R. Sauers, S. Husain, K. Krogh-Jespersen, J.D. Westbrook, G.R. Bird, *J. Phys. Chem.* 96 (1992) 7988.
20. M. Adachi, Y. Murata, S. Nakamura, *J. Phys. Chem.* 99 (1995) 14240.
21. G. Klebe, F. Graser, E. Hädicke, J. Berndt, *Acta Cryst. B* 45 (1989) 69.
22. V.M. Agranovich, Theory of Excitons, Nauka, Moscow, 1968 (in Russian).
23. J. Frenkel, *Phys. Rev.* 37 (1931) 17.
24. J. Frenkel, *Phys. Rev.* 37 (1931) 1276.
25. A.S. Davydov, *Zh. Eksperim. i Teor. Fiz.* 18 (1948) 210.
26. D.P. Craig, *J. Chem. Soc.* (1955) 2302.
27. V. Agranovich, *Fizika Tverdogo Tela* 3 (1961) 811, *Sov. Phys.-Solid State* 3 (1961) 592.
28. V.M. Agranovich, Y.V. Konobeev, *Phys. Stat. Sol.* 27 (1968) 435.
29. D. Haarer, M.R. Philpott, "Excitons and polarons in organic weak charge transfer crystals", in: V.M. Agranovich, R.M. Hochstrasser (Eds.), Spectroscopy and Excitation Dynamics of Condensed Molecular Systems, North-Holland, Amsterdam, 1983, pp. 27–82, Chapter 2.
30. Z.G. Soos, *Ann. Rev. Phys. Chem.* 25 (1974) 121.
31. L.E. Lyons, *J. Chem. Soc.* (1957) 5001.
32. R.E. Merrifield, *J. Chem. Phys.* 34 (1961) 1835.

33. J.P. Hernandez, S.-I. Choi, *J. Chem. Phys.* 50 (1969) 1524.
34. W.L. Pollans, S.-I. Choi, *J. Chem. Phys.* 52 (1970) 3691.
35. L. Sebastian, G. Weiser, *Chem. Phys.* 61 (1981) 125.
36. L. Sebastian, G. Weiser, G. Peter, H. Bässler, *Chem. Phys.* 75 (1983) 103.
37. P. Petelenz, M. Slawik, K. Yokoi, M.Z. Zgierski, *J. Chem. Phys.* 105 (1996) 4427.
38. M. Slawik, P. Petelenz, *J. Chem. Phys.* 107 (1997) 7114.
39. M. Slawik, P. Petelenz, *J. Chem. Phys.* 111 (1999) 7576.
40. P.J. Bounds, P. Petelenz, W. Siebrand, *Chem. Phys.* 63 (1981) 303.
41. M. Hoffmann, Z.G. Soos, *Phys. Rev. B* 66 (2002) 024305.
42. A. Szabo, N.S. Ostlund, Modern Quantum Chemistry, Dover Publications, Inc., Mineola, New York, 1996.
43. R.S. Mulliken, *J. Amer. Chem. Soc.* 74 (1952) 811.
44. R.S. Mulliken, W.B. Person, Molecular Complexes. A Lecture and Reprint Volume, Wiley-Interscience, New York, 1969.
45. H. Haken, H.C. Wolf, Molekülphysik und Quantenchemie, Springer-Verlag, Berlin, 1994.
46. D.B. Fitchen, "Zero-phonon transitions", in: W. Fowler (Ed.), Physics of Color Centers, Academic Press, New York, 1968, pp. 294–350, Chapter 5.
47. V.L. Broude, E.I. Rashba, E.F. Sheka, Spectroscopy of Molecular Excitons, Springer-Verlag Berlin, 1985.
48. H. Haken, Quantum Field Theory of Solids, North-Holland, Amsterdam, 1976.
49. T. Holstein, *Ann. Phys.* 8 (1959) 325.
50. T. Holstein, *Ann. Phys.* 8 (1959) 343.
51. J. Appel, in: F. Seitz, D. Trunbull, H. Ehrenreich (Eds.), Polarons, in: Solid State Physics. Advances in Research and Applications, Vol. 21, Academic Press, New York, 1968, pp. 193–391, Chapter 3.
52. M.I. Klinger, Problems of Linear Electron (Polaron) Transport Theory in Semiconductors, Pergamon Press, Oxford, 1979.
53. H. Böttger, V.V. Bryskin, Hopping Conduction in Solids, VCH, Weinheim, 1985.
54. E.A. Silinsh, V. Čápek, Organic Molecular Crystals. Interaction, Localization, and Transport Phenomena, AIP Press, New York, 1994.
55. D.R. Yarkony, R. Silbey, *J. Chem. Phys.* 67 (1977) 5818.
56. P.O.J. Scherer, E.W. Knapp, S.F. Fischer, *Chem. Phys. Lett.* 106 (1984) 191.
57. Y. Zhao, D.W. Brown, K. Lindenberg, *J. Chem. Phys.* 106 (1997) 5622.
58. A.H. Romero, D.W. Brown, K. Lindenberg, *Phys. Rev. B* 60 (1999) 4618.
59. A.H. Romero, D.W. Brown, K. Lindenberg, *Phys. Rev. B* 60 (1999) 14080.
60. A.H. Romero, D.W. Brown, K. Lindenberg, *Phys. Rev. B* 59 (1999) 13728.
61. A.H. Romero, D.W. Brown, K. Lindenberg, *J. Lum.* 83–84 (1999) 147.
62. A.S. Alexandrov, V.V. Kabanov, D.K. Ray, *Phys. Rev. B* 49 (1994) 9915.
63. G. Wellein, H. Röder, H. Fehske, *Phys. Rev. B* 53 (1996) 9666.
64. G. Wellein, H. Fehske, *Phys. Rev. B* 56 (1997) 4513.
65. G. Wellein, H. Fehske, *Phys. Rev. B* 58 (1998) 6208.
66. J. Bonča, S.A. Trugman, I. Batistić, *Phys. Rev. B* 60 (1999) 1633.
67. H. De Raedt, A. Lagendijk, *Phys. Rev. Lett.* 49 (1982) 1522.
68. H. De Raedt, A. Lagendijk, *Phys. Rev. B* 27 (1983) 6097.
69. H. De Raedt, A. Lagendijk, *Phys. Rev. B* 30 (1984) 1671.
70. P.E. Kornilovitch, E.R. Pike, *Phys. Rev. B* 55 (1997) R8634.
71. E. Jeckelmann, S.R. White, *Phys. Rev. B* 57 (1998) 6376.
72. M.R. Philpott, *J. Chem. Phys.* 54 (1971) 111.
73. R.E. Merrifield, *Radiation Research* 20 (1963) 154.
74. P.O.J. Scherer, S.F. Fischer, *Chem. Phys.* 86 (1984) 269.

75. N. Lu, S. Mukamel, *J. Chem. Phys.* 95 (1991) 1588.
76. F.C. Spano, S. Siddiqui, *Chem. Phys. Lett.* 314 (1999) 481.
77. F.C. Spano, *J. Chem. Phys.* 116 (2002) 5877.
78. F.C. Spano, *J. Chem. Phys.* 118 (2003) 981.
79. I.G. Lang, Y.A. Firsov, *JETP (USSR)* 43 (1962) 1843.
80. P.M. Morse, H. Feshbach, Methods of Theoretical Physics, McGraw-Hill, New York, 1953.
81. W.T. Simpson, D.L. Peterson, *J. Chem. Phys.* 26 (1957) 588.
82. Z.G. Soos, S. Ramasesha, D.S. Galvão, S. Etemad, *Phys. Rev. B* 47 (1993) 1742.
83. S. Ramasesha, D.S. Galvão, Z.G. Soos, *J. Phys. Chem.* 97 (1993) 2823.
84. M. Möbus, N. Karl, T. Kobayashi, *J. Crystal Growth* 116 (1992) 495.
85. E. Hädicke, F. Graser, *Acta Cryst. C* 42 (1986) 189.
86. F. Wooten, Optical Properties of Solids, Academic Press, New York, 1972.
87. R. Brendel, *J. Appl. Phys.* 71 (1992) 1.
88. M. Hoffmann, Z.G. Soos, K. Leo, *Nonlinear Optics* 29 (2002) 227.
89. T.-W. Wu, D.W. Brown, K. Lindenberg, *Phys. Rev. B* 47 (1993) 10122.
90. A.Y. Kobitski, R. Scholz, I. Vragović, H.P. Wagner, D.R.T. Zahn, *Phys. Rev. B* 66 (2002) 153204.
91. R. Scholz, I. Vragović, A.Y. Kobitski, M. Schreiber, H.P. Wagner, D.R.T. Zahn, *Phys. Stat. Sol. B* 234 (2002) 402.
92. E. Umbach, W. Gebauer, A. Soukopp, M. Bässler, M. Sokolowski, *J. Lum.* 76–77 (1998) 641.
93. H. Fidder, J. Knoester, D.A. Wiersma, *J. Chem. Phys.* 95 (1991) 7880.
94. K. Schmidt, Electronic excited states in quasi-one-dimensional organic solids with strong coupling of Frenkel and charge-transfer excitons, Ph.D. thesis, TU Dresden, 2002.
95. Y. Kayanuma, *Phys. Rev. B* 38 (1988) 9797.
96. V. Agranovich, K. Schmidt, K. Leo, *Chem. Phys. Lett.* 325 (2000) 308.
97. K. Schmidt, *Phys. Lett. A* 293 (2002) 83.
98. K. Schmidt, K. Leo, V.M. Agranovich, "Surface states and quantum confinement in linear molecular chains with strong mixing of Frenkel and charge-transfer excitons", in: V.M. Agranovich, G.C.L. Rocca (Eds.), Organic Nanostructures: Science and Applications. Proceedings of the International School of Physics "Enrico Fermi", Course CXLIX, Varenna, 2001, IOS Press, Amsterdam, 2002, pp. 521–541.
99. E.I. Rashba, "Self-trapping of excitons", in: E.I. Rashba, M.D. Sturge (Eds.), Excitons, North-Holland, 1982, p. 543.
100. V.M. Agranovich, A.M. Kamchatnov, *Chem. Phys.* 245 (1999) 175.
101. J.M. Turlet, P. Kottis, M.R. Philpott, *Advan. Chem. Phys.* 54 (1983) 303.
102. E.V. Tsiper, Z.G. Soos, W. Gao, A. Kahn, *Chem. Phys. Lett.* 360 (2002) 47.
103. F.F. So, S.R. Forrest, Y.Q. Shi, W.H. Steier, *Appl. Phys. Lett.* 56 (1990) 674.
104. V. Agranovich, R. Atanasov, G. Bassani, *Chem. Phys. Lett.* 199 (1992) 621.
105. M. Leonhardt, O. Mager, H. Port, *Chem. Phys. Lett.* 313 (1999) 24.
106. T.W. Canzler, Ultrafast dynamics in quasi-one-dimensional organic molecular crystals. Self-assembled monolayers of photochromic molecules, Ph.D. thesis, TU Dresden, 2002.
107. HyperChem, Computational Chemistry (Manual), Hypercube, Inc., Waterloo, Ontario, Canada, 1996.

## Chapter 6

# Two-Dimensional Charge-Transfer Excitons at a Donor–Acceptor Interface

### V.M. AGRANOVICH

*Institute of Spectroscopy, Russian Academy of Sciences, Troitsk, Moscow Reg., 142190, Russia*

### G.C. LA ROCCA

*Scuola Normale Superiore and I.N.F.M., Piazza dei Cavalieri 7, 56126 Pisa, Italy*

1. Phase Transition from Dielectric to Conducting State (Cold Photoconductivity) . . . . . . . 293
2. Cumulative Photovoltage in Asymmetrical Donor–Acceptor Organic Superlattices . . . . . 304
3. Nonlinear Optical Response of Charge-Transfer Excitons at Donor–Acceptor Interface . . 311
   References . . . . . . . . . . . . . . . . . . . . . . . . . . . . . . . . . . . . . . . . . 315

In the modern literature on collective and nonlinear properties of excitons (see, for instance, [1–3]) one may find only investigations devoted to Wannier–Mott or Frenkel excitons [4,5]. The studies of collective properties of charge transfer excitons (CTEs) are only in the very beginning. On the other hand, such type of excitons also play an important role in the understanding of the optical and photoelectric properties of many organic materials, including nanostructures (see [6] and also [7,8]).

In this chapter, we describe some unusual properties of CTEs. In particular, the phenomena which may arise in organic multilayers at donor–acceptor interfaces in the system of 2D charge-transfer excitons at high pumping. We discuss the dielectric-conductor phase transition at donor–acceptor interface, the photo-voltaic effect in asymmetrical superlattices and the resonant as well as off-resonant $\chi^3$ optical nonlinearity.

## 1. Phase Transition from Dielectric to Conducting State (Cold Photoconductivity)

In this section, following the papers by Agranovich and Ilinskii [9] and Kiselev et al. [10], we consider the stability of interacting CTEs at a (donor–acceptor) D-A interface and the possibility of a transition to a conducting state.

The realistic possibility to consider such organic crystalline structures has only recently appeared due to progress in the development of the organic molecular beam deposition (OMBD) and other related techniques [7]. Such progress has led to the monolayer control over the growth of organic thin films and superlattices with extremely high chemical purity and structural precision. This opens a wide range of possibilities for creating new types of ordered organic multilayer structures including ordered interfaces. It is well known that the necessity for lattice matching places strong restrictions on the materials which can be employed to produce high quality interfaces using inorganic semiconductor materials. This occurs since inorganic semiconductor materials are bonded by short-range covalent or ionic forces. On the contrary, the organic materials are bonded by weak van der Waals forces. This fact relaxes the above described restrictions and broadens the choice of materials that can be used to prepare organic crystalline layered structures with the required properties (for more details and many examples, see [7, 11,12]). Note, that the D-A interfaces can be created also in Langmuir–Blodgett films [13,14]. We can mention the paper [15] where in order to study the nonlinear optical properties of multilayers organic superlattices have been grown with a structure of the type ... $AAA|DDD|AAA|DDD$ ..., the vertical dashes indicating the donor–acceptor interfaces. In this paper, the molecule $C_{60}$ was used as acceptor ($A$) and molecules perylene, coronene and others as donors of electrons ($D$).

### 1.1. ANALYTICAL APPROACH

Consider the CTEs on a single D-A interface with a highly ordered structure. To explain the main collective effects in the physics of CTEs at a D-A interface, we assume that the static dipoles of the CTEs are aligned approximately normal to the interface plane, resulting in mutual repulsion. For example, if the static CTE dipole moment is equal to 20 D and the distance between them is 5 Å (the lattice constant at the interface) the repulsion energy is near 1 eV. If the distance between CTEs increases to 10 Å the repulsion energy decreases to about 0.1 eV. It is important to note that for crystals in which the lowest energy electronic excitations are CTEs, these repulsion energies are of the order of the energy difference $B$ from the CTE level to the lowest conduction band ($B < 0.5$ eV, see [6]). Thus, at high CTE concentrations we can expect that due to the repulsion energy the higher energy states are populated with free carriers, thus producing photoconductivity even at very low temperature (cold photoconductivity). Of course, for example, multiphoton ionization or other optical nonlinear processes can produce photoconductivity at low temperature. However, such processes are universal, they take place in condensed matter of any nature and they are not relevant to CTEs and their interaction which we discuss in this paper.

We assume below that the lowest conductivity band, which is responsible for the conductivity along the D-A interface, has a lower energy than the Frenkel excitations in the donor and acceptor materials. In this particular case, the interface at low temperature provides the lowest energy site for the CTE and interface free carriers. In [9] cold photoconductivity at the D-A interface was considered under the assumption that the time required for a phase transition to the conducting state is smaller than the CTE life time and a phase transition was obtained by minimizing the total energy of bond (CTE) and dissociated excitations (free carriers). Following [9] let's calculate at first the energy of a 2D array of self trapped CTEs at $T = 0$.

The energy of CTEs (of concentration $n_1$) and the energy of dissociated e–h pairs (of concentration $n_2$), can be calculated by assuming that the total number of excitations determined by the optical pumping intensity is constant:

$$n_1 + n_2 = n. \qquad (1)$$

The energy of the CTE array is therefore:

$$E_1 = n_1 \Delta + E_{\text{int}}, \qquad (2)$$

where $\Delta$ is the energy of a single CTE and $E_{\text{int}}$ is the total repulsion energy of their interaction. This energy can be estimated from the average distance $\rho$ between CTEs and their dipole $p$[1] as:

$$E_{\text{int}} = A \frac{p^2}{\rho^3} \frac{n_1}{2}, \qquad (3)$$

where $A$ is a geometric constant depending on the CTE distribution in the interface plane. For example, for a square lattice $A \approx 10$. Since the CTE concentration by definition is $n_1 = 1/\rho^2$ the total electrostatic energy of the interaction between the dipole moments is (see also the results of numerical simulation shown in Figure 4 below):

$$E_{\text{int}} = \frac{A p^2 n_1^{5/2}}{2}. \qquad (4)$$

We can approximate the energy of the dissociated pairs as $E_2 = (\Delta + B) n_2$, where the kinetic energy of the free carriers has been neglected (due to the self trapping and narrow electronic bands). Assuming that we consider the region near the threshold where the concentration $n_2 \ll n_1$, we can also neglect the interaction of the free carriers with the CTEs. The total energy of the system is then written as

$$E = E_1(n_1) + E_2(n_2) = n\Delta + \frac{A p^2 (n - n_2)^{5/2}}{2} + B n_2. \qquad (5)$$

---

[1] We notice that the dipole–dipole interaction between two molecules embedded in a medium of dielectric constant $\varepsilon$ is given by that in vacuum multiplied by the factor $((\varepsilon + 2)/3)^2/\varepsilon$ [5], which accounts for screening and local field effects when the distance between the molecules is larger than the lattice constant. For a typical value of $\varepsilon \simeq 3$, this correction is not important and will never be included in this chapter.

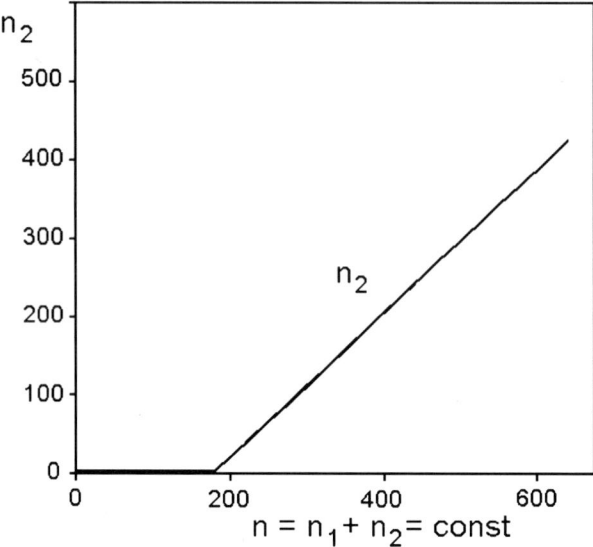

Fig. 1. The number of the dissociated pairs ($n_2$) as a function of the total number of excitations ($n_1 + n_2$) at the donor–acceptor interface (according to the simplified analytical model of Eq. (6)) and in units consistent with the numerical simulations discussed below.

Minimizing the above expression with respect to $n_2$, gives:

$$n_2 = n - \left(\frac{4B}{5Ap^2}\right)^{2/3}. \qquad (6)$$

It is clear from Eq. (6) that $n_2$ is positive at $n > n_{cr} = [4B/(5Ap^2)]^{2/3}$ (see Figure 1). The appearance of free carriers at $n > n_{cr}$ is considered to be a phase transition from the dielectric to conducting state. This transition corresponds to photoconductivity at low temperatures (i.e., to cold photoconductivity) and is due to long range dipole–dipole interactions between CTEs. In this simplified picture, we neglect the randomness in the CTE distribution and the transient to establish an equilibrium steady state. The establishment of a steady state depends on the pump intensity and the CTE lifetime which above was considered as infinite.

The phase transition in the system of interacting CTEs at finite temperature was considered in the paper [9] for a 1D donor–acceptor interface. Such 1D interfaces can be found, for example, among CT crystals containing segregated stacks. In such crystals, the 1D interfaces separate the columns of donors and acceptors. However, all the segregated stacks organic solids have an ionic ground electronic state and the excited CT state in this case corresponds to a transition of two neighboring ions (positive donor and negative acceptor) to the neutral state [6].

For 1D structures, the interaction between CTEs can be taken into account only for nearest neighbors. Besides, as it follows from [6], due to strong self-trapping of CTEs the band-width of CTEs can be neglected. Due to these assumptions an exactly solvable model was obtained in [9].

As it could be expected, it follows from the results of this model that the phase transition region broadens with increasing temperature and disappears at high temperatures. The same picture was obtained for 2D structure in the paper [16] in the Weiss molecular field approximation.

The transition from a dielectric to a conducting state can be also investigated using computer simulations. Following [10] we describe below such numerical results for a more realistic model where the random distribution in space and also the finite lifetime of CTEs are explicitly taken into account.

### 1.2. NUMERICAL SIMULATIONS

To numerically simulate the time evolution of CTEs distributed over a two-dimensional donor–acceptor interface the D-A sites were arranged in a square lattice. It was assumed that the D-A interface is uniformly irradiated with a time independent source of intensity, $I$. Only one CTE can be generated at any site, so every D-A site can be either occupied or not. The CT exciton generated at a given lattice site will stay there and it cannot move to another D-A sites because of self trapping.

Once generated, there are two mechanisms for the CTE to disappear. First, recombination occurs because of the finite lifetime of the CTE, $\tau$. The second mechanism is via dissociation. The CTE exciton dissociates when, due to the dipole–dipole interaction, the energy of the particular exciton exceeds some threshold. If there are $n_1$ CTEs occupying the D-A interface, the electrostatic energy of the $i$th exciton in the electric field of the other excitons surrounding this site is:

$$V_i = \sum_{j=1}^{n_1} \frac{p^2}{r_{ij}^3} \quad (j \neq i). \tag{7}$$

The $i$th CT exciton dissociates when the repulsion energy, $V_i$, is larger than the energy $B$. This condition is satisfied and dissociation occurs when more CTEs appear on sites adjacent to that occupied by the $i$th exciton. The electrostatic potential energy of the exciton strongly increases when a few CTEs occupy the nearest lattice sites. If this occurs, one or more CTEs will dissociate, thereby reducing the total system energy. Such a mechanism should result in correlations between exciton positions, and an ordering of the system of non-mobile CT excitons can be expected. Such a spatial ordering suggests the existence of a critical pump light intensity above which there is an onset of cold photoconductivity.

In contrast to the thermodynamical theory, in simulations the process of recombination of free carriers which can results in the creation of CTEs was neglected.

Near the threshold where the concentration of free carriers is small the contribution of this process to the number of CTEs indeed will be negligible. However, even at higher concentration, the effect of free carrier recombination can be reduced by applying along the interface an electrical field. This field will separate electrons and holes and thus will create the photocurrent which has to be measured (see also Section 1.4 below).

Computer simulations were performed for a two-dimensional square lattice containing $600 \times 600$ sites. Under continuous pumping of the sample with a constant intensity, the CTEs are generated in the process described above. In order to avoid the influence of boundary conditions, we simulate the evolution of only the central part of the lattice. This square, central sublattice consists of $200 \times 200$ D-A sites, $N_{sites} = 40,000$. Next, we replicate the central sublattice by adding 8 more square sublattices surrounding the central one. That is, the exciton positions calculated for the central $200 \times 200$ sites square lattice is reflected via a mirror symmetry operation to the other surrounding 8 squares.

To simulate the time-evolution, we run the system through equally spaced time steps separated by the interval, $\Delta t$. The value of $\Delta t$ is chosen to be much shorter than the CTE lifetime, i.e., $(\Delta t \ll \tau)$. Here, we choose $\Delta t = \tau/50$.

We start the simulations when there are no CTEs at the interface. Under the influence of the pumping the excitons begin to appear. After the time $\approx \tau$, the number of CT excitons occupying the lattice reaches the steady state value. In our current work we take a time interval of $5\tau$ to ensure that the steady state is reached. From this time on the necessary statistical information is collected.

The time-evolution of the system is simulated as follows. At every time step a few CTEs (depending on the pumping intensity, $I$) are created at randomly chosen positions at the central sublattice with $N_{sites} = 40,000$ sites. Then we go over the central sublattice sites and check every D-A molecule. With some probability the exciton at this site can recombine, as explained above. It also can dissociate if its electrostatic energy is high enough. The rules for these events to happen at one particular D-A site are:

1. If the site is empty the charge-transfer exciton can be created with the probability $P_{create} = I \Delta t / N_{sites}$.
2. If a charge-transfer exciton already occupies this site it can recombinate with the probability $P_{rec} = \Delta t / \tau$.

Next, during the same time step, we calculate the energy of every CTE in the electrostatic field produced by the dipole moments of all other excitons. The energy of the $i$th CT exciton can be found using Eq. (7). If this energy is greater than the dissociation threshold, $B$, the CT exciton dissociates. Finally, we recalculate the energies of all CT excitons that remain at the D-A interface.

## 1.3. RESULTS OF NUMERICAL SIMULATIONS

All results reported below are collected after steady state is achieved. Figure 2 shows the dependence of the number of CTEs ($n_1$) on the value $S$ which is the product of generation intensity of the CTEs $I$ and the CTEs lifetime $\tau$: $S = I\tau$. The steady state number of dissociated pairs is determined by its own lifetime but we do not estimate here the concentration of carriers and conductivity. Nevertheless, on Figure 2 is plotted the value $S_2$ which is equal to the number of dissociation which take place at given $S$ in steady state during time, $\tau$. We find qualitative agreement with the analytical theory that the CTEs populate the D-A interface only up to some saturation concentration. Further increase of the pumping $S$ results mainly into the dissociation of CTEs into electron–hole pairs. When the number of the CTEs ($n_1$) reaches the saturation density the number of dissociations during interval $\tau$ into electron–hole pairs $S_2$ increases linearly with the pumping $S$. It is interesting to compare the critical concentration of CTEs from the analytical simplified model (see above) in which the CTEs are assumed to be in thermal equilibrium and having a spatially ordered structure (square lattice) with the results of the numerical simulations. In the simulations with a random

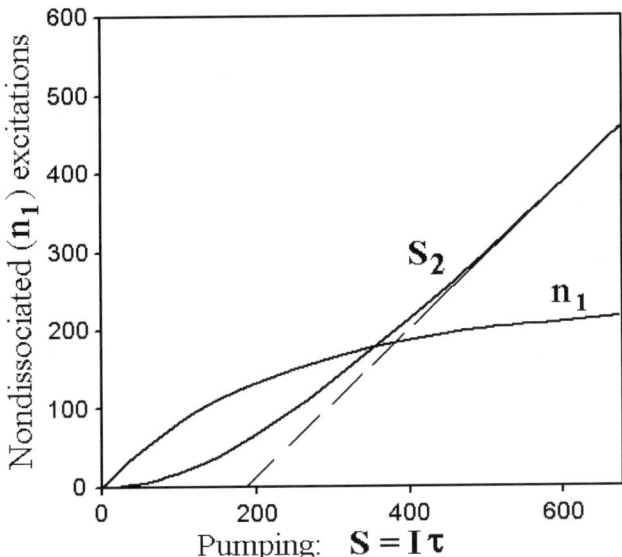

Fig. 2. The steady state number of the CTEs ($n_1$) occupying the donor–acceptor interface and the number of the dissociated pairs ($S_2$) as a function of the pumping intensity. The pumping intensity $S$ is equal to the number of the charge-transfer excitons produced at the interface during a CTE lifetime in the absence of dissociation processes. The results are from the numerical simulations of the CTE system described in the text.

CTE distribution, dissociated pairs appear even at low pumping. Nevertheless, following qualitatively the results of the analytical model (Figure 1), as a critical concentration of CTEs we can take the concentration corresponding to the saturation of CTEs at interface or, what is nearly the same, the concentration of CTEs which corresponds to intersection of the linear $S_2$ asymptote with the horizontal axis. In Figure 2 the value of $n_1$ is approximately 200 and thus the corresponding critical dimensionless concentration $C_{cr} = 200/40000 = 0.5\%$. The curve on the Figure 1 corresponds to value $M = Ba^3/p^2 = 0.01$. From the analytical theory it follows that for the same value $M$ the critical dimensionless concentration $C_{cr} = (4Ba^3/5Ap^2)^{2/3} = (4M/5A)^{2/3} = 0.85\%$. Thus, a random CTE distribution decreases the critical concentration for the transition to the conducting state. This effect could be expected, because, for a random distribution, in contrast to the analytical model of ordered CTEs, the occurrence of small distances between CTEs is allowed even at low CTE concentration. In both approaches, the critical concentration strongly depends on the values of $B$, $p$ and $a$ and below we compare the critical concentrations for different values of these parameters. For example, for $B = 0.2$ eV, $p = 20$ D and $a = 5$ Å, correspond to $M = 0.1$, the analytical model gives $C_{cr} = 4\%$, computer simulations give $C_{cr} = 2.5\%$. For $M = 0.05$, analytical approach gives $C_{cr} = 2.5\%$, computer simulations gives $C_{cr} = 1.5\%$ and so on. Thus, for a random CTE distribution at the D-A interface, the critical concentration is almost twice as small as that predicted by the analytical model of ordered CTEs with infinite lifetime.

As we already mentioned above the dissociation of nearest CTEs results into the change of the correlation function of their spatial distribution. This affects the repulsion energy distribution of the CTEs. If the CTE dissociation is absent their energy distribution would have a peak (associated with the average distance between the excitons) and a tail extending to high energies (such a tail is associated with the CTEs occupying nearby lattice sites). The dissociation prevents the creation of clusters of the closely placed CT excitons and, especially, it prohibits CTEs from occupying adjacent sites at the D-A interface and it cuts off that high energy tail. This is shown in Figure 3 which presents the distribution of the CTE repulsion energies for different pumping intensities $S$. On this figure the dissociation threshold is represented by the energy $B$ and the origin corresponds to the situation when the CTEs occupy infinitely remote sites. It is seen from this figure that the peak of the energy distribution increases with the number of CTE and so does the width of the distribution. It is interesting to note also that the position of the peak of the repulsion energy distribution (corresponding to the energy of highest probability) varies with the steady state number of CTEs approximately in the way which the theoretical model of ordered CTEs lattice predicts. As it follows from Eq. (5) the CTE energy as a function of the number of CTEs should vary as $n_1^{3/2}$. Figure 4 demonstrates that in computer simulations such dependence take place with high accuracy. Figure 5 shows the distribution of the nearest-neighbor

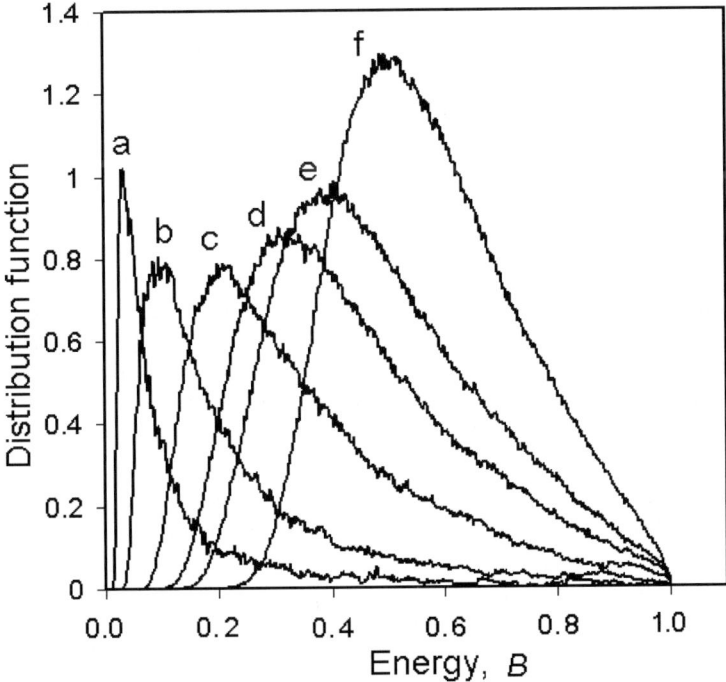

Fig. 3. Steady state distribution of the CTE electrostatic energies for different pumping intensities. The pumping intensity $S$ is the number of CTEs produced at the interface during a CTE lifetime in the absence of dissociation processes. The values of $S$ and the corresponding ones of $n_1$ are given by: (a) $S = 50$, $n_1 = 45$; (b) $S = 100$, $n_1 = 82$; (c) $S = 200$, $n_1 = 133$; (d) $S = 350$, $n_1 = 177$; (e) $S = 500$, $n_1 = 202$; (f) $S = 1500$, $n_1 = 247$. The repulsion energy of the CTEs is given in units of the dissociation energy $B$.

(n-n) distances between the CTEs in the steady state when the pumping intensity $S$ varies. We took $B = 0.02$ eV in Figure 5, with $c_{cr} = 0.5\%$ as discussed above; the closest approach of any two CTEs is then about five lattice spacings. At low intensity [curve (a)] the CTE distribution is broad and there are isolated CTEs at sites up to $20a$ from any other CTE. At higher $S$ the distribution sharpens. At high $S$ the CTE coverage of the D-A interface is roughly uniform (n-n distances vary from $5a$ to $16a$). For the parameters used, CTEs closer than $5a$ dissociate and there is always a lower limit on separations.

### 1.4. Concluding Remarks

In this section, we have considered a transition to a conducting state at $T = 0$ due to the CT exciton–exciton repulsion. Using computer simulations in which

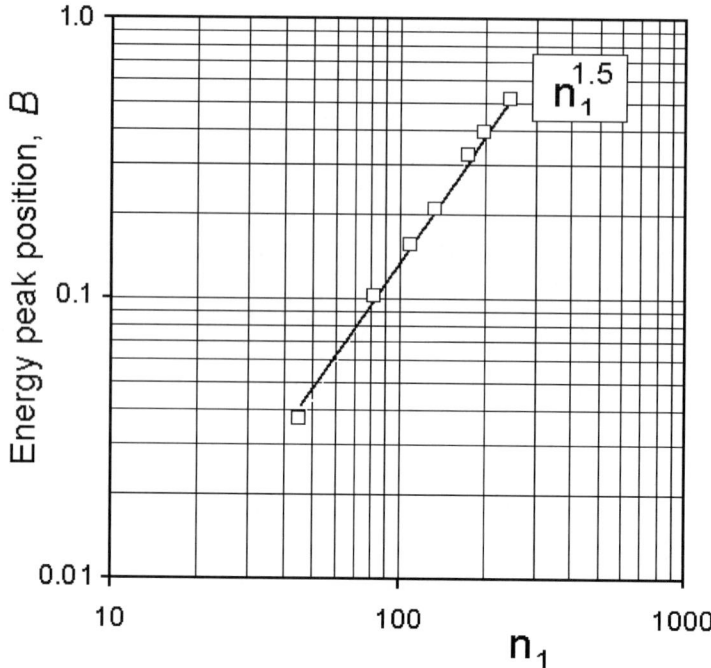

Fig. 4. Position of the energy distribution peak as a function of the number of the CT excitons, $n_1$. The position of the energy peak obtained from the numerical simulations appears to be proportional to $n_1^{1.5}$.

the randomness of the CTE distribution and their finite lifetime were taken into account, we have found how the repulsive interactions between CTEs populates higher energy states of dissociated e–h pairs and thus creates free carriers. It is clear that at a finite temperature, the repulsion can also be important because it decreases the activation energy. This decrease of activation energy depends on the concentration of CTEs.

The computer simulation demonstrates also that the critical concentration of CTEs should depend on their mobility. If this mobility is small and the CTE distribution at a D-A interface is random, the critical concentration leading to a phase transition to the conducting state will be smaller than in the case of mobile CTEs. In the case in which their lifetime is long and their repulsion is strong an ordered state is realized.

An interesting problem is how to observe the predicted photoconductivity. It is evident that for crystalline D-A multilayers the conductivity along the interfaces can be measured. Alternatively, the optical properties of the interface near the conductivity transition can be observed. For such experiments, methods for observing

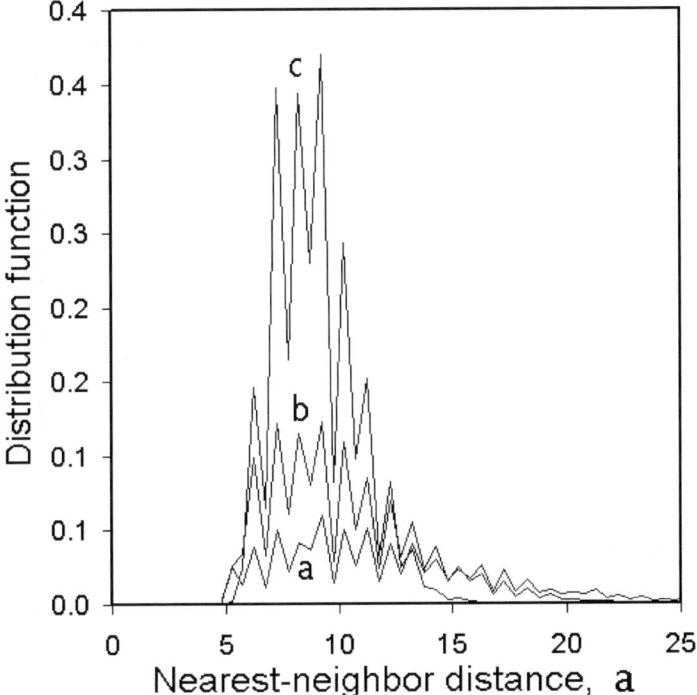

Fig. 5. Distribution of the nearest-neighbor distances, in units of the lattice constant $a$, between the charge-transfer excitons for three different pumping intensities. The pumping intensity $S$ and the corresponding number $n_1$ are: (a) $S = 100$, $n_1 = 82$; (b) $S = 200$, $n_1 = 133$; (c) $S = 1500$, $n_1 = 247$.

photoconductivity parallel to the plane of dipoles developed in the investigations of Langmuir–Blodgett films can be used. However, for such measurements we need to have nearly perfectly ordered crystalline D-A multilayers with a large interface area.

The states and mobility of free carriers at D-A interfaces are also important for the observation of photoconductivity. In the discussion of these problems, it is tempting to use the analogy with Tamm states of electrons or holes at the crystal interface. In the case of Tamm states, the wave vector is a good quantum number. The states are coherent and they form the conduction and valence bands of surface states. This is typical for surface states of inorganic semiconductors or metals, since the electron–phonon coupling is too weak to destroy the bands. On the other hand, for D-A organic structures as well as for organic crystals in general, the width of energy bands at room temperature is small (order 0.1 eV), and electron–phonon coupling can be rather strong, giving rise to self trapping of free carriers and hopping conductivity. We expect that the same is true for free carriers at an organic interface. In the discussion of carrier mobility at the D-A interface we have to take into account not only self trapping, but also Coulomb interactions between free carriers, and with CTEs and recombination processes. In semiconductors the recombination of electron–hole pairs usually is slow due to spatial delocalization. To suppress the recombination of free carriers at a D-A interface, an electric field can be applied parallel to the interface. This electric field can also be used for observation of photoconductivity along the D-A interface. Thus to observe the phenomena considered here it is necessary to have high quality D-A structure, as well as resolve many difficult technical problems usually connected to this type of experiments.

## 2. Cumulative Photovoltage in Asymmetrical Donor–Acceptor Organic Superlattices

### 2.1. INTRODUCTION

Photo-voltaic energy conversion is an important component of a future network of renewable energy sources that could provide a sustainable energy supply without greenhouse gas emission [17]. That is the reason behind the intensive investigations in the field of photo-voltaics which also demonstrate a very peculiar competition between the use of inorganic and organic materials for the realization of solar cells with increasing efficiency. No inorganic material matches the absorption coefficients of organic dyes, which are in the range of $10^5$ cm$^{-1}$, and give rise to the hope of producing organic based very thin solar cells with low energy and material consumption.

From the point of view of macroscopic electrodynamics the appearance of a constant current under the influence of light (photo-voltaic effect) can be understood in the framework of the theory of perturbation (for a small light intensity) from the nonlinear relation (see also [18]):

$$J_i(0) = \sigma_{ijl}^{(2)}(0; \omega, -\omega) E_j(\omega) E_l(-\omega)$$

$$+ \sigma_{ijlmn}^{(4)}(0; \omega_1, -\omega_1, \omega_2, -\omega_2) E_j(\omega_1) E_l(-\omega_1) E_m(\omega_2) E_n(-\omega_2)$$

$$+ \cdots, \qquad (8)$$

where $\vec{J}$ is a constant electrical current, $\vec{E}(\omega)$ is the amplitude of the electrical field of light with frequency $\omega$ and $\sigma^{(2),(4)}$ are the tensors of nonlinear conductivity. These tensors are constant in space for infinite homogeneous media and are different from zero only for non-center-symmetrical structures. The calculation of these tensors is the problem of microscopical theory. Such theory needs the analysis of the main mechanisms which can be responsible for the appearance of the photo-voltaic effect. In the case considered below of asymmetrical organic donor–acceptor superlattices with periods of the order of 100 Å it is necessary to carry out the analysis of possible mechanisms of photo-voltaics in organic nanostructured materials. It is worth to mention also that in some cases the electronic excitations which appear under the influence of light absorption may be responsible for the structural and chemical rearrangement of organic materials. In this case, of course, the theory of perturbation is not applicable and as a result the tensors of nonlinear conductivity depend themselves on the light intensity.

Below in this section following [19], we discuss the electro-optical properties of an asymmetrical stack of organic donor–acceptor (D-A) interfaces. As we mentioned, the technological progress in molecular organic beam deposition is very fast and there is little doubt that a variety of such systems will be synthesized in the near future. With this in mind, we discuss the properties of a superlattice of the type $\ldots DDD|AAA|NNN|DDD|AAA|NNN\ldots$, where $N$ stands for a material which is neither a good donor nor a good acceptor, and all molecules in the ground state are neutral. In such a non-center-symmetrical structure, when CTEs are generated by the pumping light, all the interface CTE dipoles point in the same direction and the potential differences due to the dipole layers at each D-A interface add up to a macroscopic voltage across the superlattice creating a macroscopic potential drop. The corresponding electric field will drive the free electrons and holes produced by the absorption of light and, thus, will provide a photo-voltaic current. It is interesting to mention that in the case of D-A interfaces with negative charge-transfer exciton energy or, in other words, in the case of donor and acceptor for which the charge transfer exists already in ground state, a macroscopic potential drop also has to appear in the asymmetrical structure which we consider here but in this case it will be independent on the intensity of incident light. For both cases the appearing electrical field will be directed from donor side to acceptor side as we also will show below by direct calculations.

Before discussing the origin of electro-optical effects in organic asymmetrical multilayer structures, it is worth to mention that in 1996 Professor Rolf Landauer called our attention to the fact that the photo-voltaic mechanism which we proposed in [19] for organic D-A asymmetrical superlattices is similar to those considered to generate cumulative photovoltages in some inorganic crystals. He had in mind the papers by Cheroff and Keller [20], Pensak [21] and Goldstein [22] which independently discovered the large photovoltages in ZnS and CdTe crystals with periodic intrinsic inhomogeneities, and also the paper by Swanson [23]. In the latter, it is theoretically demonstrated that the large photovoltages observed in those crystals are plausible and can be due to structures without a center of inversion by a wide class of different mechanisms. Conversely, the conditions under which a photoconductor with periodic inhomogeneities does not show a cumulative photovoltage are shown to be very restrictive and improbable. In the Introduction of the paper [23] it was mentioned, that "the large photovoltages observed in [20] in some insulators are due to numerous internal electrostatic barriers which under illumination act as $p-n$ junctions connected in series. The crystal structure of these materials does not exhibit inversion symmetry, and it is conceivable that the directionality of the crystal induces directionality of the internal barriers. Such an intrinsic directionality is necessary, since a structure of randomly alternating conductivity type could provide no basis for a preferred direction in which the voltage accumulates." It follows from this remarks that, indeed, the structures which have been investigated in [20,21] and [22] have many similar features with the asymmetrical D-A organic structure which is under discussion in this section, particularly in the case where charge transfer takes place already in the ground state of a donor–acceptor pair. In our paper [19], we assume that the charge transfer takes place only in the excited state of a D-A pair and that the main role in the appearance of the photovoltage is not played by an intrinsic ground state asymmetry of the asymmetrical D-A multilayer structure, but rather by the asymmetry of the sheets of dipoles at all donor–acceptor interfaces which arise only due to the generation of CTEs via light absorption. The density of CTEs on the sheets is dependent on the light intensity and this should transform the photo-voltaic effect in asymmetrical organic donor–acceptor multi-layers into a strongly nonlinear function of the light intensity.

It is known that a photocurrent may be measured from the response to external voltages under illumination. In materials having a large enough photovoltage, this voltage itself can be used without an external bias to measure a photocurrent (short-circuit photocurrent). One is not only interested in measuring the short-circuit photocurrent, but it is of interest to measure independently the open-circuit photovoltage which is the other main characteristic of photo-voltaics (see, for example, [20,21] and [22]). Both these characteristics of photo-voltaics, generally speaking, are dependent on the concentration and mobility of charge carriers. We will demonstrate below that in an asymmetrical stack of organic donor–acceptor (D-A) interfaces the open-circuit photovoltage may arise even in the absence of free carriers.

## 2.2. On the Mechanisms of the Photo-Voltaic Effect in Organics

To discuss in more details the possible peculiarities of photo-voltaics in an asymmetrical D-A multi-layer structure, we will make a few remarks concerning possible mechanisms of photo-voltaic effect in multilayered structures containing organic dyes (see also [29]).

One model (see, for example, [24]) is based on the assumption that the contacting organic layers have dark $n$- or $p$-type conductivity and that the band structure of the organic D-A interface is similar to that of a $p$–$n$ junction between inorganic semiconductors. In this case in the region near the interface between these layers a depletion region is formed, where the internal electric field results in the dissociation of an excited electron–hole pair into free carriers. In an alternative model [25–28], the photo-voltaic effect occurs exclusively at the interface of an organic dye with a second dye, the charge separation takes place at the interface and an extended region with an electrical field is not required and may be even absent (see also [29]). Natural photosynthesis may be considered as the important example of such type of carrier generation. Indeed, exciton are formed by light absorption in the antenna pigments and diffuse to the reaction center where they dissociate with almost unit quantum efficiency by injecting an electron and hole in opposite spacial directions down a chain of acceptors and donors, respectively (see [30,31]). It is important to mention that in this mechanism of photo-voltaics the electrons and holes after the process of interface charge separation diffuse away from the D-A interface in opposite directions as if driven by an effective electric field directed from the acceptor A to the donor D, or in other words along the same direction of the CTE dipole. This direction, as will be shown below, is opposite to the direction of the electrical field created by a sheet of CTE dipoles mentioned above.

In our discussion below, we assume that the binding energy of a CTE is large, and the CTEs are rather stable and do not participate in the photo-generation of free carriers. We will show that the asymmetrical stack of D-A interfaces under pumping of CTEs provides a macroscopic potential drop that can be considered as a cumulative photovoltage.

### 2.3. CUMULATIVE PHOTOVOLTAGE IN AN ASYMMETRICAL STACK OF D-A INTERFACES

To estimate the potential profile determined by the interface CTEs, we consider first a single D-A interface with a two-dimensional (2D) density of CTEs $n$ of order $10^{12}$ cm$^{-2}$, each one having an electric dipole moment $\mu$ of about 20 Debye. Such large dipole moments are not unusual for CTEs and the 2D density above, taking a superlattice period of a few tens of monolayers, would correspond to a bulk concentration of excited molecules of order $10^{-4}$, which is not problematic in relation to the photochemical stability of the organic materials. In a first approximation, this CTE configuration corresponds to a uniform static dipole moment per unit area $\mu n$ perpendicular to the D-A interface. As the CTEs repel each other through the dipole–dipole interaction $\Delta H = \mu^2/\rho^3$ ($\rho$ being the exciton–exciton distance) and as their mobility is not negligible, they tend indeed to be uniformly spaced along the interface; for instance, a similar system of dipoles moving classically along a plane and oriented perpendicular to it have been shown [32] to order in a 2D lattice at low temperatures and to form a homogeneous liquid at high temperatures. The CTE repulsive interaction here considered is, of course, the same that at a higher concentration would give rise to the dielectric-conductor transition as discussed in the previous section.

If we assume that the dipole layer is uniform and that we can neglect its thickness the corresponding polarization per unit volume is $\vec{P}(z) = \vec{p}_0 \delta(z)$, where $\vec{p}_0$ is the polarization per unit area and the $z$-axis is directed along the normal to the interface planes from the donor to the acceptor side. In our case, the vector $\vec{p}_0 \equiv (0, 0, -p_0)$, $p_0 = \mu n$. Using the equation div $D(\vec{r}) = 0$ where $D(\vec{r})$ is Maxwell displacement vector, $\vec{D} = \vec{E} + 4\pi \vec{P}$, and taking into account that all these values depend only on $z$ we have:

$$\frac{dD_z(z)}{dz} = 0, \qquad E_z(z) - 4\pi p_0 \delta(z) = 0, \qquad E_z = -\frac{dV}{dz}, \qquad (9)$$

where $V(z)$ is the potential. It is follow from these equations that

$$V(z) = V(-\infty) - \int_{-\infty}^{z} 4\pi p_0 \delta(z)\, dz. \qquad (10)$$

Thus, in the approximation of a uniform dipole layer, the resulting electrostatic potential has the form of a sharp step of hight $\Delta V = 4\pi \mu n$. This effect is well-known and is used in the surface double layer model for the work function of metals (see [33,34]). In our case the height of the step $\Delta V = 4\pi \mu n \simeq 0.1$ V at the interface. Thus, the electrical field in the region of a step with an effective thickness $a$ is equal to $E_z = 4\pi \mu n/a$, opposite to the direction of the CTE dipole moment. Probably, it is impossible to create a density of CTEs much larger than that used in our estimation above because the insulator to metal transition [9] leading to the ionization of the CTEs sets a limit on $n$, as discussed above.

For a typical intermolecular spacing $a$ of about 5 Å and the exciton densities considered here, the average distance $\rho_0$ between two excitons along the interface plane ($n \simeq 1/\rho_0^2$) is much larger than the interface thickness; as a consequence, the potential does not exhibit an abrupt jump and the electric field corresponding to $\Delta V$ is not restricted to the interface, but extends on either side over a layer of width comparable to $\rho_0$, as shown in Figure 6. Taking for the sake of simplicity the CTEs located at the sites of a square lattice, this figure shows $\phi(z) = V(z)/(2\pi \mu n)$ where $V(z)$ is the electrostatic potential along a direction perpendicular to the D-A interface and passing through the center of a square unit cell. Even in the case of a homogeneous disordered 2D distribution of CTEs, the average electric field profile is not expected to be much different than for a 2D square lattice of equal density. When considering an asymmetrical superlattice in which all the interface voltage drops add up, the qualitative shape of the total electrostatic potential profile will be determined by the ratio between the distance between successive D-A interfaces, i.e., the superlattice period $L$ and $\rho_0$: if $L$ is larger than $\rho_0$ (i.e., at a high density $n$) the potential will resemble a staircase as shown in Figure 7, otherwise (i.e., at a low density $n$) it will have a rather uniform

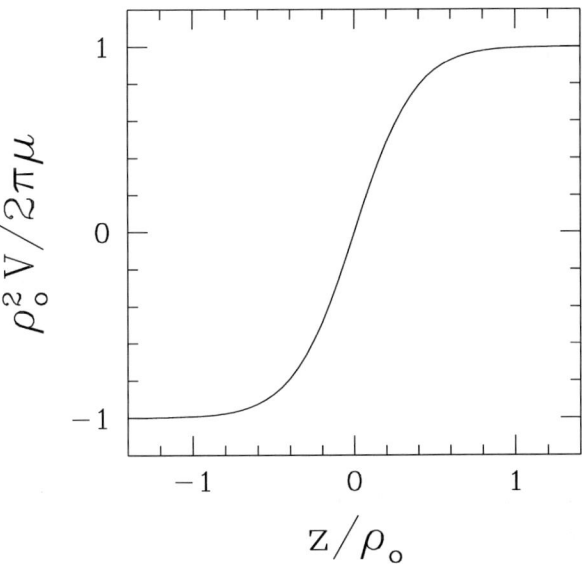

Fig. 6. Normalized potential profile across a D-A interface in the presence of a CTE density $n \simeq 1/\rho_0^2$.

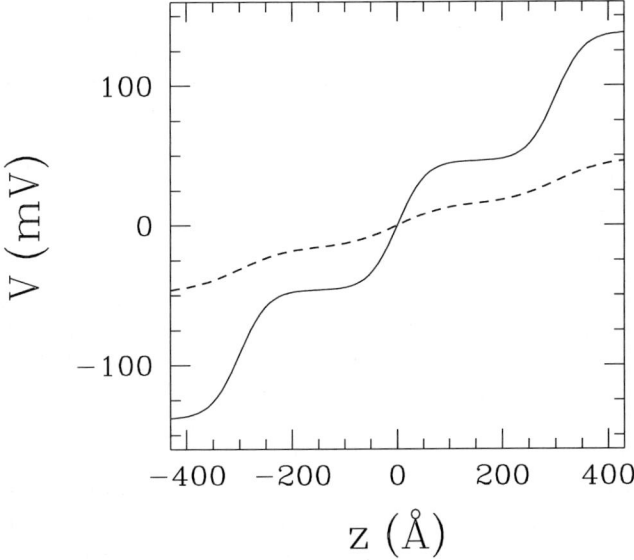

Fig. 7. Electrostatic potential along a D-A-N superlattice in the presence of 2D CTEs at the D-A interfaces with different densities: $n \simeq 10^{12}$ cm$^{-2}$ (solid line) and $n \simeq 3.5 \times 10^{11}$ cm$^{-2}$ (dashed line).

slope. In either case, the average electric field in the direction of growth will be given by $E_0 = \Delta V/L$ and can be comparable to the electric field in the depletion layer of a typical semiconductor $p - n$ junction (for the values of $\mu$ estimated above, $L \simeq 300$ Å and $n \simeq 10^{12}$ cm$^{-2}$, for instance, $E_0$ is about $3 \cdot 10^4$ V/cm). Of course, the uniformity along the superlattice planes will never be perfect in a real structure and, in general, a rather complicated spatial pattern of electric field force lines (and therefore current filaments) can be expected.

Under steady illumination, a condition of dynamical equilibrium will be reached with a constant 2D density of interface CTEs. Free carriers will also be present, either photogenerated directly or, for instance, as a result of the thermal ionization of the CTEs (see, for example, [25–27]) at room temperature. The electric field above will effectively separate the electrons and holes and drive a current along the superlattice crossing the D-A interfaces from the donor to the acceptor side. We are here neglecting the effects of the periodic changes of the band edges due to the superlattice compositional changes (such discontinuities could be minimized by a proper choice of materials) with respect to the additive effect of the potential variations at the D-A interfaces. Of course, these effects would be important to estimate the electron and hole mobilities along the growth axis. In the structures here considered, electrons could be effectively injected into the valence band on the acceptor side through a contact with a large work function metal and extracted from the conduction band on the donor side through a contact with a small work function metal (as done in the reverse direction in LED devices); such a current would deliver power to the external circuit load at the expense of the light absorbed.

The expected efficiency of photo-voltaic conversion is strongly dependent on the dark conductivity, on the processes of carrier photo-generation, on the kinetics of excitons and charge carriers and can be estimated only in the framework of a complete theory which properly takes into account also the structure of the heterojunction.

However, the qualitative features of its dependence on the light intensity can be established on the ground of more simple considerations. At low intensity of light the steady state CTE concentration, which is dependent on the pumping intensity $I$ of CTEs and their lifetime, will be small and the macroscopical potential drop will be negligible. The macroscopic potential drop $V \sim I$ will be important with increasing CTE concentration. For a constant density of free carriers, the photo-voltaic power will be proportional to $I$. At still higher pumping intensities, when the presence of the CTE leads via their "cold" ionization to an increase in the density of free carriers as described above, the photovoltage will no longer increase, but the photo-voltaic power dependence on $I$ will still be approximately linear due to the increase in free carrier density.

## 3. Nonlinear Optical Response of Charge-Transfer Excitons at Donor–Acceptor Interface

In this section following the paper [35], we discuss the resonant and off-resonant optical nonlinearities of a system of charge-transfer excitons (CTEs) at a donor–acceptor (D-A) interface. We continue to consider an interface between organic materials (neutral in the ground state) for which the lowest electronic excitations are CTEs corresponding to the displacement of an electron from the donor to the acceptor side. We assume that the photogenerated CTEs have large static electric dipole moments perpendicular to the D-A interface plane. We show that this system may exhibit strong resonant optical nonlinearities induced by the dipole–dipole repulsion among the CTEs. We show that on account of this long range interaction, the excitation intensity dependence of the Kerr nonlinearity is non-analytic: it depends on the two-dimensional density $n$ of CTEs as $n^{3/2}$. Therefore, the dependence of the CW nonlinear polarization on the laser electric field is beyond the usual power expansion. We also point out that the static electric field produced by the CTEs modifies the hyperpolarizabilities of nearby molecules. Therefore, in media with CTEs the intensity dependence of the nonlinear optical response can be stronger than usually expected, and this theoretical prediction could be easily experimentally tested. First we consider the effects of the dipole–dipole repulsion among CTEs on the nonlinear optical response and, then, the influence on the nonlinear hyperpolarizabilities of nearby molecules exerted by the large static field associated with a CTE.

### 3.1. Resonant Optical Nonlinearity of CTEs: the Role of the Exciton–Exciton Repulsion

#### 3.1.1. Exciton–exciton interaction at a D-A interface

We have already discussed some peculiarities of the exciton–exciton interaction at a D-A interface above. Nevertheless, for the convenience of the readers we repeat here some results of this discussion to make also this section self-contained.

The dipole–dipole interaction energy between two CTE having a dipole moment $\mu$ (which is typically 20 Debye) at a distance $\rho$ along the D-A interface plane is $U = \mu^2/\rho^3$ and is rather large for small distances and decreases with a long range as $\rho^{-3}$; for instance, for $\rho \simeq 5$ Å, $U \simeq 1$ eV and for $\rho \simeq 10$ Å, $U \simeq 0.3$ eV. As the number of excitons at a distance $\rho$ along the interface also scales as $\rho$ the interaction energy of a given exciton with all other excitons which are at a distance $\rho$ decreases more slowly and scales as $\rho^{-2}$, rather than as $\rho^{-3}$.

The average exciton–exciton distance $\rho_0$ is related to the two-dimensional (2D) density of CTEs $n$ by $n \simeq 1/\rho_0^2$. The ensuing repulsion increases the energy of the CTEs and the corresponding energy shift $\Delta\varepsilon$ is given by the average interaction of one exciton with all the others. We can expect that CTEs having a non-negligible mobility will tend to order in such a way as to minimize the energy of repulsion. As discussed above, we assume a simple 2D square lattice structure for the CTE spatial distribution, then

$$\Delta\varepsilon \simeq 10\,\mu^2/\rho_0^3 = 10\,V_0\,(n/N)^{(3/2)}, \tag{11}$$

with $V_0 = \mu^2/a^3$ and $N$ the total 2D density of molecules given by $N = 1/a^2$ where $a$ is the molecular crystal lattice constant along the interface (which is typically 5 Å). It is clear that such an additional repulsion energy is required to create a CTE at an empty site (i.e., at a vacancy in the CTE lattice).

To estimate the energy shift $\Delta\varepsilon$ we can use the expression

$$\Delta\varepsilon = 10\,V_0\,(a/\rho)^3. \tag{12}$$

Assuming that the mean distance between CTEs $\rho \simeq 10\,a$, e.g., that the concentration of CTEs is order of $10^{-2}$ we obtain that $\Delta\varepsilon \simeq 150\,\text{cm}^{-1}$, which is significant on the scale of the homogeneous width of a CTE transition.

As discussed above, the scaling dependence $\Delta\varepsilon \simeq n^{3/2}$ follows also from the results of numerical simulations and, thus, the estimate made is expected to be valid even for a disordered homogeneous distribution of CTEs of comparable 2D density $n$. We wish to stress here the non-analytic dependence of $\Delta\varepsilon$ on $n$: a perturbation theory expansion in terms of $n$, $n^2$, $n^3$, etc., would be inadequate as the leading correction scales like $n^{3/2}$.

### 3.1.2. Nonlinear D-A interface polarizability

The D-A interface polarizability due the CTE contribution can be written as $\chi(\omega) \simeq A/((\varepsilon_0 + \Delta\varepsilon)^2 - \omega^2)$, where $\varepsilon_0$ is the CTE energy for $n = 0$ and $A$ is a constant proportional to the CTE oscillator strength. Expanding $\chi(\omega)$ in series of $\Delta\varepsilon/\varepsilon_0$ we find that $\chi(\omega) = \chi_0^{(1)}(\omega)(1 - 2\varepsilon_0\Delta\varepsilon(n)/(\varepsilon_0^2 - \omega^2)) = \chi_0^{(1)} + \Delta\chi(n)$ where $\chi_0^{(1)}$ is the polarizability for $n = 0$. Thus, for the nonlinear correction to the polarizability corresponding usually to the Kerr nonlinearity, we have $\Delta\chi/\chi_0^{(1)} = -2\varepsilon_0\Delta\varepsilon/(\varepsilon_0^2 - \omega^2) \simeq -\Delta\varepsilon/(\varepsilon_0 - \omega)$, where for resonant pumping $|\varepsilon_0 - \omega| \simeq \delta$, $\delta$ being the exciton linewidth. Assuming the concentration $n \simeq N/100$ and using the previous estimates, we have $\Delta\varepsilon \simeq 150\,\text{cm}^{-1}$ which, even for $\delta \simeq 500\,\text{cm}^{-1}$ gives for the resonant nonlinearity $\Delta\chi/\chi_0^{(1)} \simeq 0.3$. Such a large change in polarizability is not due to the two-level-system-like anharmonicity which is the main nonlinear mechanism for Wannier–Mott excitons in semiconductors (phase space filling [1,36–39]), but it is caused by the exciton–exciton interaction which in the present case is particularly large; in fact, we have that $\Delta\chi/\chi_0^{(1)}$ is one order of magnitude larger than $n/n_S$ where the CTE saturation density $n_S$ is given by $N$ itself. A 2D concentration along each interface of 0.01 can hardly be dangerous for organic crystals as, taking into account the thickness of the layer between successive interfaces, it may correspond to a 3D concentration of order $10^{-3}$. We note again that the continuous wave (CW) optical nonlinearity of 2D CTEs here considered is also interesting as it goes beyond the usual perturbation theory [40–42]. As a matter of fact, because in steady state equilibrium conditions $n \propto I \propto |E(\omega)|^2$ ($I$ being the pump light intensity), the shift of the CTE energy $\Delta\varepsilon(n)$ and the nonlinear correction to the polarizability turn out to be proportional to $|E|^3$. Thus, the nonlinear part of the polarization is $\Delta\vec{P}(\omega) \propto |E(\omega)|^3 \vec{E}(\omega)$ and such a term cannot be found in the usual expansion of the polarization $\vec{P}$ in powers of the components of the electric field $\vec{E}$. This peculiarity stems from the long-range exciton–exciton interaction that shifts the CTE energy in a non-analytic way with respect to $n$. The unusual dependence of $\Delta\vec{P}$ on $\vec{E}$ here considered should be easily experimentally observed for CW resonant pumping.

### 3.2. Photogenerated Static Electric Field: Influence on the Nonresonant Optical Response

In the above discussion, we have only considered the effects due to the CTE–CTE repulsion, which contribute to the resonant nonlinear absorption (as well as to other resonant nonlinearities) by the CTEs themselves. Here, however, we want to mention a more general mechanism by which the nonlinear optical properties of media containing CTEs in the excited state can be enhanced. This influence is due to the strong static electric field arising in the vicinity of an excited CTE. If, for example, the CTE (or CT complex) static electric dipole moment is 20 Debye, at a distance of 5 Å it creates a field $E^{CTE}$ of order $10^7$ V/cm. Such strong electric fields have to be taken into account in the calculation of the nonlinear susceptibilities, because they change the hyperpolarizabilities $\alpha, \beta, \gamma$, etc., of all molecules close to the CTE. For instance, in the presence of this CTE induced static fields, the microscopic molecular hyperpolarizabilities are modified as follows $\alpha_{ij} = \alpha_{ij}^{(0)} + \alpha_{ijl}^{(1)} E_l^{CTE} + \alpha_{ijlk}^{(2)} E_l^{CTE} E_k^{CTE} + \cdots, \beta_{ijl} = \beta_{ijl}^{(0)} + \beta_{ijlk}^{(1)} E_k^{CTE} + \cdots$, etc. The changes in the molecular hyperpolarizabilities are reflected in all the macroscopic nonlinear optical constants of a medium. As the CTE induced electrical fields increase with the concentration of CTEs, a stronger dependence of the nonlinear optical response on the intensity of light is obtained. For example, $\chi^{(3)}$ will include a contribution from $\chi^{(4)}$ and the corresponding polarization will be given by $\Delta P_i \simeq \chi_{ijlm}^{(3)} E_j E_l E_m + \chi_{ijlmq}^{(4)} E_j E_l E_m E_q^{CTE}$ where $E$ is the pump light electric field, therefore, if the second term is not negligible, $|\Delta P|^2$ will depend on the pump intensity $I \propto |E(\omega)|^2$ more strongly than like $I^3$, through the dependence of $E^{CTE}$ on $I$. Of course, the macroscopic susceptibilities will depend on an average over the positions and orientations of the molecular species involved with respect to the static electric dipoles of the CTEs photogenerated in the medium. For the 2D distribution of CTEs at a D-A interface considered above, the presence of a static macroscopic electric field extending on either side of the interface from the donor to the acceptor side may affect the molecular hyperpolarizabilities enough as to reduce the symmetry of the optical response tensors; for instance, centrosymmetric molecules may acquire a second order polarizability $\beta_{ijl} = 0 + \beta^{(1)}_{ijlk} E^{CTE}_k + \cdots$. The influence of static electric fields on the molecular hyperpolarizabilities has long been know, but in our case the static field effects are controlled by the pumping light as they are associated to the presence of the CTEs: this can lead to a novel class of *all optical* nonlinearities. The calculation of second harmonic generation induced by charge-transfer excitations in centrosymmetric medium can be found in [43]. These calculations can be easily adapted to the case of D-A interfaces.

All these considerations may also apply to noncrystalline polymeric media with impurities having a CT complex in the excited state. If the polymers are approximately oriented in one direction and the structure is such that the CT complexes along a polymer are nearly parallel to each other and perpendicular to the polymer direction, we have a more or less ordered 1D system of CTEs having a strong repulsive interaction and we expect again, as for 2D CTEs, a large enhancement of the optical nonlinearities and a stronger intensity dependence of the nonlinear optical response. Indeed, for a 1D system the correction to the energy is already proportional to $n^3$ ($\Delta \varepsilon \propto \mu^2/\rho_0^3$, $\rho_0 \simeq 1/n$, $n$ being the 1D density) and the nonlinear contribution to the polarizability due to the CTE-CTE repulsion is proportional to $|E(\omega)|^6$. Such a contribution corresponds to a nonlinear polarization $\Delta \vec{P}(\omega) \propto |E(\omega)|^6 \vec{E}(\omega)$ and this effect, in contrast to what we had for the 2D case, can be formally included in the usual analytical expansion of the polarization $\vec{P}$ in powers of the components of the pumping light electric field $\vec{E}(\omega)$. In any case, the appearance of CTEs under the influence of a strong illumination can change drastically the intensity dependence of the nonlinear optical properties and just this theoretical prediction could be checked easily by experiments. For the 2D case and, particularly, the 1D case, this dependence becomes stronger: the CTE contribution to the nonlinear absorption coefficient is proportional to $I^{3/2}$ for the 2D case and to $I^3$ for the 1D case, instead of the usual linear dependence on $I$.

In conclusion, we have studied in this subsection the nonlinear optical response of 2D CTEs at a D-A interface: such a structure belongs to a class of novel systems of current interest to material scientists [7]. The long range exciton–exciton interaction leads to a large resonant nonlinear polarization exhibiting an unusual dependence on the light electric field which goes beyond the standard perturbation expansion. The static electric fields induced by the photogenerated CTEs affect the off resonant hyperpolarizabilities of nearby molecules giving rise to new and more strong all optical nonlinearities.

## References

1. H. Haug, S.W. Koch, Quantum Theory of the Optical and Electronic Properties of Semiconductors, World Scientific, Singapore, 1990.
2. D.S. Chemla, J. Zyss, Nonlinear Optical Properties of Organic Molecules and Crystals, Academic Press, Orlando, 1987.
3. J. Zyss (Ed.), Molecular Nonlinear Optics, Materials, Physics, and Devices, Academic Press, 1994.
4. E. Rashba, Sturge (Eds.), Excitons, North-Holland, Amsterdam, 1982.
5. V.M. Agranovich, M.D. Galanin, Electronic Excitations Energy Transfer in Condensed Matter, North-Holland, Amsterdam, 1982.
6. D. Haarer, M. Philpott, in: V.M. Agranovich, R.M. Hochstrasser (Eds.), Spectroscopy and Excitation Dynamics of Condensed Molecular Systems, North-Holland, Amsterdam, 1983, pp. 27–82.
7. S. Forrest, *Chemical Reviews* 97 (1997) 1793.
8. M. Pope, C.E. Swenberg, Electronic Processes in Organic Crystals and Polymers, Oxford University Press, 1999.
9. V.M. Agranovich, K.N. Ilinski, *Phys. Lett. A* 191 (1994) 309.
10. S.A. Kiselev, E. Hartung, Z.G. Soos, S.R. Forrest, V.M. Agranovich, *Chemical Physics* 238 (1998) 365.
11. F.F. So, S.R. Forrest, Y.Q. Shi, W.H. Steier, "Quasi-epitaxial growth of organic multiple quantum well structures by organic molecular beam deposition", *Appl. Phys. Lett.* 56 (1990) 674–676;
    Y. Imanishi, S. Hattori, A. Nakuta, S. Numata, "Direct observation of an organic superlattice structure", *Phys. Rev. Lett.* 71 (1993) 2098–2101;
    T. Nanaka, Y. Mori, N. Nagai, Y. Nakagawa, N. Saeda, T. Nakahagi, A. Ishitani, "Organic superlattice film by alternate deposition of single molecular layers", *Thin Solid Films* 239 (1994) 214–219.
12. E. Umbach, C. Seidel, J. Taborski, R. Li, A. Soukopp, "Highly-ordered organic adsorbates: commensurate superstructures, OMBE, and 1D nanostructures", *Phys. Stat. Sol. (B)* 192 (1995) 389;
    E. Umbach, R. Fink, "How to control the properties of interfaces and thin films of organic molecules?", in: V.M. Agranovich, G.C. La Rocca (Eds.), Organic Nanostructures: Science and Applications, SIF, IOS Press, 2002, p. 233.
13. M.C. Petty, Langmuir–Blodgett Films, Cambridge University Press, 1996, p. 134.
14. T. Bjornholm, T. Geiser, J. Larsen, M. Jorgensen, K. Brunfeldt, K. Schaumburg, K. Bechgaard, "Donor-acceptor interfaces in Langmuir–Blodgett films: structural and optical features related to nonlinear electronic properties", *Synthetic Metals* 55–57 (1993) 3813–3818.
15. Y. Okada-Shudo, F. Kajzar, C. Meritt, Z. Kafafi, "Second and third order nonlinear optical properties of multilayer and composite C60 thin films", *Synthetic Metals* 94 (1998) 91–98.
16. K.-D. Zhu, T. Kabayashi, *Phys. Lett. A* 201 (1995) 439.
17. A.L. Fahhhrenbruch, A.H. Bube, Fundamentals of Solar Cells. Photovoltaic Solar Energy Conversion, Academic Press, New York, 1983.
18. V.I. Belinicher, B.I. Sturman, *Sov. Phys.-Uspekhy* 130 (1980) 415.
19. V.M. Agranovich, G.C. La Rocca, F. Bassani, *JETP Lett.* 62 (1995) 405.
20. G. Cherroff, S.P. Keller, *Phys. Rev.* 111 (1958) 98.
21. L. Pensak, *Phys. Rev.* 109 (1958) 601.
22. B. Goldstein, *Phys. Rev.* 109 (1958) 601.
23. J.A. Swanson, *IBM Journal of Research and Development* 5 (1961) 210.
24. D. Woehrle, B. Tennigkeit, J. Elbe, L. Kreienhoop, G. Schnupfeil, *Mol. Cryst. Liquid Cryst.* 230 (1993) 221;
    S. Gunster, S. Siebentritt, D. Messner, *Mol. Cryst. Liquid Cryst.* 230 (1993) 351;

K. Takahashi, S.I. Nakatani, T. Matsuda, H. Nanbu, T. Kamura, K. Murata, *Chem. Letters* 11 (1994) 2001.
25. B.A. Gregg, Y.I. Kim, *J. Phys. Chem.* 98 (1994) 2412.
26. A. Kay, R. Humphrybaker, M. Graetzel, *J. Phys. Chem.* 98 (1994) 952.
27. N. Karl, A. Bauer, J. Holzaepfel, J. Marktanner, M. Moebius, F. Stoelzle, "Exciton diffusion and charge separation at donor–acceptor interfaces in thin film organic photo-voltaic cells", *Mol. Cryst. Liquid. Cryst.* 252 (1994) 243.
28. A.A. Zakhidov, K. Yoshino, *Synthetic Metals* 64 (1994) 155.
29. T.J. Savenije, R.B.M. Koehorst, T.J. Schaafsma, *Chem. Phys. Lett.* 244 (1995) 363.
30. M. Baltscheffsky (Ed.), Current Research in Photosynthesis, Vol. 1, Kluwer Academic, Dordrecht, 1990.
31. J. Deisenhofer, J.R. Norris (Eds.), The Photosynthetic Reaction Center, Academic Press, New York, 1993.
32. P. Pieranski, "Two-dimensional interfacial colloidal crystals", *Phys. Rev. Lett.* 43 (1980) 569–573; R.K. Kalia, P. Vashishta, "Interfacial colloidal crystals and melting transition", *J. Phys. C* 14 (1981) L643–L648.
33. E. Wigner, J. Bardeen, "Theory of work function of monovalent metals", *Phys. Rev.* 48 (1935) 84–87.
34. J. Bardeen, "Theory of the work function: surface double layer", *Phys. Rev.* 49 (1936) 653–663.
35. V.M. Agranovich, G.C. La Rocca, F. Bassani, *Chem. Phys. Lett.* 247 (1995) 525.
36. S. Schmitt-Rink, D.S. Chemla, D.A.B. Miller, *Phys. Rev. B* 32 (1985) 6601.
37. S. Schmitt-Rink, D.S. Chemla, D.A.B. Miller, *Adv. in Phys.* 38 (1989) 89.
38. B.I. Green, J. Orenstein, S. Schmitt-Rink, *Science* 247 (1990) 679.
39. G. La Rocca, in: V.M. Agranovich, G.F. Bassani (Eds.), Electronic Excitations in Organic Based Nanostructures, in: Thin Films and Nanostructures, Vol. 31, Elsevier, San Diego, 2003.
40. N. Blombergen, Nonlinear Optics, Benjamin, New York, 1965.
41. C. Flytzanis, in: H. Rabin, C.L. Tang (Eds.), in: Quantum Electronics, Vol. 1, Academic Press, New York, 1975, p. 1.
42. Y.R. Shen, The Principles of Nonlinear Optics, Wiley, New York, 1984.
43. P. Reineker, V.M. Agranovich, V.I. Yudson, *Chem. Phys. Lett.* 260 (1996) 621.

Chapter 7

# Hybridization of Frenkel and Wannier–Mott Excitons in Organic-Inorganic Heterostructures. Strong Coupling Regime

V.M. AGRANOVICH AND V.I. YUDSON

*Institute for Spectroscopy, Russian Academy of Sciences, Troitsk, Moscow obl., 142190 Russia*

P. REINEKER

*Department of Theoretical Physics, University of Ulm, 89089 Ulm, Germany*

1. Introduction . . . . . . . . . . . . . . . . . . . . . . . . . . . . . . . . . . . . . . . . 317
2. Hybrid 2D Frenkel Wannier–Mott Excitons at the Interface of Organic and Inorganic Quantum Wells . . . . . . . . . . . . . . . . . . . . . . . . . . . . . . . . . . . . . . . . . 321
3. Nonlinear Optics of 2D Hybrid Frenkel–Wannier–Mott Excitons . . . . . . . . . . . . . . 331
4. Hybrid Excitons in Parallel Organic and Inorganic Semiconductor Quantum Wires . . . . . 339
5. On the Hybridization of "Zero-Dimensional" Frenkel and Wannier–Mott Excitons . . . . . 344
6. Hybridization of Excitons in Microcavity Configurations . . . . . . . . . . . . . . . . . 345
   References . . . . . . . . . . . . . . . . . . . . . . . . . . . . . . . . . . . . . . . . . 352

## 1. Introduction

The need for systems having better opto-electronic properties to be used in applications has been driving researchers in materials science to develop novel compounds and novel structures. The progress in the field has been impressive, mainly due to the use of innovative growth techniques such as molecular beam epitaxy (MBE) and the realization of systems in two-dimensional (2D), one-dimensional (1D) and zero-dimensional (0D) confined geometries. We have now many newly developed organic or inorganic structures with very interesting properties. We mention here as a typical example the request for efficient second harmonic generation (SHG) where we can see a very peculiar "competition" in the use of organic or inorganic materials. Inorganic semiconductors (e.g., GaAlAs, ZnCdSe) have been used to design MBE asymmetric quantum wells (QWs) having values of $\chi^{(2)}$ much larger than the corresponding bulk materials. Organic materials have also been used for the same purpose: molecular charge transfer excitations lead to a strong enhancement of SHG. From the theoretical point of view, scientists working independently with covalent or molecular crystals have exploited actually the same basic idea, i.e., achieving a large change of static dipole moment upon excitation. In the following chapters of this book we discuss the possibility of obtaining qualitatively new physical effects, potentially useful, as we expect, also for technological applications. In our considerations we follow another strategy by ingeniously combining organic with inorganic materials in one and the same hybrid structure.

In this chapter as a first example of such a strategy we discuss the properties of electronic excitations in nanostructures based on combinations of organic materials as well as inorganic semiconductors, having respectively Frenkel excitons and Wannier–Mott excitons with nearly equal energies. We show that in this case the resonant coupling between organic and inorganic quantum wells (or wires or dots) may lead to several interesting effects of linear optics, such as splitting of the excitonic spectrum and drastic change of exciton dispersion, but also to strong enhancement of the resonant optical nonlinearities.

The electronic excitations known as excitons correspond to a bound state of one electron and one hole and can be created by light or can appear as a result of relaxation processes of free electrons and holes, which, for example, may be injected electrically. As discussed in the first and second introductory chapters there are two models conventionally used to classify excitons – the small radius Frenkel exciton (FE) model and the large radius Wannier–Mott exciton (WE) model.

The internal structure of Wannier–Mott excitons [1] (see also the chapter by G. La Rocca) can be represented by hydrogen-like wave functions. Such a representation results from the two-particle, Coulombic electron–hole states in a crystalline periodic potential. The mean electron–hole distance for this type of exciton is typically large (in comparison with the lattice constant). On the other hand, the Frenkel exciton is represented as an electronic state of a crystal in which electrons and holes are placed on the same molecule. We can say that Frenkel excitons in organic crystals have radii $a_F$, comparable to the lattice constant $a_F \sim a \sim 5$ Å. In contrast, weakly bound Wannier excitons in semiconductor QWs have large Bohr radii ($a_B \sim 100$ Å in III–V materials and $a_B \sim 30$ Å in II–VI ones, in both cases $a_B \gg a$). The oscillator strength of a Frenkel exciton is close to a molecular oscillator strength $F$ and in some cases may be very large, whereas the oscillator strength $f$ of a Wannier exciton is usually much weaker: in a quantum well $f \sim \frac{Fa^3}{a_B^2 L}$ where $L$ is the QW width ($a_B > L > a$). Both types of excitons interact with lattice vibrations through exciton–phonon coupling.

In high quality semiconductors as well as in organic crystalline materials, the optical properties near and below the band gap are dominated by the exciton transitions and the same situation takes place also for organic and inorganic QWs (or wires or dots). The excitonic optical nonlinearities in semiconductor QWs can be large because the ideal bosonic approximation for Wannier excitons breaks down as soon as they start to overlap with each other, i.e., when their 2D density $n$ becomes comparable to the saturation density $n_S \sim 1/(\pi a_B^2)$ (due to large Bohr radius $n_S$ is rather small and is, typically, $10^{12}$ cm$^{-2}$). Then, due to phase space filling (PSF), exchange and collisional broadening, the exciton resonance is bleached. However, for the $\chi^{(3)}$ optical nonlinearities a generic figure of merit scales (see also below) like $I_P^{-1}(\Delta\chi/\chi)$ where $\Delta\chi$ is the nonlinear change in the susceptibility in the presence of the pump of intensity $I_P$. As $\Delta\chi/\chi \sim n/n_S$ and as $n/n_S \sim n a_B^2$ and $n \propto f I_P \propto a_B^{-2} I_P$, such a figure of merit is nearly independent of the exciton Bohr radius [2].

As for the Frenkel excitons in organic crystals, just because they have small radii, they have a very large saturation density. Thus, pronounced PSF nonlinearities of the exciton resonance in molecular crystals are practically impossible to achieve as very high excitonic concentrations are needed. Of course, other mechanisms may effectively enhance the optical nonlinearities of organic materials (the discussion of some of them can be found, for example, in the chapter by G. La Rocca of this book) but as a rule the resonance $\chi^{(3)}$ optical nonlinearity in organics, in contrast to the $\chi^{(2)}$ optical nonlinearity, is much smaller than in inorganic semiconductors. In hybrid structure which we consider in this chapter the optical nonlinearity of organics is neglected.

As mentioned above it was found in [2] that the figure of merit of resonance $\chi^{(3)}$ optical nonlinearity for semiconductor quantum well is independent of the exciton Bohr radius. Within the rather good approximation of neglecting of optical nonlinearities in organics we will show in this chapter that this independence does not hold for the organic-inorganic hybrid structures. On the contrary, we show that in hybrid structures the dependence of the resonance $\chi^{(3)}$ optical nonlinearity on exciton Bohr radius may be very strong.

In such a structure the Frenkel exciton in the organic material and the Wannier–Mott exciton in the semiconductor can be coupled through their dipole–dipole interaction at the interface. Due to this coupling one may expect (see also [3]) the formation of new eigenstates given by appropriate coherent linear combinations of large radius exciton states in the inorganic material and small radius exciton states in the organic one. These hybrid electronic excitations will be characterized by a radius dominated by their Wannier component and by an oscillator strength dominated by their Frenkel component contribution. Thus, they can have at the same time a small saturation density $n_S$ and a large oscillator strength $F$. In this way, the desirable properties of both the inorganic and organic material conspire to overcome the basic limitation mentioned above for the figure of merit of the exciton resonance nonlinearities.

One of the most natural choices to implement this idea is a layered structure with an interface between a covalent semiconductor and a crystalline molecular semiconductor. In such heterojunctions, there is obviously some cause for concern about the detrimental effects that lack of material purity and structural quality would have on the formation and the functional properties of the hybrid excitons. As we mentioned, the realistic possibility to consider such organic-inorganic crystalline structures has only recently appeared due to progress in the development of the organic molecular beam deposition (OMBD) and other related techniques.

This progress has led to a monolayer level control in the growth of organic thin films and superlattices with extremely high chemical purity and structural precision. This opens a wide range of possibilities for the creation of new types of ordered organic multilayer structures including highly ordered interfaces. It is well known that the requirement of lattice matching places strong restrictions on the materials which can be employed to produce high quality interfaces using inorganic semiconductor materials. This is due to the fact that they are bound by short-range covalent or ionic forces. On the contrary, organic materials are bound by weak Van der Waals forces. This fact lifts such restrictions and broadens the choice of materials that can be used to prepare organic crystalline layered structures with the required properties (for more details and many examples, see Ref. [4]). In the following sections, we will discuss in detail the electronic excitation spectra arising from the Frenkel–Wannier exciton hybridization in different geometrical configurations: quantum wells, quantum wires and quantum dots. At the same time, the nonlinear optical properties of hybrid excitons will be also considered in detail: we predict a strong dependence of the figure of merit of the excitonic resonance optical nonlinearity on the Bohr radius and its strong enhancement, in some cases by two orders of magnitude as compared to traditional systems. A few other results on the physics of hybrid excitons taken from the current literature will also be presented.

In this chapter, first, we discuss the properties of hybrid Frenkel–Wannier–Mott excitons in the case of strong coupling, which appear when the energy splitting of the excitonic spectrum is large as compared to the width of the exciton resonances. Just in this case the new and rather peculiar excitations share at the same time both the properties of the Wannier excitons (e.g., the large radius) and of the Frenkel excitons (e.g., the large oscillator strength). We discuss two-dimensional configurations (interfaces or coupled quantum wells) which are studied most extensively and also one-dimensional (quantum wires) and zero-dimensional (quantum dots) configurations. In particular, we show that hybrid excitons are expected to have resonant optical nonlinearities significantly enhanced with respect to traditional inorganic or organic systems. We also consider analogous phenomena in organic microcavities where the Frenkel exciton resonances are close to the cavity photon mode resonance. The experimental observation of different optical effects in such microcavities will be described in the chapter by David Lidzey. The case of the weak coupling regime is discussed in the chapter by Denis Basko.

## 2. Hybrid 2D Frenkel Wannier–Mott Excitons at the Interface of Organic and Inorganic Quantum Wells

### 2.1. CONFIGURATION OF HETEROSTRUCTURE AND GENERAL RELATIONS

Here we study the effects of resonant interaction between an organic quantum well (OQW) and an inorganic one (IQW) and demonstrate how new hybrid states arise [3]. The configuration we consider is the following. A plane semiconductor IQW of thickness $L_w$ occupies the region $|z| < L_w/2$, the $z$-axis being chosen along the growth direction. All the space with $z > L_w/2$ is filled by the barrier material and that with $z < -L_w/2$ by the organic material in which the OQW is placed (Figure 1).

For simplicity, we treat the interaction of excitations IQW with organic molecules in the dipole approximation, neglecting the contribution of higher multipoles to the interaction, and we consider the OQW as a single monolayer, i.e., as a 2D lattice of molecules at discrete sites $\vec{n}$, placed at $z = -z_0 < -L_w/2$ (the

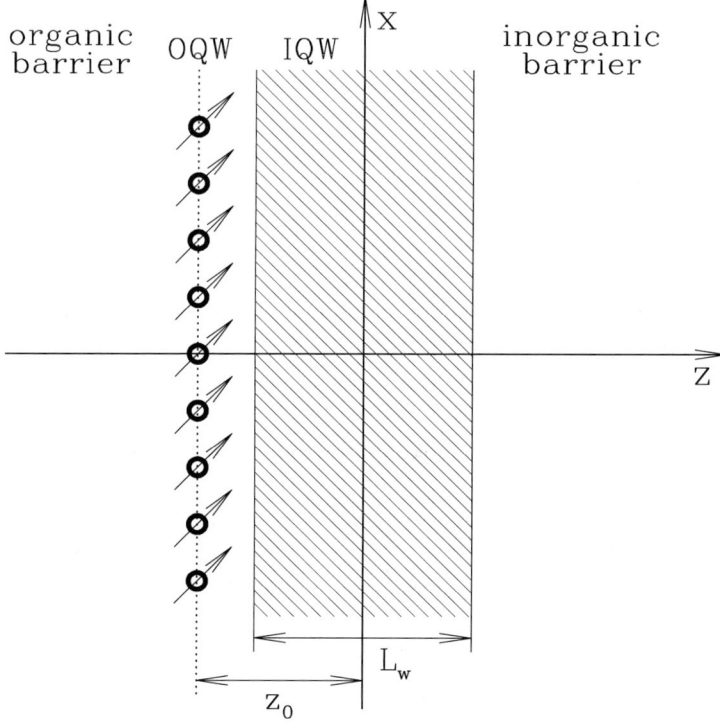

Fig. 1. The physical configuration under study.

generalization to the case of several monolayers is easy). All the semiconductor well-barrier structure ($z > -L_w/2$) is assumed to have the same background dielectric constant $\varepsilon$, while the organic half-space ($z < -L_w/2$) has the dielectric constant $\tilde{\varepsilon}$ (corresponding to the organic substrate). For example, the role of the OQW can be played by the outermost monolayer of the organic crystal. Due to gas-condensed matter shift (see the chapters by J. Knoester and V.M. Agranovich and by V.M. Agranovich and A.M. Kamchatnov) the exciton transition in such a monolayer can be blue-shifted with respect to the bulk excitonic transition (as, for instance, this takes place for anthracene crystals), thus giving rise to 2D Frenkel excitonic states [17]. In this case the difference $L_w/2 - z_0$ is of the order of the organic crystal lattice constant.

Due to the different electronic structure of the two QWs under consideration and the rather large organic crystal lattice constant, the OQW and the IQW states are assumed to have zero wave function overlap. It is known that this is a rather good approximation for organic crystals in the bulk for the ground and for the lowest energy excited states. Thus, we assume that the same takes place also at the interface between organic and inorganic QWs. Assuming perfect 2D translational invariance of the system, we classify the excitons by their in-plane wave vector $\vec{k}$. Supposing that for some bands of Frenkel excitons in the OQW and Wannier–Mott excitons in the IQW the energy separation is much less than the distance to other exciton bands we take into account only the hybridization between these two bands. We choose as a basis set the "pure" Frenkel and Wannier states, i.e., the state (denoted by $|F, \vec{k}\rangle$) when the OQW is excited, while the IQW is in its ground state, and vice versa (denoted by $|W, \vec{k}\rangle$), their energies being $E_F(\vec{k})$ and $E_W(\vec{k})$. We seek the new hybrid states in the form

$$|\alpha, \vec{k}\rangle = A_\alpha(\vec{k})|F, \vec{k}\rangle + B_\alpha(\vec{k})|W, \vec{k}\rangle, \quad (1)$$

where $\alpha =$ "$u$", "$l$" labels the two resulting states (upper and lower branches). The Schrödinger equation for the coefficients $A$, $B$ is then written as:

$$(E_F(\vec{k}) - E)A(\vec{k}) + \langle F, \vec{k}|\hat{H}_{\text{int}}|W, \vec{k}\rangle B(\vec{k}) = 0,$$
$$\langle W, \vec{k}|\hat{H}_{\text{int}}|F, \vec{k}\rangle A(\vec{k}) + (E_W(\vec{k}) - E)B(\vec{k}) = 0, \quad (2)$$

where $\hat{H}_{\text{int}}$ is the Hamiltonian of the interaction between the QWs. Solution of (2) gives the energies of the upper and lower branches and the splitting $\Delta(\vec{k})$:

$$E_{u,l}(\vec{k}) = \frac{E_F(\vec{k}) + E_W(\vec{k}) \pm \Delta(\vec{k})}{2},$$

$$\Delta(\vec{k}) \equiv \sqrt{(E_F(\vec{k}) - E_W(\vec{k}))^2 + 4\Gamma^2(\vec{k})}, \quad (3)$$

where we use the notation $\Gamma(\vec{k}) \equiv |\langle W, \vec{k}|\hat{H}_{\text{int}}|F, \vec{k}\rangle|$ for the coupling matrix element. For the orthonormalized new states the weighting coefficients are given by

$$|A_u(\vec{k})|^2 = |B_l(\vec{k})|^2 = \frac{1}{2}\left(1 + \frac{E_F(\vec{k}) - E_W(\vec{k})}{\Delta(\vec{k})}\right), \quad (4)$$

$$|A_l(\vec{k})|^2 = |B_u(\vec{k})|^2 = \frac{1}{2}\left(1 - \frac{E_F(\vec{k}) - E_W(\vec{k})}{\Delta(\vec{k})}\right). \quad (5)$$

## 2.2. THE COUPLING MATRIX ELEMENT

To evaluate the matrix element $\Gamma(\vec{k})$ determining the resonance interaction between Frenkel and Wannier–Mott excitons we write down the interaction Hamiltonian as

$$\hat{H}_{\text{int}} = -\sum_{\vec{n}} \hat{\vec{p}}^F(\vec{n}) \cdot \hat{\vec{\mathcal{E}}}(\vec{n}), \quad (6)$$

where $\hat{\vec{p}}^F(\vec{n})$ is the operator of the dipole moment of the organic molecule situated at the lattice site $\vec{n}$, and $\hat{\vec{\mathcal{E}}}(\vec{n})$ is the operator of the electric field at the point $\vec{n}$, produced by the IQW exciton. If we introduce the operator of the IQW polarization $\hat{\vec{P}}^W(\vec{r})$, then the operators $\hat{\vec{\mathcal{E}}}(\vec{n})$ and $\hat{\vec{P}}^W(\vec{r})$ are related to each other exactly in the same way as the corresponding classical quantities in electrostatics:

$$\hat{\mathcal{E}}_i(\vec{r}) = \int d^3\vec{r}' \mathcal{D}_{ij}(\vec{r}_\| - \vec{r}'_\|, z, z') \hat{P}^W_j(\vec{r}'), \quad (7)$$

where $i, j = x, y, z$, $\vec{r}_\| \equiv (x, y)$ and $\mathcal{D}_{ij}(\vec{r}, \vec{r}')$ is the Green's function appearing in the analogous problem of classical electrostatics. It is equal to the $i$th Cartesian component of the classical static electric field at the point $\vec{r}$, produced by the $j$th component of the classical point dipole, situated at the point $\vec{r}'$ and is connected to the Green's function $G$ of the Poisson equation in an inhomogeneous medium with the dielectric constant $\varepsilon_{ij}(\vec{r})$:

$$\mathcal{D}_{ij}(\vec{r}, \vec{r}') = -\frac{\partial}{\partial x_i} \frac{\partial}{\partial x'_j} G(\vec{r}, \vec{r}'), \quad (8)$$

$$\frac{\partial}{\partial x_i} \varepsilon_{ij}(\vec{r}) \frac{\partial}{\partial x_j} G(\vec{r}, \vec{r}') = -4\pi \delta(\vec{r} - \vec{r}'). \quad (9)$$

Since our system is translationally invariant in two dimensions, it is convenient to consider the Fourier transform:

$$\mathcal{D}_{ij}(\vec{r}_\| - \vec{r}'_\|, z, z') = \int \frac{d^2\vec{k}}{(2\pi)^2} \mathcal{D}_{ij}(\vec{k}, z, z') e^{i\vec{k}(\vec{r}_\| - \vec{r}'_\|)}, \quad (10)$$

and analogously for $G(\vec{r}_\| - \vec{r}'_\|, z, z')$. Then $G(\vec{k}, z, z')e^{i\vec{k}\vec{r}_\|}$ is the potential, produced by a charge density wave $\rho(\vec{r}) = \delta(z - z') e^{i\vec{k}\vec{r}_\|}$. In our case the dielectric constant is a simple step function

$$\varepsilon_{ij}(\vec{r}) = \begin{cases} \tilde{\varepsilon}\,\delta_{ij}, & z < -L_w/2, \\ \varepsilon\,\delta_{ij}, & z > -L_w/2, \end{cases} \quad (11)$$

and the potential may be readily found from Poisson's equation

$$\left(\frac{d^2}{dz^2} - k^2\right) G(\vec{k}, z, z') = -\frac{4\pi\,\delta(z - z')}{\varepsilon(z)}, \quad (12)$$

with the usual electrostatic boundary conditions at the interface $z = -L_w/2$ (continuity of the tangential component of the electric field $-i\vec{k}G$ and the normal component of the electric displacement $-\varepsilon(z)\partial G/\partial z$. The Green's function $\mathcal{D}_{ij}$ for $z < -L_w/2$, $z' > -L_w/2$ is then given by:

$$\mathcal{D}_{ij}(\vec{k}, z, z') = \frac{4\pi}{\varepsilon + \tilde{\varepsilon}} k e^{k(z-z')} \left(\frac{ik_i}{k} + \delta_{i,z}\right)\left(\frac{ik_j}{k} + \delta_{j,z}\right). \quad (13)$$

Thus, the matrix element of $\hat{H}_{\text{int}}$ we are interested in can be written as

$$\langle F, \vec{k} | \hat{H}_{\text{int}} | W, \vec{k} \rangle$$
$$= -\sum_{\vec{n}} \int d^3\vec{r} \langle F, \vec{k} | \hat{p}_i(\vec{n}) | 0 \rangle \mathcal{D}_{ij}(\vec{n} - \vec{r}_\|, -z_0, z) \langle 0 | \hat{P}_j^W(\vec{r}) | W, \vec{k} \rangle. \quad (14)$$

The matrix element of the IQW polarization between the ground state $|0\rangle$ and $|W, \vec{k}\rangle$ for 1s-exciton with the Bohr radius $a_B$ is equal to [5,6]

$$\langle 0 | \hat{P}^W(\vec{r}) | W, \vec{k} \rangle = \sqrt{\frac{2}{\pi a_B^2}} \frac{\vec{d}^{vc} e^{i\vec{k}\vec{r}_\|}}{\sqrt{S}} \chi^e(z) \chi^h(z), \quad (15)$$

where $\sqrt{2/(\pi a_B^2)}$ is the value of the 1s-wave function of the relative motion of the electron and hole, taken at $r_\| = 0$; $\chi^e(z)$, $\chi^h(z)$ are the envelope functions for the electron and hole in the IQW confinement potential (we assume the IQW to be thin, so that the transverse and the relative in-plane motion of the electron and hole are decoupled) and $S$ is the in-plane normalization area. Finally,

$$\vec{d}^{vc} = \int_{u.c.} u_v^*(\vec{r})(-e\vec{r}) u_c(\vec{r}) d^3\vec{r} \quad (16)$$

is the matrix element of the electric dipole moment between the conduction and valence bands ($\vec{d}^{vc}$ is taken to be independent of $\vec{k}$, $u_{c(v)}$ are the Bloch functions for the conduction (valence) band extremum and the integration in (16) is performed over the unit cell). Its Cartesian components $d_i^{vc}$ ($i = x, y, z$) may be expressed in terms of the Kane's energy $E_0$ [5]:

$$|d_i^{vc}|^2 = \frac{e^2\hbar^2 E_0 c_i^2}{2m_0 E_g^2}, \tag{17}$$

where $m_0$ is the free electron mass, $E_g$ is the energy gap between the conduction and valence bands and $c_i$ is the appropriate symmetry coefficient. In semiconductors of the zinc-blende structure $c_x^{hh} = c_y^{hh} = 1/\sqrt{2}$, $c_z^{hh} = 0$ (heavy holes) and $c_x^{lh} = c_y^{lh} = 1/\sqrt{6}$, $c_z^{lh} = \sqrt{2/3}$ (light holes). We see that only light holes can contribute to the $z$-component of the IQW polarization. For the Frenkel exciton the dipole moment matrix element, contributing to the matrix element (14), is given by

$$\langle F, \vec{k}|\hat{p}_i(\vec{n})|0\rangle = \frac{e^{-i\vec{k}\vec{n}}}{\sqrt{N}} \vec{d}^{F*} = \frac{e^{-i\vec{k}\vec{n}}}{\sqrt{S}} a_F \vec{d}^{F*}, \tag{18}$$

where $\vec{d}^F$ is the transition dipole moment for a single organic molecule (analogous to $\vec{d}^{vc}$ in the semiconductor), $N$ is the total number of sites in the lattice, $a_F$ is the lattice constant, which may be considered as the radius of the Frenkel exciton.

Now we can write the final expression for the coupling matrix element:

$$\langle F, \vec{k}|\hat{H}_{int}|W, \vec{k}\rangle = -\sqrt{\frac{2}{\pi}} \frac{d_i^{F*}}{a_F} \frac{d_j^{vc}}{a_B} \int dz\, \mathcal{D}_{ij}(\vec{k}, -z_0, z)\chi^e(z)\chi^h(z). \tag{19}$$

From Eqs. (13), (19) we see that the only contributing polarizations for the semiconductor are those along $\vec{k}$ (L-modes) and along the growth direction $z$ (Z-modes, only for the light holes, according to Eq. (17)). For simplicity we take the electron and hole confinement wave functions for the lowest subbands in the approximation of an infinitely deep IQW:

$$\chi^e(z)\chi^h(z) = \frac{2}{L_w}\cos^2\left(\frac{\pi z}{L_w}\right), \tag{20}$$

and assume the transition dipole moment in the organics $\vec{d}^F$ to be real (which is always possible with an appropriate choice of molecular wave functions). Without loss of generality we may take the vector $\vec{k}$ along the $x$ axis. Evaluating the integral in (19), we obtain the interaction parameter $\Gamma_{L,Z}$ for the L- and Z-modes:

$$\Gamma_{L(Z)}(k) = \frac{8\sqrt{2\pi}}{\varepsilon + \tilde{\varepsilon}} \frac{e^{-kz_0}\sinh(kL_w/2)}{1 + (\frac{kL_w}{2\pi})^2} \frac{|d_{x(z)}^{vc}|\sqrt{(d_x^F)^2 + (d_z^F)^2}}{a_F a_B L_w}. \tag{21}$$

It is seen that $\Gamma(k)$ has a maximum $\Gamma_{max}$ at $k = k_{max}$. The value of $k_{max}$ for arbitrary $z_0$ and $L_w$ may be found numerically, for $z_0 - L_w/2 > 0.1 L_w$ it is well

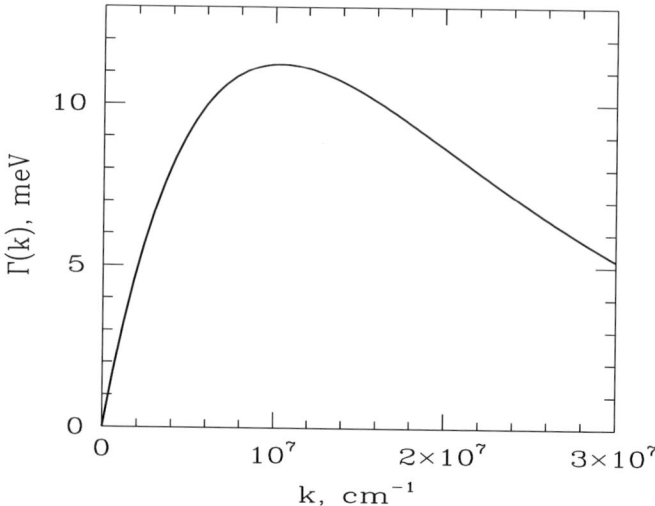

Fig. 2. The interaction parameter $\Gamma(k)$ for $d^{vc} = 12$ Debye, $d^F = 5$ Debye, $a_B = 25$ Å, $a_F = 5$ Å, $L_w = 10$ Å, $z_0 = 10$ Å, $\varepsilon_\infty = 6$, $\tilde{\varepsilon}_\infty = 4$.

described by the formula (see also Figure 2)

$$k_{max} \simeq \frac{1}{L_w} \ln\left(\frac{2z_0 + L_w}{2z_0 - L_w}\right), \quad (22)$$

while in the limit $z_0 \simeq L_w/2$ we have $k_{max} \simeq 2.4/L_w$.

### 2.3. Dispersion Relations of Hybrid States

To calculate the dispersion relation of the hybrid excitons we approximate the WE energy by a parabola with the in-plane effective mass $m_W = m_e + m_h$, $m_{e(h)}$ being the electron (hole) mass, and neglect the FE dispersion since the typical masses are $(5–100)m_0$:

$$E_W(\vec{k}) = E_W(0) + \frac{\hbar^2 k^2}{2m_W}, \quad E_F(\vec{k}) = E_F(0), \quad E_F(0) - E_W(0) \equiv \delta. \quad (23)$$

We will measure all energies with respect to $E_W(0)$. The dispersion of the hybrid states (3) can be written as

$$E_{u,l}(\vec{k}) - E_W(0) = \frac{\delta}{2} + \frac{\hbar^2 k^2}{4m_W} \pm \sqrt{\left(\frac{\delta}{2} - \frac{\hbar^2 k^2}{4m_W}\right)^2 + \Gamma^2(\vec{k})}. \quad (24)$$

To perform numerical estimates we choose the following values of the parameters. For the IQW those representative of II–VI semiconductor (e.g., ZnSe/ZnCdSe)

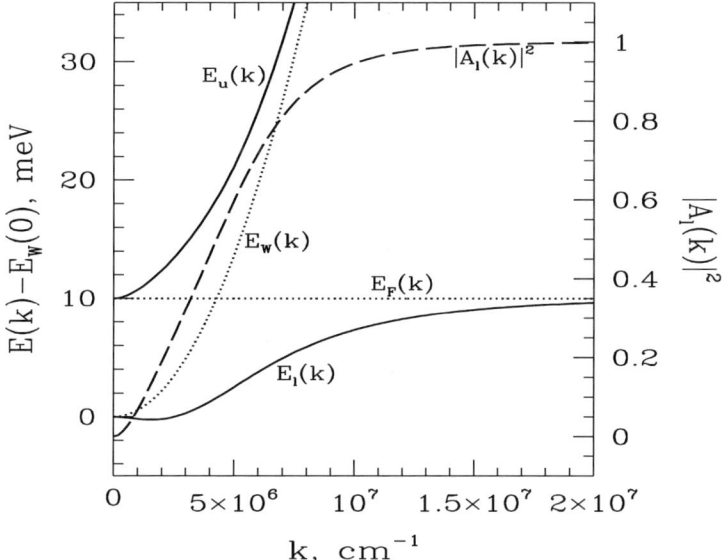

Fig. 3. The dispersion $E_{u,l}(k)$ of the upper and lower hybrid exciton branches (solid lines) and that of the unperturbed Frenkel and Wannier excitons (dotted lines). The "weight" of the FE component in the lower branch $|A_l(k)|^2$ is shown by the dashed line. The parameters are the same as on the Figure 2 ($m_W = 0.7 m_0$), the detuning $\delta = 10$ meV.

quantum wells are taken [7]: $\varepsilon = \varepsilon_\infty = 6$, $d^{vc}/a_B \approx 0.1\,e$ (which corresponds to $d^{vc} \simeq 12$ Debye and a Bohr radius of 25 Å), the exciton mass $m_W = 0.7 m_0$ and the well width $L_w = 10$ Å. For the organic part of the structure, we take parameters typical for such media (e.g., see [4,8,9]): $\tilde{\varepsilon} = \tilde{\varepsilon}_\infty = 4$, the transition dipole for the molecules in the monolayer $d^F = 5$ Debye, $a_F = 5$ Å and $z_0 = 10$ Å. We plot $\Gamma(k)$ for these values of parameters on Figure 2.

We see that $\Gamma_{\max} \simeq 11$ meV. The dispersion curves $E_{u,l}(k)$ along with the FE weight in the lower branch $|A_l(k)|^2$ for three different detunings $\delta = 10$ meV, $\delta = 0$ and $\delta = -10$ meV are plotted on Figures 3–5.

For $\delta > 0$ the properties of the excited states are changed drastically. In this case the zero approximation dispersion curves for FE and WE cross at the point $k = k_0 = \sqrt{2m_W \delta/\hbar^2}$. At $k = 0$ the upper states are purely F-like and the lower states W-like, at $k \sim k_0$ they are strongly mixed and a large splitting of their dispersion curves is present, $\Delta(k_0) \sim 2\Gamma(k_0)$, and for large $k$ ($k \gg k_0$) they "interchange": the upper branch becomes W-like with the quadratic dispersion and excitations of the lower branch tend to FE. If $\delta < 0$ then $E_W(k) > E_F(k)$ for all $k$ and no crossing occurs, $E_u(k)$ closely follows the WE dispersion and $|A_u(k)|^2 \ll 1$, the lower state is FE-like.

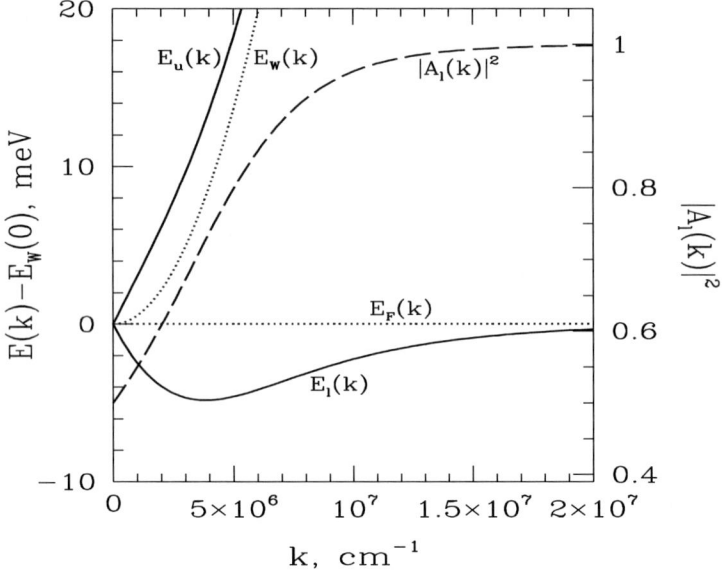

Fig. 4. The same as on Figure 3, but $\delta = 0$.

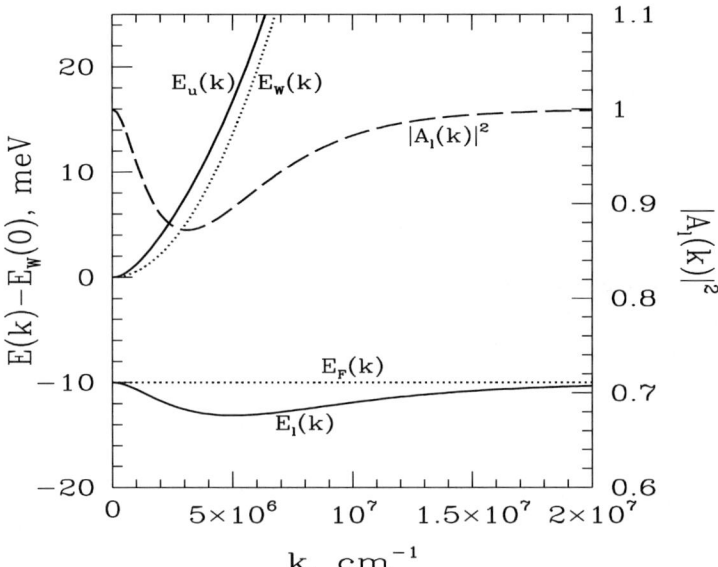

Fig. 5. The same as on Figure 3, but $\delta = -10$ meV.

A nontrivial feature of the lower branch dispersion is a minimum away from $k = 0$, which is always present for $\delta \leq 0$ as well as for some positive values of $\delta$, $0 < \delta < \delta_{cr}$ and is the deepest for $\delta = 0$. The critical value of $\delta$ may be found if one looks at the values of the derivatives of $E_l(k)$ at $k = 0$. It turns out that

$$\frac{dE_l(0)}{dk} = 0 \; (\delta \neq 0), \qquad \frac{d^2 E_l(0)}{dk^2} < 0 \; (\delta < \delta_{cr}),$$

$$\frac{d^2 E_l(0)}{dk^2} > 0 \; (\delta > \delta_{cr}), \tag{25}$$

and $\delta_{cr}$ when the minimum "splits" off $k = 0$ is given by

$$\delta_{cr} = \left(\frac{d\Gamma(0)}{dk}\right)^2 \frac{2m_W}{\hbar^2}. \tag{26}$$

For our parameters $\delta_{cr} \simeq 16$ meV. For large negative values of $\delta \ll -\Gamma_{max}$ the lower branch dispersion at $k \ll \sqrt{2m_W \delta/\hbar^2}$ may be approximated by

$$E_l(k) - E_W(0) \simeq -|\delta| - \frac{\Gamma^2(k)}{|\delta|}. \tag{27}$$

So, the depth of the minimum for large $|\delta|$ is $\Gamma_{max}^2/|\delta|$ while for small $\delta$ it is of the order of $\Gamma_{max}$ and we see that the effective range of $\delta$, when the minimum is the most pronounced, is $-\Gamma_{max} \lesssim \delta < \delta_{cr}$. As a consequence, at low temperatures and under optical pumping at frequencies above the excitonic resonance excitons will accumulate in this minimum, which can be detected, for example, by pump-probe experiments. The fluorescence from these states should increase with temperature since states with small $k$ become populated.

## 2.4. LINEAR OPTICAL RESPONSE OF HYBRID STATES

If an incident electromagnetic wave with the electric field $\vec{\mathcal{E}}(\vec{r}) = \vec{\mathcal{E}}_0 e^{i\vec{Q}\vec{r}}$ is present, then the interaction with the hybrid structure is described by the Hamiltonian (neglecting local field corrections)

$$\hat{H}_{em} = -\vec{\mathcal{E}}_0 \cdot \left( \sum_{\vec{n}} \hat{\vec{p}}^F(\vec{n}) e^{i\vec{Q}_\parallel \vec{n}} + \int dz \int d^2\vec{r}_\parallel \, \hat{\vec{P}}^W(\vec{r}) e^{i\vec{Q}_\parallel \vec{r}_\parallel} \right), \tag{28}$$

where we have neglected the $z$-dependence of the incident field since the thickness of our structure is much less than the light wavelength. The corresponding matrix element is different from zero only if $\vec{k} = \vec{Q}_\parallel$ and in this case equal to

$$\langle \alpha, \vec{k} | \hat{H}_{em} | 0 \rangle \equiv -\vec{\mathcal{E}}_0 \cdot \vec{M}_{\vec{k}}^\alpha = -\vec{\mathcal{E}}_0 \cdot \left( A_\alpha^*(\vec{k}) \vec{M}^F + B_\alpha^*(\vec{k}) \vec{M}_{\vec{k}}^W \right), \tag{29}$$

where

$$\vec{M}^F = \sqrt{N}\,\vec{d}^{F*} = \frac{\sqrt{S}}{a_F}\,\vec{d}^{F*}, \qquad (30)$$

$$\vec{M}^W_{\vec{k}} = \sqrt{\frac{2}{\pi}}\,\frac{\sqrt{S}}{a_B}\,\vec{d}^{vc*}\int \chi^{e*}(z)\,\chi^{h*}(z)\,dz \qquad (31)$$

are the optical matrix elements for the isolated OQW and IQW respectively, which are independent of $\vec{k}$. Usually we have $M^F \gg M^W$ since $a_F \ll a_B$, and in the region of strong mixing the oscillator strengths $f^\alpha$ of a hybrid state is determined by its FE component:

$$f^\alpha(\vec{k}) \simeq |A_\alpha(\vec{k})|^2 f^F. \qquad (32)$$

At the crossing point $k = k_0$ (for $\delta > 0$) we have $|A_\alpha(k_0)|^2 = 1/2$ and the FE oscillator strength is equally distributed between the two hybrid states. For the hybrid exciton radii the opposite relation holds. Calculating the expectation value of the exciton radius squared $\hat{r}^2$ in the state $|\alpha, \vec{k}\rangle$ we obtain

$$\langle \alpha, \vec{k}|\hat{r}^2|\alpha, \vec{k}\rangle = |A_\alpha(\vec{k})|^2 \langle F, \vec{k}|\hat{r}^2|F, \vec{k}\rangle + |B_\alpha(\vec{k})|^2 \langle W, \vec{k}|\hat{r}^2|W, \vec{k}\rangle$$
$$\simeq |B_\alpha(\vec{k})|^2 a_B^2, \qquad (33)$$

since $a_B \gg a_F$ and we neglect the latter. Cross terms do not appear since we neglect the single-particle wave function mixing between the two QWs. We see that the new states can possess both large oscillator strengths and exciton radii. This effect is especially pronounced if the crossing of the FE and WE dispersion curves occurs for a value of the wave vector close to that of the maximum of the coupling strength: $k_0 \simeq k_{\max}$. Since $k_0$ is determined by the detuning $\delta$, and $k_{\max}$, in turn, depends on $L_w$ and $z_0$ (Eq. (22)), a special choice of these parameters should be made for maximizing the effect. Also, in order to take advantage of the hybrid states in optics, the wave vector of light in the medium $q = n_\infty \omega/c$ ($n_\infty$ being the background refraction index) should not be far from $k_0$. Usually near excitonic resonances, $q < k_0$ and special care should be taken to overcome this difficulty (e.g., using a coupled diffraction grating with period $2\pi/k_0$ [10] or a prism). We mention, however, that even in the region of small wave vectors in which the 2D excitons are radiative, the hybridization may be realized not due to the instantaneous dipole–dipole interaction, but due to the retarded interaction stemming from the exchange of photons. Such a situation has been analysed (even in the nonlinear regime) with an appropriate transfer matrix approach, which is equivalent to the solution of the full Maxwell equations [11].

Concerning the choice of materials for the implementation of the system considered here, examples of molecular substances having small radius ($\leqslant 5$ Å) excitons with energies of a few eV, among those already successfully grown [4] as crystalline layers on a variety of inorganic (including semiconductor) crystals, are the acenes, such as tetracene (2 eV) or pentacene (1.5 eV), the metal phtalocyanines, such as VOPc (1.6 eV) or CuPc (1.8 eV), and the tetracarboxilic compounds, such as NTCDA (3.1 eV) or PTCDA (2.2 eV). Semiconductors having large radius excitons with matching energies are, for instance, the III–V and II–VI ternary solid solutions such as GaAlAs, ZnCdSe and ZnSe [12]; beside a judicious choice of alloy composition and well thickness, a fine tuning of the resonance condition could be achieved applying an external static electric field along the growth direction (quantum confined Stark effect [13]; for hybrid excitons it has been considered in Ref. [14]). A major experimental problem is the control of the interface quality: the inhomogeneous broadening should remain small and the in-plane wavevector $\vec{k}$ a (sufficiently) good quantum number; organic superlattices with high quality interfaces have been demonstrated [4].

The necessary condition for the hybrid states to be observable is that the exciton linewidths must be smaller then the splitting $\Delta(k)$. This is the case in the present calculations, where for $k_0 = k_{max}$ we have $\Delta(k_0) = 2\Gamma_{max} \simeq 20$ meV, while in inorganic QWs the homogeneous linewidth at low temperatures is $\sim 1$ meV [15, 16]. The nonradiative linewidth of a 2D Frenkel exciton in an OQW can also be small: in the case of a 2D-exciton in the outermost monolayer of anthracene this linewidth at low temperatures is $\sim 2$ meV [17]. In principle, apart from the resonance condition and the large difference in excitonic radii, the present model demands no specific requisite and the rapid progress in the growth of organic crystalline multilayers justifies some optimism about its concrete realization.

We also mention here the work [18], where the effects of the exciton–phonon interaction in hybrid systems were studied. In this work the resonant Raman spectroscopy is also suggested as a tool for studying hybrid organic-inorganic QWs.

## 3. Nonlinear Optics of 2D Hybrid Frenkel–Wannier–Mott Excitons

### 3.1. THE RESONANT $\chi^{(3)}$ NONLINEARITY

From the results of the previous subsection we may expect that the exciton hybridization should strongly modify the nonlinear optical properties of the structure under consideration. Indeed, hybrid excitons can combine both a large oscillator strength, which makes it easy to produce large populations, and a large radius, which, in turn, leads to low saturation densities. In this subsection we analyze the situation quantitatively [19], calculating the response of the interband polarization $\vec{P} = \vec{P}^W + \vec{P}^F$ on the external driving electric field (corresponding to a cw experiment)

$$\vec{\mathcal{E}}(\vec{r}, t) = \vec{\mathcal{E}}_0 e^{i\vec{Q}\vec{r} - i\omega t} + \text{c.c.} \tag{34}$$

in the presence of a large density of excitations using the standard technique of semiconductor Bloch equations [13,20]. Since we are considering a cw experiment, the populations are stationary and may be treated as parameters in the equation for the time-dependent interband polarization.

First, we express the operator of the electron–hole interband polarization $\hat{\vec{P}}^W(\vec{r})$ in terms of the electron and hole creation and annihilation operators in the envelope function approximation, following the standard procedure [5,20]:

$$\hat{\vec{P}}^W(\vec{r}) = \frac{\vec{d}^{vc}}{S} \chi^e(z) \chi^h(z) \sum_{\vec{k},\vec{q}} e^{i\vec{k}\vec{r}_\parallel} \hat{h}_{-\vec{q}} \hat{c}_{\vec{k}+\vec{q}} + \text{h.c.} \tag{35}$$

Here $\chi^e(z)$, $\chi^h(z)$ are electron and hole wave functions in the given IQW subbands (resonant with the FE), $\hat{c}_{\vec{k}}$ and $\hat{h}_{\vec{k}}$ are annihilation operators for an electron and hole with the in-plane wave vector $\vec{k}$ in the subbands under consideration, $S$ is the in-plane normalization area and $\vec{d}^{vc}$ is the matrix element (16). We do not take into account the spin degeneracy, considering thus the polarization produced by electrons and holes with a given spin (thus, the final expression for the susceptibility should be multiplied by two). An analogous expression for the OQW polarization is

$$\hat{\vec{P}}^F(\vec{r}) = \frac{\vec{d}^F}{a_F \sqrt{S}} \delta(z+z_0) \sum_{\vec{k}} e^{i\vec{k}\vec{r}_\parallel} \hat{B}_{\vec{k}} + \text{h.c.}, \tag{36}$$

where $\hat{B}_{\vec{k}}$ is the annihilation operator for the Frenkel exciton, which is assumed to be tightly bound. Besides the term of the Hamiltonian describing free Frenkel excitons and free electron–hole pairs (with the single-particle energies $E_F(\vec{k})$, $\varepsilon_e(\vec{k})$ and $\varepsilon_h(\vec{k})$, respectively) the Hamiltonian we consider here includes the following.

(i) The Coulomb interaction between electrons and holes

$$\hat{H}_{\text{coul}} = \frac{1}{2S} \sum_{\vec{q} \neq 0} v(\vec{q})$$

$$\times \sum_{\vec{k},\vec{k}'} (\hat{c}^\dagger_{\vec{k}+\vec{q}} \hat{c}^\dagger_{\vec{k}'-\vec{q}} \hat{c}_{\vec{k}'} \hat{c}_{\vec{k}} + \hat{h}^\dagger_{\vec{k}+\vec{q}} \hat{h}^\dagger_{\vec{k}'-\vec{q}} \hat{h}_{\vec{k}'} \hat{h}_{\vec{k}} - 2\hat{c}^\dagger_{\vec{k}+\vec{q}} \hat{h}^\dagger_{\vec{k}'-\vec{q}} \hat{h}_{\vec{k}'} \hat{c}_{\vec{k}}), \tag{37}$$

$$v(\vec{q}) = \frac{2\pi e^2}{\varepsilon_0 q}, \tag{38}$$

$\varepsilon_0$ being the static dielectric constant of the IQW.

(ii) The dipole–dipole interaction between the QWs, as follows from Eqs. (6), (7)

$$\hat{H}_{\text{hyb}} = \sum_{\vec{k}} V_{\text{hyb}}(\vec{k}) \hat{B}^\dagger_{\vec{k}} \sum_{\vec{q}} \hat{h}_{-\vec{q}} \hat{c}_{\vec{k}+\vec{q}} + \text{h.c.}, \tag{39}$$

$$V_{\text{hyb}}(\vec{k}) = -\frac{d_i^{F*} d_j^{vc}}{a_F \sqrt{S}} \int dz \, \mathcal{D}_{ij}(\vec{k}, -z_0, z) \, \chi^e(z) \, \chi^h(z), \tag{40}$$

which corresponds to (19) with $\sqrt{2/(\pi a_B^2)}$ is replaced by $1/\sqrt{S}$ since we use plane waves as the basis for the semiconductor states. Of course, this interaction is also of Coulomb nature, but since we treat the OQW and the IQW as completely different systems and neglect all effects of electronic exchange between them, these pieces of the Hamiltonian come separately.

(iii) The interaction with the driving electric field (34)

$$\hat{H}_{\text{dr}} = -(\vec{\mathcal{E}}_0 \cdot \vec{M}^F) e^{-i\omega t} \hat{B}_{\vec{Q}_\|}^\dagger - (\vec{\mathcal{E}}_0 \cdot \vec{M}^{eh}) e^{-i\omega t} \sum_{\vec{q}} \hat{c}_{\vec{Q}_\| + \vec{q}}^\dagger \hat{h}_{-\vec{q}}^\dagger + \text{h.c.}, \tag{41}$$

$$\vec{M}^{eh} = \vec{d}^{vc*} \int \chi^{e*}(z) \, \chi^{h*}(z) \, dz, \qquad \vec{M}^F = \frac{\sqrt{S}}{a_F} \vec{d}^{F*}, \tag{42}$$

where we again neglect the $z$-dependence of the field and the wave vector dependence of $\vec{M}^{eh}$.

Given the Hamiltonian, we can write the equations of motion for the Heisenberg operators. The polarization is obtained by averaging the expressions (35), (36) over the equilibrium density matrix. The result is expressed in terms of the polarization functions

$$\langle \hat{h}_{-\vec{q}}(t) \, \hat{c}_{\vec{k}+\vec{q}}(t) \rangle = \mathcal{P}_{\vec{k}}^W(\vec{q}), \qquad \langle \hat{B}_{\vec{k}}(t) \rangle = \mathcal{P}_{\vec{k}}^F. \tag{43}$$

Average values of the four-operator terms are factorized in the Hartree–Fock approximation and are expressed in terms of the polarization functions and the populations defined by

$$\langle \hat{c}_{\vec{q}}^\dagger(t) \, \hat{c}_{\vec{q}'}(t) \rangle = \delta_{\vec{q}\vec{q}'} \, n_{\vec{q}}^e, \qquad \langle \hat{h}_{\vec{q}}^\dagger(t) \, \hat{h}_{\vec{q}'}(t) \rangle = \delta_{\vec{q}\vec{q}'} \, n_{\vec{q}}^h. \tag{44}$$

Here the averages with different wave vectors correspond to the intraband polarization, which is far off resonance and may be neglected. Since the electric field excites only states with the given total in-plane wave vector $\vec{Q}_\|$, from now on we set $\vec{k} = \vec{Q}_\|$. As a result, we obtain the equations for the polarization functions:

$$i\hbar \frac{d\mathcal{P}_{\vec{k}}^F}{dt} = E_F(\vec{k}) \mathcal{P}_{\vec{k}}^F + V_{\text{hyb}}(\vec{k}) \sum_{\vec{q}} \mathcal{P}_{\vec{k}}^W(\vec{q}) - (\vec{\mathcal{E}}_0 \cdot \vec{M}^F) e^{-i\omega t}, \tag{45}$$

$$i\hbar \frac{d\mathcal{P}_{\vec{k}}^W(\vec{q})}{dt} = \hat{\mathcal{H}}_0 \mathcal{P}_{\vec{k}}^W(\vec{q}) + \hat{\mathcal{H}}_1 \mathcal{P}_{\vec{k}}^W(\vec{q})$$
$$+ (1 - n_{\vec{k}+\vec{q}}^e - n_{-\vec{q}}^h) [V_{\text{hyb}}^*(\vec{k}) \mathcal{P}_{\vec{k}}^F - (\vec{\mathcal{E}}_0 \cdot \vec{M}^{eh}) e^{-i\omega t}], \tag{46}$$

$$\hat{\mathcal{H}}_0 \mathcal{P}_{\vec{k}}^W(\vec{q}) \equiv [\varepsilon_e(\vec{k}+\vec{q}) + \varepsilon_h(-\vec{q})]\mathcal{P}_{\vec{k}}^W(\vec{q}) - \sum_{\vec{q}'} \frac{v(\vec{q}-\vec{q}')}{S}\mathcal{P}_{\vec{k}}^W(\vec{q}'),$$

$$\hat{\mathcal{H}}_1 \mathcal{P}_{\vec{k}}^W(\vec{q}) \equiv -\left[\sum_{\vec{q}'} \frac{v(\vec{q}-\vec{q}')}{S}(n_{\vec{k}+\vec{q}'}^e + n_{-\vec{q}'}^h)\right]\mathcal{P}_{\vec{k}}^W(\vec{q})$$

$$+ (n_{\vec{k}+\vec{q}}^e + n_{-\vec{q}}^h) \sum_{\vec{q}'} \frac{v(\vec{q}-\vec{q}')}{S}\mathcal{P}_{\vec{k}}^W(\vec{q}').$$

Here the "Hamiltonian" $\hat{\mathcal{H}}_0$ describes the evolution of the polarization in an isolated IQW in the absence of electron–hole populations and corresponds to the Wannier equation [20]. The resonant Wannier exciton wave function in the momentum space $\Phi_{\vec{k}}(\vec{q})$ is its eigenfunction with the eigenvalue $E_W(\vec{k})$. The "Hamiltonian" $\hat{\mathcal{H}}_1$ describes the nonlinear many-particle corrections. It is proportional to the populations $n^e$, $n^h$ and we treat it perturbatively, keeping only the first-order corrections to the eigenfunction $\delta\Phi_{\vec{k}}(\vec{q})$ and to the eigenvalue $\delta E_W(\vec{k})$. Since populations are proportional to the intensity of the applied field $|\mathcal{E}_0|^2$, our calculation describes a third-order nonlinearity.

We seek the solutions depending on time as $e^{-i\omega t}$. The solution for $\mathcal{P}_{\vec{k}}^W(\vec{q})$ may be expressed in terms of the orthonormal basis of eigenfunctions of $\hat{\mathcal{H}}_0 + \hat{\mathcal{H}}_1$. Picking up only the resonant term, we may write

$$\mathcal{P}_{\vec{k}}^W(\vec{q}) = u_{\vec{k}}^W(\Phi_{\vec{k}}(\vec{q}) + \delta\Phi_{\vec{k}}(\vec{q}))e^{-i\omega t}, \tag{47}$$

$$u_{\vec{k}}^W e^{-i\omega t} = \sum_{\vec{q}}(\Phi_{\vec{k}}^*(\vec{q}) + \delta\Phi_{\vec{k}}^*(\vec{q}))\mathcal{P}_{\vec{k}}^W(\vec{q}), \tag{48}$$

$$\mathcal{P}_{\vec{k}}^F = u_{\vec{k}}^F e^{-i\omega t}. \tag{49}$$

Then Eqs. (45), (46) are reduced to

$$(\hbar\omega - E_F(\vec{k}))u_{\vec{k}}^F = V_{FW}(\vec{k})u_{\vec{k}}^W - J_F,$$

$$(\hbar\omega - E_W(\vec{k}) - \delta E_W(\vec{k}))u_{\vec{k}}^W = \beta_{\vec{k}} V_{FW}^*(\vec{k})u_{\vec{k}}^F - \beta_{\vec{k}} J_W, \tag{50}$$

where we have introduced the coupling matrix element

$$V_{FW}(\vec{k}) + \delta V_{FW}(\vec{k}) = V_{\text{hyb}}(\vec{k})\sum_{\vec{q}}(\Phi_{\vec{k}}(\vec{q}) + \delta\Phi_{\vec{k}}(\vec{q})), \tag{51}$$

the effective driving forces

$$J_F = \vec{\mathcal{E}}_0 \cdot \vec{M}^F, \qquad J_W + \delta J_W = \vec{\mathcal{E}}_0 \cdot \vec{M}^{eh} \sum_{\vec{q}}(\Phi_{\vec{k}}^*(\vec{q}) + \delta\Phi_{\vec{k}}^*(\vec{q})), \tag{52}$$

and the Pauli blocking factor which is given by

$$\beta_{\vec{k}} = \frac{\sum_{\vec{q}} (1 - n^e_{\vec{k}+\vec{q}} - n^h_{-\vec{q}}) \Phi^*_{\vec{k}}(\vec{q})}{\sum_{\vec{q}} \Phi^*_{\vec{k}}(\vec{q})}. \qquad (53)$$

In the low-density limit and with $\mathcal{E}_0 = 0$ these equations correspond to the eigenvalue equation (2) with the coupling matrix elements given by (19) since for 1s-exciton

$$\Phi_{\vec{k}}(\vec{q}) = \sqrt{\frac{8\pi a_B^2}{S}} \frac{1}{(q^2 a_B^2 + 1)^{3/2}}, \qquad \sum_{\vec{q}} \Phi_{\vec{k}}(\vec{q}) = \sqrt{2S/(\pi a_B^2)}. \qquad (54)$$

Solving the system (50), we obtain for the polarization of the structure under consideration (per unit area):

$$P^s_i(\vec{r}_\parallel) \equiv \int \langle \hat{P}^F_i(\vec{r}) + \hat{P}^W_i(\vec{r}) \rangle dz \simeq \frac{u^F_{\vec{k}} M^{F*}_i}{S} e^{-i\omega t + i\vec{k}\vec{r}_\parallel} + \text{c.c.}$$

$$= \chi_{ij}(\omega, \vec{k}) \mathcal{E}_{0j} e^{-i\omega t + i\vec{k}\vec{r}_\parallel} + \text{c.c.}, \qquad (55)$$

where we have retained only the term proportional to $|\vec{M}^F|^2$ since $|J_W| \ll |J_F|$. Finally, we obtain for the susceptibility (not forgetting the factor of 2 originating from spin degeneracy as mentioned in the beginning of this section):

$$\chi_{ij}(\omega, \vec{k}) = 2 \frac{d^{F*}_i d^F_j}{a_F^2}$$

$$\times \frac{E_W(\vec{k}) + \delta E_W(\vec{k}) - \hbar\omega}{(E_W(\vec{k}) + \delta E_W(\vec{k}) - \hbar\omega)(E_F(\vec{k}) - \hbar\omega) - \beta_{\vec{k}} |V_{FW}(\vec{k}) + \delta V_{FW}(\vec{k})|^2}. \qquad (56)$$

In Eq. (56) the nonlinearities appear through the blue shift $\delta E_W$, the blocking factor $\beta$ and the modification of the hybridization $\delta V_{FW}$ due to the correction $\delta \Phi$; all these effects are typical of Wannier excitons [13], but here they belong to the hybrid excitons which also have a large oscillator strength characteristic of Frenkel excitons. When only excitons are present (i.e., under resonant excitation at low temperature), the nonlinear corrections can be calculated to first order in the $n$'s with

$$n^e_{\vec{k}+\vec{q}} = n^h_{-\vec{q}} \simeq \frac{n_T S}{4} |\Phi_{\vec{k}}(\vec{q})|^2, \qquad (57)$$

where $n_T$ is the total density of electron–hole pairs and the factor $1/4$ takes into account electron (and hole) spin degeneracy of two and an equal population of resonant FE and WE. In terms of the previous subsection, this corresponds to the situation when $k \simeq k_0$, thus $E_F(\vec{k}) \simeq E_W(\vec{k})$, $|A_\alpha|^2 \simeq |B_\alpha|^2 \simeq 1/2$, and $|u_{\vec{k}}^F|^2 \simeq |u_{\vec{k}}^W|^2 \simeq (n_T S)/4$. The blue shift $\delta E_W$ is given by the expectation value of $\mathcal{H}_1$ on $\Phi_{\vec{k}}(\vec{q})$ and reduces to

$$\delta E_W \simeq 0.48 \, E_b \, \pi a_B^2 \, n_T, \tag{58}$$

where $E_b$ is the binding energy of a 2D Wannier exciton. The blocking factor is calculated from Eq. (53) and results as

$$\beta_{\vec{k}} \simeq 1 - 0.57 \, \pi a_B^2 n_T. \tag{59}$$

The effect of $\delta V_{FW}$ can be estimated [13] writing $\delta\Phi_{\vec{k}}(\vec{q})$ as a sum over all continuous and discrete excitonic states which are then approximated by plane-waves in the expression for $|V_{FW} + \delta V_{FW}|^2$ obtaining

$$|V_{FW} + \delta V_{FW}|^2 \simeq \left(1 - 0.48 \, \pi a_B^2 n_T\right)|V_{FW}|^2. \tag{60}$$

Close to resonance (denoting the detuning $\hbar\omega - E_W(\vec{k})$ by $\Delta E$) Eq. (56) can be approximated by

$$\chi_{ij}(\omega, \vec{k}) = -2 \frac{d_i^{F*} d_j^F}{a_F^2} \frac{\Delta E}{\Delta E^2 - |V_F W|^2}$$

$$\times \left[1 - \pi a_B^2 n_T \left(\frac{1.05\,|V_{FW}|^2 - 0.48\,E_b\,\Delta E}{\Delta E^2 - |V_{FW}|^2} + \frac{0.48\,E_b}{\Delta E}\right)\right]$$

$$= \chi_{ij}^{(1)}(\omega, \vec{k})\left(1 - \frac{n_T}{n_S}\right), \tag{61}$$

$\chi_{ij}^{(1)}(\omega, \vec{k})$ being the susceptibility of the hybrid structure at $n_T = 0$ (the linear susceptibility) and $n_S$ is the saturation density. The characteristic feature of the expression (61) is the presence of the factor $(d^F/a_F)^2$ in $\chi^{(1)}$ instead of $(d^{vc}/a_B)^2$ in the analogous expression for an isolated IQW. This leads to the enhancement of absorption, determined by $\Im\chi^{(1)}$. Thus, while the saturation density is comparable to that of Wannier excitons ($n_S \sim 1/a_B^2$), the density of photogenerated electron–hole pairs, for a given light intensity, can be two orders of magnitude larger (by a factor $\sim (a_B/a_F)^2$); for the same reason, also the linear susceptibility $\chi^{(1)}$ can be two orders of magnitude larger. Therefore, the present theory substantiates the intuitive expectation of very pronounced nonlinear optical properties of the hybrid excitons.

While the range of validity of Eq. (61) (with respect to variations of $\Delta E$ and $n_T$) is rather limited, the expression for $\chi$ given by Eq. (56) holds true as long as the basic approximations of the present approach are tenable. These are, in addition to the first-order perturbation theory with respect to the excitation density $n_T$, the usual Hartree–Fock decoupling in the equations of motion adopted in

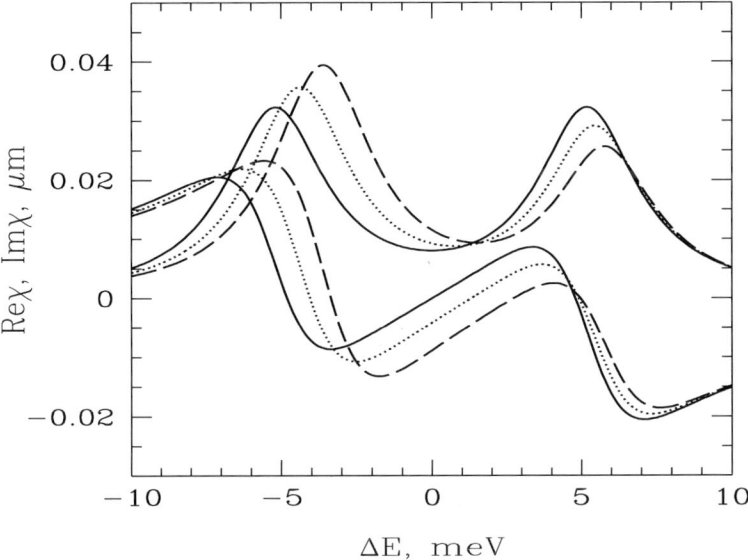

Fig. 6. Real and imaginary parts of the 2D susceptibility $\chi$ near the hybrid exciton resonances in the linear regime (solid lines), medium excitation density ($n_T = 3 \cdot 10^{10}$ cm$^{-2}$, dotted lines) and high excitation density ($n_T = 10^{11}$ cm$^{-2}$, long-dashed lines). Other parameters are $d^{vc} = 20$ Debye, $a_B = 60$ Å $E_b = 40$ meV, $\varepsilon_\infty = 11$, all the rest are the same as in the previous subsection. Line-widths $\hbar\gamma_W = \hbar\gamma_F = 2$ meV.

Eqs. (45), (46), the subsistence of well defined individual excitons (valid only for $n_T \lesssim n_S$) and the neglect of screening due to the reduced screening efficiency of a two-dimensional exciton gas [13,20]. Numerical examples of the predictions of Eq. (56) have been obtained using the values of semiconductor parameters representative of III–V semiconductor (e.g., GaAs/AlGaAs) quantum wells, since the necessary information on homogeneous linewidths of excitons in II–VI QWs is not presently available to the authors. Namely, we set $\varepsilon_\infty = 11$, $d^{vc} = 20$ Debye, the Bohr radius $a_B = 60$ Å and the binding energy is taken to be $E_b \simeq 20$ meV, all the rest are the same as in the previous subsection. This gives $|V_{FW}| \simeq 4$ meV at $k = 10^7$ cm$^{-1}$. Assuming a phenomenological linewidth $\hbar\gamma_W = \hbar\gamma_F = 2$ meV for both excitons Figure 6 shows the split resonance of the hybrid excitons at different excitation densities (linear regime, $n_T = 3 \cdot 10^{10}$ cm$^{-2}$, and $n_T = 10^{11}$ cm$^{-2}$); it is noticeable, in particular, that for vanishing excitation density the mixing is complete and the oscillator strength is equally shared between the two peaks, whereas for high excitation density, due to the small blue shift of the WE, the stronger line corresponds to the lowest (more Frenkel-like) hybrid exciton. Figure 7 shows the effect of a density dependent broadening of the Wannier exciton: $\hbar\gamma_W = 1$ meV

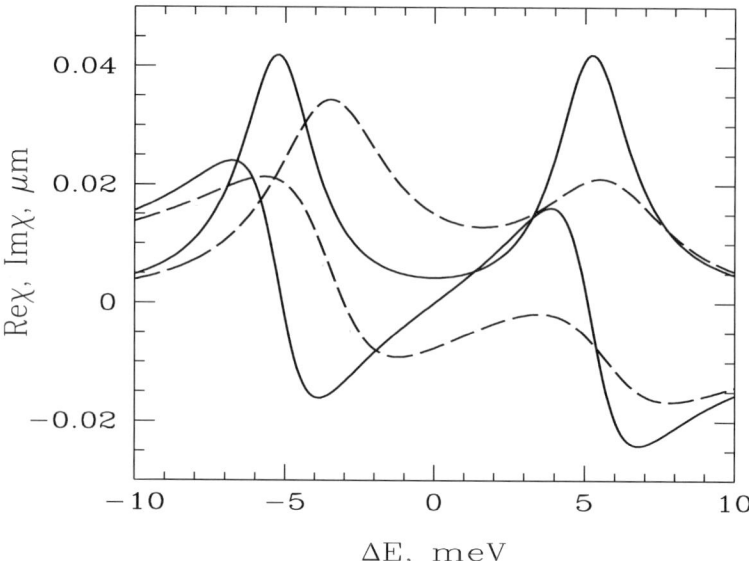

Fig. 7. Real and imaginary parts of $\chi$ in the linear regime (solid lines), and high excitation density ($n = 10^{11}$ cm$^{-2}$, long-dashed lines), in the first case $\hbar\gamma_W = 1$ meV, in the second case $\hbar\gamma_W = 3$ meV.

at low excitation densities and $\hbar\gamma_W = 3$ meV at high excitation density [21] ($\hbar\gamma_F$ being fixed at 2 meV).

From numerical estimates such as those shown in Figures 6 and 7, we obtain for the relative nonlinear change in the absorption coefficient close to resonance $|\Delta\alpha|/\alpha \sim 10^{-11}$ cm$^2$ $n_T$, which is analogous to the case of a semiconductor multiple quantum well. However, for a given pump intensity the 2D density of photogenerated excitons $n_T$ in our case of hybrid excitons is about two orders of magnitude larger because the oscillator strength of hybrid excitons is comparable to the one of Frenkel excitons rather the one of Wannier excitons. A similar theoretical approach can be used to calculate the dynamical Stark effect for hybrid excitons which shows qualitative and quantitative differences with respect to the case of the usual inorganic semiconductor QWs [22].

## 3.2. Second-Order Susceptibility $\chi^{(2)}$

As was already mentioned, the calculations, performed here, correspond to the third-order nonlinearity. But the hybrid system considered here has also a nonzero second-order susceptibility $\chi^{(2)}$. For such a structure $\chi^{(2)} \neq 0$ even if the original OQW and IQW are centro-symmetric and the second-order processes are forbidden by parity conservation. Such a phenomenon can take place because the resonant dipole–dipole coupling breaks the symmetry along the growth direction. Of course, any interaction between the OQW and the IQW can be responsible for symmetry breaking. However, the resonant dipole–dipole coupling considered here is probably the strongest among others. In a geometrical sense, this system is analogous to an asymmetric semiconductor QW. The calculation of $\chi^{(2)}$ for such a system can be found in Ref. [23] and the calculation of $\chi^{(2)}$ for the hybrid system may be performed following the lines of the latter work and here we restrict ourselves only to some qualitative remarks.

The general microscopic expression for the $n$th-order susceptibility contains $n+1$ dipole moment matrix elements, involving $n$ intermediate states. For the linear susceptibility there is only one intermediate state, and if the latter is a hybrid one, the corresponding dipole matrix elements are determined mainly by the Frenkel component of the hybrid state. Thus, the linear susceptibility of the hybrid structure contains the factor $(d^F/a_F)^2$, as is seen from Eq. (61). For the second-order nonlinear susceptibility $\chi^{(2)}$ one must have two intermediate states or three virtual transitions. One of them may be a hybrid one, and as long as the materials under consideration have no static dipole moments, the other intermediate state has to be an excited state of the IQW, which is not resonant with the Frenkel exciton. Hence, the result will be proportional to $d^F/a_F$, the other two virtual transitions will give a factor, coinciding with that for an isolated IQW. One may apply analogous arguments to the case of the third-order nonlinearity: of three needed intermediate states one may be the hybrid one, the second may be the ground state, and the third one—again the hybrid state (such a scheme corresponds to the Kerr nonlinearity). Thus, one should obtain a factor $(d^F/a_F)^4$. Indeed, in Eq. (61) we have $(d^F/a_F)^2$ in $\chi^{(1)}$ and another $(d^F/a_F)^2$ comes from $n_T$ when the latter is expressed in terms of the incident electric field. There exists another mechanism for the second harmonic generation. It does not require the parity breaking, since the optical quadratic nonlinearity appears due to the contribution of spatial derivatives of the electric field to the nonlinear response [24,25]. It works also in the case of an isolated symmetric QW and corresponds to the higher multipole contribution rather then the dipole one, which is usually considered. The hybrid system will again have an advantage here because of the increase in the oscillator strength due to the Frenkel exciton component.

## 4. Hybrid Excitons in Parallel Organic and Inorganic Semiconductor Quantum Wires

In this section (see also [26]) we study hybrid exciton states not in a two-dimensional (2D) but in a one-dimensional (1D) system of parallel organic and inorganic semiconductor quantum wires. In particular, we will show that the interwire hybridization strength is nonzero even for small exciton wave vectors along the wires and decays rather slowly with increasing interwire spacing. This is in contrast with 2D quantum wells considered in the preceding subsection, where the dipole–dipole coupling decays fast with increasing interwell distance and is nonzero only for nonzero exciton wave vectors.

Although there is already a considerable progress in the preparation of organic-inorganic semiconductor nanostructures, the achievements are mostly connected with the fabrication of planar heterostructures. As to laterally modulated composite organic-semiconductor heterostructures, we realize that the fabrication of such systems is a more formidable task. However, the expected unique physical properties of such systems may justify the efforts. A motivation for studying laterally modulated hybrid systems, such as a system of organic-inorganic wires, is the following. As has been shown in the preceding subsection for a 2D organic molecular layer in contact with a neighboring semiconductor 2D quantum well, the electrostatic coupling of Frenkel and Wannier–Mott excitons vanishes in the range of small two-dimensional exciton wave vectors (see Eq. (21)). This is a consequence of the well-known fact that the electric field of a uniformly polarized layer vanishes outside the layer. Thus, in 2D systems, the conditions for the manifestation of hybridization effects are not favorable just for the most interesting case of small wave vector excitons which interact actively with light. To lift this restriction, here we consider another organic semiconductor system where the hybridization of Frenkel and Wannier–Mott excitons is especially effective just for excitons with small wave vectors. Namely, we consider a system of parallel organic and semiconductor quantum wires. The Frenkel $|F, k, l\rangle$ and Wannier–Mott $|W, k, l\rangle$ exciton states in the wires are characterized by a one-dimensional (1D) wave vector $k$ along the wires and by the label $l$ counting quantized states of the transverse motion of excitons within the wires. To simplify the consideration, below we restrict our analysis to the lowest transverse state of excitons and omit the label $l$.

Our task is to calculate the hybridization parameter

$$\Gamma(k) = \langle F, k | H_{\text{int}} | W, k \rangle, \tag{62}$$

that determines the resonance coupling of the Frenkel and Wannier–Mott excitons, see Eqs. (1)–(5). The Hamiltonian $H_{\text{int}}$ of the dipole–dipole interaction between the two systems is given by Eq. (6) which may be represented also in the form

$$H_{\text{int}} = -\int P_i^F(\mathbf{r}) \mathcal{D}_{ij}(\mathbf{r}, \mathbf{r}') P_j^W(\mathbf{r}') d\mathbf{r} d\mathbf{r}'', \tag{63}$$

where the integrations goes over the organic and semiconductor wires, $\mathcal{D}_{ij}(\mathbf{r}, \mathbf{r}')$ is the Green's function determined by Eqs. (8) and (9); indices $i, j$ denote vector components in Cartesian coordinates with the $z$-axis chosen along the wires, and the $x$- and $y$-axes perpendicular to the wires; the $y$-axis is perpendicular to the plane determined by the wires. The transition polarization operator $\mathbf{P}^F$ for Frenkel excitons is given by

$$P_i^F(\mathbf{r}) = \sum_{\mathbf{n}} \mu_i^F \left(A_{\mathbf{n}}^\dagger + A_{\mathbf{n}}\right) \delta(\mathbf{r} - \mathbf{r_n}), \quad (64)$$

where the summation runs over sites $\mathbf{n}$ (with the radius-vector $\mathbf{r_n}$) of the molecular lattice, $A_{\mathbf{n}}^\dagger$ is the creation operator of the Frenkel exciton at the site $\mathbf{n}$, and $\mu^F$ is the transition dipole moment for Frenkel excitons. The transition polarization operator $\mathbf{P}^W$ for Wannier–Mott excitons is given by

$$\mathbf{P}^W(\mathbf{r}) = \boldsymbol{\mu}^W \Psi_e(\mathbf{r}) \Psi_h(\mathbf{r}) + \text{h.c.}, \quad (65)$$

where $\Psi_{e(h)}(\mathbf{r})$ is the electron (hole) annihilation operator and $\boldsymbol{\mu}^W$ is the (intracell) optical transition dipole moment.

The state corresponding to the Frenkel exciton is represented as

$$|F, k\rangle = \frac{1}{\sqrt{N_F}} \sum_{\mathbf{n}} \exp(ikz_{\mathbf{n}}) A_{\mathbf{n}}^\dagger |0\rangle, \quad (66)$$

where $N_F$ is the total number of molecular sites and $|0\rangle$ is the exciton vacuum state. For the state of the Wannier–Mott exciton we take the following representation:

$$|W, k\rangle = \frac{1}{\sqrt{L}} \int dz_e\, dz_h \exp\left(ik\frac{m_e z_e + m_h z_h}{m_e + m_h}\right) \Phi_0(z_e - z_h) \Psi_{e0}^\dagger(z_e) \Psi_{h0}^\dagger(z_h) |0\rangle. \quad (67)$$

Here $L$ is the length of the wires, $m_e$ and $m_h$ are effective electron and hole masses, respectively; the function $\Phi_0(z_e - z_h)$ describes the relative 1D motion of the bound electron and the hole. The operators $\Psi_{e(h)0}^\dagger(z)$ in Eq. (67) correspond to the lowest ($l = 0$) state of the transverse motion in the operator expansion over transverse modes $\phi_l$:

$$\Psi_{e(h)}^\dagger(\mathbf{r}) = \sum_{l} \Psi_{e(h)l}^\dagger(z) \phi_{e(h)l}^*(\boldsymbol{\rho}), \quad (68)$$

where $\mathbf{r} = (\boldsymbol{\rho}, z)$.

In calculating the hybridization parameter $\Gamma$ (see Eq. (62)) we meet the following matrix elements of the polarization operators between the ground ($|0\rangle$) and corresponding excited states ($|F, k\rangle$ or $|W, k\rangle$). With the use of Eqs. (64) and (66) we obtain:

$$\langle F, k | P_i^F(\mathbf{r}) | 0 \rangle = \frac{\mu_i^F}{\sqrt{N_F}} \sum_{\mathbf{n}} \exp(-ikz_{\mathbf{n}}) \delta(\mathbf{r} - \mathbf{r_n})$$

$$\approx \frac{\mu_i^F}{\sqrt{N_F} v_F} \exp(-ikz), \quad (69)$$

where we have used the long-wavelength approximation ($ka_F \ll 1$, $a_F$ is the organic lattice constant) and substituted the summation over lattice sites $\mathbf{n}$ by integration over $d\mathbf{r}_n/v_F$, where $v_F$ is the volume of the organic lattice elementary cell. Similarly, with the use of Eqs. (65), (67), and (68) we find:

$$\langle 0|P_j^W(\mathbf{r}')|W,k\rangle = \frac{\mu_j^W}{\sqrt{L}}\phi_{e0}(\boldsymbol{\rho}')\phi_{h0}(\boldsymbol{\rho}')\Phi_0(0)\exp(ikz'). \tag{70}$$

Combining Eqs. (62), (69), and (70) we arrive at the following expression for the hybridization parameter $\Gamma$ from Eq. (62):

$$\Gamma(k) = -\mu_i^F \mu_j^W \Phi_0(0)\sqrt{\frac{N_F}{L}}\int \frac{d^2\rho\, d^2\rho'}{S_F}\mathcal{D}_{ij}(\boldsymbol{\rho},\boldsymbol{\rho}';k)\phi_{e0}(\boldsymbol{\rho}')\phi_{h0}(\boldsymbol{\rho}'), \tag{71}$$

where $S_F$ is the organic wire cross section area, $\mathcal{D}_{ij}(\boldsymbol{\rho},\boldsymbol{\rho}';k)$ is the Fourier transform of $\mathcal{D}_{ij}(\boldsymbol{\rho},z;\boldsymbol{\rho}',z')$ with respect to the difference $z-z'$ (the surrounding medium is assumed to be homogeneous in the direction parallel to the wires). Eq. (71) determines the hybridization parameter of interest for an arbitrary system geometry and dielectric tensor of the surrounding medium. Below we consider in more detail the particular case of thin wires ($d_w \ll R$, $d_w \ll 1/k$, where $d_w$ is the thickness of wires, $R$ is the distance between the wires) embedded into a medium with an isotropic dielectric tensor $\varepsilon_{ij} = \varepsilon\delta_{ij}$. In this case the Green's function $\mathcal{D}_{ij}(\mathbf{r};\mathbf{r}') = \varepsilon^{-1}\partial_i\partial_j G_0(\mathbf{r}-\mathbf{r}')$, where $G_0(\mathbf{r}-\mathbf{r}') = 1/|\mathbf{r}-\mathbf{r}'|$. The Fourier transform $G_0(\boldsymbol{\rho}-\boldsymbol{\rho}';k)$ of $G_0(\mathbf{r}-\mathbf{r}')$ with respect to $z-z'$ is given by

$$G_0(\boldsymbol{\rho}-\boldsymbol{\rho}';k) = 2K_0(k|\boldsymbol{\rho}-\boldsymbol{\rho}'|), \tag{72}$$

where $K_0$ is the modified Bessel function of zeroth order [27]. For thin wires, we may neglect the variation of the transverse coordinates $\boldsymbol{\rho}$ and $\boldsymbol{\rho}'$ in the argument of the function $\mathcal{D}_{ij}(\boldsymbol{\rho}-\boldsymbol{\rho}';k)$, substituting $\boldsymbol{\rho}-\boldsymbol{\rho}'$ by the vector $(R,0)$ (of course, this can be done only after taking spatial derivatives of $G_0(\boldsymbol{\rho}-\boldsymbol{\rho}';k)$).

As a result, Eq. (71) for the hybridization parameter $\Gamma(k)$ takes the form:

$$\Gamma(k) = f_{eh}\Phi_0(0)\sqrt{\frac{N_F}{L}}\frac{\mu_i^F \mu_j^W}{\varepsilon}C_{ij}, \tag{73}$$

where

$$f_{eh} = \int \phi_{e0}(\boldsymbol{\rho})\phi_{h0}(\boldsymbol{\rho})\,d^2\rho, \tag{74}$$

$$C_{ij} = -2[(\nabla_i^\perp - ik\delta_{iz})(\nabla_j^\perp - ik\delta_{jz})K_0(k\rho)]_{(\rho_x,\rho_y)=(R,0)}. \tag{75}$$

Here $\nabla_i^\perp$ denotes the derivative with respect to the transverse variables ($i=x,y$). The function $K_0$ possesses the following limiting behavior [27]:

$$K_0(x) = \begin{cases} \sqrt{\frac{\pi}{2x}}\exp(-x), & x \gg 1, \\ -\ln(x/2), & x \ll 1. \end{cases} \tag{76}$$

As follows from Eq. (76), the interwire coupling is suppressed exponentially for excitons with wave vectors $k \gtrsim 1/R$, i.e., for a major part of the Brillouin zone. On the contrary, coupling of excitons with relatively small wave vectors $k \lesssim 1/R$ is quite efficient. This is in contrast to the case of a 2D system of quantum wells where the coupling at small wave vectors is suppressed because the electric field outside of a uniformly polarized layer vanishes.

The range of small wave vectors $k \sim 1/\lambda \ll 1/R$ is of special interest as excitons with such wave vectors may be created straightforwardly by light of wavelength $\lambda$. In the leading order in $kR \ll 1$, the hybridization parameter $\Gamma(k)$ of Eq. (73) has the following form

$$\Gamma(k) = \frac{f_{eh}}{\varepsilon R^2}\sqrt{\frac{2S_F}{a_{1B}v_F}}(\mu_y^F \mu_y^W - \mu_x^F \mu_x^W), \qquad (77)$$

where $S_F$ and $v_F$ are the cross-section and the volume of an elementary lattice cell for the molecular wire; the 1D exciton ground state wave function $\Phi_0(0) = 1/\sqrt{2a_{1B}}$ in the strong-confinement limit has been expressed in terms of the 1D Bohr radius $a_{1B} = (a_0/2)\sqrt{E_0/E_1}$, with $a_0$ and $E_0$ being the Bohr radius and the ground state energy of the bulk exciton, respectively; $E_1$ is the ground state energy of the 1D exciton, see [28,29].

Note that the excitonic polarization component along the wires does not contribute to $\Gamma$ in the leading order in $kR$. This is due to the obvious fact that a uniform longitudinal polarization is not accompanied by the appearance of an electric charge. To estimate the value of $\Gamma$ we use the following parameter values: $a_{1b} = 30$ Å, $\mu^F = 5$ Debye, $\mu^W = 10$ Debye, $S_F = (50 \text{ Å})^2$, $v_F = 100 \text{ (Å)}^3$, $R = 50$ Å, $f_{eh} = 1$, $\varepsilon = 3$, and we obtain $\Gamma \approx 5.4$ meV. Similar to the 2D case, the resonance coupling of 1D Frenkel and Wannier–Mott excitons results in the appearance of hybrid states described by Eqs. (1)–(5).

The coupling is strong if the energies of Frenkel and Wannier–Mott excitons are in resonance: $|E_F(k) - E_W(k)| \sim |\Gamma(k)|$. In this case the size of the hybrid state is comparable with that for Wannier–Mott excitons, i.e., it is much larger than the radius of Frenkel excitons. This causes a high sensitivity of the hybrid states to external fields. Outside the resonance range, the coupling is governed by the parameter $\Gamma^2/(E_F - E_W)$ and is rather small. The condition of resonance is rather strict for the considered range of parameters and requires a careful choice of materials for both wires. Naturally, the exciton linewidth should be small as compared to $\Gamma$. In general, the excitonic linewidth is determined by radiative and nonradiative processes and by other dephasing processes.

At low temperatures, the nonradiative linewidth may be $\sim 1$ meV for Wannier–Mott excitons, and in some cases rather small for Frenkel excitons. These linewidths may be thus in some particular structures smaller than the resonant splitting $2\Gamma \approx 11$ meV of the hybrid excitations. In perfect translationally invariant wires there is no radiative decay of excitons with wave vectors $k > 2\pi/\lambda$ ($\lambda$ is the wavelength of the exciton luminescence in a medium with the refraction index $n = \sqrt{\varepsilon}$). The radiative decay channel opens for excitons with rather small wave vectors $k < 2\pi/\lambda$. For the transition dipole moment oriented perpendicular to the wire, the radiative decay rate $\gamma_{\text{rad}}$ of Frenkel excitons in a 1D wire is given by

$$\gamma_{\text{rad}}(k) = \frac{\pi |\mu^F|^2 S_F}{\varepsilon v_F} [k^2 + (2\pi/\lambda)^2], \quad k \leqslant 2\pi/\lambda. \tag{78}$$

Using Eqs. (77) and (78) we obtain the following estimate (for small $k$ range):

$$\frac{2\Gamma(k)}{\gamma_{\text{rad}}(k)} \sim \frac{f_{eh} \lambda^2 \mu^W}{2\pi^3 R^2 \mu^F} \sqrt{\frac{v_F}{2 S_F a_{1B}}}. \tag{79}$$

For the above chosen parameters values and for the exciton radiation wavelength to be of the order of 5000 Å in vacuum, we find $2\Gamma/\gamma_{\text{rad}} \sim 3$. This means that in perfect wires the collective exciton radiative decay channel may be competitive with the hybridization.

One of the ways of increasing the ratio (79) is to decrease the molecular wire cross section $S_F$. This is due to the square-root dependence of $\Gamma$ on $S_F$ versus the linear dependence of $\gamma_{\text{rad}}$. Another circumstance that is favorable for an increase of the ratio $\Gamma/\gamma_{\text{rad}}$, is the fact that the expression (78) is valid for very perfect wires with the exciton coherence length $l_{\text{coh}}$ being longer than $\lambda$. On the contrary, if $l_{\text{coh}} \ll \lambda$ (that is more realistic situation), the radiative exciton decay rate will be diminished by the factor $l_{\text{coh}}/\lambda$. In this case, the radiative decay would be much smaller than the hybridization parameter $\Gamma$. The latter does not suffer from the finite exciton coherence length as long as $l_{\text{coh}} > R$. This is due to the fact that the leading contribution to the quantity Eq. (77) stems from the distances of the order of $R$.

To summarize, we have demonstrated the possibility of strong resonance hybridization of 1D Frenkel and Wannier–Mott excitons in parallel organic and semiconductor wires. Like in 2D case, the new states possess the properties of both types of excitons. They have a relatively large size (along the wires) like Wannier–Mott excitons, but they have also a large transition dipole moment which is typical for Frenkel excitons. Thus, one may expect strong nonlinear optical effects in such systems.

## 5. On the Hybridization of "Zero-Dimensional" Frenkel and Wannier–Mott Excitons

In the previous sections we have considered the hybridization of Frenkel and Wannier–Mott excitons in two-dimensional (quantum wells) and one-dimensional (quantum wires) geometries. For the sake of completeness, in this subsection we shall briefly and qualitatively discuss the zero-dimensional (0D) case that corresponds to a quantum dot (QD) geometry. We have in mind a configuration where a semiconductor QD is located near a small size organic cluster or is just covered by a thin shell of an organic material.

There are the following qualitative differences in the hybridization scenario in the considered geometry as compared to those in 2D and 1D cases. The presence of a translational symmetry along planes or wires has resulted in a selection rule restricting the hybridization to states $|F, k\rangle$ and $|W, k\rangle$ with the same wave vectors. The absence of the translational symmetry in 0D case leads, generally speaking, to the coupling of all exciton states. This circumstance is not favorable for the efficient hybridization: as was discussed earlier, the condition for the strong hybridization is an energy resonance between the mixed exciton states. Therefore, it is desirable to deal with a situation where there are only two resonant (Frenkel and Wannier–Mott) exciton states strongly coupled to each other and only weakly connected with other states. This may be achieved in small size QDs and clusters where the exciton motion is quantized. As soon as the resonance between two (say, the lowest) exciton states $|F\rangle$ and $|W\rangle$ has been achieved, the further description in terms of an effective two-level model is similar to what has been done in the previous sections. The hybridization parameter $\langle F|H_{\text{int}}|W\rangle$ is determined by properties of the resonant states $|F\rangle$ and $|W\rangle$ for a concrete system configuration, i.e., on the shape and symmetry of the semiconductor QD and organic cluster, on the orientation of organic molecules, etc. Just as one example of a great variety of possible geometries, we mention here the case of a spherical semiconductor QD covered with a thin shell of organic molecules (see [30a] for more details; on the system of quantum dot array embedded in an organic host see [30b] and on hybrid exciton state in a quantum dot-dendrite system see [30c]).

A feature of this situation is that in the case of a too symmetric orientation of the transition dipole moments of organic molecules (for instance, if all of them are oriented along the same direction) the hybridization parameter vanishes. This is because the electric field that corresponds to an excitation of the organic shell, coincides with the field inside a uniformly polarized shell, and is zero for a spherical shell. To provide a nonzero hybridization one should allow for, e.g., a spatial variation of the shell thickness or of the angular orientation of the molecular transition dipole moments.

## 6. Hybridization of Excitons in Microcavity Configurations

In the last decade planar semiconductor microcavities attract much attention, since they provide the possibility to enhance and control the interaction between light and electronic excitations. When the microcavity mode is resonant with the excitonic transition, two different regimes of interaction can be distinguished, depending on the ratio between the coupling strength and the damping in the system. In the weak coupling regime, when the damping prevails over the light-matter interaction, the interaction just modifies the radiative decay rate and the emission angular pattern of the cavity mode. On the contrary, in the strong coupling regime, the damping is small in comparison with the interaction, and the true eigenstates of the system are the doublet of the cavity polaritons separated by the Rabi splitting [31]. The Rabi splitting is proportional to the square root of the oscillator strength of the transition, and its typical value in inorganic materials is about 5–10 meV. In inorganic semiconductors, the strong coupling regime was extensively investigated both experimentally and theoretically (see [31] for reviews), and the dynamics of microcavity polaritons was well understood [32]. Recently, high-quality crystalline *organic* nanostructures became available for the fabrication and inclusion into a microcavity. Their advantage is that they possess Frenkel excitons, which may have an oscillator strength much stronger than the one in inorganic structures. Thus, as it was noticed in [33], the strength of their interaction with light (and, consequently, the resulting Rabi splitting) is expected to be much larger (up to in dozens of times) than the one in the inorganic analogues. On the other hand, for many known cases of disordered organics the relation between the inhomogeneous broadening of the excitonic resonance and its oscillator strength is such that the strong coupling between the molecular electronic excitation and the cavity photon is washed out. That is why disordered organic materials are usually associated with the weak coupling regime. Quite recently, however, the strong coupling just in disordered organic semiconductor microcavities has been observed using materials with large oscillator strength and relatively narrow (40–90 meV) bare exciton absorption linewidths. First, porphyrin dyes with a single narrow optical absorption line were used as a semiconductor material [34]. A giant Rabi splitting (160 meV at room temperature) was observed. Soon after, the strong coupling in organic semiconductors was observed using another type of a semiconductor material, namely, cyanine dyes [35,36]. The typical polaritonic branches separated by a giant Rabi splitting ($\Delta = 80$–$300$ meV at room temperature) were observed in these structures (Chapter 8 by D. Lidzey).

The possibility to create organic microcavities with large Rabi splitting is very interesting for the problem of hybridization of Frenkel and Wannier–Mott excitons which we discuss in this chapter. Indeed, the structure described in the previous section raises the technologically challenging problem of growing high quality organic-inorganic heterostructures only a few nanometers apart and a more promising way of realizing a hybrid exciton system is to couple Frenkel and Wannier excitons through a microcavity (MC) electromagnetic field [33]. We can expect hybridization in microcavities with the large Rabi splitting to arise not due to the Coulombic short-range interaction, but due to the strong long range interaction stemming from virtual cavity photon exchange. For cavity embedded QWs, the

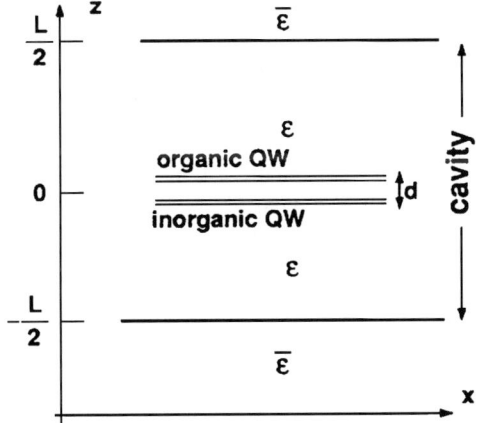

Fig. 8. Scheme of microcavity embedded organic and inorganic quantum wells. The mirrors are simply described by a very high dielectric constant $\bar{\varepsilon} \gg \varepsilon$.

fabrication problems would be much alleviated as their separation can be of the order of an optical wavelength. For the sake of simplicity, however, in the following discussion we assume that both QWs lie at the center ($z \simeq 0$) of a single MC at a distance $d \ll \lambda$ from each other (see Figure 8). This situation is qualitatively equivalent to that of two coupled microcavities for which the growth conditions could be separately optimized for the organic and inorganic well [33].

Microcavity embedded organic QWs in the weak coupling regime have already been realized [37] and effects such as spectral narrowing and increased directionality of light emission have been demonstrated. To achieve the strong coupling regime with organic material, as observed for inorganic QWs [31], we need molecular compounds combining a large oscillator strength of the lowest energy electronic transition with an absorption linewidth smaller than the cavity mode splitting. Good candidates for such structures are thin film crystals of aromatic molecules like anthracene, tetracene, terrylene and many others. For example, five monolayers of terrylene ($d \simeq 50$ Å) exhibit an oscillator strength per unit area as large as $10^{15}$ cm$^{-2}$, more than a hundred times the one of a GaAs QW exciton. Other possibilities to observe the hybridization of Frenkel and Wannier–Mott excitons are opened by using the strong coupling regime in microcavities with disordered organic material. These results we discuss below in Section 2.

### 6.1. Illustrative Estimates

To illustrate the theoretical results derived for such a system [33], we use material parameter data available from experiments or from realistic estimates. We assume that $E_F(k) = E_C(k=0)$ and $E_W(k=0) = E_C(k=0)(1+\eta)$, i.e., a Frenkel exciton resonant with the cavity mode $E_C$ (we neglect the dispersion of the FE) and a Wannier exciton with a fractional detuning of $\eta$ at $k=0$. Using the reduced variable $\bar{k} = k/k_{cav}$ with $k_{cav} = \pi/L$, we have for this case $E_C(k)/E_C(0) = \sqrt{1+\bar{k}^2}$ and $E_W(k)/E_W(0) = 1 + \eta + a\bar{k}^2$ with $a = \hbar^2 k_{cav}^2/2M E_C(0)$. For resonance at $E_C(0) = 1.5$ eV and $\varepsilon \simeq 10$, we have $k_{cav} = 2.4 \cdot 10^5$ cm$^{-1}$ and, using an exciton mass $M = 0.3 m_0$ ($m_0$ is the free electron mass), $a = 10^{-5}$. The inorganic QW Rabi splitting $\Delta_1$ is taken to be 3 meV, then, assuming a ratio of the organic to inorganic QW oscillator strength $F/f \simeq 60$, we have for the organic QW Rabi splitting $\Delta_2 \simeq 23$ meV. The ratio $\Delta_2/\Delta_1 \simeq 8$ is by no means unusually large and, as a matter of fact, even larger oscillator strengths can easily be attained with many organic materials. For example, from the standard LT splittings of 0.08 meV in GaAs ($\varepsilon \simeq 12$) and $\simeq 50$ meV for the lowest singlet exciton in tetracene ($\varepsilon \simeq 9$) [38], their oscillator strength ratio is about 500. The large splittings $\Delta_2 \approx 100$ meV expected from such estimates give reasonable hope for reaching the strong coupling regime even at room temperature since the absorption linewidth can be as low as a few tens of meV in selected organic systems. We assume such a situation in our calculations and neglect dissipation for both bare excitonic states. The dispersion of cavity polaritons $E_j(k)$ and of weighting coefficients $N_j^{F,W,C}(k)$ (analogous to $A$ and $B$ of Eq. (1)) are shown in Figure 9.

From Figure 9(c) it is seen that the branch 1 (which at large wave vectors turns into a pure WE) contains a big part of the FE state ($|N_1^F|^2$) for $\bar{k} < 0.1$. As seen from Figure 9(d) the branch 2 (which at large wave vectors turns into a pure FE) for $\bar{k} < 0.25$ also retains a large part of the FE state ($|N_1^F|^2$) while exhibiting a large cavity photon component. The FE component is crucial in assisting the inelastic relaxation that will be considered whereas the cavity photon component obviously has a large radiative width. For $\bar{k} \ll 1$ even for high mirror reflection $(1-R = 10^{-3})$, the cavity mode radiative lifetime is of order $\tau \simeq 1$ ps. The better mixing of branch 2 with the cavity photon means faster radiative decay in a larger phase space. Such a short lifetime is only effective in a very narrow region of phase space ($k < 0.05\, k_{cav}$) in the case of typical inorganic QW splittings; such a region can only be reached in about 100 ps due to slowed-down relaxation [32] in the flat part of the dispersion curve, poorly coupled to the cavity mode. In our case, to populate the states of branch 2 with a large radiative width (i.e., those with $k < 0.2\, k_{cav}$), we can assume that the parameters of the MC with two QWs are such that for $k, k' < 0.2\, k_{cav}$ an inelastic resonance condition is realized, i.e., that the energy difference $E_1(k) - E_2(k')$ is close to the energy of some intramolecular optical phonon strongly coupled to excitons. For this case, the relaxation rate can be of the order of 10 ps or less [33], i.e., at least one order of magnitude faster than for MC with an inorganic QW.

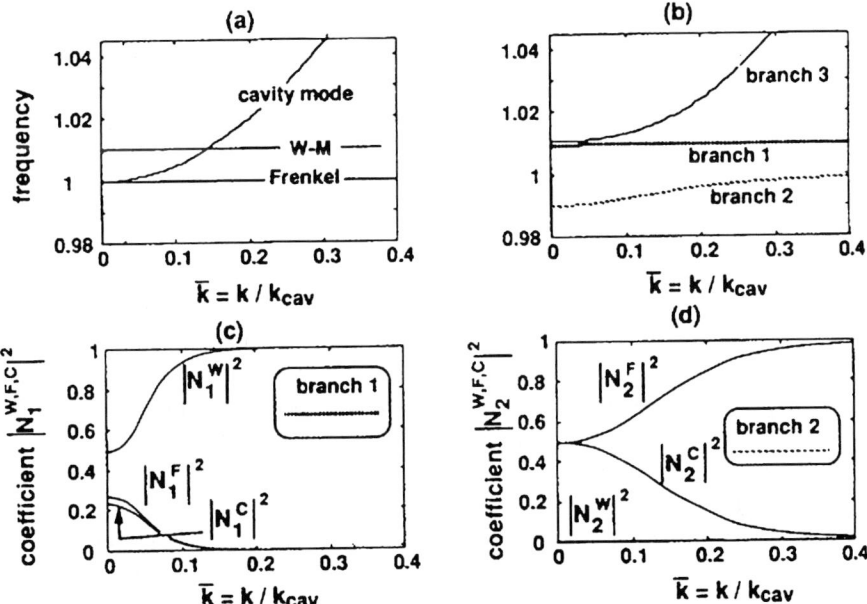

Fig. 9. (a) Bare dispersion curves of cavity photon, WE and FE normalized to the cavity mode at $k = 0$. The FE exciton is resonant with the cavity mode and the WE has a positive detuning; (b) Cavity polariton dispersion curves: for large wavevectors, branches 1, 2 and 3 turn into WE, FE and cavity photon, respectively; (c) Weighting coefficients of branch 1; (d) Weighting coefficients of branch 2.

Summarizing, we have considered the new possibilities which may appear in microcavities containing resonating organic and inorganic QWs. Although our estimates are preliminary and we, for example, did not take into account the dissipation of exciton states, we can expect in such structures a drastic shortening of the relaxation time of excitons into states having a large radiative width and a short fluorescence decay time. We can also expect that the combination of electrical pumping of excitons in inorganic QWs with the fast relaxation and fluorescence of excitons in organic QWs will open up a new scenario of excitonic processes in microcavities, which is of interest for both basic science and device applications.

### 6.2. Microcavities with Disordered Organics: Coexistence of Coherent and Incoherent States

We already mentioned that in microcavities with disordered organic materials a very large Rabi splitting can be achieved. This potentially opens an interesting possibility to create the hybridization of Frenkel and Wannier excitons in such

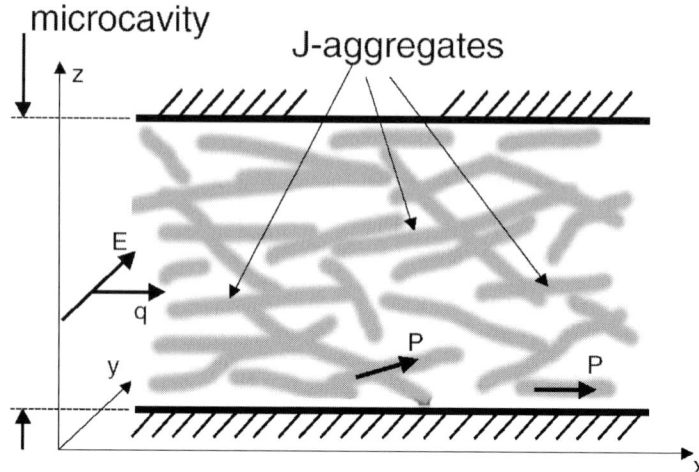

Fig. 10. A draft of a microcavity with J-aggregates formed from a cyanine dye suspended in a transparent polymer matrix. The dipole moments of the J-aggregates are indicated for two particular chains.

structures. However, in microcavities with disordered organics some peculiar features in the spectra of cavity polaritons arise which has to be taken into account.

The observation of Rabi splitting in microcavities with disordered organics, the investigations of photoluminescence and resonance Raman scattering as well as investigations of reflection of light convincingly demonstrate the existence in such microcavities, at least for small in-plane wave vectors, of the coherent states (polaritons) (see the chapter by D. Lidzey). For this region of the wave vectors the situation is similar to what has been observed in microcavities with ordered inorganic semiconductor. However, when we go to consideration of larger wave vectors the picture changes drastically (more details can be found in the paper [39]). Qualitatively this change can be understood on the basis of consideration of polariton spectra with, for example, particular case of J-aggregates as active material (see Figure 10).

However, such a consideration is applicable also for any organic or inorganic, ordered or disordered material with weak intermolecular resonance interaction, where the electronic excitations have negligible dispersion and strong damping.

Indeed, the cavity photons are coherent excited states of the microcavity, and they all are characterized by a two-dimensional in-plane wave vector ($q$). The resonance between the cavity photon and the molecular electronic excitations occurs at relatively small values of the cavity photon wave vector. Thus, in the region of resonance the cavity photon wavelength is large in comparison with the mean distance $\bar{R}$ between the J-aggregates, and the system of J-aggregates can be treated as effectively homogeneous. The cavity photon creates a coherent polarization in

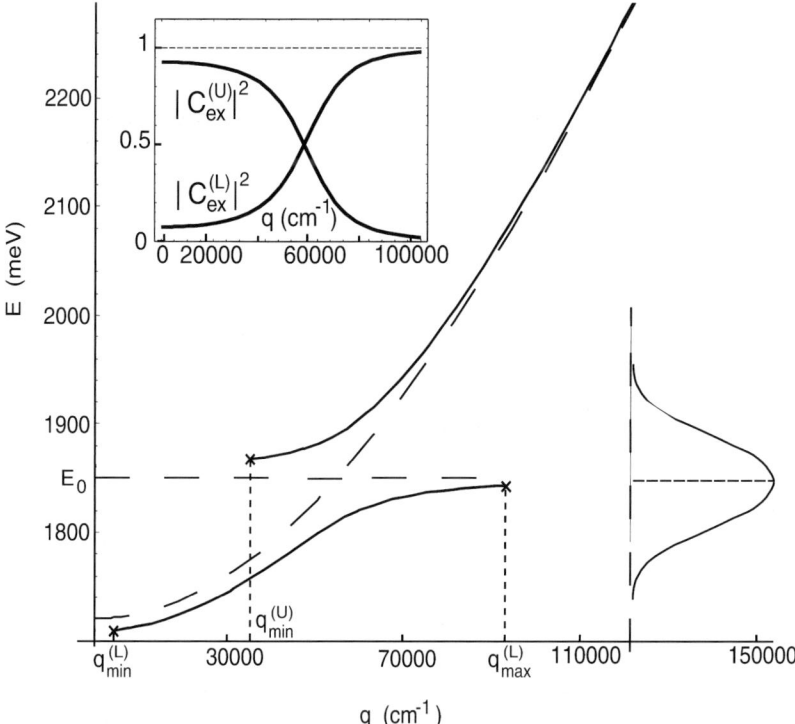

Fig. 11. The dispersion curves for the coherent polaritonic states (solid lines) and for uncoupled cavity photon and the molecular excitation (dashed lines). The cross shows the end-point of the lower polariton dispersion curve. On the right, the inhomogeneously broadened exciton line is drafted. The inset shows the excitonic weights for upper ($|c_{ex}^{(U)}|^2$) and lower ($|c_{ex}^{(L)}|^2$) polaritonic branches.

the organic medium and interacts strongly with the molecular resonance. If the interaction is strong enough, it leads to the strong hybridization of the exciton and cavity photon modes, which results in a Rabi-splitting and the formation of upper and lower branches of coherent (polaritonic) states. In the case of ordered organic material the number of states in the lower polariton branch is equal to the number of electronic excited states. In the two-level model approximation for the organic molecule this number is equal to the number of organic molecules. A key result which we want to outline here is that the number of the lower branch coherent states in the case of disordered organics is much smaller than the total number of electronic excited states. The rest of organic excited states are incoherent, being similar to the excited molecular states in a bulk (non-cavity) film. Qualitatively the smallness of the number of lower polariton branch coherent states in such a situation can be understood if we take into account the fact that cavity photons with large wave vectors are not resonant with the molecular excitations and a strong dephasing of molecular excitations destroys any short wavelength coherent polarization. Thus, for incoherent states, the exciton–photon coupling is not important, and we can say that the system of these excited states is in the regime of weak light-matter coupling. In the case of crystalline inorganic semiconductors, the interaction of small wavelength cavity photons with exciton states is also negligible, however in this case the coherence of short wavelength excitons occurs as a result of resonance Coulomb intermolecular interactions and not as a result of exciton–photon interactions. Coulomb resonance interactions in disordered organic materials are responsible for the inhomogeneous broadening which has to be taken into account. However, such interactions in organic materials with large dissipative width of electronic transitions is too weak to create coherent exciton-like states. Thus, in microcavities with such materials the coherent (polaritonic) states and incoherent states (which are uncoupled to light) coexist (see Figure 11).

We may expect that the results of these consideration will be applicable for any microcavity containing either disordered or even crystalline organic or inorganic materials and this circumstance has to be taken into account in the discussion of linear and nonlinear resonance optical properties of microcavities with optically active material in the region of exciton resonances where the resonance intermolecular interaction is weak in comparison with the dissipative width.

## References

1. E.I. Rashba, M.D. Sturge (Eds.), Excitons, North-Holland, Amsterdam, 1982.
2. B.I. Green, J. Orenstein, S. Schmitt-Rink, *Science* 247 (1990) 679.
3. V.M. Agranovich, R. Atanasov, F. Bassani, *Solid State Commun.* 92 (1994) 295.
4. S.R. Forrest, *Chem. Rev.* 97 (1997) 1793, *Chem. Rev.* 92 (1994) 295.
5. G. Bastard, Wave Mechanics Applied to Semiconductor Heterostructures, CNRS, Paris, 1988.
6. R. Atanasov, F. Bassani, V.M. Agranovich, *Phys. Rev. B* 49 (1994) 2658.
7. R. Cingolani, in: Semiconductors and Semimetals, Vol. 44, Academic Press, 1997, pp. 163–226.
8. T. Yamada, H. Hoshi, T. Manaka, K. Ishikawa, H. Takezoe, A. Fukuda, *Phys. Rev. B* 53 (1996) R13314.
9. A.F. Hebard, R.C. Haddon, R.M. Fleming, A.R. Kortan, *Appl. Phys. Lett.* 59 (1991) 2109.
10. M. Kohl, D. Heitmann, P. Grambow, K. Ploog, *Phys. Rev. B* 42 (1990) 2941.
11. G.C. La Rocca, *Physica Scripta T* 66 (1996) 142.
12. O. Madelung (Ed.), Landolt-Börnstein, Vol. III/22a, Springer-Verlag, Berlin, 1987.
13. S. Schmitt-Rink, D.S. Chemla, D.A.B. Miller, *Phys. Rev. B* 32 (1985) 6601, *Adv. Phys.* 38 (1989) 89.
14. S. Jaziri, S. Romdhane, H. Bouchriha, R. Bennaceur, *Phys. Lett. A* 234 (1997) 141.
15. L. Schulteis, et al., *Phys. Rev. B* 34 (1986) 9027.
16. A. Vinattieri, et al., *Solid State Commun.* 88 (1993) 189.
17. J.M. Turlet, et al., *Adv. Chem. Phys.* 54 (1983) 303.
18. A. D'Andrea, R. Muzi, *Solid State Commun.* 95 (1995) 493.
19. G.C. La Rocca, F. Bassani, V.M. Agranovich, in: O. Keller (Ed.), Notions and Perspectives of Nonlinear Optics, World Scientific, Singapore, 1996; *Nuovo Cimento D* 17 (1995) 1555.

20. H. Haug, S.W. Koch, Quantum Theory of the Optical and Electronic Properties of Semiconductors, 3rd edn., World Scientific, Singapore, 1994.
21. A. Honold, et al., *Phys. Rev. B* 40 (1989) 6442;
    B. Devead, et al., *Phys. Rev. Lett.* 67 (1991) 2355.
22. G.C. La Rocca, F. Bassani, *Phys. Lett. A* 247 (1998) 365.
23. R. Atanasov, F. Bassani, V.M. Agranovich, *Phys. Rev. B* 50 (1994) 7809.
24. N. Blombergen, R.K. Chang, S.S. Jha, C.H. Lee, *Phys. Rev.* 174 (1968) 813.
25. V.M. Agranovich, V.L. Ginzburg, Crystal Optics with Spatial Dispersion and Excitons, Springer-Verlag, Berlin, Heidelberg, New York, Tokyo, 1984.
26. V.I. Yudson, P. Reineker, V.M. Agranovich, *Phys. Rev. B* 52 (1995) R5543.
27. I.S. Gradstein, I.M. Ryzhik, Table of Integrals, Series, and Products, Academic Press, San Diego, 1980.
28. R. Loudon, *Am. J. Phys.* 27 (1959) 649.
29. L. Bányai, I. Galbraith, C. Ell, H. Haug, *Phys. Rev. B* 36 (1987) 6099.
30. (a) A. Engelmann, V.I. Yudson, P. Reineker, *Phys. Rev. B* 57 (1998) 1784;
    (b) N.Q. Huong, J.L. Birman, *Phys. Rev. B* 61 (1999) 13131;
    (c) N.Q. Huong, J.L. Birman, *Phys. Rev. B* 61 (2003) 075319.
31. E. Burstein, C. Weisbuch (Eds.), Confined Electrons and Photons: New Physics and Applications, Plenum, New York, 1995;
    M.S. Skolnick, T.A. Fisher, D.M. Whittaker, *Semicond. Sci. Technol.* 13 (1998) 645, and references therein.
32. F. Tassone, C. Piermarocchi, V. Savona, A. Quattropani, P. Schwendimann, *Phys. Rev. B* 56 (1997) 7554.
33. V. Agranovich, H. Benisty, C. Weisbuch, *Solid State Commun.* 102 (1997) 631.
34. D.G. Lidzey, D.D.C. Bradley, M.S. Skolnik, T. Virgili, S. Walker, D.M. Whittaker, *Nature* 395 (1998) 53.
35. D.G. Lidzey, D.D.C. Bradley, T. Virgili, A. Armitage, M.S. Skolnik, *Phys. Rev. Lett.* 82 (1999) 3316.
36. A.I. Tartakovskii, M. Emam-Ismail, D.G. Lidzey, M.S. Skolnik, D.D.C. Bradley, S. Walker, V.M. Agranovich, *Phys. Rev. B* 63 (2001) 121302.
37. B. Zhang, L. Zhuang, Y. Lin, Z. Xia, Y. Ma, X. Ding, S. Wang, D. Zhou, C. Huang, *Solid State Commun.* 87 (1996) 445;
    A. Dodabalapur, L.J. Rothberg, T.M. Miller, E.W. Kwoch, *Appl. Phys. Lett.* 64 (1994) 2486;
    N. Takada, T. Tsutsui, S. Saito, *Appl. Phys. Lett.* 63 (1993) 2032.
38. G. Vaubel, H. Baessler, *Mol. Cryst. Liq. Cryst.* 12 (1970) 39;
    J.M. Turlet, M.R. Philpott, *J. Chem. Phys.* 62 (1975) 4260.
39. V.M. Agranovich, M.L. Litinskaya, D.G. Lidzey, *Phys. Stat. Sol. (B)* 234 (2002) 130–138; *Phys. Rev. B* 67 (2002) 085311.

Chapter 8
# Strong Optical Coupling in Organic Semiconductor Microcavities

DAVID G. LIDZEY

*Department of Physics and Astronomy, University of Sheffield, Hicks Building, Hounsfield, Sheffield, S37RH, UK*

1. Introduction . . . . . . . . . . . . . . . . . . . . . . . . . . . . . . . . . . . . 355
2. Organic Semiconductor Microcavities . . . . . . . . . . . . . . . . . . . . . 356
3. Organic Semiconductors for Strong Optical Coupling . . . . . . . . . . . . 367
4. Optical Measurement Techniques . . . . . . . . . . . . . . . . . . . . . . . 373
5. Dispersion of Cavity Polaritons . . . . . . . . . . . . . . . . . . . . . . . . 374
6. Cavity Emission Following Non-Resonant Laser Excitation . . . . . . . . . 381
7. Photon Emission Following Resonant Excitation . . . . . . . . . . . . . . . 389
8. Photon-Mediated Hybridisation between Frenkel Excitons . . . . . . . . . 393
9. Future Prospects . . . . . . . . . . . . . . . . . . . . . . . . . . . . . . . . 397
   Appendix . . . . . . . . . . . . . . . . . . . . . . . . . . . . . . . . . . . . 398
   References . . . . . . . . . . . . . . . . . . . . . . . . . . . . . . . . . . . 400

## 1. Introduction

A semiconductor microcavity is a structure in which a wavelength-thickness semiconductor layer is positioned between two closely separated mirrors. The cavity mirrors quantize the local electromagnetic field into a set of discreet confined photon modes. If the energy of one of the modes is resonant with an optical transition of the semiconductor, it is possible to modify both the semiconductor absorption and emission characteristics. Such structures are of fundamental and practical interest, with applications in lasers and other light-emitting devices [1–7].

There are two regimes into which the interactions between a semiconductor and the electromagnetic field can be classified, namely the weak and strong-coupling regimes. In the strong-coupling regime, a cavity photon couples to an exciton having the same energy and in-plane momentum. The coherent coupled state thus formed is termed a cavity-polariton, and can be considered as an admixture of the exciton and cavity-photon modes. Strong-coupling in microcavities was first observed in 1992 by Weisbuch and colleagues, who fabricated heterostructures containing a series of inorganic (III–V) quantum-wells (QWs) [8]. Since then strongly-coupled inorganic semiconductor microcavities have been extensively studied both experimentally and theoretically by a large number of groups. The reader who wishes to delve deeper into the subject of optical strong coupling in microcavities, is encouraged to consult a number of comprehensive review papers listed at the end of this chapter [9–13]. Even though the subject has now reached a degree of maturity, the observation of new effects in strongly-coupled microcavities continues to surprise and delight researchers. One important advance that has emerged within the past few years has been the observation that organic semiconductors can also undergo strong-coupling in suitably designed optical resonators. This advance is particularly important if strong-coupling is to find applications, as the binding energy of organic (Frenkel) excitons in conjugated polymers is usually between 0.1 and 0.5 eV [14] and in the range 0.5 to 1.0 eV in alkali halide crystals [15]. This allows the direct observation and manipulation of Frenkel excitons at room temperature. This is in contrast to inorganic (Mott–Wannier) quantum-well excitons, whose binding energy is typically around 10 meV (dependent on well-width and barrier composition), which thus requires the use of low temperatures to facilitate their observation.

In this chapter, we review our research on strong-coupling in microcavities using organic semiconductors. In Section 2 we discuss the optical properties of microcavities both in the weak and strong-coupling regimes. In Section 3, we summarise the requirements that a semiconductor material must possess to undergo strong-coupling in a microcavity, and discuss the optical properties of a number of organic materials that we have used to achieve strong coupling. In Section 4 we summarise the experimental methods that we have used to study such structures. Section 5 presents some of our observations made via white light reflectivity measurements, whilst in Sections 6 and 7, the photon emission from organic strongly-coupled microcavities is presented following both non-resonant and then resonant excitation respectively. In Section 8, we present our results on microcavities containing two different organic semiconductors, and demonstrate how such structures support new types of hybridised and delocalised optical excitations. Finally, in Section 9 we outline the possible future developments and applications that may emerge from this exciting research area.

## 2. Organic Semiconductor Microcavities

### 2.1. The Optical Properties of Microcavities

A microcavity is a planar Fabry–Perot cavity in which two mirrors are placed either side of a dielectric medium. The presence of the two mirrors quantizes the cavity photon modes, such that only photons having a certain energy and in-plane momentum can be supported. Such photons are confined by the cavity until they either escape from the cavity by penetration through one of the mirrors, or until some optical loss mechanism occurs, such as scattering or absorption. For the simplest case of two infinitely reflecting mirrors (Figure 1) a series of cavity modes are supported having a wavelength $\lambda$ measured outside the cavity given by

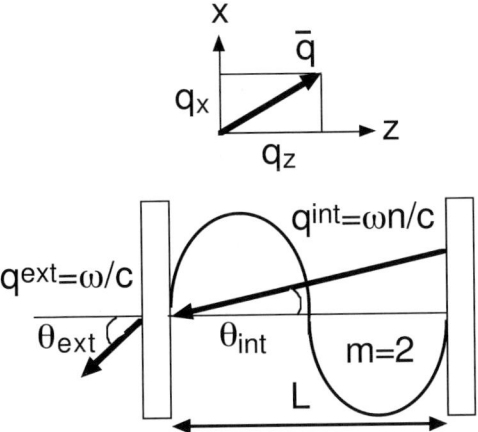

Fig. 1. A schematic diagram of a microcavity, showing the two cavity mirrors and the confined optical field. This particular cavity mode corresponds to an $m = 2$ mode. The internal cavity angle $\theta_{int}$ is also defined.

the expression

$$\frac{m\lambda}{2} = nL\cos\theta_{int} \qquad (1)$$

where $m$ is a mode number, $n$ is the refractive index of the medium between the mirrors and $L$ is their physical separation. In this equation, we have the usual definition of wavelength

$$\lambda = \frac{2\pi c}{\omega} \qquad (2)$$

where $\omega$ is the angular frequency of light, and $c$ is the speed of light in vacuum. The frequency of the cavity mode is dependent on its in-plane wavevector. The in-plane wavevector of the cavity mode ($|q_x^{cav}|$) determines the angle at which it is detected from outside the cavity. The in-plane wavevector of a cavity mode corresponding to an angle $\theta_{int}$ is given by

$$q_x^{cav} = \frac{\omega n}{c} \sin\theta_{int}. \qquad (3)$$

Light with a frequency $\omega$ escaping from the cavity will have the same in-plane component of wavevector ($q_x^{ext}$), where

$$q_x^{ext} = \frac{\omega}{c} \sin\theta_{ext}. \qquad (4)$$

From the equality $q_x^{int} = q_x^{ext}$ it follows

$$\theta_{int} = \arcsin\left\{\frac{\sin\theta_{ext}}{n}\right\}. \qquad (5)$$

For each cavity mode frequency, the component of the wavevector parallel to the cavity growth direction ($q_z$) is constant. For each frequency of a given cavity mode, the $z$ component of the wavevector is given by

$$q_z^{cav} = \frac{2\pi}{\lambda} \cos\theta_{int} \qquad (6)$$

and is constant and this component is independent of frequency, and is given by $q_z^{cav} = \frac{m\pi}{L}$. At $\theta_{int} = 0$, the wavelength of the cavity mode is defined as the cut-off wavelength ($\lambda_{cutoff}$)—see the chapter by F. Bassani in this book. We can thus relate the wavelength of a light measured outside a cavity to the external viewing angle via

$$\lambda = \lambda_{cutoff} \cos\left(\sin^{-1}\left\{\frac{\sin\theta_{ext}}{n}\right\}\right). \qquad (7)$$

If the two mirrors are *closely* separated (i.e., a *micro*cavity) it is possible to arrange that there is only one optical mode present that can interact with any semiconductor material placed within the cavity. For example, a cavity fabricated from two mirrors separated by 150 nm, containing a material having $n = 1.5$, the fundamental cavity mode ($m = 1$) will have a wavelength of 450 nm at normal incidence (corresponding to the blue end of the visible spectrum). The next optical mode ($m = 2$) will be positioned at 225 nm (in the deep ultra-violet). The energy separation between these optical modes is 2.75 eV, which is significantly larger than the absorption or emission bandwidth of many organic and inorganic materials.

The structure shown in Figure 1 represents an ideal cavity consisting of two completely reflecting surfaces. However such a structure cannot be easily realised in practice (at least at visible wavelengths) as there are no materials available that have unity reflectivity. Microcavities have been fabricated based on two metallic mirrors [16–19] (which in the case of silver approximates a perfect reflector), however there are limitations to this approach. To be able to usefully 'use' the light emitted from a source inside the cavity, it must escape from the cavity by penetration through one of the cavity mirrors. If the metallic mirrors are thick compared to their optical skin depth, then light cannot escape from the cavity and there is very weak output coupling. If one of the mirrors are thin, then some light can escape from the cavity through the semi-transparent mirror. However the reflectivity of a thin metallic film is significantly lower compared to a bulk film, which results in a microcavity that has a rather low finesse, and only weakly confines photons. In addition, the absorption of light passing through even a thin (20 nm) metallic film is significant, resulting in a microcavity that has quite high losses.

To achieve effective photon confinement within a microcavity, it is more common to utilise dielectric mirrors (distributed Bragg reflectors or DBRs) as the cavity reflectors. Such dielectric reflectors can have a much higher reflectivity

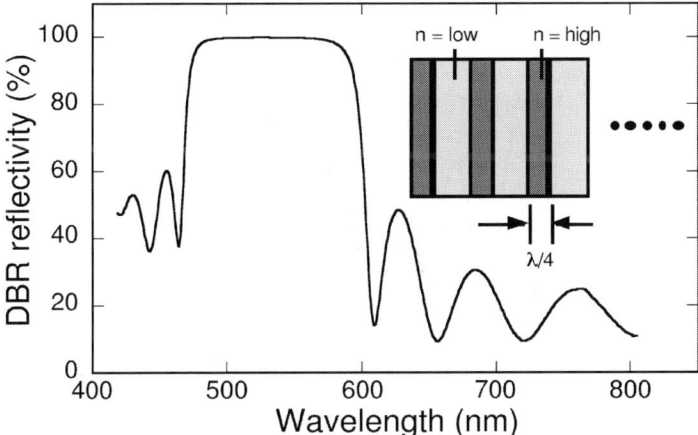

Fig. 2. A reflectivity spectrum of a dielectric mirror (DBR) composed of 12 repeat silicon oxide/silicon nitride $\lambda/4$ pairs. The inset shows a schematic diagram of a DBR.

than a metallic surface, and have the significant advantage of very low absorption loss. A DBR is a multilayer structure composed of a series of 'pairs' of dielectric layers [20]. Each mirror pair is composed of two different dielectric materials, having either a high, or a low refractive index, but with each layer having an optical thickness of $\lambda/4$. DBRs can have a very high optical reflectivity (> 99%) over a range of wavelengths (termed the 'stop-band'). The high reflectivity of a DBR results from in-phase optical reflections from each of the layers within the dielectric stack. Figure 2 shows the measured reflectivity spectrum from a 12 pair DBR composed of alternate layers of silicon oxide ($n = 1.45$) and silicon nitride ($n = 1.95$), on a polished glass substrate. The mirror has a maximum reflectivity of 99.8% at 525 nm, and a stop-band extending from 475 to 580 nm.

Inorganic semiconductor microcavities are usually composed of two DBRs deposited either side of the active semiconductor layer (which is often comprised of a series of quantum wells). As the semiconductor mirrors are deposited using the same molecular beam epitaxy (MBE) process as the quantum wells, the whole structure can, in principal be grown in a single deposition run. This is usually not possible when creating organic semiconductor microcavities, as the process used to deposit the (usually inorganic) DBR is often very different from that used to deposit the organic semiconductor. For example, the DBR shown in Figure 2 was grown by plasma enhanced chemical vapour deposition, which involves reacting two flowing gas streams above a glass substrate heated to 300 °C. Such a deposition process is not possible *on top of* an organic semiconductor, as most organic materials suffer significant degradation at temperatures in excess of 250 °C. This incompatibility between different processing techniques to some extent restricts

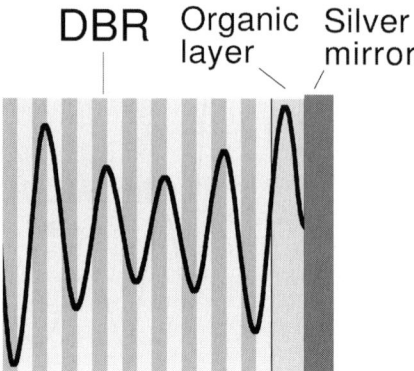

Fig. 3. A schematic diagram of a microcavity, composed of a DBR, an organic layer and a silver mirror. The optical field confined by the cavity (calculated using a TMR model) is shown superimposed on the structure.

the type of structures that can be grown using organic materials. Because of this, most of the organic semiconductor microcavities studied so far have comprised of one DBR and one metallic mirror [21–27]. Metallic thin films can be successfully deposited onto an organic thin film by thermal evaporation, with the organic substrate held at room temperature. This process causes little or no degradation to the optical or electronic properties of the organic material.

Figure 3 shows a schematic figure of an organic semiconductor microcavity based on a metallic mirror and a DBR along with the amplitude of the optical field confined within the cavity as calculated using a transfer matrix reflectivity (TMR) model [20]. This type of structure has been used in all of the experiments described in this chapter. Such microcavities support a $\lambda/2$ optical mode between the two mirrors, whose wavelength can be adjusted by simply changing the thickness of the organic film. It can be seen that the optical field penetrates a significant distance into DBR. Because there is a finite amplitude of the optical field at the DBR surface, light can couple out of the cavity to the outside world. In contrast, light cannot penetrate through the 'thick' ($> 100$ nm) silver mirror, and so all experiments on this structure to study the optical properties of the cavity must be performed through the DBR. When fabricating microcavities of this type, it must be remembered that if the reflectivity of the DBR is very much greater than that of the metallic mirror, then the photons within the cavity are more likely to be absorbed within the metallic mirror than escape through a very high reflectivity DBR. Hence with single DBR cavities, there is a practical limitation to the finesse that can be realised. In our work, we have used DBRs consisting of 9 dielectric pairs, having a peak reflectivity of around 98%. Recent reports have however highlighted new techniques used to grow DBR mirrors onto organic semiconductors, and have realised very high finesse microcavities operating in the weak-coupling limit [28,29]. It will be very interesting to create strongly coupled organic semiconductor microcavities using such procedures.

The linear optical properties of a cavity can be determined by measuring its reflectivity as a function of wavelength. We discuss the practicalities of cavity reflectivity measurement in Section 4. Figure 4(a) shows a reflectivity spectrum measured for a metal-DBR microcavity containing a single layer of a transparent dielectric polymer (having a thickness of $\sim 220$ nm and an average refractive index of 1.6). It can be seen that a single sharp 'dip' is seen in the reflectivity spectrum at 1.74 eV having a linewidth FWHM (full width a half maximum) of 14 meV. The sharp dip in the reflectivity indicates the presence of an optical mode within the cavity to which photons from the outside world can couple. The dip in reflectivity apparently indicates that such photons do not reappear form the cavity. This is because there are other 'guided' optical modes within the cavity that cannot couple to the outside world. If a photon scatters into one of these modes, then it cannot escape from the cavity and can become effectively 'lost'.

The FWHM of the optical mode ($\Delta\lambda$) is an important parameter in determining the $Q$-factor of the cavity. The $Q$-factor is defined as

$$Q = \frac{\lambda}{\Delta\lambda} \qquad (8)$$

where $\lambda$ is the wavelength of the optical mode. The $Q$-factor relates the energy stored by a (damped) oscillator to the energy dissipated per oscillation cycle, and thus cavities with high $Q$-factors confine photons for many 'round-trips' of the photon. As the linewidth of the cavity mode is homogeneous in nature, it can be used to estimate the lifetime of the confined photons within the cavity via the energy-time representation of the uncertainty principle. For example, a cavity with an optical linewidth of 14 meV will trap photons for around 50 fs. Much higher $Q$-factors can be realised using cavities based on two DBRs. The highest finesse microcavities to date have been fabricated by Stanley et al. [30] and have had linewidths of around 130 µeV ($Q = 6000$), corresponding to photon lifetimes of approximately 5 ps. As we will show below, the lifetime of the photon within a cavity is an important factor in determining whether strong coupling can occur.

The optical modes supported by microcavities have a strong angular dispersion (see Eq. (7). This can be readily seen in Figure 4(b), which plots the energy of a confined photon mode (determined from the reflectivity spectra), as a function of external viewing angle. The measured photon energy is shown as solid points and the line is a calculation of the photon energy using the analytical expression given in Eq. (2). The best fit to the data is achieved using an average refractive index of $n = 1.6$, with photon energy at normal incidence being $E_0 = 1.763$ eV. Figure 4(c) re-plots the data in terms of frequency ($\omega$) versus in plane momentum ($q_x$). From this data we can deduce an effective photon mass of $10^{-5}$ of the mass of an electron. We show in the following sections, that the dispersion in the photon

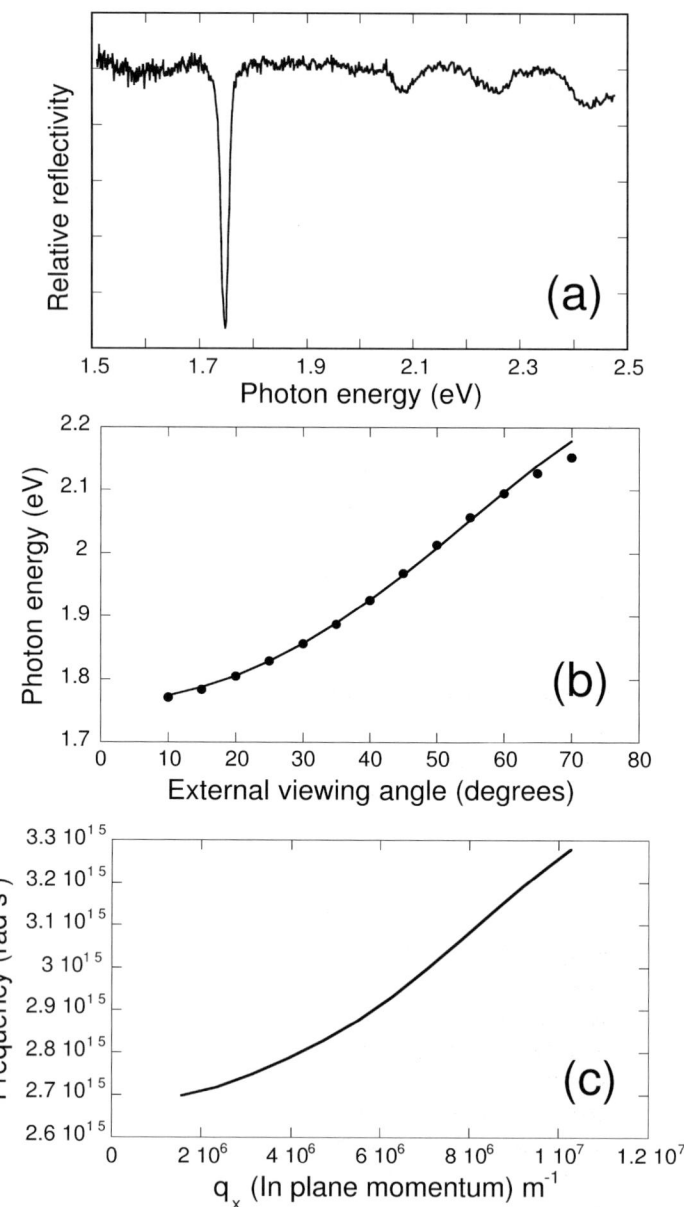

Fig. 4. (a) A reflectivity spectrum measured from a microcavity composed of a DBR, an optically transparent polymer polyvinyl alcohol (PVA), and a silver mirror. The linewidth of the cavity mode is 14 meV. (b) The energy dispersion of the cavity mode as a function of external viewing angle (solid points). The solid line is a calculation of the cavity mode dispersion using Eq. (2).

energy can be used to tune the energy of the photons with respect to an exciton mode within the cavity and thus explore their mutual coupling.

In our discussion of the optical properties of microcavities, we have so-far only considered 'empty' microcavities—i.e., structures that are composed of simple dielectric materials that have no resonant interactions with the cavity photons. We now consider a microcavity containing a semiconductor material, and the interactions that can occur between optically active transitions in the semiconductor and confined cavity photons. There are two distinct interaction regimes that can occur within a microcavity; the weak and the strong coupling regimes. The subject of this chapter is the strong-coupling regime, however for completeness we briefly summarise the salient point of the weak coupling regime.

## 2.2. Weak Coupling

Within the weak coupling regime, the spontaneous emission of a dipole source placed within a cavity can be described via Fermi's Golden Rule:

$$W_{i \to f} \propto \langle f | E.\mu | i \rangle \rho \qquad (9)$$

Here the transition rate ($W$) between initial ($i$) and final ($f$) states is proportional to the density of final optical states $\rho$. The Hamiltonian for the transition is given by $E.\mu$ where $E$ is the electric field experienced by an emitter having a transition dipole moment $\mu$. In the weak coupling regime, the microcavity modifies the density of optical states, enhancing them at the cavity mode wavelength, and suppressing them elsewhere. The modification of the frequency dependent density of optical states in the cavity also changes the distribution of vacuum field fluctuations that effectively 'stimulate' spontaneous emission: At the cavity mode wavelength, the density of vacuum field fluctuations is enhanced, leading to an increase in the spontaneous emission rate.

If a broad-band emitter is placed into a microcavity, the cavity enhances the emission intensity at the cavity mode wavelength and suppresses it elsewhere, producing a significantly spectrally narrowed source. A number of authors have demonstrated that the intensity of emission can be enhanced at the cavity mode wavelength, in some cavities by up to 60 times [25,31]. Fluorescence lifetime emission measurements show however that the *overall* spontaneous emission rate from an excited atom or exciton within a 1-dimensional microcavity is rather small—of the order of 20% at most [5,32]. Such small changes occur because there are a large number of 'leaky' modes in a 1D microcavity, including many guided optical modes in a DBR. These leaky modes limit the overall effect of the cavity on an exciton, with the result being that changes in the spontaneous emission rate are small. A microcavity in the weak coupling regime can thus be thought of a structure that redistributes the emission from a source placed between

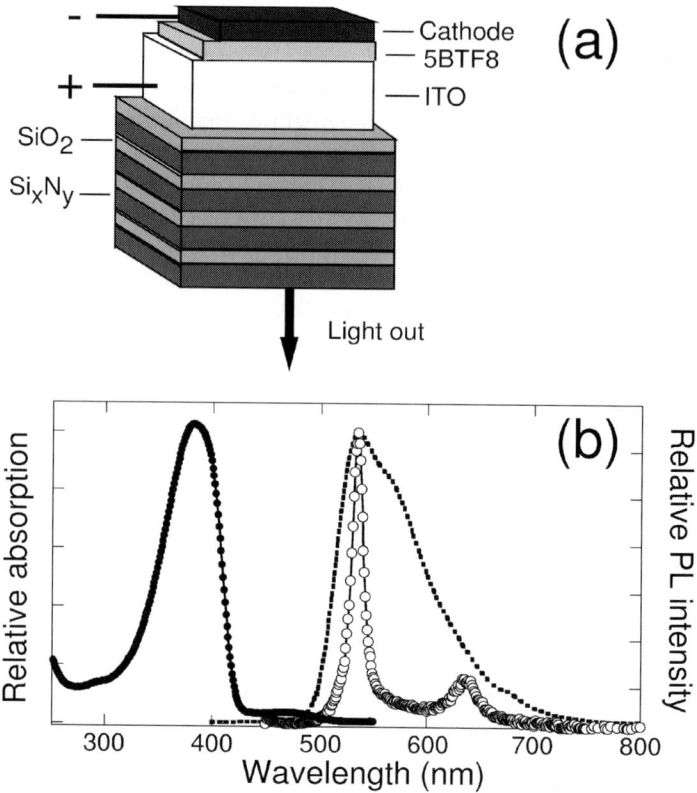

Fig. 5. (a) Schematic diagram of a resonant cavity LED (RCLED) containing a conjugated polymer as the active charge transporting and emissive material. (b) Absorption (solid dots) and emission (dotted line) from the polymer system used in the RCLED shown in part (a) in a non-cavity device. The spectra shown using open circles is the electroluminescence emission from the RCLED.

the mirrors, allowing light to escape from the structure that would otherwise be trapped by total internal reflection.

Such properties are in fact highly desirable as they can be used to enhance the external quantum efficiency of light emitting devices. The strong spectral narrowing that can be achieved is also very useful in display devices, where pure, spectrally narrow emission colours are required to achieve a full colour palette. Figure 5(a) shows a schematic diagram of a light emitting diode based on a conjugated polymer that has been engineered into a microcavity (a resonant-cavity LED). The structure is based on a DBR, an indium tin-oxide anode (used to inject holes into the conjugated polymer) and a top cathode-mirror, composed of a thin (10 nm) layer of calcium to facilitate the injection of electrons into vacant states in the polymer. A thick layer of aluminium is deposited onto the calcium, to create a reasonably highly reflective cavity mirror.

Figure 5(b) shows the emission from the cavity (open circles) along with the absorption (full circles) and emission (dotted line) from the same conjugated polymer in a non-cavity electroluminescent device. The spectral narrowing of emission from the cavity is very clear, with a strong emission peak visible at 535 nm, having a linewidth of 12 nm. This compares to the electroluminescence emission from a non-cavity LED that has a FWHM of around 90 nm. It can be seen that the emission from the microcavity LED is not completely 'pure': there is also weak emission from another feature at 635 nm. This emission comes from a second cavity mode, whose wavelength coincides with the long wavelength emission tail of the polymer. Such bi-modal emission can be easily eliminated by reducing the overall length of the cavity [3,26].

## 2.3. Strong Coupling

In the strong-coupling regime, cavity photons and excitons can couple, forming a coherent superposition of states termed a cavity polariton. Such quasi-particle states can be thought of as being part-light and part-matter. In order to couple an exciton and a photon to form a cavity polariton, both states must have the same energy and in-plane momentum ($q_x^{\text{cav}}$). In traditional inorganic semiconductors, the semiconductor crystal lattice ensures that the system has translational symmetry, and thus the exciton wavevector is considered a good quantum number. In our microcavities however, we have used organic semiconductors as the active medium. Such organic materials (in all but a highly crystalline form) can be considered as very disordered, and thus such systems do not have long-range translational symmetry. This lack of symmetry does not mean that strong-coupling cannot be achieved: If the mean distance ($r$) between the (organic semiconductor) chromophores within the cavity is small in comparison with the wavelength of the light ($\lambda$), then the medium can be considered as effectively optically homogeneous. In our molecular systems, $r$ is typically between 1 and 20 nm, whilst $\lambda$ is around 600 to 700 nm, allowing us to satisfy the relation $1/k \gg r$. As resonance between the cavity photon and the exciton occurs for $k$ vectors of the order of $1/\lambda$, the wavevector can be considered a good quantum number. This, we will demonstrate, allows us to create the same type of cavity polaritons that are observed in inorganic microcavities containing ordered semiconductors. However as we couple the cavity photon to organic Frenkel excitons (rather than inorganic Mott–Wannier excitons), we can explore a much larger exciton–photon coupling strengths, due to the very large oscillator strength of Frenkel excitons.

In a microcavity, the strong-coupling regime is evidenced by an anticrossing between photon and exciton modes, and on resonance the appearance of two equal intensity transitions, separated by the Rabi-splitting energy $\hbar\Omega_{\text{Rabi}}$. Such a splitting can only be realised when the interaction strength (expressed as an angular frequency $\Omega_{Rabi}$) between a photon and an exciton is greater than (i) the inverse cavity photon lifetime, and (ii) the inverse exciton dephasing time [13]. Condition (i) dictates that the photon is confined in the cavity for a time that is longer than the period of the Rabi oscillations ($2\pi/\Omega_{Rabi}$). Condition (ii) ensures that an ensemble of excitons retain their mutual coherence in their wavefunctions for a time which is also longer than the Rabi oscillation period. If either of these conditions is not met, then a coherent superposition of states cannot be formed because either the photon leaks from the cavity, or the exciton undergoes dephasing before one period of oscillation between the exciton and photon modes can occur. In this case, the system is in the weak coupling limit and the exciton decays by spontaneous emission. It is common to express such inequalities in terms of the homogeneous energy linewidths of both the exciton ($\Gamma_{ex}$) and the photon modes ($\Gamma_p$); thus

$$\Gamma_{ex} < \hbar\Omega_{Rabi}, \tag{10}$$

$$\Gamma_p < \hbar\Omega_{Rabi}. \tag{11}$$

It is also important to consider the *inhomogeneous* linewidth when deciding whether strong coupling can be achieved. To be able to resolve a splitting between a photon and an inhomogeneously broadened semiconductor transition, the inhomogeneous linewidth also needs to be narrower than the Rabi-splitting energy. To a first approximation, the Rabi-splitting energy can be expressed using

$$\hbar\Omega_{Rabi} \propto \sqrt{\frac{f}{n_c^2 L_{tot}}} \tag{12}$$

where $f$ is the oscillator strength of the semiconductor excitons per unit area, $n_c$ is the average refractive index of the semiconductor and $L_{tot}$ is the *physical* path length of the cavity. The oscillator strength $f$ per unit area of a thin film is given by

$$f \propto \alpha L \tag{13}$$

where $L$ is the film thickness, and $\alpha$ is the optical attenuation coefficient (per unit path length) at the peak of the absorption. If the inhomogeneous linewidth of the semiconductor material is very broad, then the oscillator strength of the transition is effectively spread over frequency space, and thus the peak value of $f$ that is achievable is relatively small. A reduced value of $f$ results in a reduced Rabi-splitting, with a possible consequence being that the Rabi-splitting becomes smaller than the inhomogeneous linewidth of the excitons. In this case the broad inhomogeneous linewidth of the excitons (which encompasses a large distribution of narrower homogeneous states), results in a distribution of reduced coupling strengths, effectively masking the anticrossing behaviour. The eventual consequence of using states with very broad inhomogeneous linewidths is that the system operates in the weak-coupling regime.

The operation of a material system in the weak coupling regime can be seen in Figure 5(b). Here the organic semiconductor used has rather broad and featureless transitions, which are characterised by transition linewidths of the order of 500 meV. Such a material system would not be anticipated to undergo strong-coupling, as it effectively constitutes a continuum of states. In addition, in this structure, the cavity mode has been deliberately positioned at a wavelength that coincides with a region of high electroluminescence emission intensity from the conjugated polymer. Because of the large Stokes shift in this material system, there is a very low residual absorption (and thus no transition having significant oscillator strength) at the cavity mode wavelength, again indicating that the system operates in the weak coupling limit.

## 3. Organic Semiconductors for Strong Optical Coupling

### 3.1. Background

Before we discuss strong coupling of organic semiconductors, it is useful to briefly describe the magnitude of the Rabi-splitting observed in microcavities containing inorganic (III–V) QW excitons. The inhomogeneous linewidth of a QW exciton (which is often broadened by fluctuations in the alloy composition of the inorganic crystal, and also by roughness in the thickness of the quantum well) is typically around 1 meV. By utilising highly reflective semiconductor DBRs, it is relatively straightforward to create microcavities supporting confined photon states having linewidths of 1 meV. In microcavities based on III–V semiconductor quantum wells (utilising GaAs/InGaAs), Rabi splittings of around 5 meV have been observed (at 20 K) [9]. Therefore the homogeneous and inhomogeneous linewidths of both the photon and exciton are smaller than the Rabi splitting, satisfying the inequalities given in Eqs. (5) and (6).

### 3.2. Molecular Dyes

As we have discussed in Section 2.3, to achieve strong optical coupling, it is necessary to use semiconductor materials whose absorption linewidths are narrow compared to the Rabi-splitting energy. The materials we have used to achieve strong-coupling have had inhomogeneous absorption linewidths between 40 and 90 meV. This linewidth is well over an order of magnitude larger than the Rabi-splitting observed in III–V QW microcavities. However this relatively broad linewidth has not precluded organic semiconductors from reaching the strong-coupling regime; this is because the oscillator strength of a thin film of an organic semiconductor can be at least two orders of magnitude larger than that of a series of inorganic QWs. It can be seen from Eq. (7), that the Rabi-splitting energy is proportional to $\sqrt{f}$, and thus the Rabi-splittings that we observe can be significantly larger than the relatively broad linewidth of the organic semiconductors that we have used in our microcavities.

In our microcavities, we have used organic materials that we have used have the distinct advantage that they are relatively easy to fabricate into high-quality, defect-free, thin films suitable for inclusion into a microcavity. In our first observation of strong coupling in a microcavity [21], we used a porphyrin dye having a single narrow optical absorption. The dye used (tetra-(2,6-*tert*-butyl)phenol-porphyrin zinc) which we term 4TBPPZn is one example of a large family of porphyrin macrocycles whose photophysical and photochemical properties have been extensively studied [33]. The chemical structure, absorption spectrum and electronic energy-level scheme of the 4TBPPZn molecule are shown in Figure 6(a). This molecule has an intense, narrow ($90 \pm 5$ meV at room temperature) transition in the form of the so called "Soret band" absorption at 2.88 eV. The relatively narrow linewidth of the optical absorption comes from the high structural rigidity of the porphyrin molecule. The Q-band absorbance is seen as a weak double peak structure between 2.0 and 2.26 eV.

To utilise the 4TBPPZn molecular dye in a microcavity, it was dissolved into a solution of toluene containing the polymer polystyrene. The solution was then spin-coated onto a quartz substrate. On spin-coating, the toluene solution rapidly evaporates, leaving the 4TBPPZn molecules suspended in a solid polystyrene matrix. This deposition method allows the creation of films having areas of up to several cm$^2$ with a thickness anywhere between 30 nm and 1 μm. The point-to-point fluctuations in the film thickness over an area of around 1 cm$^2$ can be relatively small (of the order of 5–10 nm). By careful calibration of the viscosity of the solution and spin-speed, the thickness of the film deposited can be controlled with a precision of about 10 nm. The attainment of very smooth organic films is critical for the fabrication of high-quality microcavities. Any surface roughness or scattering within the film will lead to a significant broadening of the cavity optical mode.

The absorption spectrum of the polystyrene film containing the 4TBPPZ molecules is shown in Figure 6(a). The polystyrene matrix has an average refractive index of approximately 1.5, and has a very high degree of optical transparency over the whole of the visible spectrum. The use of a matrix polymer to suspend the active molecules is an important feature of our approach. The matrix physically separates the molecules from one another by suspending them in a 'solid solution'. It is of course possible to create a 'pure' thin film of 4TBPPZn molecules directly via spin-coating or by thermal evaporation without the use of a matrix film, however the strong intermolecular interactions that occur in such molecular films tend to significantly broaden the Soret-band absorption and reduce the pos-

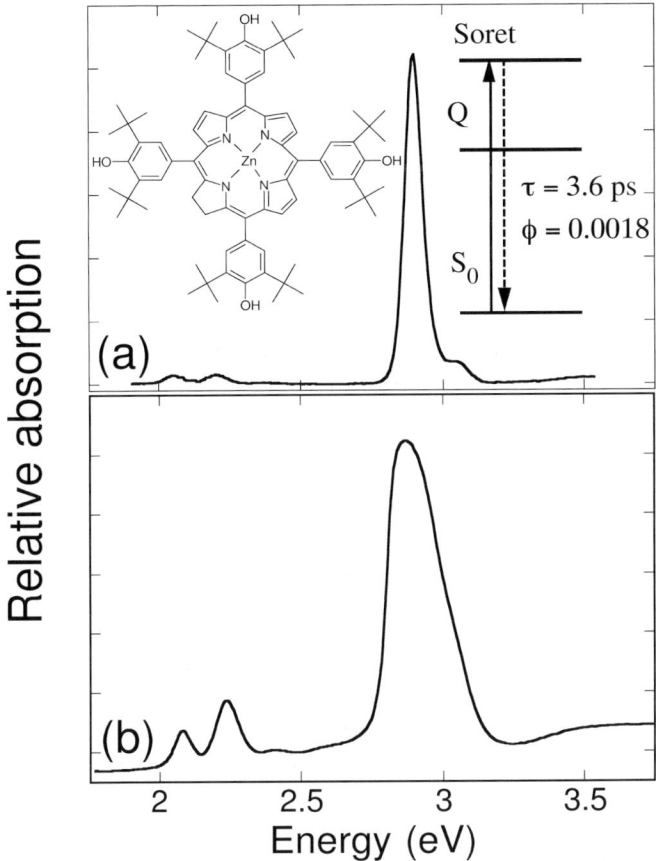

Fig. 6. (a) The absorption of a thin film of 4TBPPZn molecules in a polystyrene matrix. The insets show the chemical structure of 4TBPPZn, and its energy level scheme. (b) The absorption of a drop-cast film of 4TBPPZn on a quartz substrate.

sibility of achieving strong coupling. Figure 6(b) shows the absorption of a thin film of 'pure' 4TBPPZn molecules that have been deposited by spin-coating. It can be seen that the Soret-band absorption has now broadened, having a linewidth of approximately 220 meV. There is also an absorption background visible in the spectrum, whose origin comes from scattering. This scattering results from microcrystallinity within the film, again resulting from the strong intermolecular interactions that occur when a matrix polymer is not used. Such scattering would significantly reduce the finesse of the microcavity, and make strong-coupling much more difficult to achieve.

### 3.3. J-Aggregates and Other Self-Assembled Molecular Systems

A second type of organic semiconductor system that we have used for strong coupling is based on J-aggregates of cyanine dyes, which are a class of material that has found diverse uses in photography [34] and as laser dyes [35]. Cyanine dyes carry a net charge, which drives a self-assembly of the molecules in a polar solution to form 1-dimensional aggregates termed J-aggregates [36–38]. Intermolecular interactions in J-aggregates is responsible for the appearance of an excitonic band. Due to the 'head to tail' packing of the molecules in the aggregate the lowest energy state in the band has zero wavevector. To a first approximation we can say that optical transitions are only permitted to the lowest energy point in the band. For this reason and due to phenomena resulting from motional narrowing [39], J-aggregate absorption spectra are often characterised by a single, relatively narrow and intense optical transition, significantly red-shifted from the absorption of the un-aggregated monomer. Following optical excitation, the primary fundamental excitations that are created in the aggregate are singlet excitons. Such excitons have relatively large binding energies, which allow them to be created and studied at room temperature. In J-aggregates, the excitons are highly mobile and can be viewed as being delocalised over a relatively large number of molecular units.

To process cyanine dyes into microcavities, the J-aggregates were suspended in a polymer matrix. Cyanine dyes are usually soluble in polar solvents such as water and methanol, and thus we have used the matrix material poly(vinyl alcohol) [PVA] which is also soluble in aqueous solvents. Strong-coupling has been achieved with two different J-aggregate forming dyes (whose chemical structures are shown in the insets in Figures 7(a) and 7(b) [22,23]). To fabricate a thin film of J-aggregates, the cyanine dye is dissolved at a concentration of $\sim 1$ mg/ml into a 50/50 water-methanol mix containing the PVA polymer. Thin films of the composite organic film were formed by spin-coating. During the spin-coating process, the water and methanol evaporate, which rapidly raises the concentration of the cyanine dye. This drives an association of the molecules to form the J-aggregates. Figure 7(a) and (a) shows the optical absorption and photoluminescence of two different PVA/J-aggregate thin films. The J-aggregates shown in Figure 7(a) have an absorption peaking at 1.84 eV, with a linewidth of 40 meV. The absorption is slightly asymmetric, having a weak tail that extends to higher energies. The PL is the mirror image of the absorption, and is Stokes shifted down in energy by 5 meV.

This solution based deposition method is critical for the formation of the J-aggregates. If a thin film of the cyanine dye is deposited by a non-aqueous method (for example by thermal evaporation), the molecules are unable to associate with one another and the formation of the 1-dimensional aggregates is blocked. This is highlighted by the absorption spectra shown in Figure 8. For ease of comparison, we plot the normalised absorption of an amorphous cyanine dye film formed

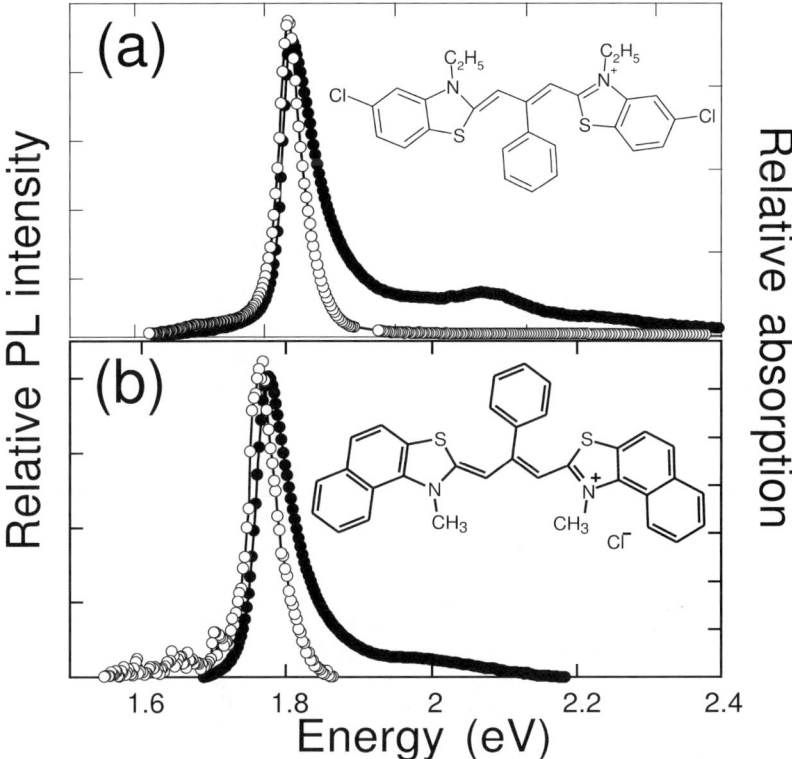

Fig. 7. Absorption (solid dots) and photoluminescence emission (open circles) of a thin film of J-aggregated cyanine dyes in a PVA matrix. Note, the chemical structure of each of the dyes are shown as insets.

by vapour deposition along with the same material in a PVA matrix which has self assembled in solution to form J-aggregates. However, the peak absorption of J-aggregated film would be much more intense than that of an amorphous film containing the same number of monomers per unit volume. To a good first approximation, the oscillator strength of each of the monomers is concentrated into the J-aggregate transition band. It can be seen that the J-aggregate absorption is located at the extreme low-energy end of the monomer absorption. Whilst the aggregated molecules are a good candidate for strong-coupling, the amorphous material clearly is not. There is thus a subtle difference in the use of J-aggregates for strong coupling as apposed to the use of highly rigid molecular dyes as exemplified by the material system shown in Figure 2: Here, we do not rely on the structural rigidity of the molecule to create a narrow optical transition. Instead the narrow transition arises because of the natural tendency of the molecules to

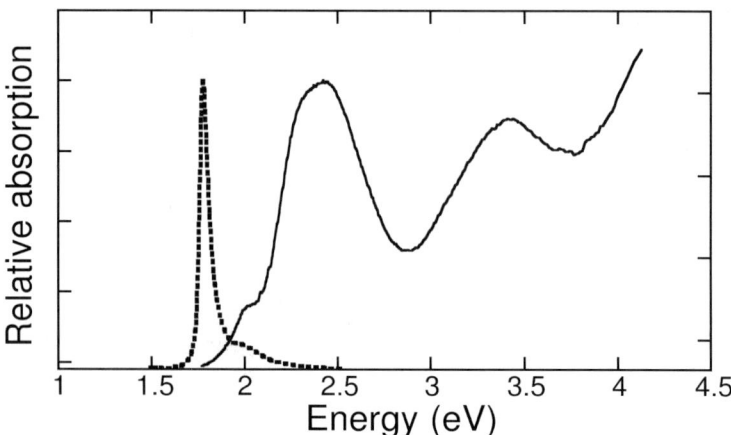

Fig. 8. The absorption of an amorphous film (solid line) and a J-aggregated film (dotted line) of the cyanine dye shown in Figure 7(b).

self organise into molecular superstructures which have delocalised and narrow optical transitions, determined by the structure of the molecular packing.

One consequence of the self-assembly of the cyanine dye molecules during the spin-coating process is a significant degree of inhomogeneity in their spatial distribution. We have made preliminary investigations of such J-aggregate films using a near-field optical microscope, and find strong variations in the density of aggregates within the PVA film on length-scales of a few hundred nanometers and below. We can in fact estimate the number of molecules within each aggregate on the basis of the reduction in the linewidth of the J-aggregate absorption compared to the monomer absorption. It has been shown [39], that the inhomogeneous linewidth of an aggregate composed on $N$ molecules scales with $N^{-1/2}$. For the material system shown in Figure 8, the linewidth of the J-aggregate is approximately 10 times smaller than that of the monomer, implying that each aggregate is comprised of about 100 coupled-molecules. We estimate the length of a single cyanine-dye molecule to be around 2 nm, and allowing for some overlap between the neighbouring molecules in the aggregate, we calculate the average length of a J-aggregate to approximately 100 to 150 nm. The cyanine dye molecules are added to the PVA matrix at a concentration of approximately $4 \times 10^{19}$ cm$^{-3}$, from which we calculate that there are approximately $2 \times 10^{17}$ aggregates per cubic centimetre, implying an average *aggregate* separation of 20 nm.

Strong-coupling effects have also been observed using perovskites. Perovskites are self assembled systems, that form alternate nano-structured organic and inorganic layers [40]. Such materials can be thought of as analogous to inorganic quantum well heterostructures: in a perovskite, the organic component of the complex [$C_6H_5C_2H_4NH_3$] acts as the barrier for the inorganic well material [a $PBI_4$ salt]. The excitons are localised within the inorganic well and have absorption linewidths of around 100 meV. Both synthetic opals [42] and distributed feedback gratings (DFBs) [43] have been infiltrated with perovskites, and strong-coupling behaviour has been evidenced.

## 4. Optical Measurement Techniques

One of the simplest techniques used to characterise microcavities is the measurement of reflectivity as a function of angle. This allows the direct determination of the energy and dispersion of the optically accessible states within the cavity. 'Photon-like' cavity modes can be readily identified by a sharp dip in the cavity reflectivity spectra. Such photonic modes in microcavities have a strong angular dispersion, with their wavelength at external viewing angle $\theta$ given by Eq. (2).

The angular dependent reflectivity spectra of our microcavities can be measured using the apparatus shown schematically in Figure 9. The microcavity (MC) is mounted on a central stage, to which are fixed 3 optical rails, each having free independent rotation around the central axis. Light from a tungsten projector lamp

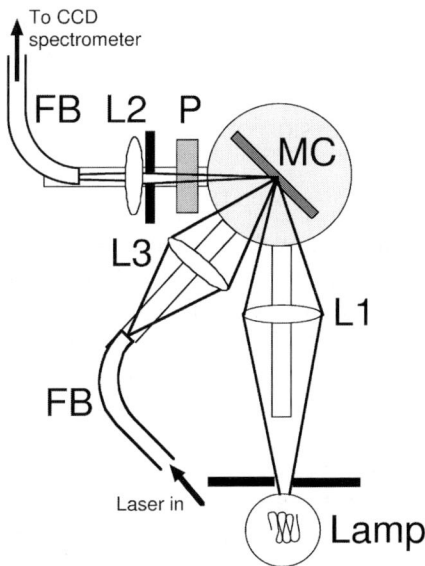

Fig. 9. The apparatus used to measure both white light reflectivity, and photon emission from an organic semiconductor microcavity. Abbreviations used in this figure: Microcavity (MC), lens (L), Fibre bundle (FB), P (Polarizer).

is imaged onto a 500 µm spot on the microcavity surface via lens L1. The specular reflection from the cavity is collected by a lens L2 and imaged into a fibre bundle (FB) that is also mounted on the optical rail. The fibre-bundle is then used to deliver the reflected light to a cooled CCD spectrograph. Using this equipment, the white light reflectivity of the microcavity can be measured at angles from 12° to 85°. A polarizer (P) placed in front of lens L2 allows either TE or TM polarisations of reflected light to be detected. By comparing the reflectivity of a microcavity to that of an aluminium mirror (which is assumed to reflect light with approximately equal efficiency over the whole visible range) the absolute reflectivity of the cavity can be determined.

Using the system shown in Figure 9, we are also able to generate and detect photoluminescence emission from the microcavity as a function of angle. Light from a laser is focussed into a fibre bundle, one end of which is mounted on an optical rail. The laser light delivered by the fibre is then focussed by lens L3 onto a 500 µm spot on the cavity surface. This system allows the photoluminescence and reflectivity to be measured from exactly the same spot on the cavity surface. This apparatus is designed to allow measurements to be made at room temperature and in air. For most purposes this is sufficient, as the organic materials that we use have reasonable photostability. However for some measurements (particularly those involving the use of lasers with high excitation density) it is preferable to mount the microcavity in a cryostat at low temperature. A second similar system is available having a wide angular access cryostat mounted on a rotation stage.

## 5. Dispersion of Cavity Polaritons

### 5.1. REFLECTIVITY MEASUREMENTS OF STRONGLY-COUPLED CAVITIES

As discussed above, the energy of the cavity photon mode varies strongly as a function of angle, shifting to higher energies at off axis viewing angles. This effect can be conveniently used to tune of the interaction between the exciton and photon modes: the exciton transition energy is to a very good approximation angle independent and hence changing the angle of incidence allows one to adjust the relative separation of the photon with respect to the exciton. We illustrate this point with measurements on microcavities containing a layer of polystyrene doped with 4TBPPZn molecules (whose structure is shown in Figure 6(a)). Microcavities were fabricated such that the cavity mode was at a lower energy than the exciton at normal incidence by some 100 to 150 meV and became resonant with it at approximately 35 to 40°. The room temperature reflectivity of a microcavity measured at normal incidence to the cavity axis is shown in Figure 10, curve A. The cavity photon mode is visible as a sharp dip in the reflectivity at 2.68 eV. The 4TBPPZn exciton absorption lies at 2.88 eV (marked by the dashed vertical

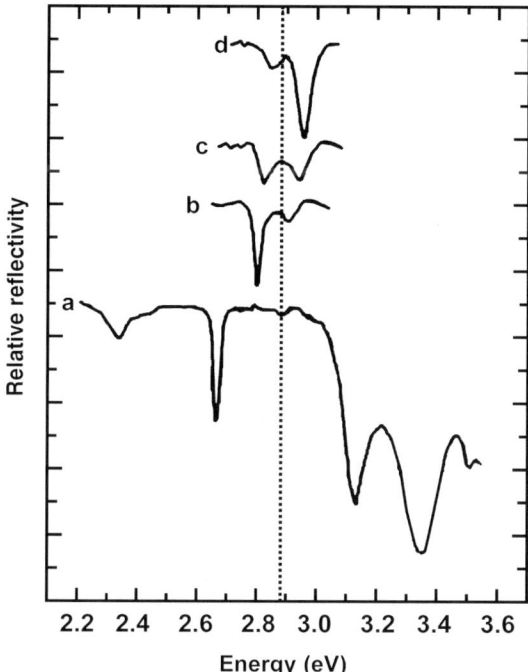

Fig. 10. A series of reflectivity spectra measured at progressively larger viewing angles. The vertical dotted line marks the resonance energy for the 4TBPPZn excitons within the cavity.

line). The exciton can only just be detected at normal incidence in the reflectivity spectrum as there is very little coupling between the photon and exciton modes as they are relatively far apart in energy. As the cavity mode is tuned closer to the exciton energy, strong-coupling is observed (curve B). The two modes of the system can now no longer be described as 'pure photon' or 'pure exciton'; rather each mode must be described as a linear superposition of photon and exciton. The exciton-like mode thus becomes increasingly visible in reflectivity, as it contains an increasing large cavity-photon like character. Such mixed modes are termed cavity polaritons. On resonance (curve C) the expected pair of equal intensity Rabi-split transitions are seen both containing equal admixtures of a photon and an exciton. For larger angles, beyond resonance (curve D) the coupling reduces, again in line with expectation. Such spectral features are entirely consistent with other measurements made on strongly-coupled inorganic-semiconductor microcavities [9].

The energies of the two transitions are plotted in Figure 11 as a function of the angle of incidence. The horizontal dashed line at 2.88 eV is the 4TBPPZn exciton energy in the polystyrene blend film. This energy is angle independent

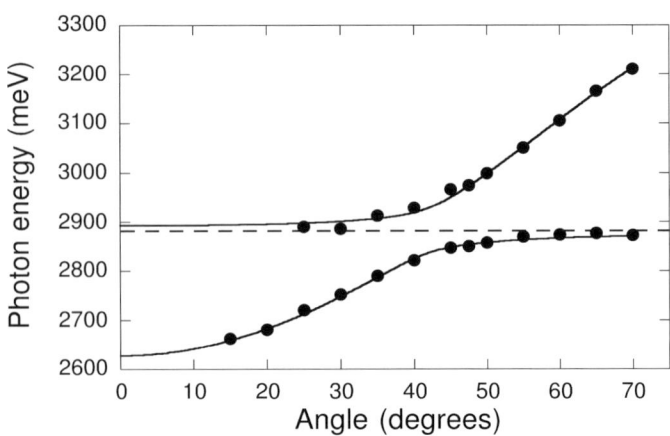

Fig. 11. Dispersion curve measured from a microcavity containing 4TBPPZn molecules. The horizontal dashed line is the peak absorption energy of the 4TBPPZn molecules, about which anticrossing occurs.

and hence defines the resonance energy for the coupled exciton–photon system. The cavity photon and 4TBPPZn exciton modes exhibit a very clear anti-crossing behaviour and the Rabi-splitting energy, of 110 meV between the two transitions at the 40° resonance angle is *exceptionally large*.

## 5.2. Macroscopic and Microscopic Analysis of Dispersion

The structure that we consider is very disordered. However as the distance between the 4TBPPZn dye molecules in the microcavity is approximately 3 nm, it is thus much smaller than the wavelength of the optical mode. This allows us to use an equation of macroscopic electrodynamics, with some dielectric constant to describe the optical properties of the cavity. We can write following Maxwell's equations for a light wave in a three dimensional isotropic medium as

$$\frac{k^2 c^2}{\omega^2} = \varepsilon(\omega) \quad (14)$$

where $\omega$ and $k$ are the frequency and wavevector of a light wave. This equation determines the dispersion relation of the frequency to the wavevector. However let us firstly consider a microcavity having a material that has an optical resonance that is very far from the cavity cut-off frequency. In this case, the dielectric constant $\varepsilon_0$ can be considered a constant, thus

$$\frac{k^2 c^2}{\omega^2} = \varepsilon_0 \quad (15)$$

For the $m$th cavity mode $k_z$ is fixed, and in this case

$$k^2 = k_x^2 + \left(\frac{m\pi}{L}\right)^2, \quad m = 1, 2, \ldots, \quad (16)$$

where $k_x$ is the in-plane component of the total wavevector. From Eqs. (14) and (15), it is easy to show that for small $k_x$, the dependence of the frequency of the cavity mode on its in-plane wavevector is given by

$$\omega^{\text{cav}}(k_x) = \omega_{\text{cutoff}} + \frac{\hbar k_x^2}{2\mu} \quad (17)$$

where $\omega_{\text{cutoff}}$ is given by

$$\omega_{\text{cutoff}} = \frac{c}{\sqrt{\varepsilon_0}}\left(\frac{\pi m}{L}\right) \quad (18)$$

and $\mu$ is the effective mass given by

$$\mu = \frac{\hbar \omega_{\text{cutoff}} \varepsilon_0}{c^2}. \quad (19)$$

In organic materials $\varepsilon_0$ is of the order of 3, and as the cutoff energy in our cavity is given by $\hbar\omega_{\text{cutoff}} \sim 2$ eV, we calculate the cavity photon effective mass to be of $10^{-5}$ of a free electron.

We assume that the material placed within the cavity, has an (excitonic) resonance given by

$$\varepsilon(\omega) = \varepsilon_0 + \frac{2E_0 f}{E_0^2 - E^2 - 2i\gamma(E)E} \quad (20)$$

where $f$ is some constant proportional to the oscillator strength, $E_0$ is the resonance energy of the exciton, and $\gamma$ is the homogeneous broadening. In the case where $E \approx E_0$ the resonance term can be simplified and it can be shown that

$$\varepsilon(\omega) = \varepsilon_0 + \frac{f}{E_0 - E} \quad (21)$$

where we neglect the term accounting for homogeneous broadening. Substituting Eq. (21) into Eq. (14), and using Eqs. (16) and (17), it can be shown that the dispersion of the cavity polaritons is given by

$$[E_c(k_x) - E](E_0 - E) = V^2 \quad (22)$$

where the energy of an optical mode in an bare cavity is given by

$$E_c(k_x) = \hbar\omega^{\text{cav}}(k_x) \quad (23)$$

and $V$ is given by

$$V = \sqrt{\frac{fE_0}{2\varepsilon_0}} \qquad (24)$$

and is half the Rabi-splitting energy. For the derivation of the above equations, we used a phenomenological Maxwell equation. The results of this derivation, summarised by Eq. (22) are identical to the results of microscopic theory, which are usually used in the analysis of inorganic semiconductor microcavities. Our derivation presents a straightforward approach to determine the main features of the Rabi-splitting if the dielectric constant of the organic semiconductor within the cavity is known. At the same time, it can be generalised to take account of both homogeneous and inhomogeneous broadening.

However we now proceed in our analysis of the strongly coupled microcavity using a microscopic theory. This method is widely used [9] in the description of inorganic semiconductor microcavities. The cavity is described as a compound oscillator consisting of the exciton and photon modes. The compound oscillator is described with the matrix equation

$$\begin{pmatrix} E^{\text{cav}}(k_x) - E & V \\ V & E_0 - E \end{pmatrix} \begin{pmatrix} \alpha \\ \beta \end{pmatrix} = 0, \qquad (25)$$

$E_0$ and $V$ are assumed to be constant at all angles. The matrix can be diagonalized to obtain the eigenvalues of the system ($E$) that represent the energy of the two polariton branches. The advantage of this approach is that the coefficients of the bare photon and exciton which describe the weighting of each of the states in the coupled modes (which are represented in Eq. (25) as $\alpha$ and $\beta$) can be directly determined. For the important case of the photon resonant with the exciton ($E^{\text{cav}} = E_0 = \varepsilon$), it is straightforward to show that the polariton branches have energies given by

$$E_\pm = \varepsilon \pm V \qquad (26)$$

where $2V = \hbar\Omega_{\text{Rabi}}$. Using the fact that $\alpha^2 + \beta^2 = 1$, it is straightforward to demonstrate that at resonance $\alpha = \beta = 1/\sqrt{2}$. This simple analysis confirms that at resonance, the two polariton branches are separated equally in energy around the resonance energy of the system ($\varepsilon$), and that both branches contain equal amplitude of the exciton and the photon modes.

To compare with the experimental data, the photon energy dispersion $E_{\text{ph}}(\theta)$ is calculated via a transfer matrix model in which the only free variables are the photon energy at $0°$ and the average (non-dispersive) refractive index of the cavity. The solid line displayed in Figure 11 is the energy of the two polariton branches calculated using Eq. (25). As it can be seen, the agreement between the data obtained from experiment and the coupled oscillator model is excellent. Figure 12

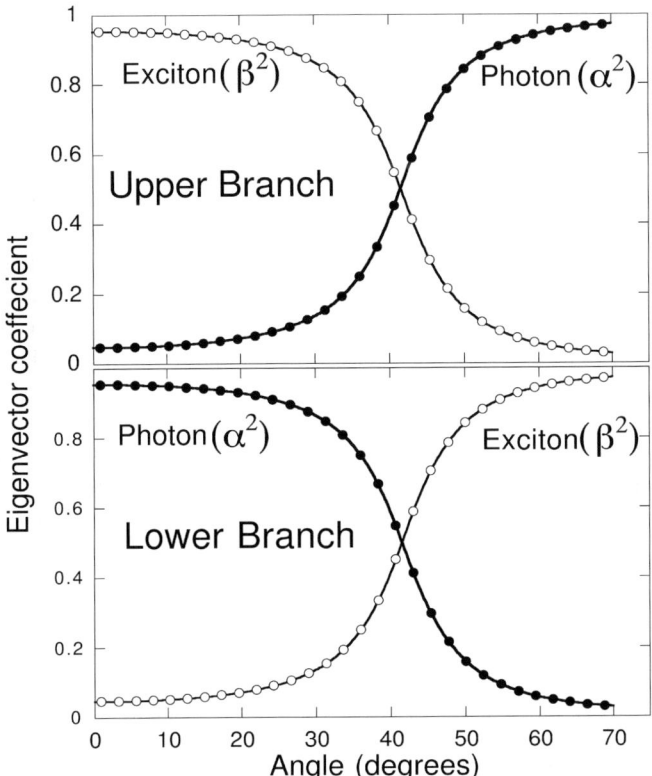

Fig. 12. Predicted exciton (open circles) and photon (filled circles) coefficients for the upper and lower polariton branches of the cavity whose dispersion curve is shown in Figure 12.

plots the values of $\alpha^2$ and $\beta^2$ for the upper and lower branches. At small angles the upper branch only contains a small component of the cavity photon, and is thus weak in reflectivity (see curve A, Figure 10). As the viewing angles increases, the upper-branch photon component grows, and thus it becomes more visible in reflectivity (curve B). At resonance (40°), the cavity photon is contained equally in both branches, and thus they are detected with equal intensity in reflectivity (curve C). Beyond resonance, the cavity photon is largely contained within the upper polariton branch, which thus becomes increasingly dominant in the reflectivity spectrum (curve D).

The excitons within the cavity can be viewed as an ensemble being driven by the confined optical field. The coupling between the exciton mode and the photon mode necessarily implies a mutual *coherence* between the excitons in the microcavity. The polarisability of the organic layer per unit area is therefore dependent

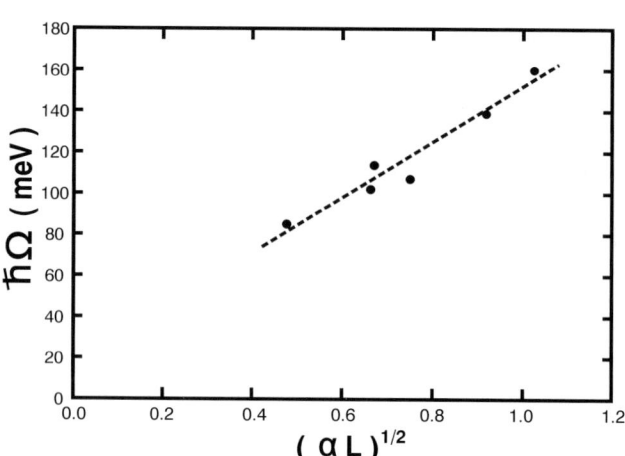

Fig. 13. The measured Rabi splitting versus the square-root of the peak absorption coefficient measured from cavities containing a thin film of 4TBPPZn molecules in a polystyrene matrix. Note the peak absorption coefficient was determined in each case from an absorption measurement made from a control (non-cavity) film, which had the same thickness and number of 4TBPPZn molecules per cm$^3$ as were used in the cavity.

on the sum of the oscillator strength of all the molecules within the cavity. By adjusting the concentration of the 4TBPPZn dye in the blend film it is possible to control oscillators strength of the organic layer, and thus adjust the energetic splitting between the two branches. This is summarised by Eqs. (12), (13) and (24), which show that the Rabi-splitting varies as the square root of the absorbance. Figure 13 shows the Rabi-splitting energy of a series of microcavities plotted as a function of $(\alpha L)^{1/2}$: The observed proportionality is in good agreement with expectation. The largest room temperature Rabi-splitting that we have measured, is 160 meV; more than 30 times the typical splittings observed for III–V quantum well microcavities, which are of the order of 5 meV [9]. In the appendix, we show that the oscillator strength of the 4TBPPZn film that gave a Rabi-splitting of 160 meV is approximately 100 times larger than that of a series of three III–V QWs. This increase in oscillator strength enhances the Rabi-splitting by a factor of 10. Two important additional effects also contribute to the enhanced Rabi splitting observed in the organic semiconductor microcavity compare to its inorganic analogue. These are the low refractive index of the organic layer and the use of a metallic reflector. We show in the appendix that these factors increase the local optical field experienced by the semiconductor, and enhance the Rabi-splitting (compared to a III–V QW microcavity) by a factor of approximately 3.3 times.

## 6. Cavity Emission Following Non-Resonant Laser Excitation

In the previous section, it was demonstrated that organic semiconductor microcavities can show giant Rabi-splittings due to the very large oscillator strength of organic excitons. In this section we turn our attention to the photon emission from an organic semiconductor microcavity following non-resonant optical excitation. In our work [22,23,44,45], we have used J-aggregates of a cyanine dye instead of the 4TBPPZn material discussed in the previous section. This change in molecular system does not result in any particularly different linear optical properties of the cavity, apart from a change in the resonance energy of the system. However as shown in Figure 7, the J-aggregates that we have studied can emit strong fluorescence following optical excitation. This fluorescence emission is a convenient probe by which to study the effect of strong-coupling on the excitons within the cavity.

### 6.1. Experimental Observations

The apparatus shown in Figure 9 was used to excite and then detect emission from the cavity following non-resonant excitation. In a non-resonant excitation experiment, photons having an energy greater than that of the polariton branches are shone into the cavity. The photons directly excite excitons in the cavity, which relax in energy and populate the 'exciton-reservoir'. The exciton reservoir describes the population of excitons within the microcavity, which have either an energy or in-plane momentum that is greater than that of the polariton modes. Such excitons cannot directly couple to a cavity photon. Instead, the population of the polariton states can only occur once the excitons have lost an appropriate amount of energy or momentum through phonon emission. A number of authors have studied the emission from cavity polaritons following non-resonant excitation [41,46–50]. We will show that our results are in accord with the observations made in microcavities containing inorganic QWs, giving us further confidence of the strong-coupling picture that we use to describe our microcavities.

To generate *cw* photoluminescence emission from the cavity, light from a red HeNe laser (having an energy of 1.96 eV) was focussed into the cavity through lens L3 (see Figure 9) at normal incidence (having a power density at the cavity surface of 50 kW m$^{-2}$). It can be seen from Figure 7(a), that the energy of the excitation photons lie within the high energy tail of the J-aggregates. The excitation is non-resonant, as the laser photons have an energy 110 meV greater than the peak absorption energy of the excitons (1.84 eV). The photon emission from the cavity was collected using lens L2, and was imaged into a fibre-bundle connected to a CCD spectrograph.

Figure 14(a) shows the relative reflectivity of the cavity measured at a series of increasing viewing angles. At 20°, two modes are visible: a sharp dip at 1.72 eV

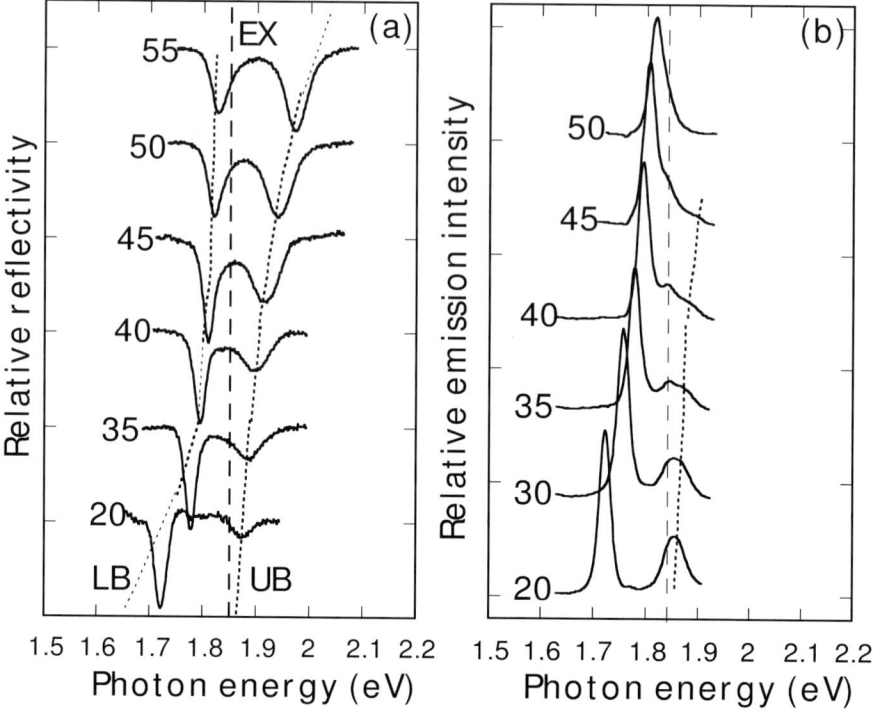

Fig. 14. (a) Reflectivity spectra measured as a function of angle for a microcavity containing a thin film of the cyanine dye J-aggregates in a PVA matrix, whose absorption and chemical structure are shown in Figure 7(a). The dotted lines are a guide for the eye showing the dispersion of the polariton branches. (b) Photoluminescence emission from the cavity as a function of angle following non-resonant optical excitation. The dispersion of the upper polariton branch is shown by a dotted line. In both figures, the peak absorption energy of the J-aggregates are shown by a vertical dashed line.

(a photon-like mode), and a second weaker dip at 1.85 eV (an exciton-like mode). Resonance between the exciton and photon occurs at 45° degrees, where two equal intensity modes are visible, split around the peak absorption energy of the excitons (which is marked with a vertical dashed line). At viewing angles greater than 45°, the higher-energy mode moves to higher energies. This behaviour is consistent with absorption and reflectivity spectra measured from microcavities containing 4TBPPZn described in the previous section.

Figure 14(b) shows the photon-emission from the cavity, recorded at a number of different viewing angles. The emission from the lower-branch is more intense than the upper-branch at all angles. Very similar behaviour has been reported from microcavities containing III–V semiconductor QWs [50]. The upper polariton branch (located at 1.85 eV at 20°), decreases in intensity as a function of

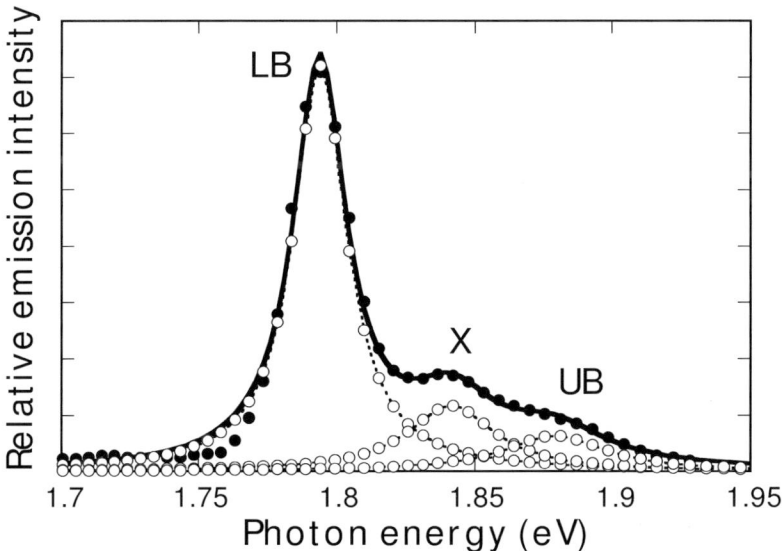

Fig. 15. Photoluminescence emission from the microcavity measured at a viewing angle of 35 degrees. Emission from the upper branch (UB), lower branch (LB) and exciton (X) are clearly identified.

increasing viewing angle. At angles of 35° and above, two features are detected in emission in addition to that from the lower-branch. This point is further illustrated in Figure 15, which plots the emission measured at 35°. The spectrum can be well described using 3 Lorentzian functions as shown. The higher and lower energy features can be positively identified as emission from the upper and lower branches, as their energy coincides with the energy of the features recorded in reflectivity (see Figure 14(a)). The energy of the central feature appears to coincide with the peak emission energy of the J-aggregates (see Figure 7(a)). It therefore appears that some of the J-aggregates emission 'escapes' from the cavity without coupling to a cavity-polariton mode. At present, the reason for such 'uncoupled' emission is not fully understood, however similar observations have been made in inorganic semiconductor microcavities, and have been attributed to emission from localised states [51] and inhomogeneous broadening [52].

Figure 16(a) shows a plot of the peak photon-emission energy from the two branches as a function of angle (solid points). The open circles mark the energy of the feature that we identify as uncoupled exciton emission. It can be seen, that the polariton branches undergo anticrossing around an energy coinciding with the peak of the J-aggregate absorption (which is marked by a dashed line). The energy of the direct exciton emission remains approximately constant as a function of angle as expected. A Rabi-splitting of 80 meV is detected between the polariton branches at an angle of 42.5°. Figure 16(b) shows a plot of the pho-

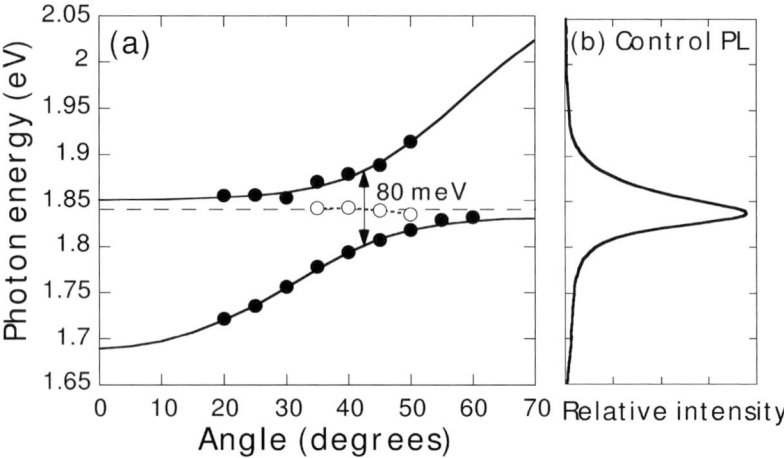

Fig. 16. Dispersion curve constructed from the measured photoluminescence emission energy of the cavity shown whose reflectivity and emission spectra are shown in Figure 14(a) and (b). Solid points indicate the measured emission energies, whilst the filled lines are predictions from a model that uses a coupled oscillator model to describe the interaction between the exciton and photon modes. The open circles are the energies of the uncoupled exciton emission (see text for details).

toluminescence emission intensity from a control film of J-aggregates. It can be seen that the J-aggregate emission spectrum is approximately symmetrical with respect to the energy at which anticrossing occurs (1.84 eV). This symmetry is not however reflected in the emission intensity from the two polariton branches which are strongly asymmetric (see Figure 14(b)). At resonance, the emission intensity from lower branch is 14 times larger than that from the upper branch.

### 6.2. A Model for Non-Resonant Excitation and Emission

We now describe a model that we have used [44] to describe the angular dependent emission intensity from the microcavity. The model considers the scattering of excitons from the exciton reservoir into the upper and lower polariton branches, followed by their radiative decay. Our model provides a good fit to the experimental data, provided that a term is included which allows a transfer of the polariton population to occur from the upper to the lower branch. The model is summarised schematically in Figure 17.

We first calculate the weight of the exciton and photon modes in the polariton branches as a function of angle using Eq. (10). The best-fit theoretical dispersion for the energy of the polariton branches is shown as a solid line in Figure 15(a) and is in good agreement with the experimental data. We also introduce the labels

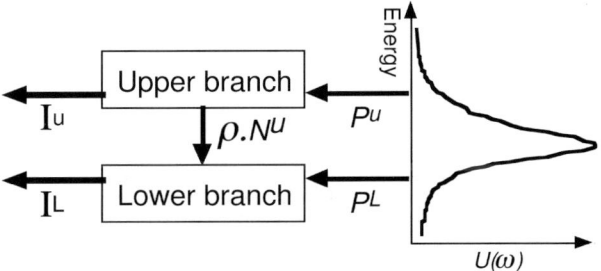

Fig. 17. Schematic diagram showing the main components of the model used to describe the emission from the non-resonantly excited microcavity. Here the relative population of the exciton reservoir is plotted as $U(\omega)$. The population of the upper and lower branches is labelled as $P$ (see Eq. (27), the emission from the branches is labelled as I (see Eq. (32)), and the transfer of population between the branches is labelled as $\rho$ (see Eq. (28)).

u and L to denote the upper and lower branches respectively. Unless stated, all terms are expressed as function of viewing angle $(\theta)$.

We assume that the polariton branches are populated by direct scattering of excitons from the reservoir. Furthermore we make the common assumption [53] that the scattering rate of excitons from the reservoir to a polariton state is directly proportional to the relative exciton fraction $(\beta_{u,L})$ of the polariton state. The population rate of the branches $P_\theta^{u,L}(\omega)$ is thus given by

$$P_\theta^{u,L}(\omega) = \Phi U_{u,L}(\omega)\beta_{u,L}^2 \qquad (27)$$

where $U_{u,L}(\omega)$ is the relative number of uncoupled excitons in the reservoir having the appropriate energy to scatter into the state, and $\Phi$ is a scaling constant, dependent on the intensity of the excitation laser. We assume that the distribution of $U(\omega)$ is the same as the photoluminescence emission spectra – i.e., the spectra this provides a measure of the distribution of energetically relaxed excitonic states within the J-aggregate. We assume that exciton scattering occurs from the reservoir to a polariton state, and there is no return path back to the reservoir. This is because the radiative rate of polaritons from the cavity is likely to be much faster than the relatively slow scattering process.

We introduce a population transfer term *between* the branches. We propose that population-transfer occurs by the emission of energy in the form of vibrational quanta. Such a transfer process will occur when $E_u - E_L = h\nu_{phonon}$. Figure 18 shows a resonance Raman spectrum recorded for the J-aggregate material used in this experiment. It can be seen that there is a strong Raman-active mode having an energy of 74 meV. In this particular cavity, the Rabi-splitting between the cavity branches on resonance is 80 meV; an energy almost resonant with that of the Raman mode. The slight difference between these two energies is smaller than

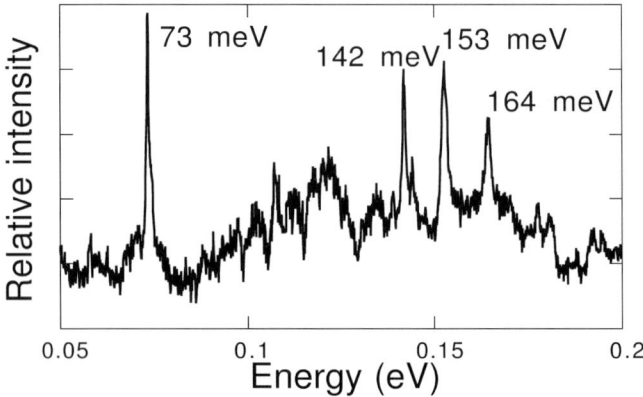

Fig. 18. Resonant Raman spectrum recorded from the J-aggregated cyanine dye in a PVA matrix used in the microcavity experiments.

the linewidth of each of the polariton branches, which have a FWHM of approximately 30 meV. This broadening also means that phonon-mediated inter-branch transitions are permitted over most of the detuning range studied in this experiment.

In our model, we assume that polariton transitions between the branches occur between initial and final states having the same in-plane momentum. We neglect momentum transfer to the optical phonon, with only energy in the $u \rightarrow L$ transition being conserved. This approximation would not be valid in high quality crystals as the uncertainty of the polariton wave vector arising due to scattering by defects or phonons is small in comparison with wave vector. The wave vector of such quasi-particles is a good quantum number and in a scattering process we would have to take into account not only energy conservation, but also momentum conservation. In high quality inorganic single crystals or quantum well systems, we meet just this case and the conservation of momentum for instance in scattering of electrons, excitons and phonons plays just as an important role as the conservation of energy.

Nevertheless, in some cases the conservation of momentum can be neglected [54]. This can be done if at least one of the quasi-particles that participates in a collision has a very small energy band width (or very large effective mass). In this case this the heavy quasi-particle can be considered as localized and we can effectively neglect its movement in the interaction processes. Such an adiabatical approximation only requires the conservation of energy. Such an approximation can be used when considering intramolecular vibrations in organic crystals. In organic crystals the band width of the optical phonons (which originate from high frequency intramolecular vibrations having an energy of the order of 50–100 meV) is usually small, and of the order of 1 meV. This means that when we consider scattering of polaritons by intramolecular optical phonons, we can neglect the bandwidth of optical phonons and just include the conservation of energy of the quasi-particles that participate in the collision. This approximation is rather good because the change of polariton energy in the region of Rabi splitting is strongly-dependent on wavevector, and is of the order 100 meV, which is much larger than the bandwidth of intramolecular phonons.

The interaction with molecular vibrations is maximised when the polaritons in both branches are exciton like, and thus the transition rate between the branches $\rho(\theta)$ is given by

$$\rho(\theta) = K_{u \to L} \beta_u^2 \beta_L^2 \tag{28}$$

where $K_{u \to L}$ is a rate-constant, assumed to be constant at all angles and branch separations. As the Rabi-splitting between the branches is much larger than $kT$, we assume that population transfer occurs from the upper to the lower branch only, and that there are no back-transfer processes.

We now consider the radiative decay of the polariton states. The radiative decay time of a polariton state $\tau_{u,L}(\theta)$ is given by

$$\tau_{u,L}(\theta) = \frac{\tau_{cav}}{\alpha_{u,L}^2(\theta)} \tag{29}$$

where $\tau_{cav}$ is the escape time of an uncoupled photon from the cavity.

The relative populations of the upper $[N_u(\theta)]$ and lower $[N_L(\theta)]$ polariton branches can be described with the two rate equations:

$$\dot{N}_L(\theta) = -\frac{N_L(\theta)}{\tau_L(\theta)} + \rho(\theta) N_u(\theta) + P_\theta^L(\omega), \tag{30}$$

$$\dot{N}_u(\theta) = -\frac{N_u(\theta)}{\tau_u(\theta)} - \rho(\theta) N_u(\theta) + P_\theta^u(\omega). \tag{31}$$

Here, the first term on the right-hand-side describes the radiative depopulation of the branches. The second term represents the population transfer from the upper to the lower branch, and the last term describes the filling of the upper and lower branches from the exciton reservoir. As our experiments measure the *cw* emission from microcavity, we set $\dot{N}_u = \dot{N}_L = 0$. The emission intensity from the cavity $[I_{u,L}]$ can be expressed in terms of the relative polariton population and its radiative decay time:

$$I_{u,L} = \frac{N_{u,L}}{\tau_{u,L}}. \tag{32}$$

It can be shown that

$$I_u(\theta) = \frac{\Phi U_u(\omega) \beta_u^2}{1 + \rho(\theta) \tau_{cav} \alpha_u^{-2}}, \tag{33}$$

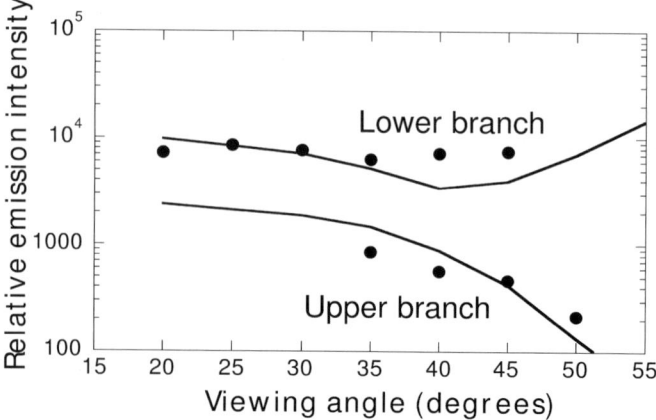

Fig. 19. Measured photoluminescence emission intensity as a function of angle for the upper and lower polariton branches. Note filled dots are the measured data, whilst the solid lines are the predictions of the model, summarised in Eqs. (33) and (34).

$$I_L(\theta) = \Phi\left[\frac{U_L \beta_u^2}{1 + \alpha_u^2 \rho^{-1}(\theta)\tau_{cav}^{-1}} + U_L(\omega)\beta_L^2\right]. \quad (34)$$

Our model only includes population transfer between branches, it does not account for any redistribution *along* each polariton branch by intra-branch scattering. Such intra-branch scattering processes have been shown to only become significant at laser fluxes much larger than that used in this experiment (i.e., > 50 kW m$^{-2}$) [49].

Figure 19 shows a fit of Eq. (18) and (19) to the measured angular dependent intensity from the upper and lower branches. The only free variables used in the fit was the product of $K_{u\to L} \cdot \tau_{cav}$ (which represents the relative competition between the inter-branch transition rate and radiative decay) and $\Phi$. We determine a best fit using a value of $K_{u\to L} \cdot \tau_{cav} = 4.5$. The measured data points are shown as solid circles, and the best fit to the data is shown as full lines. The agreement between the data and the fit is quite reasonable and in particular it replicates the strong asymmetry in emission intensity observed between the upper and lower branches. Using a bare cavity linewidth of 20 meV, we calculate a photon escape-time from the cavity of $\tau_{cav} \sim 35$ fs. The transition rate between the branches [$\rho(\theta)$] is given by Eq. (13). This term takes its maximum value at resonance, where both $\beta_u^2 = \beta_L^2 = 0.5$, with their product being equal to 0.25. In this microcavity, resonance between the photon and exciton occurs at 42.5°, thus the maximum transition rate from the upper to the lower branch is given by $\rho(42.5°) = 0.25 K_{u\to L}$. From the best fit, we determine an inter-branch transfer time at resonance to be $\tau_{trans} \approx 30$ fs. The decay time of the cavity polaritons can be calculated using Eq. (14). Using a cavity escape time of $\tau_{\text{cav}} = 35$ fs, we calculate[1] that at resonance, the upper or lower branch polariton states decay with a lifetime of $\tau_{u,L}(42.5°) \approx 70$ fs. As the transfer time between branches is approximately *half* that of the polariton decay time from either branch, it is apparent that inter-branch transfer process will dominate over radiative decay. In fact, we calculate that approximately 80% of all of the upper branch polaritons transfer to lower branch states.

The time-scale for the inter-branch transition predicted by our model is $\sim 30$ fs. In many molecular systems, both in solution and in the solid state, there are intramolecular vibrational energy redistribution processes that occur over time scales of a few tens of femtoseconds or less [55]. Thus the inter-branch population transfer proposed here could be thought of being analogous to the ultra-fast relaxation processes that occur through coupled vibrational states.

## 7. Photon Emission Following Resonant Excitation

We now review our work on studying the emission from a strongly-coupled organic semiconductor microcavity that has been excited resonantly using a laser [56]. Under resonant excitation, photons having energy and in-plane momentum equal to one of the polariton modes are incident on the cavity. These photons can then directly excite a cavity polariton state. In microcavities containing inorganic quantum wells, a number of effects have been studied following resonant excitation, including the temporal dynamics of polariton emission [57–59], polariton mediated Raman scattering [60,61], polariton relaxation mechanisms [62] and (bosonic) stimulated scattering [63,64].

### 7.1. Resonant Excitation Measurements

We have again explored the properties of strongly-coupled microcavities containing the J-aggregates shown in Figure 7(a). The resonant excitation of the microcavities was achieved using a tunable *cw* Ti-sapphire laser. In these experiments, the cavity was placed in a cryostat with wide angular access. Experiments were carried out over the temperature range 60–300 K. The experimental setup is shown schematically in the inset of Figure 20. Here, the laser is incident on the sample surface at an angle $\Theta$. The energy of the incident laser photons is fixed to the energy of the polariton state *at that angle*. The photon emission is then collected at

---

[1] This time is twice the cavity escape time, as at resonance each polariton can be thought of as 'spending' half its time as a photon, and half its time as an exciton. As escape only occurs during its photon-like part of its cycle, the polariton lifetime is approximately twice that of a cavity photon. This of course assumes that exciton dephasing is not faster than the photon lifetime.

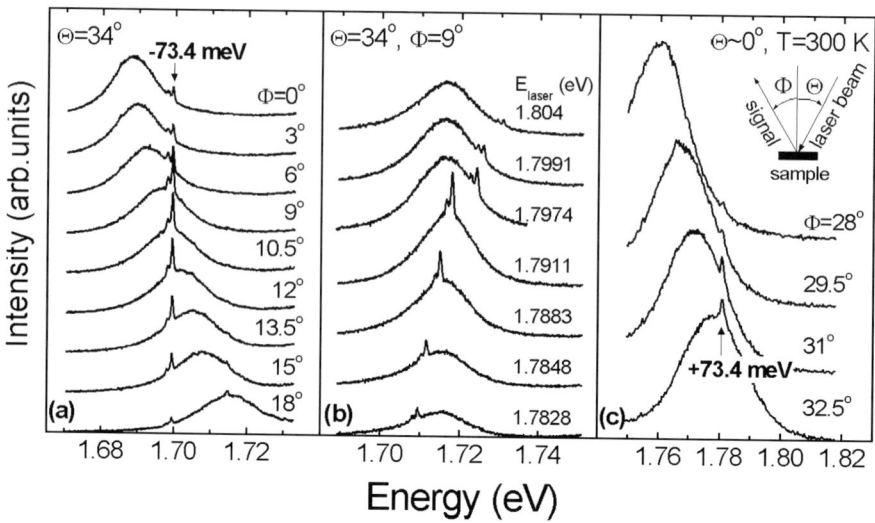

Fig. 20. Photon emission from the microcavity following resonant excitation. In part (a), the excitation angle and energy remains fixed whilst the angle of detection is varies. In part (b), the angles of excitation and emission remain fixed, whilst the energy of excitation is varied. Part (c) shows the room temperature analogue of the data shown in part (a), which were recorded at 120 K. The inset to part (c) defines the angles of excitation and emission.

an angle $\Phi$. Figure 20(a) shows a series of spectra recorded at $T = 120$ K for excitation at $\Theta = 34°$, and detection at a series of angles between $\Phi = 0°$ to $18°$. It can be seen that each spectrum is composed of a relatively broad emission peak, with additional sharp features superimposed at 1.698 and 1.699 eV. The broad emission occurs from radiative-decay of polaritons that have scattered down in energy along the lower branch. We find that the energy of the sharp features are independent of collection angle, with the strongest sharp line occurring at an energy of 73.4 meV below the laser line. Figure 21 shows a schematic sketch of the experimental methodology. The solid arrows indicate transitions to different states that can be probed by angle-tuning.

To determine the origin of the sharp peaks, spectra were recorded for fixed angles of excitation ($\Theta$) and detection ($\Phi$). However now, the energy of the laser was varied, tuning it through resonance with the lower polariton branch. A series of spectra recorded at $T = 120$ K are presented in Figure 20(b). As expected the lower polariton branch emission energy is constant, as the angle of detection is fixed. However, the energy of the sharp lines superimposed on the broader PL peak depend strongly on the laser energy. The energy separation between these peaks and the laser is constant with a separation of $-73.4$ meV below the laser line being found for the strongest line. The laser tuning experiment is represented

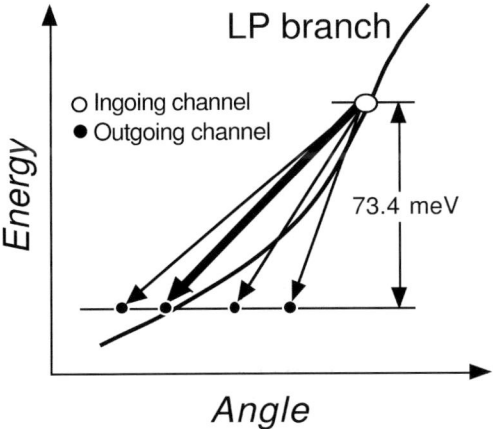

Fig. 21. Schematic diagram of the resonant excitation experiment, illustrating angle tuning. Only when the ingoing (excitation) and outgoing (emission) channels both lie on the lower polariton branch is the condition of exact double resonance achieved.

in the diagram of Figure 21 by the dashed arrows. The constant energy separation from the laser energy, independent of laser energy and angle of detection, provides strong evidence that the sharp features arise from an inelastic scattering process, namely Raman scattering. It can be seen in the Raman spectra shown in Figure 17, that there is indeed a strong Raman line (that originates from a vibrational mode of the cyanine monomer) at 73 meV.

We present further evidence to support the assignment of Raman scattering to the sharp features shown in Figures 20(a) and (b) in Figure 20(c). Here, polariton emission spectra recorded at 300 K are presented, generated under conditions of resonant excitation into the lower polariton branch at $\Theta \approx 0°$ at an energy of 1.707 eV. The polariton emission is now observed at energies up to nearly 80 meV above the excitation energy, due to the large thermal energy of equilibrium polaritons at 300 K. A relatively weak sharp feature is also observed at 1.781 eV, corresponding to an energy of +73.4 meV above the laser energy. This sharp feature arises from anti-Stokes Raman scattering. This process occurs as the phonon population at 300 K is sufficient to allow the anti-Stokes companion of the Stokes features observed at lower temperature in Figures 20(a) and (b) to be detected. It is important to note that in contrast to inorganic microcavities where strong coupling effects are only easily observed at $T < 100$ K due to the small exciton binding energies and Rabi splittings, the polaritons in organic structures are stable at room temperature and above. This high temperature stability permits the observation of the anti-Stokes resonant Raman process for the first time in a strongly coupled microcavity.

## 7.2. THE GENERATION OF RAMAN EMISSION IN A MICROCAVITY

In can be seen from Figures 20(a) and (b), that the Raman intensities are maximised when they are in resonance with the polariton emission peaks. In Figure 21(a), the excitation is in resonance with the lower polariton branch at 34°. Thus for detection at 10.5° where the Raman intensity is maximum, both incident and scattered photon energies coincide with lower polariton branch states. In this case the double resonance condition is achieved (the thick arrow in the diagram of Figure 21). Similarly in Figure 20(b), where the laser energy is tuned, the maximum in Raman intensity is found when both the laser and the scattered photons are resonant with the polariton dispersion curves.

Raman scattering in a microcavity can be though of as the result of three steps: (1) The transmission of a photon into the sample and conversion to a polariton, (2) phonon-mediated scattering from one polariton state to another, (3) the subsequent propagation and transmission of the scattered polariton and conversion to an external photon. The transmission of a photon into the cavity at any particular angle (1) is maximised when its energy coincides with the energy of a photon-like cavity mode. Similarly in process 3, the escape of a photon from the cavity at any particular angle is also maximised when the energy of the photon coincides with the energy of a photon-like cavity mode. On the basis of this, we can understand the observed energy and angular dependence of the Raman signal: In angle tuning experiments, the energy of the cavity mode (which can be identified by the polariton emission) is tuned through the energy of the Raman scattered photon. The escape of the Raman photons from the cavity is therefore maximised when their energy coincides with the energy of the polariton emission (schematically shown in Figure 21). In energy tuning experiments, both the efficiency of transmission of the photons in and out of the cavity are varied. Again, the Raman scattering efficiency is maximised when the energy of the incident laser and the Raman photon coincide with points on the lower branch.

Control experiments have demonstrated that the Raman scattering efficiency in the microcavities (characterised by a finesse of $Q \approx 50$) under double resonance conditions is about 300 times larger than that for the a non-cavity control sample of J-aggregates in PVA. It has been shown that the enhancement of the Raman efficiency is a strong function of the magnitude of the confined optical field within the cavity, which is itself a function of cavity finesse. High finesse inorganic cavities with $Q \sim 2000$ have been reported to show enhancements up to $10^4$ [65]. We therefore conclude that our measurements are in good accord with the optical processes that occur in inorganic semiconductor microcavities.

The cavity-induced enhancement of Raman scattering is particularly interesting for investigations of non-linear processes. Recent studies of resonant Raman scattering in conjugated polymer thin films have exhibited non-linear character under high power pulsed laser excitation [66]. The significant enhancements of Raman scattering found in this work therefore suggest interesting opportunities for the investigation of non-linear processes in strongly coupled microcavities [67] using high levels of resonant pulsed laser excitation.

## 8. Photon-Mediated Hybridisation between Frenkel Excitons

Recently a number of authors have studied the interaction between different optical and electronic excitations in a microcavity. In the weak-coupling regime, it has been claimed that the optical structure of a microcavity can be used to enhance dipole–dipole interactions [68–70]. These observations have not yet been validated theoretically [71], and it is clear that more research in this area is needed. However it is clear that modifying electronic processes by control over optical nanostructure would be an attractive and exciting prospect. Coupling between different optical and electronic excitations has also been studied in the strong-coupling regime. Here, the interactions between two inorganic QW exciton states with a single photon state [53] two photon states with a single (inorganic) exciton state [72], and two (inorganic) exciton states interacting with two photon states [73] have been reported.

In this section, we review our recent work [23,45] on the fabrication of microcavities containing two different types of organic semiconductor, both of which are strongly coupled to the same cavity mode. As discussed in Section 6, strong-coupling in a microcavity necessarily implies that the excitonic states within the cavity are all driven coherently by the same optical field. As we will show, simultaneous strong coupling of the individual exciton species to a single cavity-photon mode leads to new eigenmodes that can be described as exciton–photon–exciton admixtures of the three states. One may think of this coupled system in terms of a photon-mediated hybridisation of the two exciton states. This hybridisation therefore *coherently* couples the two different organic semiconductor materials. We will show that such systems are analogous with biological light-harvesting complexes, which are able to mediate long-range energy transfer.

### 8.1. Hybrid Semiconductor Microcavities

To achieve a photon mediated hybridisation between two organic semiconductors, it is necessary to choose a combination of materials, whereby (i) each individual material can strongly-couple to a cavity photon, and (ii) the energy separation ($\Delta E$) between the transitions of the two semiconductors must not be much greater than the Rabi-splitting energy. To achieve these conditions, we have utilised the two cyanine dyes shown in Figures 7(a) and (b). For simplicity, we label these materials $Ex_1$ and $Ex_2$. The difference in their peak absorption energy is approximately 60 meV that is comfortably less than the Rabi-splittings of 80 meV that we

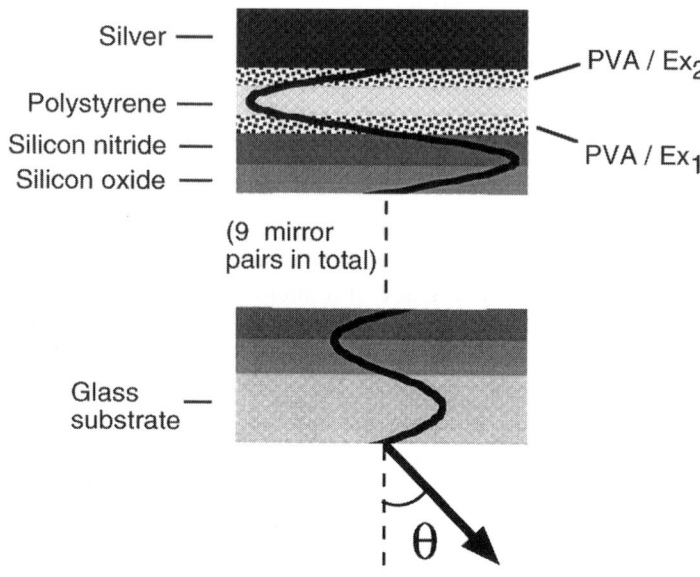

Fig. 22. Schematic diagram of a microcavity containing two different cyanine dyes. $Ex_1$ corresponds to the dye shown in Figure 7(b), and $Ex_2$ corresponds to the dye shown in Figure 7(a).

observe in single component microcavities. The generic structure of the microcavities that we have fabricated is shown in Figure 22. The cavity medium between the DBR and the metallic mirror is composed of two J-aggregate films separated by a 100 nm thick layer of the transparent dielectric polymer polystyrene. This ensures that the only coupling that can occur between the two exciton species is that mediated by a cavity photon, as short-range dipole–dipole interactions are characterised by Förster transfer radii of typically less than 10 nm.

Figure 23(a) shows a series of reflectivity spectra measured at increasing viewing angle (spectra are displaced vertically for the sake of clarity). The vertical dotted lines correspond to the peak absorption energy of the $Ex_1$ and $Ex_2$ excitons in a non-cavity control film. At 15°, a strong dip can be seen at 1.72 eV, which corresponds to the cavity photon. Two other features are also apparent at 1.79 and 1.87 eV, which correspond to the $Ex_1$ and $Ex_2$ exciton-like modes. The energy of three modes as a function of angle is plotted in Figure 23(b) with the solid dots being the measured data points. The energy of the uncoupled excitons is shown as horizontal dotted lines. As in the single component cavity, the polariton branches undergo anti-crossing around the absorption energy of the two different exciton modes.

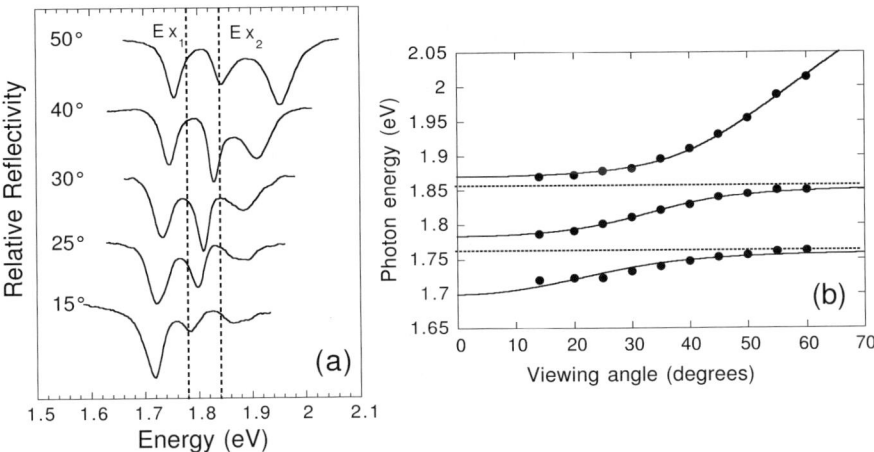

Fig. 23. (a) A series of reflectivity spectra measured at different angles from a microcavity containing two different J-aggregated dyes. The peak absorption energies of the J-aggregates are marked with vertical dashed lines. (b) A dispersion curve of the three polariton branches determined from the spectra shown in part (a). The peak absorption energies of the two J-aggregates around which anticrossing occurs are also shown as horizontal dotted lines.

We describe this system using a model based on the interaction of two excitons and one photon mode, and we write the matrix equation

$$\begin{pmatrix} E^{cav}(k_x) - E & V_1 & V_2 \\ V_1 & E_{Ex_1} - E & 0 \\ V_2 & 0 & E_{Ex_2} - E \end{pmatrix} \begin{pmatrix} \alpha \\ \beta \\ \gamma \end{pmatrix} = 0 \qquad (35)$$

where $E^{cav}$, $E_{Ex_1}$ and $E_{Ex_2}$ are the energies of the non-interacting cavity photon and the two excitons and $E$ are the eigenvalues of the coupled system. $V_1$ and $V_2$ are the interaction potentials between the photon and each of the two excitons and $\alpha$, $\beta$ and $\gamma$ are the coefficients that correspond to the bare photon, and the Ex$_1$ and Ex$_2$ exciton modes respectively. The same result can be obtained using the macroscopic approach (summarised by Eq. (20)) using a dielectric constant with two different resonances. We therefore write

$$\varepsilon(\omega) = \varepsilon_0 + \frac{f_1}{\omega_1^2 - \omega^2 + i\gamma} + \frac{f_2}{\omega_2^2 - \omega^2 + i\gamma} \qquad (36)$$

where $\omega_1$ and $\omega_2$ are the frequencies of the two resonances, and $f_1$ and $f_2$ are their corresponding oscillator strengths.

However we proceed using Eq. (35) to describe the energy of the polariton branches. This allows us to readily calculate the mixing between the cavity photon and the two exciton states in each of the polariton branches. The matrix is diagonalised, and then fit to the experimental data as was demonstrated for a single component cavity. The solid lines in Figure 23(b) are the eigenvalues predicted using the coupled oscillator model. Our best fit used values for the exciton energies of $E_{Ex_1} = 1.762$ eV, $E_{Ex_2} = 1.857$ eV, and predicts Rabi-splittings of $\hbar\Omega_{Ex_1} = 77.5$ meV, $\hbar\Omega_{Ex_2} = 78.9$ meV. The exciton energy used in the model varies slightly from the measured values (1.778 eV, 1.842 eV), however such shifts can be explained by the asymmetric linewidths of the J-aggregate excitons [74,75]. As the energy separation between the excitons (60 meV) is smaller than the Rabi-splittings between the branches, it is possible for the photon to interact with both exciton modes *simultaneously*, forming a coherent superposition of particles.

Figure 24 shows the eigenvector coefficients used to describe the 3 cavity branches. It can be seen, that the upper and lower polariton branches are mainly composed of a superposition of the cavity photon ($\alpha$), and either $Ex_1(\beta)$ or $Ex_2(\gamma)$. However, there is a mixing between all three modes in the central branch, and at 30° the middle branch contains approximately equal amplitudes of the cavity photon and the two exciton species. A hybrid state has thus been created, delocalised throughout the cavity, which is composed of coherently coupled excitons spatially separated by some 100 nm. It is interesting to consider whether it will prove possible to utilise our system as the basis for efficient long-range energy transfer. We can perhaps view the hybrid exciton state supported by this cavity as linking two cavity-polariton states, each of whose excitonic components are widely physically separated. In Section 6 we demonstrated that the emission intensity from a microcavity containing a single organic semiconductor could be explained using a phonon-mediated energy-transfer process between the upper and lower polariton branches. In this hybrid semiconductor microcavity, a similar energy transfer process between (for example) the upper polariton branch to the lower branch would involve the movement of energy, as the excitonic components associated with these branches are physically separated. If such a system could be demonstrated, it could, in principle provide a method to move energy via a cavity photon over distances much larger than is permitted by direct dipole–dipole coupling.

It is intriguing to consider whether there are similarities between the phenomena we have been studying and the processes that occur in natural photosynthesis. The observed behaviour of the light-harvesting complex in funneling energy to the reaction centre and the subsequent electron transfer that initiates the energy conversion process pose many intriguing questions. These processes are extremely fast and efficient and are believed to involve coherent excitations of several different molecules [76]. In the light harvesting complex, there are expected to be many energy transfer processes between closely separated and efficiently coupled chromophores before trapping on the reaction centre, yet the overall efficiency is still near unity. In this situation, the Förster dipole–dipole description of energy trans-

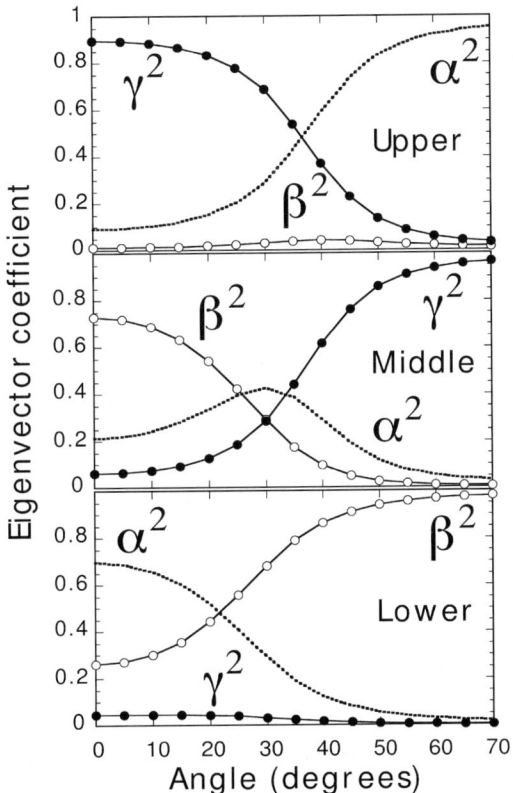

Fig. 24. Predicted exciton and photon coefficients for the upper middle and lower polariton branches of the cavity whose dispersion curve is shown in Figure 23(b). The coefficients $\alpha$, $\beta$ and $\gamma$ correspond to the bare photon, and the $Ex_1$ and $Ex_2$ exciton modes respectively. Note for the middle branch at 30 degrees, the state formed is composed of almost equal amplitudes of the photon and the two different excitons.

fer is expected to be limited because the point dipole approximation breaks down and transfer occurs from excited vibrational levels. In addition, the chromophore units are considered to be strongly-coupled with exciton splittings of $\approx 25$ meV that are larger than the inhomogeneous linewidths. Short time energy transfer may well then be akin to the scattering processes between coupled exciton levels that are discussed, albeit in a different context, above.

## 9. Future Prospects

In this chapter, I hope that I have communicated some of the excitement of research on strongly-coupled organic semiconductor microcavities. The interest and novelty comes in part from the significant difference between the properties of Mott–Wannier excitons, (which have been studied in detail in microcavities), and Frenkel excitons. It is already clear that such differences result in the formation of optical structures that have very different optical properties. So far, we have shown that because of the large oscillator strength of Frenkel excitons, we can achieve strong coupling at room temperature, and can observe significantly enhanced Rabi-splittings. We have also shown that the large binding energy of Frenkel excitons allows the observation of anti-stokes Raman scattering in a microcavity for the first time. We anticipate that there will almost certainly be more new effects waiting to be discovered. Such experimental work will, by necessity, have to be coupled with high-level theoretical investigation to allow a full picture of the physics of organic semiconductor microcavities to developed.

One area that is likely to be particularly promising is the study of stimulated scattering effects. At present, this area is generating considerable excitement in the inorganic semiconductor microcavity community. It has been shown [4,62, 77–79] that under resonant pumping, highly non-linear processes occur. Such effects originate because of the bosonic character of cavity polaritons. At a certain pump power, the number density of a particular polariton state (usually one with zero in plane momentum) can stimulate the scattering of other polaritons into the same state. The strong increase of final state population corresponds to a "condensation" to a polariton mode with macroscopic occupancy, the phenomena possessing a number of similarities to Bose–Einstein condensation. It has been shown [4] that the maximum temperature at which such a process will occur is a function of the binding energy of the excitons within the cavity, and thus strongly-coupled organic excitons are likely to display such scattering effects at room temperature. It has been proposed that such structures could find applications as ultra-fast optical switches and amplifiers.

Finally, recent theoretical work has discussed the optical properties of hybrid organic-inorganic structures [80,81]. It is predicted that a microcavity which contained strongly-coupled Mott–Wannier excitons and Frenkel excitons would allow the creation of hybridised systems, where the characteristics of the organic and inorganic excitons would both in some measure be present in a hybrid-polariton state. Such structures may be of significant importance in creating new types of semiconductor optoelectronic devices and in generating optical structures having enhanced optical non-linearity.

## Appendix

Our experimental measurements have demonstrated that organic semiconductor microcavities can show Rabi-splittings at least an order of magnitude larger than those typically found in organic semiconductor microcavities. Such enhancements in splitting can be understood using Eq. (7). The total optical path length in a microcavity ($L_{\text{eff}}$) is given by

$$L_{\text{eff}} = L_{\text{c}} + L_{\text{DBR}}^{\text{Tot}} \qquad (A.1)$$

where $L_{\text{c}}$ is the distance between the two cavity mirrors, and $L_{\text{eff}}^{\text{Tot}}$ is the total penetration of the optical field into the cavity DBRs. The optical penetration into a *single* dielectric mirror $L_{\text{DBR}}$ is given by

$$L_{\text{DBR}} = \frac{\lambda}{4n_{\text{c}}} \frac{n_{\text{L}} n_{\text{H}}}{n_{\text{H}} - n_{\text{L}}} \qquad (A.2)$$

where $\lambda$ is the wavelength of the light in the cavity, $n_{\text{c}}$ is the refractive index of the material within the cavity, and $n_{\text{H}}$ ($n_{\text{L}}$) is the refractive index of the high (low) layers which comprise the dielectric mirror.

In Table A.1 below, we compare numerically the optical properties of an inorganic microcavity and the organic 4TBPPZn cavity that demonstrated a 160 meV Rabi-splitting.

In Table A.1, the effective optical cavity length of the organic microcavity is calculated assuming the penetration of the optical field into the metallic mirror is negligible. The physical length of the cavity is then obtained by dividing the optical cavity length by the cavity refractive index ($n_{\text{c}}$). The effective oscillator strength of the inorganic QWs used above takes into account that not all of the QWs are located at an antinode of the confined optical field. This reduces the effective oscillator strength of the 3 QWs to 2.52 $f_{\text{inorg}}$, where $f_{\text{inorg}}$ is the oscillator strength of a single QW. The effective oscillator strength of the organic 4TBPPZn

Table A.1.

| Parameter | Inorganic III–V QW Cavity | Organic 4TBPPZn Cavity |
|---|---|---|
| Cavity structure | 20 pair DBR/cavity/18 pair DBR | 9 pair DBR/cavity/Silver |
| DBR Mirror composition | $Al_x Ga_{1-x} As$ ($n_{\text{L}} = 3.08$) | $SiO_2$ ($n_{\text{L}} = 1.45$) |
| | AlAs ($n_{\text{H}} = 3.66$) | $Si_x N_y$ ($n_{\text{H}} = 1.95$) |
| Cavity material and ref. index | GaAs, 3 $In_y Ga_{1-y} As$ QWs, | 4TBPPZn in PS, |
| | $n_{\text{c}} = 3.08$ | $n_{\text{c}} = 1.63$ |
| Penetration in each DBR ($L_{\text{DBR}}$) | 1.6 $\lambda$ | 0.87 $\lambda$ |
| Cavity path-length ($L_{\text{c}}$) | $\lambda$ | $\lambda/2$ |
| Total optical path-length ($L_{\text{eff}}$) | 4.2 $\lambda$ | 1.37 $\lambda$ |
| Cavity operational wavelength | 855 nm | 430 nm |
| Total physical cavity length | 1165 nm | 360 nm |
| Rabi-splitting energy | 5 meV | 160 meV |
| Oscillator strength per QW | $4.2 \times 10^{12}$ cm$^{-2}$ | -N/A- |
| Effective oscillator strength | $1.05 \times 10^{13}$ cm$^{-2}$ | $\sim 10^{15}$ cm$^{-2}$ |

layer within the cavity can then be estimated using

$$f_{\text{org}} = f_{\text{Inorg}} \left( \frac{n_{\text{org}}^2 L_{\text{eff}}^{\text{org}}}{n_{\text{inorg}}^2 L_{\text{eff}}^{\text{inorg}}} \right) \left( \frac{\Omega_{\text{org}}}{\Omega_{\text{inorg}}} \right)^2. \quad (A.3)$$

It can be seen from Table A.1, that the effective oscillator strength of the material used in the cavity that had a 160 meV splitting is approximately 100 times larger than the effective oscillator strength of 3 III–V QWs. This enhanced oscillator strength is anticipated to increase the Rabi-splitting by a factor of 10 times. An additional enhancement of the Rabi-splitting also arises because the optical field is more effectively concentrated in the region of the organic semiconductor material. In Eq. (A.3), the term describing the spatial extent of the optical field ($n^2 L_{\text{eff}}$) is 11 times larger in the inorganic cavity compared to the organic cavity. This occurs because of the low refractive index of the organic semiconductor and because the cavity utilises a metallic mirror in the place of one of the DBRs. The effect of the increase in the optical field in the region of the semiconductor increases the Rabi-splitting by a factor of $\sqrt{11} = 3.3$.

## References

1. R.B. Fletcher, D.G. Lidzey, D.D.C. Bradley, M. Bernius, S. Walker, *Appl. Phys. Lett.* 77 (9) (2000) 1262–1264.
2. Ünlü M.S., Strite S., *J. Appl. Phys.* 78 (1995) 607–638.
3. P.K.H. Ho, D.S. Thomas, R.H. Friend, N. Tessler, *Science* 285 (1999) 233–236.
4. M. Saba, C. Ciuti, J. Bloch, V. Thierry-Mieg, R. Andre, L. Si Dang, S. Kundermann, A. Mura, G. Bongiovanni, J.L. Staehli, B. Deveaud, *Nature* 414 (6865) (2001) 731–735.
5. R.H. Jordan, L.J. Rothberg, A. Dodabalapur, R.E. Slusher, *Appl. Phys. Lett.* 69 (14) (1996) 1997–1999.
6. R. Piron, E. Toussaere, D. Josse, J. Zyss, *Appl. Phys. Lett.* 77 (16) (2000) 2461–2463.
7. D.G. Lidzey, D.D.C. Bradley, S.J. Martin, M.A. Pate, *IEEE Journal of Selected Topics in Quantum Electronics* 4 (1) (1998) 113–118.
8. C. Weisbuch, M. Nishioka, A. Ishikawa, Y. Arakawa, *Phys. Rev. Lett.* 69 (1992) 3314–3318.
9. M.S. Skolnick, T.A. Fisher, D.M. Whittaker, *Semicond. Sci. Technol.* 13 (1998) 645–669.
10. C. Weisbuch, H. Benisty, R. Houdre, *Journal of Luminescence* 85 (2000) 271–293.
11. G. Kitrova, et al., *Rev. Mod. Phys.* 71 (1999) 1591.
12. C. Weisbuch, J.G. Rarity (Eds.), Microcavities and Photonic Bandgaps: Physics and Applications, Kluwer, Dordrecht, 1996.
13. T.B. Norris, in: E. Burstein, C. Weisbuch (Eds.), Confined Electrons and Photons, Plenum Press, New York, 1995, pp. 503–521.
14. S.F. Alvarado, P.F. Seidler, D.G. Lidzey, D.D.C. Bradley, *Phys. Rev. Lett.* 81 (5) (1998) 1082–1086.
15. M. Fox, Optical Properties of Solids, Oxford University Press, New York, 2001.
16. P.T. Worthing, J.A.E. Wasey, W.L. Barnes, *J. Appl. Phys.* 89 (1) (2001) 615–625.
17. V. Cimrova, U. Scherf, D. Nehr, *Appl. Phys. Lett.* 69 (5) (1996) 608–610.
18. G. Bourdon, I. Robert, R. Adams, K. Nelep, I. Sagnes, J.M. Moison, I. Abram, *Appl. Phys. Lett.* 77 (9) (2000) 1345–1348.
19. S.E. Burns, N. Pfeffer, J. Gruner, M. Remmers, T. Javoreck, D. Neher, R.H. Friend, *Adv. Mater.* 9 (5) (1997) 395–398.

20. E. Hecht, Optics, 2nd Edition, Addison-Wesley, 1987, Chapter 9.
21. D.G. Lidzey, D.D.C. Bradley, M.S. Skolnick, T. Virgili, S. Walker, D.M. Whittaker, *Nature* 395 (1998) 53.
22. D.G. Lidzey, D.D.C. Bradley, T. Virgili, A. Armitage, M.S. Skolnick, S. Walker, *Phys. Rev. Lett.* 82 (16) (1999) 3316–3319.
23. D.G. Lidzey, D.D.C. Bradley, A. Armitage, S. Walker, M.S. Skolnick, *Science* 288 (2000) 1620–1623.
24. T. Virgili, D.G. Lidzey, M. Grell, S. Walker, A. Asimakis, D.D.C. Bradley, *Chem. Phys. Lett.* 341 (2001) 219–224.
25. D.G. Lidzey, D.D.C. Bradley, M.A. Pate, J.P.R. David, D.M. Whittaker, T.A. Fisher, M.S. Skolnick, *Appl. Phys. Lett.* 71 (6) (1997) 744–746.
26. T.A. Fisher, D.G. Lidzey, M.A. Pate, M.S. Weaver, D.M. Whittaker, M.S. Skolnick, D.D.C. Bradley, *Appl. Phys. Lett.* 67 (10) (1995) 1355–1358.
27. P. Schouwink, H.V. Berlepsch, L. Dahne, R.F. Mahrt, *Chem. Phys. Lett.* 344 (2001) 352–356.
28. M. Anni, G. Gigli, R. Cingolani, S. Patane, A. Arena, M. Allegrini, *Appl. Phys. Lett.* 79 (9) (2001) 1382–1384.
29. G.J. Denton, N. Tessler, M.A. Stevens, S.E. Burns, N.T. Harrison, R.H. Friend, *Proceedings of the SPIE* 3145 (1997) 24–35.
30. R.P. Stanley, R. Houdre, U. Oesterle, M. Gailhanou, M. Ilegems, *Appl. Phys. Lett.* 65 (15) (1994) 1883–1885.
31. A.M. Vredenberg, N.E.J. Hunt, E.F. Schubert, D.C. Jacobson, J.M. Poate, G.J. Zydzik, *Phys. Rev. Lett.* 71 (4) (1993) 517–520.
32. M. Hopmeier, U. Siegner, U. Lemmer, W. Guss, J. Pommerehne, R. Sander, A. Greiner, R.F. Mahrt, H. Bassler, J. Feldmann, E.O. Gobel, *Synth Met.* 76 (1–3) (1996) 117–119.
33. D. Dolphin (Ed.), The Porphyrins, Academic Press, New York, 1978.
34. D.M. Sturmer, D.M. Heseltine, in: T.H. James (Ed.), The Theory of Photographic Processes, Macmillan Publishing Co., New York, 1977.
35. F. Bos, *Appl. Optics* 20 (1981) 1886.
36. D. Möbius, *Advanced Materials* 7 (5) (1995) 437–443.
37. E.O. Potma, D.A. Wiersma, *J. Chem. Phys.* 108 (1998) 4894.
38. T. Kobayashi (Ed.), J-Aggregates, World Scientific, Singapore, 1996.
39. E.W. Knapp, *Chem. Phys.* 85 (1984) 73–82.
40. M. Era, S. Morimoto, T. Tsutsui, S. Saito, *Appl. Phys. Lett.* 65 (6) (1994) 676–678.
41. R. Houdré, C. Weisbuch, R.P. Stanley, U. Oesterle, P. Pellandini, M. Ilegems, *Phys. Rev. Lett.* 73 (15) (1994) 2043–2046.
42. K. Sumioka, H. Nagahama, T. Tsutsui, *Appl. Phys. Lett.* 78 (2001) 1328.
43. T. Fujita, Y. Sato, T. Kuitana, T. Ishihara, *Phys. Rev. B* 57 (19) (1998) 12428–12434.
44. D.G. Lidzey, A.M. Fox, M.D. Rahn, M.S. Skolnick, V.M. Agranovich, S. Walker, *Phys. Rev. B* 65 (2002) 195312.
45. D.G. Lidzey, D.D.C. Bradley, M.S. Skolnick, S. Walker, *Synth. Met.* 124 (2001) 37–40.
46. J. Wainstain, G. Cassabois, P.H. Roussignol, C. Delalande, M. Voos, F.T. Assone, R. Houdré, R.P. Stanley, U. Oesterle, *Superlattices and Microstructures* 22 (3) (1997) 389–392.
47. J. Wainstain, C. Delalande, M. Voos, R. Houdré, R.P. Stanley, U. Oesterle, *Solid State Communications* 99 (5) (1996) 317–321.
48. J. Wainstain, C. Delalande, M. Voos, J. Bloch, V. Thierry-Mieg, R. Planel, R. Houdré, R.P. Stanley, U. Oesterle, *Solid State Communications* 106 (11) (1998) 711–714.
49. A.I. Tartakovskii, M. Emam-Ismail, R.M. Stevenson, M.S. Skolnick, V.N. Astratov, D.M. Whittaker, J.J. Bamberg, J.S. Roberts, *Phys. Rev. B* 62 (4) (2000) R2283–R2286.
50. R.P. Stanley, R. Houdré, C. Weisbuch, U. Oesterle, M. Ilegems, *Phys. Rev. B* 53 (16) (1996) 10995–11007.

51. M. Muller, J. Bleuse, R. Andre, *Phys. Rev. B* 62 (24) (2000) 16886–16892.
52. R. Houdré, R.P. Stanley, U. Oesterle, P. Pellandini, M. Ilegemes, in: J. Rarity, C. Weisbuch (Eds.), Microcavities and Photonic Bandgaps: Physics and Applications, in: NATO ASI Series E: Applied Sciences, Vol. 324, Kluwer, Academic Press, 1996, pp. 33–42.
53. J. Wainstain, C. Delalande, D. Gendt, M. Voos, J. Bloch, V. Thierry-Mieg, R. Planel, *Physical Review B* 58 (11) (1998) 7269–7279.
54. V.M. Agranovich, R.M. Hochstrasser (Eds.), Spectroscopy and Excitation Dynamics of Condensed Molecular Systems, North-Holland, Amsterdam, 1983.
55. L.D. Book, N.F. Scherer, *J. Chem. Phys.* 111 (3) (1999) 792–795.
56. A.I. Tartakovoskii, M. Emam-Ismail, D.G. Lidzey, M.S. Skolnick, D.D.C. Bradley, S. Walker, V.M. Agranovich, *Phys. Rev. B* 63 (2001) 121302(R).
57. B. Sermage, S. Long, I. Abram, J.Y. Marzin, J. Bloch, R. Planel, V. Thierry-Mieg, *Phys. Rev. B* 53 (1996) 16516.
58. V. Savona, C. Weisbuch, *Phys. Rev. B* 54 (15) (1996) 10835–10840.
59. T.B. Norris, J.K. Rhee, C.Y. Sung, Y. Arakawa, M. Nishioka, *Phys. Rev. B* 50 (19) (1994) 14633–14666.
60. A. Fainstain, B. Jusserande, V. Thierry-Mieg, *Phys. Rev. Lett.* 75 (1995) 3764–3767.
61. A. Fainstain, B. Jusserand, V. Thierry-Mieg, *Phys. Rev. Lett.* 78 (1997) 1576–1579.
62. R.P. Stanley, S. Pau, U. Oesterle, R. Houdré, M. Ilegems, *Phys. Rev. B* 55 (1997) R4867.
63. R.M. Stevenson, V.N. Astratov, M.S. Skolnick, D.M. Whittaker, M. Emam-Ismail, A.I. Tartakovoskii, P.G. Savvidis, J.J. Baumberg, J.S. Roberts, *Phys. Rev. Lett.* 85 (17) (2000) 3680–3686.
64. P.G. Savvidis, J.J. Baumberg, R.M. Stevenson, M.S. Skolnick, D.M. Whittaker, J.S. Roberts, *Phys. Rev. Lett.* 84 (2000) 1547.
65. A. Fainstain, B. Jusserand, *Phys. Rev. B* 57 (1997) 2402.
66. M.N. Shkunov, W. Gellermann, Z.V. Vardeny, *Appl. Phys. Lett.* 73 (20) (1998) 2878–2880.
67. A.L. Bradley, J.P. Doran, T. Aherne, J. Hegarty, R.P. Stanley, R. Houdre, U. Oesterle, M. Ilegems, *Phys. Rev. B* 57 (16) (1998) 9957–9964.
68. M. Hopmeier, W. Guss, M. Deussen, E.O. Gobel, R.F. Mahrt, *Phys. Rev. Lett.* 82 (20) (1999) 4118–4121.
69. P. Andrew, W.L. Barnes, *Science* 290 (2000) 785–788.
70. Finlayson C.E., Ginger D.S., Greenham N.C., *Chem. Phys. Lett.* 338 (2–3) (2001) 83–87.
71. D.M. Basko, F. Bassani, G.C. La Rocca, V.M. Agranovich, *Phys. Rev B.* 62 (23) (2000) 15962–15977.
72. A. Armitage, M.S. Skolnick, A.V. Kavokin, D.M. Whittaker, V.N. Astratov, G.A. Gehring, J.S. Roberts, *Physical Review B* 58 (23) (1998) 15367–15370.
73. A. Armitage, M.S. Skolnick, V.N. Astratov, D.M. Whittaker, G. Panzarini, L.C. Andreani, T.A. Fisher, J.S. Roberts, A.V. Kavorkin, M.A. Kaliteevski, M.R. Vladimirova, *Phys. Rev. B* 57 (23) (1998) 14877–14881.
74. A. Armitage, D.G. Lidzey, D.D.C. Bradley, T. Virgili, M.S. Skolnick, S. Walker, *Synth. Met.* 111–112 (2000) 377–379.
75. C. Ell, J. Prineas, T.R. Nelson, S. Park, H.M. Gibbs, G. Khitrova, S.W. Koch, R. Houdre, *Phys. Rev. Lett.* 80 (1998) 4795.
76. G.R. Fleming, R. van Grondelle, *Physics Today* (February 1994) 48–55.
77. P.G. Savvidis, J.J. Baumberg, R.M. Stevenson, M.S. Skolnick, D.M. Whittaker, J.S. Roberts, *Phys. Rev. B* 62 (20) (2000) R13278.
78. R. Huang, F. Tassone, Y. Yamamoto, *Phys. Rev. B* 61 (12) (2000) R7854.
79. L. Si Dang, D. Heger, R. Andre, F. Boeuf, R. Romestain, *Phys. Rev. Lett.* 81 (18) (1998) 3920–3923.
80. V.M. Agranovich, H. Benisty, C. Weisbuch, *Solid State Commun.* 102 (1997) 631–636.
81. V.M. Agranovich, G.C. La Rocca, F. Bassani, *Phys. Stat. Sol. (a)* 164 (1997) 39.

Chapter 9
# Electronic Energy Transfer in a Planar Microcavity

D.M. BASKO

*ICTP Trieste, International Centre for Theoretic Physics, Strada Costiera 11, Trieste, I-34000, Italy*

| | |
|---|---|
| 1. Energy Transfer: an Introductory Discussion | 404 |
| 2. Semiclassical Description of the Transfer | 406 |
| 3. Modeling a Specific Structure | 413 |
| 4. Weak Absorption Regime | 422 |
| 5. Numerical Results | 426 |
| 6. Discussion | 439 |
| 7. Concluding Remarks | 444 |
| Acknowledgements | 445 |
| References | 445 |

The role of microcavities in controlling the optical properties of different materials and structures and their importance for nanoscience and nanotechnology has already been addressed in the previous chapter by D. Lidzey. In particular, for the present study the most important feature is that a microcavity changes the mode structure of the electromagnetic field [1], and thus all the interactions mediated by the electromagnetic field are also modified. In particular, the problem of the dipole–dipole interaction, modified by the cavity, was addressed both theoretically [2,3] and experimentally [4–6]. Another manifestation of the cavity effect on the electromagnetic interaction is the cavity-photon-mediated hybridization of the excitations in two spatially separated layers, predicted in Ref. [7]. For organic layers this was observed [8], as described in the previous chapter. An analogous phenomenon was observed in semiconductor structures [9].

In the present chapter we focus on a different situation, assuming the electromagnetic coupling between the two species not to be strong enough to produce a coherent superposition of states. Instead, we assume the dissipative processes to destroy completely the quantum-mechanical coherence between the two subsystems. In such a situation the excitation, initially localized on one subsystem (usually called donor), is incoherently transferred to the second subsystem (acceptor).

This problem may be viewed at a different angle. The phenomenon of the resonant electronic energy transfer has been studied for many decades [10]. Usually one considers resonant donor and acceptor coupled electromagnetically. For example, in the work by Förster [11], which has become one of the most important milestones in this field, energy transfer due to the Coulomb dipole–dipole interaction was considered. Now we ask ourselves: how will the picture of the transfer be modified if the interaction is mediated by a third resonant entity, like a cavity mode?

Below we discuss the physics of the energy transfer in a planar cavity, the role of different dissipative processes in play. After the description of the theoretical approach used to calculate the rates of different processes, we present some results for a model structure [12]. The latter is chosen to illustrate the basic ideas and to show possible important effects, rather than to reproduce a specific experimental configuration. Finally, in the end of this chapter some relevant experiments are discussed.

## 1. Energy Transfer: an Introductory Discussion

Three main characters are going to enter our play: (i) donor—a piece of matter, something consisting of atoms and molecules, which may be, e.g., one or several organic dye molecules, a molecular J-aggregate, a molecular crystalline or inorganic semiconductor structure, etc.; (ii) electromagnetic field; (iii) acceptor—again some material system, for which the same words, as for the donor, may be repeated. Donor and acceptor may be identical, like two molecules of the same dye, or may be completely different, like an inorganic semiconductor quantum well and some amorphous organic substance. What is important, is that *the energies of the electronic excitations in the donor and the acceptor are close enough*. How close is "enough" will become clear when we discuss various processes that may occur in such a system.

Each of the three is a quantum-mechanical subsystem, and the second one (field) interacts with the first (donor) and the third (acceptor). We assume the interaction between the donor and the field to be weak compared to the dissipation into the field or acceptor degrees of freedom. Quantitatively, this means that the characteristic shifts of the donor energy levels due to the interaction are much less than the characteristic spectral widths entering the problem. Note that we do not make any assumptions on the strength of the coupling between the field and the acceptors. As we shall see later on, different cases may arise, leading to different behaviour of the whole system.

Now we can pose the problem as follows. Let at time $t = 0$ the donor be in an excited state. As the system starts to evolve, the donor excitation will be transferred incoherently to the acceptor or to the field (in other words, dissipated into the acceptor or the field degrees of freedom). The question is: what are the corresponding rates for different decay channels? Which part of the donor energy will end up in the acceptor?

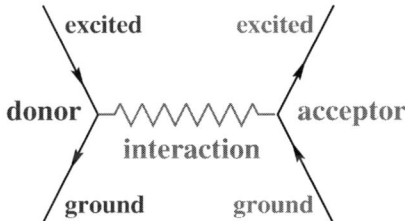

Fig. 1. A naive "Feynman diagram" for the energy transfer.

The relevant quantum mechanical picture of the energy transfer from the donor to the acceptor is that of the decay of a discrete initial state into a continuum of final states due to the electromagnetic interaction, which may be represented by a naive "Feynman diagram", as shown in Figure 1. Namely, this decay occurs via a set of intermediate states, which are the excited states of the electromagnetic field, i.e., photons, represented by the wavy line on the diagram. The simplest calculational tool, describing this phenomenon, is basically the second-order Fermi Golden Rule.

In this picture the initial state (excited donor) decays into final states (excitations of electromagnetic field and acceptor) with the energies $\hbar\omega_f$ equal to the energy $\hbar\omega_i$ of the initial state, which is guaranteed by the presence of the $\delta$-function $\delta(\hbar\omega_f - \hbar\omega_i)$ in Fermi Golden Rule. The energies of the intermediate states, on the other hand, may be arbitrary. If the energy of an intermediate state is different from $\hbar\omega_i$, then one speaks about virtual states or virtual photons. Intermediate states whose energies coincide with $\hbar\omega_i$, are called real photons and correspond to usual propagating light. A virtual photon cannot participate in any dissipative process, since it would violate the energy conservation law, and in particular, a virtual photon cannot be detected. A real photon, on the contrary, may itself decay somewhere, and thus it can carry the excitation away from both the donor and the acceptor. The contribution of a real photon to the transfer is nothing else but the emission of this photon by the donor and its subsequent reabsorption by the acceptor.

The character of the intermediate states most contributing to the transfer depends on the typical length scale of the problem. If the donor and the acceptor are separated by a distance much shorter than the wavelength of the resonant light $\lambda_D \equiv 2\pi c/\omega_D$ ($\omega_D$ being the donor frequency), then the intermediate states with large wave vectors $k \gg \omega_D/c$ contribute the most, and the transfer is dominated by the scalar photons, responsible for the Coulomb interaction.[1] Obviously, these

---

[1] Expanding the Coulomb interaction in multipoles [13], one obtains that the first term is the dipole–dipole one (since the molecules are not charged), which gives the Förster formula [11]. At very small distances higher multipole terms are also important for the transfer, as considered by Dexter [14].

states are virtual since their energy $\hbar ck \gg \hbar\omega_D$. If the separation is much larger than $\lambda_D$, then the dominant contribution comes from the transverse photons. In this case both virtual and real states may contribute to the transfer.

The simple perturbative picture described in the previous two paragraphs refers to the case when both the donor and the acceptor interact weakly with the field, and all the states preserve most of their individuality. In a more general situation, when the coupling between the field and the acceptor is arbitrarily strong, the acceptor and the photon states may become strongly mixed. Graphically, the wavy line on the diagram in Figure 1 should now represent not just a bare photon, but a photon, dressed by the interaction with the acceptors. An example of strongly mixed photon-acceptor states are the well-known cavity polaritons, like those discussed by D. Lidzey in the previous chapter. Namely, if many acceptor molecules with a sharp and narrow absorption line are placed in a high-quality microcavity, the acceptor excitations get strongly mixed with the cavity photons, giving rise to the cavity polaritons.

If the excited states of the acceptor and the field are strongly mixed, it only makes sense to speak about the donor decay into the continuum of the mixed states.[2] One may ask, if the final states of the decay are mixed, can we in principle distinguish the transfer to the acceptor from other dissipative processes? This issue will be discussed in more detail in Section 6, at this stage we can say the following. Yes, different channels for the donor decay may be identified if one assumes that each channel corresponds to a different dissipative bath, and the baths are independent from each other. By the latter we mean that after a sufficiently long time the baths may be considered non-interacting, like a photon that has escaped at infinity, or an excitation in the acceptor that has been brought to a lower energy by the vibronic relaxation (the Stokes shift), or just the heat produced in an absorbing metallic mirror.

## 2. Semiclassical Description of the Transfer

It has been known since long time ago that the quantum-mechanical problem of the radiative decay of an excited atom or molecule is equivalent to a well-known problem of an oscillating dipole in classical electrodynamics. In quantum electrodynamics the discrete excited state of the atom decays into the continuum of photon states, while a classical dipole emits electromagnetic waves and thus loses energy, which leads to the damping of the oscillations. This analogy may also be applied to the problem of the energy transfer if one calculates classically the

---

[2]Note that the interaction between the donor and the electromagnetic field is always assumed here to be weak, otherwise the problem is no longer that of just the donor decay.

power dissipated by the electromagnetic field in the absorbing acceptor, as considered by Galanin and Frank for a luminescent molecule transferring its energy to the surrounding solution [15]. The case of transfer between two point molecules was considered in Ref. [16] both in classical and quantum picture, and in Ref. [17] the classical analogy was used to consider energy transfer in more complex geometries. The deep reason for this classical analogy to be valid is that, on one hand, in quantum electrodynamics the field operators for the quantized electromagnetic field satisfy the same Maxwell's equations as their average values participating in the classical electrodynamics, which is guaranteed by the linearity of the Maxwell's equations. On the other hand, the atomic and molecular susceptibilities, introduced phenomenologically into the classical Maxwell's equations, are in fact quantum-mechanical Kubo averages.

This analogy turns out to be fruitful for the problems described in the previous section, and will be used throughout this chapter. This way of calculation gives exactly the same result as the quantum mechanical approach outlined in the previous section, while the language of the classical electrodynamics is more intuitive and more illustrative.

### 2.1. Calculation of the Transfer Rate from Maxwell's Equations

So, we are going to represent the donor by a classical oscillator. For non-magnetic materials (the only ones we are going to consider) it is convenient to express the oscillating charge density $\rho(\mathbf{r}, t)$ and the current density $\mathbf{j}(\mathbf{r}, t)$ as

$$\rho(\mathbf{r}, t) = -\operatorname{div} \mathbf{P}(\mathbf{r}, t), \qquad \mathbf{j}(\mathbf{r}, t) = \frac{\partial \mathbf{P}(\mathbf{r}, t)}{\partial t}, \tag{2.1}$$

in terms of the polarization $\mathbf{P}(\mathbf{r}, t)$. The simplest example is a molecule in its excited state $|e\rangle$, located in the point $\mathbf{r}_0$. If we adopt the dipole approximation, the corresponding classical polarization is given by

$$\mathbf{P}(\mathbf{r}, t) = \mathbf{d}^D \, \delta(\mathbf{r} - \mathbf{r}_0) \, e^{-i\omega_D t} + \text{c.c.}, \qquad \mathbf{d}^D \equiv \langle g | \hat{\mathbf{d}} | e \rangle, \tag{2.2}$$

where $|g\rangle$ is the molecular ground state, and $\hat{\mathbf{d}}$ is the operator of the molecular dipole moment. Correspondingly, $\mathbf{d}^D$ is the molecular transition dipole. Finally, $\omega_D$ is the donor transition frequency, and "c.c." stands for the complex conjugate. If one wants to take into account the quadrupole polarization, which may be relevant at short distances or for dipole-forbidden transitions, the corresponding expression is

$$P_i(\mathbf{r}, t) = -\frac{Q_{ij}^D}{6} \frac{\partial}{\partial x_j} \delta(\mathbf{r} - \mathbf{r}_0) \, e^{-i\omega_D t} + \text{c.c.}, \tag{2.3}$$

where $Q_{ij}^D$ is the molecular quadrupole moment matrix element.

If the wave function of the donor excitation is delocalized over a spatial region, not small compared to the light wavelength or to the characteristic distance between donors and acceptors, one cannot already treat it as a point dipole. However, one still can associate with it some classical polarization of the form

$$\mathbf{P}(\mathbf{r}, t) = \mathbf{P}^D(\mathbf{r}) e^{-i\omega_D t} + \text{c.c.}, \tag{2.4}$$

where $\mathbf{P}^D(\mathbf{r})$ should be obtained from some microscopic model for the donor. A simple example of such an excitation is an exciton in a planar layer (organic or inorganic quantum well), for which the spatial profile of the polarization is a plane wave.

Another possible complication is that instead of one donor one may have an ensemble of excited donors with different transition frequencies. This may result in a complicated kinetics; in particular, depending on how fast the energy redistribution is within the ensemble, one may or may not average the transfer rate over the donor frequencies. This as well as the previous issues will be addressed in more detail in the next section, where specific cases will be discussed. Anyway, the primary problem to solve is always the one for a given donor with a given transition frequency.

The donor is always embedded into some dielectric structure. Neglecting the spatial dispersion, we characterize the structure by a position and frequency dependent dielectric function $\varepsilon_{ij}(\mathbf{r}, \omega)$. In the latter we include the absorbing acceptors, which give their contribution to the imaginary part[3] of $\varepsilon$. The dielectric function describes also the cavity mirrors: whether dielectric or metallic, they can be included properly by choosing the correct value of $\varepsilon$ in the corresponding regions of space. Thus, $\varepsilon_{ij}(\mathbf{r}, \omega)$ contains all the information about the structure except the donors.

Taking the classical Maxwell's equations for monochromatic fields in an arbitrary dielectric structure, and eliminating the magnetic field from them, we obtain that the Cartesian components $E_i(\mathbf{r}, \omega)$ of the electric field vector satisfy the equation

$$\left[ \frac{\partial^2}{\partial x_i \partial x_j} - \delta_{ij} \frac{\partial^2}{\partial x_l \partial x_l} + \frac{\varepsilon_{ij}(\mathbf{r}, \omega) \omega^2}{c^2} \right] E_j(\mathbf{r}, \omega) = -\frac{4\pi \omega^2}{c^2} P_i(\mathbf{r}, \omega). \tag{2.5}$$

Here $P_i(\mathbf{r}, \omega)$ is the source (donor) polarization, and the summation over the repeated indices is assumed.

Having found the electric field, we may find the power dissipated into different channels. The rate of the pure radiative decay of the donor into the outgoing light

---

[3] By virtue of the Kramers–Krönig relations, the acceptors then necessarily contribute to the real part as well.

may be found calculating the energy flux of the irradiated field outside the structure, where the dissipation is absent. When calculating the energy-related quantities, which are bilinear in the fields, we should not forget to keep both complex conjugate terms, which corresponds to the electric and magnetic fields

$$\mathbf{E}(\mathbf{r},t) = \mathbf{E}(\mathbf{r},\omega)\,e^{-i\omega t} + \text{c.c.}, \qquad \mathbf{B}(\mathbf{r},t) = \mathbf{B}(\mathbf{r},\omega)\,e^{-i\omega t} + \text{c.c.} \quad (2.6)$$

The value of the Pointing vector for a monochromatic component, averaged over the period of the wave, is given by:

$$\overline{\mathbf{S}}(\mathbf{r},\omega) + \overline{\mathbf{S}}(\mathbf{r},-\omega) \equiv \frac{c}{4\pi}\overline{\left[\mathbf{E}(\mathbf{r},t)\times\mathbf{B}(\mathbf{r},t)\right]}$$

$$= \frac{c^2}{2\pi\omega}\,\text{Im}\!\left[\mathbf{E}^*(\mathbf{r},\omega)\times\text{rot}\,\mathbf{E}(\mathbf{r},\omega)\right]. \quad (2.7)$$

Integrating the normal component of the Pointing vector over a remote surface, we obtain the total power $Q_{\text{rad}}(\omega)$ leaving the system in the form of light. Divided by the oscillator energy, it gives the donor radiative decay rate:

$$\Gamma_D^{(\text{rad})} = \frac{Q_{\text{rad}}(\omega_D)}{\hbar\omega_D}. \quad (2.8)$$

To find the transfer rate to the acceptor, we should calculate the corresponding dissipated power. Suppose the acceptor is a molecule situated in the point $\mathbf{r}_1$, with the polarizability $\alpha_{ij}(\omega)$ which incorporates the local field effects. Than its dipole moment, induced by the field from the donor, is given by

$$d_i(t) = \alpha_{ij}(\omega) E_j(\mathbf{r}_1,\omega)\,e^{-i\omega t} + \text{c.c.}, \quad (2.9)$$

and the corresponding dissipated power, averaged over the period of the wave

$$Q_{\text{acc}}(\omega) = \overline{\mathbf{E}(\mathbf{r}_1,t)\cdot\dot{\mathbf{d}}(t)} = 2\omega\,\text{Im}\!\left[E_i^*(\mathbf{r}_1,\omega)\alpha_{ij}(\omega)E_j(\mathbf{r}_1,\omega)\right]$$

$$= 2\omega\,E_i^*(\mathbf{r}_1,\omega)\,\text{Im}\,\alpha_{ij}(\omega)\,E_j(\mathbf{r}_1,\omega), \quad (2.10)$$

where the in the last equation we have used the general symmetry property $\alpha_{ij}(\omega) = \alpha_{ji}(\omega)$, valid if the system is not placed into an external magnetic field [18]. Again, the correct quantum mechanical transfer rate is nothing else but this power divided by the donor energy:

$$\Gamma_D^{(\text{acc})} = \frac{Q_{\text{acc}}(\omega_D)}{\hbar\omega_D}. \quad (2.11)$$

If the point-like donor and the acceptor are close to each other (compared to the light wavelength), the transfer is mostly due to the Coulomb interaction (that is, nonradiative), and the resulting expression for $\Gamma_D^{(\text{acc})}$ is nothing else but the Förster formula.

Usually instead of a single acceptor molecule one deals with many molecules distributed with some density in a host matrix. If the characteristic distance between the molecules is much smaller than the characteristic length scale at which the donor field is varying, one can approximate them as a continuous medium [10], whose dielectric constant has a contribution $\varepsilon_{ij}^{\text{acc}}(\mathbf{r}, \omega)$ from the resonance in the acceptor we are interested in. In this case the losses are given by [19]:

$$Q_{\text{acc}}(\omega) = \frac{\omega}{2\pi} \int d^3\mathbf{r} \, \text{Im}\big[E_i^*(\mathbf{r}, \omega)\varepsilon_{ij}^{\text{acc}}(\omega)E_j(\mathbf{r}, \omega)\big]. \quad (2.12)$$

When calculating the electric field, one might use Eq. (2.5) with the dielectric function not containing the contribution from the acceptors, and to recall about the latter only when calculating the losses in (2.10) or (2.12). This is justified if the acceptors are dilute enough or the corresponding transition is weak, so that the contribution to the dielectric function is small. This brings us back to the discussion of the previous section, since neglection of the acceptors when calculating the electric field is nothing else than the perturbation theory with respect to the interaction between the acceptors and the electromagnetic field, corresponding to the diagram in Figure 1 with the bare photon line.

On the contrary, taking the full dielectric function in Eq. (2.5) means dealing with the photons, "dressed" by the interaction with the acceptor. This becomes crucial if one wants to see how the presence of the acceptor modifies the optical properties of the structure. Again, the relevant example is that of the strong coupling in a planar microcavity. If $\varepsilon_{ij}(\mathbf{r}, \omega)$ contains, besides well-reflecting mirrors, a strong and sharp resonance in the acceptor, Eq. (2.5) gives the two polariton modes, as we shall see below.

A question may arise: how can we speak about the transfer rate, i.e., the transition probability per unit time, when an acceptor, separated from the donor by a distance $l_{d-a}$, must not know anything about the donor at least within the time interval $l_{d-a}/c$? The thing is that the whole approach is valid only at distances $l_{d-a} \ll c/\Gamma_D$, where $\Gamma_D$ is the total donor decay rate. Indeed, in the classical electrodynamics we calculate the losses for a *quasistationary* oscillating dipole, i.e., with the almost constant amplitude. Obviously, at the distances of the order of $c/\Gamma_D$ one cannot ignore the fact that the donor oscillations are damped. Analogously, in the quantum mechanics, Fermi Golden Rule is derived looking at the evolution of the wave function at times smaller than $1/\Gamma_D$, thus it cannot pick up correctly the processes occurring at the characteristic distances exceeding $c/\Gamma_D$. But once the condition $l_{d-a} \gg c/\Gamma_D$ is satisfied, what happens in the first interval $l_{d-a}/c$ is not important, since it is only a small correction, and the most of the transfer occurs afterwards.

The donor decay may be viewed in a different way: the electric field produced by the donor acts on the donor itself, performing a negative work, thus damping the donor oscillations. From this we can find the total decay of the donor, given by:

$$\Gamma_D^{(tot)} = -\frac{1}{\hbar\omega_D} \int d^3\mathbf{r} \overline{\left[\mathbf{E}(\mathbf{r},t) \cdot \frac{\partial \mathbf{P}(\mathbf{r},t)}{\partial t}\right]}$$

$$= \frac{2}{\hbar} \int d^3\mathbf{r}\, \text{Im}[\mathbf{E}(\mathbf{r}) \cdot \mathbf{P}^*(\mathbf{r})], \quad (2.13)$$

where, as before, the overline means the average over the period of the wave. Obviously, the total donor decay rate must be equal to the sum of the contributions from all the decay channels, discussed above: the transfer to the acceptors, the decay into the outgoing light (luminescence), dissipation in the mirrors (if there is any), etc. Formally, it follows from the local energy conservation law

$$-\mathbf{E} \cdot \frac{\partial \mathbf{P}}{\partial t} = \text{div}\, \mathbf{S} + \frac{1}{4\pi}\left[\mathbf{E} \cdot \frac{\partial \mathbf{D}}{\partial t} + \mathbf{B} \cdot \frac{\partial \mathbf{B}}{\partial t}\right], \quad (2.14)$$

obtained directly from the Maxwell's equations. Here $\mathbf{S}$ is the Pointing vector, $\mathbf{D}$ is the electric displacement vector, including the polarization of all the media except donors. The latter are treated as the external source polarization on the left-hand side of Eq. (2.14), which "pumps" the energy into the field and the media on the right-hand side.

## 2.2. Response Function

As Eq. (2.5) is a linear one, the electric field may be expressed in terms of the polarization as

$$E_i(\mathbf{r},\omega) = \int d^3\mathbf{r}'\, \chi_{ij}(\mathbf{r},\mathbf{r}',\omega) P_j(\mathbf{r}',\omega), \quad (2.15)$$

or in other words, Eq. (2.5) may be rewritten using the integral operator, inverse to the differential one in the square brackets in Eq. (2.5). The kernel $\chi_{ij}(\mathbf{r},\mathbf{r}',\omega)$ determines the response of the electric field to the external polarization and satisfies the equation:

$$\left[\frac{\partial^2}{\partial x_i \partial x_j} - \delta_{ij} \frac{\partial^2}{\partial x_l \partial x_l} + \frac{\varepsilon_{ij}(\mathbf{r},\omega)\omega^2}{c^2}\right]\chi_{jk}(\mathbf{r},\mathbf{r}',\omega)$$

$$= -\frac{4\pi\omega^2}{c^2}\delta_{ik}\delta(\mathbf{r}-\mathbf{r}'). \quad (2.16)$$

This response function characterizes only the dielectric structure under consideration, independently of the donors (which do not enter Eq. (2.16) at all). Its properties in different dielectric structures were extensively studied by Agarwal [20]. As a function of $\omega$ it may have poles in the complex plane, which correspond to the frequencies of excitations of the interacting system "field + acceptors". The total donor decay rate (2.13) may be written in terms of the response function as

$$\Gamma_D^{(\text{tot})} = \frac{2}{\hbar} \int d^3\mathbf{r}\, d^3\mathbf{r}'\, \left(P_i^D(\mathbf{r})\right)^* \operatorname{Im} \chi_{ij}(\mathbf{r}, \mathbf{r}', \omega_D)\, P_j^D(\mathbf{r}'). \quad (2.17)$$

As can be shown, using the arguments given in Ref. [21], the response function $\chi$ is closely related to the retarded propagator of the quantized electromagnetic field:

$$\chi_{ij}(\mathbf{r}, \mathbf{r}', t - t') = \frac{i}{\hbar} \langle [\hat{E}_i(\mathbf{r}, t), \hat{E}_j(\mathbf{r}', t')] \rangle \theta(t - t')$$

$$- 4\pi \delta_{ij} \delta(\mathbf{r} - \mathbf{r}') \delta(t - t'). \quad (2.18)$$

Its imaginary part is proportional to the spectral function of the quantized electromagnetic field in the given dielectric structure:

$$\operatorname{Im} \chi_{ij}(\mathbf{r}, \mathbf{r}', \omega) = \sum_{\nu} \langle 0|\hat{E}_i(\mathbf{r})|\nu\rangle \langle \nu|\hat{E}_j(\mathbf{r}')|0\rangle \, \pi \delta(\hbar\omega - \hbar\omega_\nu), \quad \omega > 0,$$

$$(2.19)$$

where $\hat{E}_i(\mathbf{r})$ is the electric field operator, $|0\rangle$ is the ground state of the system, and $|\nu\rangle$ are the exact excited states. Looking at this expansion, one may notice that the expression (2.17) indeed has the structure of Fermi Golden Rule, where the donor polarization is coupled to the electric field.

If the contribution to the dielectric function of the structure from acceptors is small, it may be convenient to represent it as

$$\varepsilon_{ij}(\mathbf{r}, \omega) = \varepsilon_{ij}^{(0)}(\mathbf{r}, \omega) + \delta\varepsilon_{ij}(\mathbf{r}, \omega), \quad (2.20)$$

where $\delta\varepsilon_{ij}(\mathbf{r}, \omega)$ is the contribution from the acceptor resonance. The term $\varepsilon_{ij}^{(0)}(\mathbf{r}, \omega)$ may often be considered independent of frequency, corresponding to the background dielectric constant of the structure. Let us introduce the response function $\chi_{ij}^{(0)}(\mathbf{r}, \mathbf{r}', \omega)$ for the structure without acceptors, satisfying Eq. (2.16) with $\varepsilon_{ij}^{(0)}(\mathbf{r}, \omega)$ instead of the full $\varepsilon_{ij}(\mathbf{r}, \omega)$. In Eq. (2.16) we split the term with $\varepsilon_{ij}$ according to (2.20), move the part with $\delta\varepsilon_{ij}$ to the right-hand side, and apply the integral operator with the kernel $\chi_{ij}^{(0)}$ to both sides of the equation. Then Eq. (2.16) is equivalently rewritten as

$$\chi_{ij}(\mathbf{r}, \mathbf{r}', \omega) = \chi_{ij}^{(0)}(\mathbf{r}, \mathbf{r}', \omega) + \int d^3\mathbf{r}''\, \chi_{ik}^{(0)}(\mathbf{r}, \mathbf{r}'', \omega) \frac{\delta\varepsilon_{kl}(\mathbf{r}'', \omega)}{4\pi} \chi_{lj}(\mathbf{r}'', \mathbf{r}', \omega),$$

$$(2.21)$$

from which the perturbative expansion in $\delta\varepsilon$ may be obtained. This equation has the structure similar to that of the Dyson equation for the photon Green's function in a medium [21], obtained from the full quantum theory, which is a manifestation of the fact that for linear macroscopic electrodynamics the classical and quantum approaches actually give the same result.

## 3. Modeling a Specific Structure

The general discussion of the previous two sections may be applied to the transfer in a planar microcavity [12], like that schematically sketched in Figure 2. The donor layer, placed at $z = z'$, is assumed to be spatially separated from the acceptors by a distance $l'$. If this distance is comparable to the wavelength, the contribution of the short-range Coulomb interaction to the energy transfer in this system is negligible, in which case all the effects are due to the interaction with the transverse electromagnetic field.

As we assume the structure to be translationally invariant in two dimensions, it is convenient to separate the in-plane component $\mathbf{r}_\| \equiv (x, y)$ of the radius-vector $\mathbf{r}$ and to make the corresponding 2D Fourier transform. In particular, it is convenient to represent the response function $\chi_{ij}$, which depends on the difference $\mathbf{r}_\| - \mathbf{r}'_\|$, as:

$$\chi_{ij}(\mathbf{r}, \mathbf{r}', \omega) = \int \frac{d^2\mathbf{k}}{(2\pi)^2} \chi_{ij}(\mathbf{k}, z, z', \omega) e^{i\mathbf{k}(\mathbf{r}_\| - \mathbf{r}'_\|)}. \tag{3.1}$$

In this situation a useful technical tool for calculations is the transfer-matrix formalism [22,23].

We assume the mirrors to be well-reflecting. Then in an empty cavity (without donors and acceptors) well-defined cavity modes exist. When an absorbing medium is placed into the cavity, the properties of the cavity may be strongly modified. For example, the dipole emission from cavity may be changed in the presence of an absorbing medium [24].

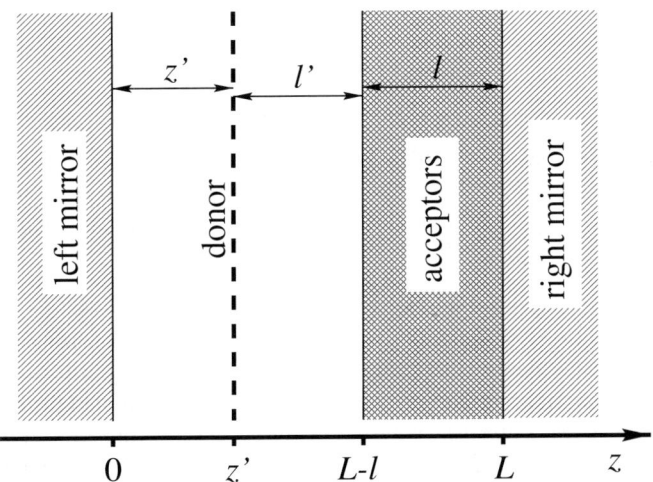

Fig. 2. The microcavity structure to be modeled.

Depending on the parameters of the system we may identify three different regimes. First, if the acceptor absorption is weak enough (i.e., $\operatorname{Im}\tilde{\varepsilon}\ll 1$, $\tilde{\varepsilon}$ being the acceptor medium dielectric function), cavity modes are preserved, correspondingly the response function $\chi$ will have poles at $\omega = \omega_k^{s,p} - i\gamma_k^{s,p}$ for each **k** for $s$- and $p$-polarizations, with small imaginary parts $\gamma_k^{s,p}$. Second, if the acceptor absorption is strong, or $\operatorname{Im}\tilde{\varepsilon} \sim 1$, and the absorption band is broad enough, the cavity mode is destroyed by the absorption, i.e., one obtains a broad smooth density of states with no sharp resonances. And third, if the acceptor absorption band is narrow, the cavity mode and the acceptor excitation may become strongly coupled, giving rise to the split upper and lower cavity polariton branches (two distinct peaks in the spectral function). Note that these three regimes are independent on the assumption about the field–donor weak coupling.

Now we proceed to formal description of different elements of the structure.

### 3.1. MIRRORS

Let the cavity occupy the region $0 < z < L$ between two mirrors which are situated at $z = 0$ and $z = L$. At this stage we do not need to specify the structure of the mirrors, knowing their reflection coefficients is sufficient. To be specific, we define the amplitude reflection coefficients analogously to Ref. [19]: as the ratio of the electric field amplitudes for the $s$-polarization, and of the magnetic field amplitudes for the $p$-polarization (in particular, for an ideal metallic surface they are equal to $-1$ and $1$ respectively). In the isotropic case each plane electromagnetic wave is characterized by its 2D wave vector **k** lying in the plane of the cavity, its frequency $\omega$, the direction of propagation in $z$ (left or right), and the polarization ($s$ or $p$). Correspondingly, the reflection coefficient of each mirror for the waves incident from inside the cavity is a function of $|\mathbf{k}|$, $\omega$, and the polarization. We denote the reflection coefficients of the left mirror by

$$r^s(k,\omega) \equiv \sqrt{R^s(k,\omega)}\, e^{i\theta^s(k,\omega)}, \qquad r^p(k,\omega) \equiv \sqrt{R^p(k,\omega)}\, e^{i\theta^p(k,\omega)},$$
(3.2)

those of the right mirror—by

$$\tilde{r}^s(k,\omega) \equiv \sqrt{\tilde{R}^s(k,\omega)}\, e^{i\tilde{\theta}^s(k,\omega)}, \qquad \tilde{r}^p(k,\omega) \equiv \sqrt{\tilde{R}^p(k,\omega)}\, e^{i\tilde{\theta}^p(k,\omega)}.$$
(3.3)

If one is not interested in the characteristics of the light outside the cavity, these amplitude reflection coefficients provide all the necessary information about the mirrors, since they determine completely the boundary conditions for the field inside the cavity (in combination with the condition that outside the cavity only outgoing waves are present). Now we make no assumptions about the mirrors, which may be dissipative like metallic mirrors, or non-dissipative like distributed Bragg reflectors (DBRs). We will keep the general form of the reflection coefficients up to the final stage.

## 3.2. Acceptors

The region $L - l < z < L$ inside the cavity is assumed to be occupied by the acceptor molecules. We assume the acceptor medium to be macroscopically homogeneous on the length scale of the order of the wavelength, and describe the acceptor layer by a complex dielectric function, as discussed in the previous section (continuous medium approximation). We also assume the acceptor medium to be isotropic. This condition does not hold always in experiments, nevertheless we make this assumption since we are not intending to describe any specific experiment, but the general trends. Anisotropy does not introduce any conceptual difficulties and may be taken into account when needed, but it would lead to much more bulky expressions.

The complex dielectric function $\tilde{\varepsilon}(\omega)$ of the acceptor medium contains two parts: the off-resonant (background) contribution $\tilde{\varepsilon}_\infty$ whose dependence on the frequency may be neglected, and the resonant contribution from the acceptor molecules:

$$\tilde{\varepsilon}(\omega) = \tilde{\varepsilon}_\infty + \tilde{\Delta}_{\text{res}}\left[\bar{s}(\xi) + is(\xi)\right], \quad \xi \equiv \frac{\hbar\omega - \hbar\omega_A}{\gamma_A}, \qquad (3.4)$$

where $\omega_A$ and $\gamma_A$ are the position and width of the acceptor absorption band, $\tilde{\Delta}_{\text{res}}$ determines the strength of the acceptor absorption at the maximum (it depends on the transition dipole moment of the acceptor molecules, and on their concentration), $s(\xi)$ is the imaginary part of the normalized resonant contribution to the $\tilde{\varepsilon}(\omega)$ (the shape of the acceptor absorption band), $\bar{s}(\xi)$ is the corresponding real part. The latter may be expressed via the former using the Kramers–Krönig relation:

$$\bar{s}(\xi) = \frac{1}{\pi} \int \frac{s(\eta)\, d\eta}{\eta - \xi}. \qquad (3.5)$$

For homogeneously broadened molecules one would have the absorption with a Lorentzian shape:

$$s(\xi) = \frac{1}{\xi^2 + 1}, \quad \bar{s}(\xi) = -\frac{\xi}{\xi^2 + 1}. \qquad (3.6)$$

However, a more realistic picture is an inhomogeneously broadened ensemble, which we model by a Gaussian absorption:

$$s(\xi) = e^{-\xi^2/2}, \quad \bar{s}(\xi) = -\sqrt{\frac{2}{\pi}} e^{-\xi^2/2} \int_0^\xi e^{\eta^2/2}\, d\eta. \qquad (3.7)$$

If needed, one can include several peaks in (3.4).

The phenomenological expression (3.4) may be obtained from a simple microscopic model, following the general procedure of Ref. [25]. We assume the acceptor molecules with the transition dipole moment $d_A$ to be distributed continuously inside the cavity with the density $n_A$ and with chaotic orientation. Then the microscopic expression for the dielectric function is

$$\tilde{\varepsilon}(\omega) = \tilde{\varepsilon}_\infty - \frac{4\pi n_A}{\hbar} \frac{d_A^2}{3} \int d\bar{\omega} \frac{2\bar{\omega} \rho_A(\bar{\omega})}{\omega^2 - \bar{\omega}^2 + i\omega\delta}, \qquad \delta \to +0. \qquad (3.8)$$

Formally the quantity $\rho_A(\omega)$ is given by the true density of states for a single acceptor convoluted with the probability for an acceptor to have a given transition frequency, but only the convolution $\rho_A(\omega)$ enters the final results and is related to the absorption spectrum measured experimentally. It is normalized according to

$$\int d\omega \, \rho_A(\omega) = 1. \qquad (3.9)$$

If we choose $\rho_A(\omega)$ to be Gaussian, which is the case when the inhomogeneous broadening dominates over the homogeneous one, we obtain the expressions (3.4), (3.7) with $\tilde{\Delta}_{\text{res}}$ related to the microscopic quantities:

$$\tilde{\Delta}_{\text{res}} = \frac{4\pi^2 n_A}{\hbar} \frac{d_A^2}{3} \frac{1}{\sqrt{2\pi} \, \gamma_A}. \qquad (3.10)$$

If the inhomogeneous broadening dominates over the homogeneous one $\gamma_A^{\text{hom}}$, the effective density of the acceptor molecules "seen" by the donor is given by $n_A^{\text{eff}} \sim n_A \gamma_A^{\text{hom}}/\gamma_A$. The description of the medium in terms of the dielectric constant is valid at characteristic distances much larger than $(n_A^{\text{eff}})^{-1/3}$.

### 3.3. Donors

The donors are supposed to be situated somewhere in the region $0 < z < L - l$, whose background dielectric constant is denoted by $\varepsilon$. We make several assumptions about donors. First, we assume that the donor absorption does not play any significant role in the frequency region of interest. This implies that either (i) donor absorption spectrum lies high above, which is the case for organics with large Stokes shift, or (ii) the donors are dilute enough. Correspondingly, the donors are assumed to have no effect on the optical properties of the cavity. Second, the donors are assumed to be incoherent between themselves, and their excitation density to be low enough for the acceptors to absorb in the linear regime, which means that the problem is reduced to that for a single donor with the subsequent averaging.

In practice, we intend to study two cases. The first one is the case when the donor is a single molecule (or a cluster of coupled molecules) with the size much smaller than the light wavelength. Such a donor may be treated as a point dipole $\mathbf{d}^D$, oscillating as given by Eq. (2.2). We assume that the transition dipole moment $\mathbf{d}_D$ includes also the local field effects. For the Lorentz model it is given by the molecular dipole, multiplied by the factor $(\varepsilon + 2)/3$.

The other case is when the donor is a thin layer, parallel to the cavity plane, whose excited states may be characterized by a definite in-plane wave vector $\mathbf{k}$. If it is much thinner than the light wavelength, its polarization may be considered to be proportional to $\delta(z)$ and may be written as

$$\mathbf{P}^D(\mathbf{r}) = \frac{\mathbf{d}_D}{a_{eh}} \frac{e^{i\mathbf{k}\mathbf{r}_\|}}{\sqrt{S}} \delta(z - z'), \tag{3.11}$$

where $z'$ is the position of the donor layer, $S$ is the normalization area, $\mathbf{d}_D$ and $a_{eh}$ are the relevant transition dipole moment, and the effective "Bohr radius", characterizing the electron and hole relative motion. Their precise definitions depend on the specific structure. For a monolayer of organic molecules with Coulomb coupling (a 2D molecular crystal) $\mathbf{d}_D$ is the monomer transition dipole (corrected for the local field effects), $a_{eh}$ is the square root of the unit cell area.

We consider in more detail the situation when the donor layer is a thin semiconductor quantum well with the background dielectric constant $\varepsilon_{sem}$, which may be different from $\varepsilon$. Then $a_{eh}$ may be identified as the 2D Bohr radius $a_{B2}$, while the dipole moment is given by

$$\mathbf{d}_{D\|} = \phi(0) a_{eh} \langle e|h\rangle_z \langle v|(-e\mathbf{r}_\|)|c\rangle, \tag{3.12}$$

$$d_{Dz} = \frac{\varepsilon_{sem}}{\varepsilon} \phi(0) a_{eh} \langle e|h\rangle_z \langle v|(-ez)|c\rangle, \tag{3.13}$$

where $\langle v|(-e\mathbf{r}_\|)|c\rangle$ and $\langle v|(-ez)|c\rangle$ are the corresponding components of matrix element of the electron dipole moment between the valence and conduction band extrema Bloch functions, $\phi(0)$ is the wave function of the in-plane relative motion of the electron and the hole when they are at the same point [for $1s$-state of the purely 2D hydrogen problem $\phi(0) = \sqrt{2/(\pi a_{B2}^2)}$], and $\langle e|h\rangle_z$ is the $z$-overlap of the electron and hole confinement wave functions in the quantum well. The factor $\varepsilon_{sem}/\varepsilon$ takes into account the boundary condition for the normal component of the electric field at the interface and plays the role of the local field correction.

Applying the general considerations of the previous section to these specific systems, we obtain the decay rate of an excited state in a plane quantum well:

$$\Gamma_{QW}(\mathbf{k}, z', \omega_D) = \frac{2}{\hbar} \frac{d_{D,i} d^*_{D,j}}{a_{eh}^2} \operatorname{Im} \chi_{ij}(\mathbf{k}, z', z', \omega_D) \tag{3.14}$$

for a state with the wave vector $\mathbf{k}$ and the energy $\hbar\omega_D$, and in the case of a point donor molecule one should integrate over the wave vectors:

$$\Gamma_{mol}(z', \omega_D) = \frac{2 d_{D,i} d^*_{D,j}}{\hbar} \int \operatorname{Im} \chi_{ij}(\mathbf{k}, z', z', \omega_D) \frac{d^2\mathbf{k}}{(2\pi)^2}. \tag{3.15}$$

The expressions (3.14) and (3.15) give the decay rate for a single donor. However, in experiments some averaged quantities are measured. If the orientations of the donor molecules are random, the relevant average is

$$\overline{d_{D,i} d^*_{D,j}} = \delta_{ij} |d_D|^2/3. \qquad (3.16)$$

Correspondingly, in the expression (3.15) the sum of the diagonal components $\chi_{ii}$ appears. The expression (3.15) may be also averaged over the spatial density $n_D(z')$ of the *excited* donors, and over the inhomogeneous distribution of the donor frequencies $\omega_D$ (corresponding to the luminescence spectrum at a given temperature $T$).

For a semiconductor quantum well, placed at a certain position $z'$, and with a narrow excitonic resonance at the frequency $\omega_D$, it is reasonable to average the decay rate over the momentum distribution $n_\mathbf{k}$ of the excitons. The simplest approximation would be to assume the excitons with the 2D spatial density $n_{\text{exc}}$ to be thermalized at some temperature $T$, which gives the Boltzmann occupation numbers:

$$n_\mathbf{k} = \frac{2\pi\hbar^2 n_{\text{exc}}}{mT} \exp\left(-\frac{\hbar^2 k^2}{2mT}\right), \qquad (3.17)$$

as it was done in Ref. [26] where Wannier–Mott exciton radiative decay in a bare semiconductor quantum well (without a cavity) was studied. As a matter of fact, the exciton intraband relaxation processes are not fast enough to maintain the Boltzmann distribution in the radiative region $k < \sqrt{\varepsilon}\,\omega_D/c$, as shown by simulations for a bare semiconductor quantum well [27], as well as for one in a microcavity [28]. As a result, the formula (3.17) overestimates the exciton population in the radiative region (and hence the radiative decay rate) by several times. Since at large donor–acceptor separations the transfer is dominated by the transverse photons, the main contribution to the energy transfer is also limited to the radiative $\mathbf{k}$, and we encounter the same problem. We shall use (3.17) as an estimate because of its simplicity. Neglecting the exciton dispersion in the radiative region, we obtain

$$\Gamma_{QW}(T, z', \omega_D)$$
$$= \frac{2\pi\hbar^2}{m a_{eh}^2 T} \frac{2 d_{D,i} d^*_{D,j}}{\hbar} \int \operatorname{Im} \chi_{ij}(\mathbf{k}, z', z', \omega_D) \frac{d^2\mathbf{k}}{(2\pi)^2}, \qquad (3.18)$$

with the integral formally coinciding with that in (3.15).

The direction of the Wannier exciton dipole moment is determined by the crystal structure. In quantum wells, made of materials with zinc-blende structure, the heavy-hole ($hh$) excitons may be polarized along $L$- and $T$-directions, the light-hole ($lh$) excitons – along $L$-, $T$-, and $Z$-directions, where we use the standard notation $L$, $T$, and $Z$ for the three basis vectors

$$\mathbf{e}_L \equiv \mathbf{k}/|\mathbf{k}|, \qquad \mathbf{e}_T \equiv \mathbf{e}_z \times \mathbf{e}_L, \qquad \mathbf{e}_Z \equiv \mathbf{e}_z \qquad (3.19)$$

at a given **k**. The decay rate (3.14) or (3.18) is then given by the corresponding diagonal component $\chi_{ii}$.

### 3.4. EXPLICIT EXPRESSIONS

In this subsection we solve Eqs. (2.5), (2.16), giving the explicit analytical expressions for the components of $\chi$ for the structure, shown in Figure 2, and the expressions for the normal component of the Pointing vector at the mirrors and for the losses in the acceptor for the classical oscillating polarization.

The nonzero components of the response function are $\chi_{TT}$, which corresponds to the $s$-polarization, and $\chi_{LL}$, $\chi_{ZZ}$, $\chi_{LZ}$, $\chi_{ZL}$, corresponding to $p$-polarization. In all the subsequent calculations we assume the donor dipole moment either to be oriented chaotically (for point donor molecules), or to be directed along $T$, $L$, or $Z$ (for $T$-, $L$-, and $Z$- excitons in a quantum well). Then, according to the above-said, we need only the diagonal components of the response function. For a point $z' < L - l$ (outside the acceptor layer) they are given by

$$\chi_{TT}(\mathbf{k}, z', z', \omega) = \frac{2\pi i}{q} \frac{\omega^2}{c^2} \frac{(1 + r^s e^{2iqz'})(1 + \tilde{r}^s_{\text{int}} e^{2iql'})}{1 - r^s \tilde{r}^s_{\text{int}} e^{2iq(L-l)}}, \qquad (3.20)$$

$$\chi_{LL}(\mathbf{k}, z', z', \omega) = \frac{2\pi i q}{\varepsilon} \frac{(1 - r^p e^{2iqz'})(1 - \tilde{r}^p_{\text{int}} e^{2iql'})}{1 - r^p \tilde{r}^p_{\text{int}} e^{2iq(L-l)}}, \qquad (3.21)$$

$$\chi_{ZZ}(\mathbf{k}, z', z', \omega) = \frac{2\pi i k^2}{\varepsilon q} \frac{(1 + r^p e^{2iqz'})(1 + \tilde{r}^p_{\text{int}} e^{2iql'})}{1 - r^p \tilde{r}^p_{\text{int}} e^{2iq(L-l)}}, \qquad (3.22)$$

where $l' \equiv L - l - z'$, and $\tilde{r}^{s,p}_{\text{int}}(k, \omega)$ are the reflection coefficients at the interface $z = L - l$ for the $s$- and $p$-polarizations respectively:

$$\tilde{r}^s_{\text{int}} = \frac{(q + \tilde{q})\tilde{r}^s e^{2i\tilde{q}l} + q - \tilde{q}}{q + \tilde{q} + (q - \tilde{q})\tilde{r}^s e^{2i\tilde{q}l}},$$

$$\tilde{r}^p_{\text{int}} = \frac{(q/\varepsilon + \tilde{q}/\tilde{\varepsilon})\tilde{r}^p e^{2i\tilde{q}l} + q/\varepsilon - \tilde{q}/\tilde{\varepsilon}}{q/\varepsilon + \tilde{q}/\tilde{\varepsilon} + (q/\varepsilon - \tilde{q}/\tilde{\varepsilon})\tilde{r}^p e^{2i\tilde{q}l}}, \qquad (3.23)$$

and we denote the $z$-component of the wave vector in the two media by

$$q(\mathbf{k}, \omega) \equiv \sqrt{\varepsilon \omega^2/c^2 - k^2}, \qquad \tilde{q}(\mathbf{k}, \omega) \equiv \sqrt{\tilde{\varepsilon}(\omega)\omega^2/c^2 - k^2}. \qquad (3.24)$$

The sign of the square root should be chosen to give non-negative real and imaginary parts of $q$, $\tilde{q}$ for $\omega > 0$, and for complex $\omega$ the square root should be understood as the analytical continuation from the real positive $\omega$ through the upper half-plane (according to the general analytical properties of response functions). Note that even in the case of a semiconductor quantum well with $\varepsilon_{\text{sem}}$ different from $\varepsilon$ the above expressions for $\chi$ are still valid, as long as the quantum well is much thinner than the light wavelength, since all the effect of the dielectric discontinuity are included in the definition of the dipole moment (3.12), (3.13).

To calculate the classical losses, we assume the external (donor) polarization be given by

$$\mathbf{P}^{\text{ext}}(\mathbf{r},t) = \mathbf{p}\,\delta(z-z')\,\frac{e^{i\mathbf{k}\mathbf{r}_\| - i\omega t}}{\sqrt{S}} + \text{c.c.}, \tag{3.25}$$

where $\mathbf{p}$ is $\mathbf{d}_D/\sqrt{S}$ for a point molecule and $\mathbf{d}_D/a_{eh}$ for a quantum well (the Fourier components of the polarization (2.2) or (3.11)).

First we consider the field in the region $L-l < z < L$, from which both the losses in the acceptors and the leakage through the right mirror can be obtained. The electric field, produced by the polarization (3.25) in the region $L-l < z < L$ may be represented as

$$\mathbf{E}(\mathbf{r}_\|,z,t) = \left(\mathbf{E}_r e^{i\tilde{q}(z-L)} + \mathbf{E}_l e^{-i\tilde{q}(z-L)}\right)\frac{e^{i\mathbf{k}\mathbf{r}_\| - i\omega t}}{\sqrt{S}} + \text{c.c.}, \tag{3.26}$$

$$\mathbf{E}_r = \tilde{A}^s\,\mathbf{e}_T - \frac{c\tilde{q}}{\bar{\varepsilon}\omega}\tilde{A}^p\,\mathbf{e}_L + \frac{ck}{\bar{\varepsilon}\omega}\tilde{A}^p\,\mathbf{e}_Z, \tag{3.27}$$

$$\mathbf{E}_l = \tilde{r}^s\tilde{A}^s\,\mathbf{e}_T + \frac{c\tilde{q}}{\bar{\varepsilon}\omega}\tilde{r}^p\tilde{A}^p\,\mathbf{e}_L + \frac{ck}{\bar{\varepsilon}\omega}\tilde{r}^p\tilde{A}^p\,\mathbf{e}_Z, \tag{3.28}$$

where $\tilde{A}^s$ is the electric field amplitude at $z = L-0$ for the $s$-polarization, $-\tilde{A}^p$ is the magnetic field amplitude at $z = L-0$ for the $p$-polarization (this parametrization of the fields corresponds to the definition of the reflection coefficients, given in the beginning of Section 3) for the right-travelling waves, and $\tilde{q}(\mathbf{k},\omega)$ is defined by (3.24). The amplitudes $\tilde{A}^{s,p}$ are given by

$$\tilde{A}^s = \frac{2\pi i}{q}\frac{\omega^2}{c^2}\frac{p_T(1+r^s e^{2iqz'})\tilde{t}^s_{\text{int}}e^{iql'}}{1-r^s\tilde{r}^s_{\text{int}}e^{2iq(L-l)}}, \tag{3.29}$$

$$\tilde{A}^p = \frac{2\pi i\omega}{c}\left[-p_L\left(1-r^p e^{2iqz'}\right) + \frac{kp_Z}{q}\left(1+r^p e^{2iqz'}\right)\right]$$

$$\times\frac{\tilde{t}^p_{\text{int}}e^{iql'}}{1-r^p\tilde{r}^p_{\text{int}}e^{2iq(L-l)}}, \tag{3.30}$$

where the reflection coefficients at the interface $z = L-l$ are defined by (3.23), and the transmission coefficients are given by

$$\tilde{t}^s_{\text{int}} = \frac{2q\,e^{i\tilde{q}l}}{q+\tilde{q}+(q-\tilde{q})\tilde{r}^s e^{2i\tilde{q}l}},$$

$$\tilde{t}_{\text{int}}^p = \frac{2(q/\varepsilon)\,e^{i\tilde{q}l}}{q/\varepsilon + \tilde{q}/\tilde{\varepsilon} + (q/\varepsilon - \tilde{q}/\tilde{\varepsilon})\,\tilde{r}^p e^{2i\tilde{q}l}}. \tag{3.31}$$

The net power absorbed in the unit area of the acceptor layer is given by

$$\begin{aligned}
W_{\text{abs}} &= \frac{\omega |\tilde{A}^s|^2 \operatorname{Im}\tilde{\varepsilon}}{2\pi S}\left[(1+|\tilde{r}^s|^2 e^{-2\tilde{q}''l})\frac{e^{2\tilde{q}''l}-1}{2\tilde{q}''} + 2\operatorname{Im}\left(\tilde{r}^s\,\frac{e^{2i\tilde{q}'l}-1}{2\tilde{q}'}\right)\right] \\
&+ \frac{\omega |\tilde{A}^p|^2 \operatorname{Im}\tilde{\varepsilon}}{2\pi S}\left[(1+|\tilde{r}^p|^2 e^{-2\tilde{q}''l})\frac{e^{2\tilde{q}''l}-1}{2\tilde{q}''}\frac{\operatorname{Re}(\tilde{q}/\tilde{\varepsilon})}{\operatorname{Re}\tilde{q}}\right. \\
&\left. + 2\operatorname{Im}\left(\tilde{r}^p\,\frac{e^{2i\tilde{q}'l}-1}{2\tilde{q}'}\right)\frac{\operatorname{Im}(\tilde{q}/\tilde{\varepsilon})}{\operatorname{Im}\tilde{q}}\right], \tag{3.32}
\end{aligned}$$

where we denote $\tilde{q}' \equiv \operatorname{Re}\tilde{q}$, $\tilde{q}'' \equiv \operatorname{Im}\tilde{q}$.

The net power, leaving the cavity through the unit area of the right mirror is given by the $z$-component of the Pointing vector:

$$\begin{aligned}
W_r &= \frac{c^2|\tilde{A}^s|^2}{2\pi S\omega}\left[(1-|\tilde{r}^s|^2)\operatorname{Re}\tilde{q} + 2\operatorname{Im}\tilde{r}^s \operatorname{Im}\tilde{q}\right] \\
&+ \frac{c^2|\tilde{A}^p|^2}{2\pi S\omega}\left[(1-|\tilde{r}^p|^2)\operatorname{Re}(\tilde{q}/\tilde{\varepsilon}) + 2\operatorname{Im}\tilde{r}^p \operatorname{Im}(\tilde{q}/\tilde{\varepsilon})\right]. \tag{3.33}
\end{aligned}$$

Near the left mirror ($0 < z < z'$) the electric field is represented as

$$\mathbf{E}(\mathbf{r}_\parallel, z, t) = \left(\mathbf{E}_r e^{iqz} + \mathbf{E}_l e^{-iqz}\right)\frac{e^{i\mathbf{kr}_\parallel - i\omega t}}{\sqrt{S}} + \text{c.c.}, \tag{3.34}$$

$$\mathbf{E}_r = r^s A^s\,\mathbf{e}_T + \frac{cq}{\varepsilon\omega}r^p A^p\,\mathbf{e}_L + \frac{ck}{\varepsilon\omega}r^p A^p\,\mathbf{e}_Z, \tag{3.35}$$

$$\mathbf{E}_l = A^s\,\mathbf{e}_T - \frac{cq}{\varepsilon\omega}A^p\,\mathbf{e}_L + \frac{ck}{\varepsilon\omega}A^p\,\mathbf{e}_Z, \tag{3.36}$$

with the amplitudes $A^{s,p}$ corresponding to the left-travelling wave and given by

$$A^s = \frac{2\pi i}{q}\frac{\omega^2}{c^2}\frac{p_T(1+\tilde{r}_{\text{int}}^s e^{2iql'})e^{iqz'}}{1-r^s\tilde{r}_{\text{int}}^s e^{2iq(L-l)}}, \tag{3.37}$$

$$A^p = \frac{2\pi i\omega}{c}\left[p_L\left(1-\tilde{r}_{\text{int}}^p e^{2iql'}\right) + \frac{kp_Z}{q}\frac{\varepsilon}{\varepsilon_0}\left(1+\tilde{r}_{\text{int}}^p e^{2iql'}\right)\right]$$
$$\times \frac{e^{iqz'}}{1-r^p\tilde{r}_{\text{int}}^p e^{2iq(L-l)}}. \tag{3.38}$$

Analogously, we obtain the net power, leaving the cavity through the unit area of the left mirror:

$$W_r = \frac{c^2|A^s|^2}{2\pi S\omega}\left[(1-|r^s|^2)\operatorname{Re} q + 2\operatorname{Im} r^s \operatorname{Im} q\right]$$

$$+ \frac{c^2|A^p|^2}{2\pi S\omega}\left[(1-|r^p|^2)\operatorname{Re}(q/\varepsilon) + 2\operatorname{Im} r^p \operatorname{Im}(q/\varepsilon)\right]. \quad (3.39)$$

For a quantum well exciton with the wave vector $\mathbf{k}$ the expressions (3.32), (3.33), (3.39) give the final result for the energy loss per unit time and divided by $\hbar\omega$ give the quantum mechanical probabilities. One may check that the sum of the decay rates obtained from (3.32), (3.33), (3.39) for each polarization is equal to the total decay rates found using (3.20), (3.21), (3.22). For a point molecule these expressions should be summed over $\mathbf{k}$:

$$\sum_{\mathbf{k}} \to S \int \frac{d^2\mathbf{k}}{(2\pi)^2}, \quad (3.40)$$

where the factor $S$ in front of the integral will cancel $1/\sqrt{S}$ in $\mathbf{p}$.

## 4. Weak Absorption Regime

In this section we consider a special case, which, on one hand, is rather relevant from the experimental point of view, and on the other hand, allows a more complete analytical treatment. Imagine a cavity with good mirrors (not necessarily the one described in the previous section), where one or several well-defined modes are present. Suppose that placing the donors and the acceptors inside does not have any dramatic effect on the cavity mode(s): they are not enormously broadened, no strong-coupling phenomena occur, the spectral function of the electromagnetic field still has distinct narrow peaks. More specifically, we assume the mode spectral width to be small compared to other characteristic frequency scales: those determined by the mirrors, by the dielectric media inside the cavity, like acceptor absorption band width, it etc. What can one say about the energy transfer in such situation?

In formal terms, the cavity modes correspond to poles in the response function. The latter may be represented near one of the poles as

$$\chi_{ij}(\mathbf{k}, z, z', \omega) \approx -\frac{u_i(\mathbf{k}, z) u_j^*(\mathbf{k}, z')}{\omega - \omega_k + i\gamma_k}, \quad (4.1)$$

where $u_i(\mathbf{k}, z)$ is proportional to the profile of the electric field of the mode.[4] The mode damping $\gamma_k$ is small compared to other frequency scales, as discussed above. It has several contributions from different sources of damping: transmission of the mirrors, absorption in the acceptors, etc. Below we shall see how these contributions may be separated. If there are several modes, the response function will be represented by a sum of several terms analogous to (4.1).

What consequences does it have for the donor, and for the energy transfer to the acceptors? Substituting (4.1) into (2.17), we see that the donor decay occurs only if the donor frequency coincides with the mode frequency (within the width $\gamma_k$). This brings us back to the discussion of real and virtual photons in Section 1. Namely, we can conclude that as long as it makes sense to speak about cavity photons as well-defined quasiparticles (weakly damped), their contribution to the donor decay and to the energy transfer is purely real: the donor emits a cavity photon, and then the latter may finish its life in several ways, corresponding to different contributions to $\gamma_k$. If it is absorbed in the acceptor, this corresponds to the energy transfer. Thus in the weak absorption regime the transfer is reduced to a somewhat trivial process of emission and reabsorption of real cavity photons (although one should not forget some important issues – see the discussion at the end of this section).

For the structure, described in the previous section, the weak absorption regime is realized if we consider $\tilde{\Delta}_{\text{res}}$, determining the resonant part of the acceptor dielectric constant (3.4), a small parameter: $\tilde{\Delta}_{\text{res}} \ll 1$. We also assume the background dielectric constant to be the same across the cavity: $\tilde{\varepsilon}_\infty = \varepsilon$. In these conditions one may neglect the reflection at the interface $\varepsilon|\tilde{\varepsilon}$, and set $\tilde{r}_{\text{int}}^{s,p} = \tilde{r}^{s,p} e^{2i\tilde{q}l}$ in (3.20)–(3.22). The spectral function $\text{Im}\,\chi$ has a sharp peak corresponding to a cavity mode. The mode dispersion and damping are different for $s$- and $p$-polarizations, thus in the denominator of (4.1) one has $\omega_k^s$, $\gamma_k^s$ for $\chi_{TT}$, and $\omega_k^p$, $\gamma_k^p$ for $\chi_{LL}$, $\chi_{ZZ}$.

Focusing on the denominators of (3.20)–(3.22), we represent them as

$$Z_{s,p} \equiv 1 - r^{s,p}\tilde{r}_{\text{int}}^{s,p} e^{2iq(L-l)} = 1 - \mathcal{R}_{s,p} e^{i\Theta_{s,p}}, \qquad (4.2)$$

$$\mathcal{R}_{s,p} \equiv \sqrt{R^{s,p}\,\tilde{R}^{s,p}}\, e^{-2\,\text{Im}\,\tilde{q}l}, \qquad (4.3)$$

$$\Theta_{s,p} \equiv 2q(L-l) + 2\,\text{Re}\,\tilde{q}l + \theta^{s,p} + \tilde{\theta}^{s,p}, \qquad (4.4)$$

where $R$ and $\theta$ were defined in (3.2), (3.3). All the quantities, appearing in (4.2)–(4.4) depend on $k$ and $\omega$. We assume that the quantities $\mathcal{R}_{s,p}(k, \omega)$ are (i) close

---

[4]The normalization of $u_i(\mathbf{k}, z)$ as it appears in Eq. (4.1), and its dimensionality in particular, may seem somewhat strange. To make it correspond to the electric field one has to introduce $\hbar$, thus invoking the quantum theory. Within the classical theory $u_i(\mathbf{k}, z)$ is just an eigenfunction of the differential operator in Eqs. (2.5), (2.16), whose normalization is what it is, as $\chi$ is obtained from the Maxwell's equations. The minus sign in (4.1) is necessary to ensure the positiveness of $\text{Im}\,\chi_{ii}(\mathbf{k}, z', z', \omega)$

to 1 (which means that the mirrors are well-reflecting and the absorption is small), and (ii) vary slowly with $k$ and $\omega$. The latter condition implies also that the cavity mode width should be small compared to the acceptor absorption band width: $\gamma_k \ll \gamma_A$. We make no assumptions about the phase $\Theta_{s,p}(k,\omega)$, whose behaviour may be complicated.

The frequencies $\omega_k^s$, $\omega_k^p$ and the dampings $\gamma_k^s$, $\gamma_k^p$ of the cavity modes for $s$- and $p$-polarizations (the real and imaginary parts of the complex zeroes of the corresponding denominators $Z_s$, $Z_p$) may be found from

$$\Theta_{s,p}(k, \omega_k^{s,p}) = 2\pi, \tag{4.5}$$

$$\frac{\partial \Theta_{s,p}(k, \omega_k^{s,p})}{\partial \omega} \gamma_k^{s,p} = \ln \frac{1}{\mathcal{R}_{s,p}(k, \omega_k^{s,p})}$$

$$\approx \frac{1}{2}\left(1 - R_k^{s,p} \widetilde{R}_k^{s,p}\right) + 2 \operatorname{Im} \widetilde{q}_k^{s,p} l, \tag{4.6}$$

where in the first equation all the dissipation has been neglected and we set $\mathcal{R}_{s,p} = 1$, while in the second equation $-i\gamma_k^{s,p}$ are considered small imaginary corrections to the solutions of the first one. We denote then

$$R_k^{s,p} \equiv R^{s,p}(k, \omega_k^{s,p}), \qquad \widetilde{R}_k^{s,p} \equiv \widetilde{R}^{s,p}(k, \omega_k^{s,p}), \qquad \widetilde{q}_k^{s,p} \equiv \widetilde{q}(\mathbf{k}, \omega_k^{s,p}). \tag{4.7}$$

The two contributions to the mode damping from the acceptor absorption and from the mirrors are clearly distinguishable in (4.6), and their ratio is given by

$$\frac{4 \operatorname{Im} \widetilde{q}_k^{s,p} l}{1 - R_k^{s,p} \widetilde{R}_k^{s,p}} \equiv \frac{\eta_k^{s,p}}{1 - \eta_k^{s,p}}, \tag{4.8}$$

where we introduce $\eta_k^{s,p}$ – the probability for the cavity photon to end up in the acceptor medium (correspondingly, $1 - \eta_k^{s,p}$ is the probability to go outside the cavity). The denominators may be expanded around their zeroes as

$$Z_{s,p}(k,\omega) \approx \frac{\partial Z_{s,p}(k, \omega_k^{s,p})}{\partial \omega}\left(\omega - \omega_k^{s,p} + i\gamma_k^{s,p}\right)$$

$$\approx -i \frac{\partial \Theta_{s,p}(k, \omega_k^{s,p})}{\partial \omega}\left(\omega - \omega_k^{s,p} + i\gamma_k^{s,p}\right)$$

$$\approx \frac{1 - R_k^{s,p} \widetilde{R}_k^{s,p} + 4 \operatorname{Im} \widetilde{q}_k^{s,p} l}{2i\gamma_k^{s,p}}\left(\omega - \omega_k^{s,p} + i\gamma_k^{s,p}\right), \tag{4.9}$$

by virtue of Eq. (4.6). One may also approximate

$$\operatorname{Re} \widetilde{q}(\mathbf{k},\omega) \approx q(\mathbf{k},\omega), \qquad \operatorname{Im} \widetilde{q}(\mathbf{k},\omega) \approx \frac{\operatorname{Im} \widetilde{\varepsilon}(\omega)}{2} \frac{\omega^2}{c^2 q(\mathbf{k},\omega)} \tag{4.10}$$

with the same degree of precision (neglecting higher order terms in $\tilde{\Delta}_{\text{res}}$).

When calculating the numerator of (4.1) one may neglect the difference between $\varepsilon$ and $\tilde{\varepsilon}$ at all and set

$$\tilde{q}(\mathbf{k}, \omega) = q\left(\mathbf{k}, \omega_k^{s,p}\right) \equiv q_k^{s,p}, \quad (4.11)$$

and the reflectances $R_k^{s,p}$, $\widetilde{R}_k^{s,p}$ equal to 1. The phases, on the contrary, are not negligible and should be kept. Using (4.5), we arrive at

$$\chi_{TT}(\mathbf{k}, z, z', \omega) \approx -\frac{4\pi}{q_k^s} \frac{(\omega_k^s)^2}{c^2} \frac{\gamma_k^s}{\omega - \omega_k^s + i\gamma_k^s}$$

$$\times \frac{(e^{-iq_k^s z} + e^{iq_k^s z + i\theta_k^s})(e^{iq_k^s z'} + e^{-iq_k^s z' - i\theta_k^s})}{1 - R_k^s \widetilde{R}_k^s + 4\,\text{Im}\,\tilde{q}_k^s l}, \quad (4.12)$$

$$\chi_{LL}(\mathbf{k}, z, z', \omega) \approx -\frac{4\pi q_k^p}{\varepsilon} \frac{\gamma_k^p}{\omega - \omega_k^p + i\gamma_k^p}$$

$$\times \frac{(e^{-iq_k^p z} - e^{iq_k^p z + i\theta_k^p})(e^{iq_k^p z'} - e^{-iq_k^p z' - i\theta_k^p})}{1 - R_k^p \widetilde{R}_k^p + 4\,\text{Im}\,\tilde{q}_k^p l}, \quad (4.13)$$

$$\chi_{ZZ}(\mathbf{k}, z, z', \omega) \approx -\frac{4\pi k^2}{\varepsilon q_k^p} \frac{\gamma_k^p}{\omega - \omega_k^p + i\gamma_k^p}$$

$$\times \frac{(e^{-iq_k^p z} + e^{iq_k^p z + i\theta_k^p})(e^{iq_k^p z'} + e^{-iq_k^p z' - i\theta_k^p})}{1 - R_k^p \widetilde{R}_k^p + 4\,\text{Im}\,\tilde{q}_k^p l}. \quad (4.14)$$

Now we can outline the main qualitative features. If the donor frequency is below the cavity mode cutoff frequency $\omega_{k=0}$ (which is obviously the same for the s- and p-polarizations), the emission process is basically switched off. If it is above $\omega_{k=0}$, then a point molecule emits the cavity photons with $|\mathbf{k}|$ determined by the energy conservation $\omega_D = \omega_k^s$, $\omega_D = \omega_k^p$ (within the corresponding widths $\gamma_k^{s,p}$). For a quantum well exciton also the momentum conservation should hold, which means that only excitons in a small region of $\mathbf{k}$ around the crossing of the exciton and photon dispersion curves can decay via photon emission. The dominant role of the real processes over the virtual processes means also that the presence of the acceptors should not modify the donor kinetics in the first approximation, since the kinetics is governed purely by the cavity photon emission, and the acceptors themselves do not represent an additional decay channel for the donors. A fraction $\eta_k$ of the emitted photons is subsequently absorbed by the acceptors, and the remaining fraction $1 - \eta_k$ eventually leaves the cavity through the mirrors (or is absorbed, if the mirrors are dissipative).

In principle, the donor decay rate into the cavity photons may be affected if the *homogeneous* broadening of the donor emission line is small compared to the cavity mode width: since the cavity mode acquires additional broadening, the photonic density of states is decreased, which changes the decay rate accordingly. However, if the cavity mode resonance peak is still sharp, the integration over **k** will cancel this effect in the first approximation.

Several other remarks should be added. In Section 3 it was assumed that the donors and the acceptors are spatially separated ($z' < L - l$). However, under the conditions of this section ($\tilde{\varepsilon}_\infty = \varepsilon$, $\tilde{\Delta}_{res} \ll 1$) the expressions (4.12)–(4.14) are valid for $L - l < z' < L$ also, i.e., if the donors are mixed with the acceptors. In particular, one may set $l = L$, which would correspond to the whole cavity, filled by the material, doped with donors and acceptors, and the expressions (4.12)–(4.14) will describe the cavity mode contribution to the donor decay rate (or, if multiplied by $\eta_k$, to the energy transfer to the acceptors). Of course, in such a situation the short-range Coulomb contribution, not picked up by (4.12)–(4.14), as it corresponds to $\chi_{ij}(\mathbf{k}, z, z', \omega)$ with $k \gg \sqrt{\varepsilon}\,\omega/c$, should be taken into account properly (since it may actually be the dominant mechanism of the donor decay).

Another remark should be made about an important case when one of the mirrors (or both) is a DBR. A characteristic feature of DBRs is the finite width of their stop band, that is the region in the $(k, \omega)$ space, where the reflectance is close to unity [23,29]. The behavior of the reflection coefficient outside this region is complicated, and actually the probability of the excitation decay into light with a wave vector outside the DBR stop band (the so-called leaky modes) is quite large [30]. To take into account this effect, as well as the Coulomb contribution, one should return to the general discussion of the previous section.

## 5. Numerical Results

To illustrate the results of the previous sections we consider a $\lambda/2$-cavity consisting of a metallic mirror at $z < 0$ and a distributed Bragg reflector (DBR) at $z > L$. As an example of such a cavity we take that described in Ref. [31], and stick to the parameters given there. The metallic mirror is assumed to be a silver layer thick enough to suppress any transmission. The dispersion of the silver complex refractive index is taken into account, approximated by

$$n_{Ag}(\omega) + i\kappa_{Ag}(\omega) \approx (0.060 + 0.35\,i)\left(\frac{\hbar\omega}{eV} - 2.6\right)^2$$
$$+ (0.055 - 1.70\,i)\left(\frac{\hbar\omega}{eV} - 2.6\right) + 0.137 + 2.72\,i,$$

which works with the absolute error of less than 0.007 for $n_{Ag}$ and 0.02 for $\kappa_{Ag}$ in the range 2 eV $< \hbar\omega <$ 3.2 eV with respect to the data of Ref. [32]. The DBR was assumed to consist of 9 periods of alternating $\lambda/4$ layers with $n_1 = 1.95$ (silicon nitride) and $n_2 = 1.45$ (silicon dioxide). We take $\sqrt{\varepsilon} = \sqrt{\tilde{\varepsilon}_\infty} \equiv n_{cav} = 1.63$

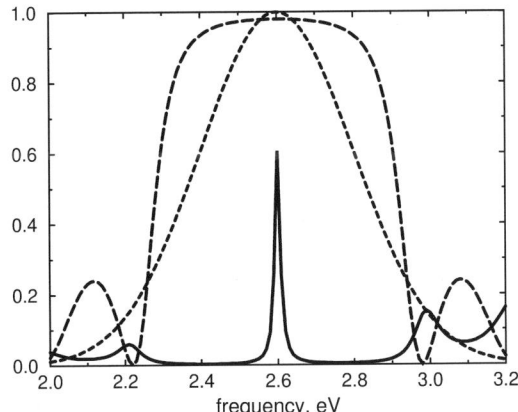

Fig. 3. The empty cavity absorbance (solid line), imaginary part of the acceptor medium dielectric function for $\hbar\omega_A = 2.6$ eV, $\gamma_A = 0.2$ eV, $\tilde{\Delta}_{res} = 1.0$ (short-dashed line) and the DBR reflectance at normal incidence (long-dashed line).

(polystyrene). The substrate refraction index is $n_{sub} = 1.55$ (quartz). These numbers yield the mirror reflectances at $\mathbf{k} = 0$, $\hbar\omega = 2.6$ eV to be $R = 0.915$, $\widetilde{R} = 0.980$. The cavity thickness was chosen to be $L = 121$ nm to give the cavity mode cutoff frequency $\hbar\omega_{k=0} = 2.6$ eV in the absence of the acceptors. The corresponding effective cavity length is $L_{eff} = 386$ nm, and the cavity mode FWHM is $2\gamma_{\mathbf{k}=0} = 17$ meV.

One of the experimentally measured quantities, characterizing the cavity, is its reflectance spectrum. Since the transmission of the metallic mirror is zero, the light, sent from the right, is either reflected back or absorbed in the metal (in the absence of acceptors it is the only dissipative channel, present in the system). In Figure 3 we plot the cavity absorbance in the absence of acceptors (solid line) together with the DBR reflectance (long-dashed line) and the imaginary part of the acceptor medium dielectric function for $\hbar\omega_A = 2.6$ eV, $\gamma_A = 0.2$ eV, $\tilde{\Delta}_{res} = 1.0$ (short-dashed line).

### 5.1. Transfer at Fixed k

Let the donor be placed at $z' = 0.5 L$, the acceptor layer thickness is chosen to be $l = 0.3 L$. At such separation between donor and acceptors the effect of the Coulomb interaction is negligible, and the main contribution to the donor–acceptor interaction comes from the transverse part. The imaginary part of the response function Im $\chi_{TT}$ at $\mathbf{k} = 0$ is plotted in Figure 4 for three values of $\tilde{\Delta}_{res}$: 3.0 (solid line), 1.0 (long-dashed line), and 0.1 (short-dashed line). The dimensionality of $\chi$ is the inverse length, so the natural units are nm$^{-1}$, which are used on the

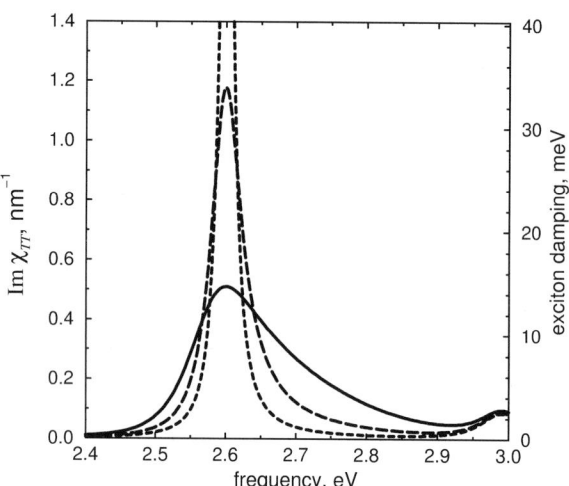

Fig. 4. The imaginary part of the transverse–transverse component of the response function $\mathrm{Im}\,\chi_{TT}$ at $\mathbf{k} = 0$ for $\tilde{\Delta}_{\mathrm{res}} = 3.0$ (solid line), $\tilde{\Delta}_{\mathrm{res}} = 1.0$ (long-dashed line), and $\tilde{\Delta}_{\mathrm{res}} = 0.1$ (short-dashed line) in nm$^{-1}$ (left axis) and rescaled to the exciton broadening (right axis, see the text for details).

left axis. However, more illustrative is to rescale the plot using the formula (3.14) for the exciton damping (broadening). We take $d_D/a_{eh} = 0.1\,e$ ($e$ being the electron charge), which is representative of II–VI semiconductors, such as ZnSe, and put the corresponding broadening in meV (the broadening of 1 meV corresponds to the lifetime of 0.658 ps).

For $\tilde{\Delta}_{\mathrm{res}} = 0.1$ the cavity mode FWHM is equal to 20 meV, which is close to the value of 17 meV for the cavity without acceptors, which means that this case falls into the category, described in the previous section: the regime of weak absorption. If one takes the acceptor transition dipole moment $d_A = 5$ D, then the corresponding density from Eq. (3.10) is 1.0 nm$^{-3}$. The peak value of $\mathrm{Im}\,\chi_{TT}$ is 2.90 nm$^{-1}$ (out of scale), which corresponds to the exciton damping $\Gamma_{QW} = 84$ meV. This value is larger then the mode FWHM, which means that the coupling of the semiconductor quantum well to the cavity mode is not weak, the formula (3.14) (Fermi Golden Rule) is inapplicable, and one has to solve the full polariton problem. However, the expression (3.15) for a point donor molecule decay rate may be still valid (the dependence of $\mathrm{Im}\,\chi$ on the wave vector will be discussed below). For $\tilde{\Delta}_{\mathrm{res}} = 1.0$ the cavity mode FWHM is equal to 54 meV, for $\tilde{\Delta}_{\mathrm{res}} = 3.0$ the spectral function is spread over several hundred meV and its shape is strongly asymmetric. This asymmetry is due to the dispersion: $\mathrm{Re}\,\tilde{\varepsilon}(\omega)$ is smaller on the high-frequency side of the resonance.

Next, we fix $\hbar\omega = 2.7$ eV (above the mode cutoff frequency), and study the exciton decay for different $k$. It is relevant for Wannier excitons in semiconductor quantum wells, since at $k \sim \omega/c$ the exciton dispersion may be neglected. We separate different contributions to the donor decay rate, as explained in Section 2 (the explicit expressions for the Pointing vector at the cavity boundaries and the losses in the acceptor medium are given in Section 3.4). Setting $\tilde{\Delta}_{\text{res}} = 3.0$, we plot in Figure 5(a) different contributions to the exciton decay rate – acceptor absorption (solid line), leakage through the right mirror (DBR, long-dashed line) and absorption in the left mirror (silver, short-dashed line) – as a function of the wave vector ($\hbar ck$ in eV). The units for Im $\chi$ are the same as on the previous plot – nm$^{-1}$ and meV. The whole range of $k$ may be divided into several regions. First, for $\hbar ck < 2.75$ eV (the DBR stop band boundary) the major part of the donor energy is transferred to the acceptor – when the mirrors are good, the light bounces back and forth, at each pass loosing some part of the energy in the acceptor layer. The decay rate is peaked around the "cavity mode" (somewhat broad and asymmetric): this peak in the spectral function follows the mode dispersion in the $(k, \omega)$-plane. The dominant contribution of the metallic mirror with respect to that of the DBR is consistent with the relation between their reflectances. At larger $k$ the dissipation in the metallic mirror decreases strongly, which is the consequence of the oblique incidence. The probability for the excitation to decay into the acceptor decreases drastically, and simultaneously the leakage through the DBR increases, becoming of the same order. The simple explanation is that in the region where the DBR reflectance is poor, the light manages to make just several passes across the cavity before escaping through the DBR, not having the possibility to lose a large part of its energy in the acceptor medium. At $ck > n_{\text{sub}}\omega$ the wave in the substrate becomes evanescent, and the leakage through the DBR is completely suppressed, the acceptors remain the only dissipative channel for the donors. At $ck > n_{\text{cav}}\omega$ the wave cannot propagate inside the cavity either, and one observes the exponentially decreasing tail in the transfer rate. Sharp features in the transfer rate correspond to the guided modes, propagating inside the DBR. An analogous plot for the case of weak absorption is shown in Figure 5(b), and the same qualitative arguments explain the main features.

Other components of the response function $\chi$, corresponding to the $p$-polarized light, behave in a similar way, except for one additional feature. Since the dielectric function of silver is mostly real and negative ($-6.48 + 0.73 i$ for $\hbar\omega = 2.7$ eV), a surface wave may propagate at the dielectric/metal interface [25], broadened due to absorption in the metal. The dispersion law for this wave is determined by

$$\frac{\omega^2}{c^2 k^2} = \frac{1}{\varepsilon_{\text{Ag}}(\omega)} + \frac{1}{\varepsilon}. \qquad (5.1)$$

For our parameters at $\hbar\omega = 2.7$ eV one obtains $\hbar ck \approx (5.7 + 0.2 i)$ eV, or $\hbar\omega = (2.7 - 0.10 i)$ eV at $\hbar ck = 5.7$ eV. Resonant excitation of the surface wave results in increasing losses in the metallic mirror. This loss mechanism becomes more

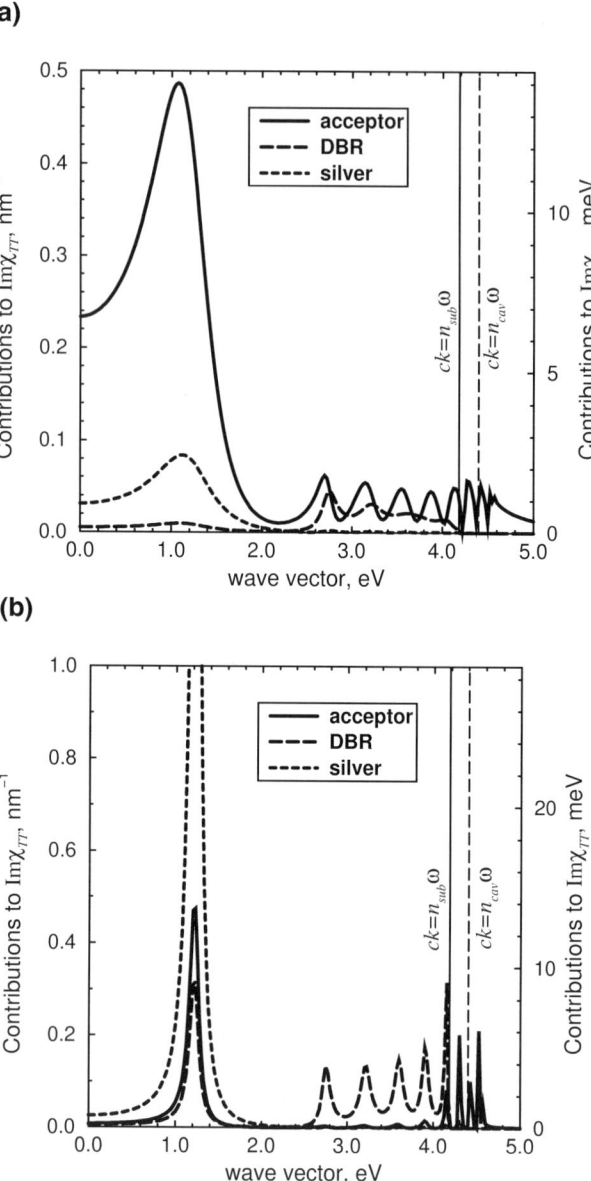

Fig. 5. Different contributions to the imaginary part of the transverse–transverse component of the response function Im $\chi_{TT}$ at $\hbar\omega = 2.7$ eV: from the acceptor (solid line), the DBR (long-dashed line), and the metallic mirror (short-dashed line) in nm$^{-1}$ (left axis) and rescaled to the exciton broadening (right axis) (a) for $\tilde{\Delta}_{\rm res} = 3.0$, (b) for $\tilde{\Delta}_{\rm res} = 0.1$. The thin vertical lines mark $ck = n_{\rm sub}\omega$ (solid line) and $ck = n_{\rm cav}\omega$ (dashed line).

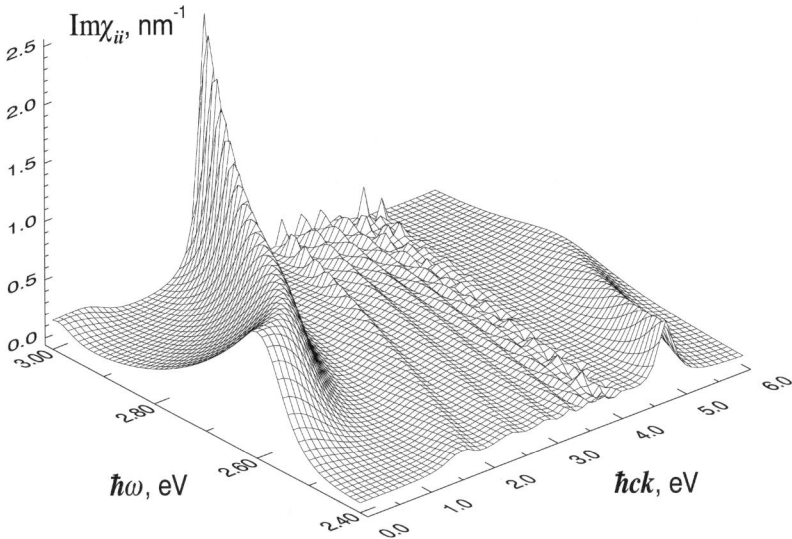

Fig. 6. The imaginary part of the sum of the diagonal components of the response function Im $\chi_{ii}$ in nm$^{-1}$ as a function of $k$ and $\omega$ for $\tilde{\Delta}_{\text{res}} = 3.0$.

efficient as the donor gets closer to the metallic mirror, since the excitation occurs via an evanescent electromagnetic wave. In Figure 6 we plot the sum of all the diagonal components Im $\chi_{ii}$ in nm$^{-1}$ as a function of $k$ and $\omega$ for $\tilde{\Delta}_{\text{res}} = 3.0$. The cavity mode is strongly broadened at small $k$, but as the frequency goes out of the acceptor absorption band, the peak gets sharper. The peak corresponding to the surface wave is seen at large $k$.

## 5.2. Transfer, Integrated over k

Next, we integrate the decay rate over $\mathbf{k}$, as in Eq. (3.15) for the decay rate $\Gamma$ of a point donor. We average (3.15) over the donor molecule orientations according to (3.16), however, keeping the donor position $z'$ and frequency $\omega_D$ as parameters. Taking the dipole moment to be $d_D = 5$ D, which is typical for organic molecules, we study different contributions to the donor damping $\Gamma$, as before, for different regimes of the acceptor–light coupling. The donor position is assumed to be $z' = 0.5 L$. If one is interested in the decay rate of thermalized Wannier excitons in a quantum well, as given by Eq. (3.18), the result for organic molecules ($\Gamma$ or the contribution from a given decay channel) should be simply multiplied by (7300 K)/$T$ (calculated using $d_D/a_{eh} = 0.1\, e$ for the semiconductor dipole moment and $m = 0.7\, m_0$ for the exciton mass, $m_0$ being the free electron mass).

First, we plot in Figures 7(a), (b) different contributions to the donor damping—acceptor absorption (solid line), leakage through the right mirror (DBR, long-dashed line) and absorption in the left mirror (silver, short-dashed line) for the same acceptor parameters as before ($\hbar\omega_A = 2.6$ eV, $\gamma_A = 0.2$ eV), corresponding to the strong/weak absorption regime: $\tilde{\Delta}_{res} = 3.0$ and $0.1$. First of all, we see that the damping never exceeds 1 μeV, which means that Fermi Golden Rule works perfectly in all cases. The difference between Figures 7(a) and (b) is the characteristic width of the transition region around the cavity mode cutoff frequency, determined by the width of the cavity mode. While for the weak absorption (the second plot) it is several ten meV and the losses depend on the donor frequency in a clear step-like manner (at higher frequencies the cavity mode is "switched on"), for the strong absorption case the transition region is washed out over several hundred meV.

The characteristic value of the differences in the donor damping $\Gamma$ for different donor frequencies is 0.1 μeV, corresponding to the lifetime of 6.6 ns. The difference in the donor relaxation kinetics may be observed if the change in $\Gamma$ is not negligible compared to the probability of nonradiative processes. Generally, if the donor is a luminescent substance, the corresponding quantum yield of the usual luminescence in the bulk typically is not very small (say, 10%). The radiative lifetime of the donor molecule in a bulk matrix with the dielectric constant $\varepsilon = 6$ for $d_D = 5$ D, $\hbar\omega_D = 2.6$ eV is evaluated as

$$\frac{4}{3\hbar} \frac{d_D^2}{\varepsilon} \left(\frac{\omega\sqrt{\varepsilon}}{c}\right)^3 \approx 8.4 \text{ ns}.$$

Thus, the expected difference in the donor kinetics may be 10–20%, which should be observable.

In agreement with Savona et al. [30], the contribution of the leaky modes to the emission through the DBR is dominant: as the donor frequency passes the cavity mode cutoff frequency (2.6 eV), the amount of energy, leaving the cavity through the DBR, does not change significantly. The same can be said about the energy, transferred to the acceptor: the change is by a factor of the order of 1 and the leaky modes contribute significantly. As for the silver mirror, the detailed analysis shows that the losses at frequencies below the mode cutoff are almost exclusively due to the surface wave, discussed above. At larger $\omega_D$ also the cavity mode contributes to the losses in the metal. In Figures 8(a), (b) we plot the same quantities as in the previous figure, but with the mirrors removed (the half-spaces at $z < 0$ and $z > L$ filled with the cavity material), intentionally preserving the axes scales. For the strong absorption case in Figure 8(a) the change is merely quantitative, which might be expected: if a substantial fraction of the light is absorbed in just one pass of the acceptor layer, the mirrors do not have a dramatic effect. On the other hand, in the weak absorption limit the step-like behaviour of

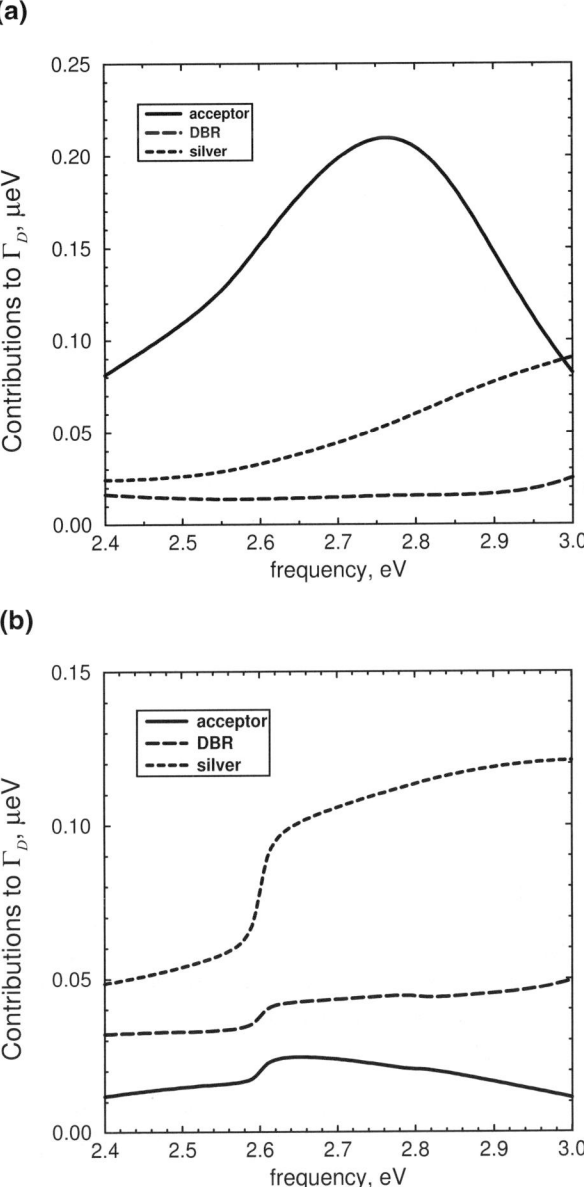

Fig. 7. Different contributions to the donor excitation damping $2\gamma_D$ (µeV) as a function of the donor frequency: from the acceptor (solid line), the DBR (long-dashed line), and the metallic mirror (short-dashed line) in µeV (a) for $\tilde{\Delta}_{\text{res}} = 3.0$, (b) for $\tilde{\Delta}_{\text{res}} = 0.1$.

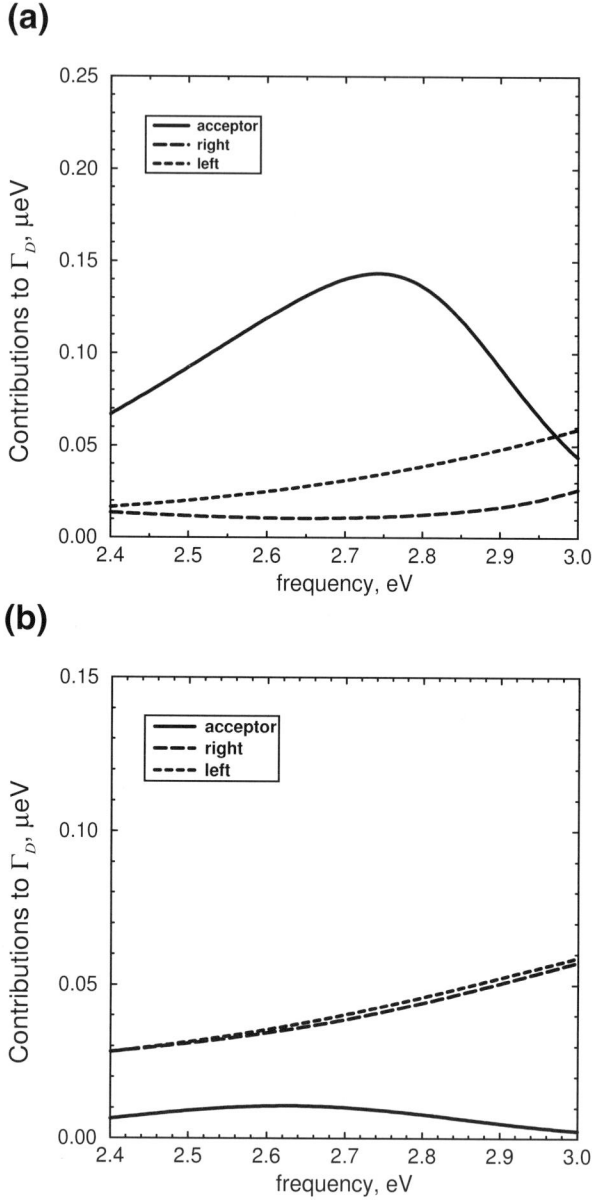

Fig. 8. Different contributions to the donor excitation damping $2\gamma_D$ (μeV) as a function of the donor frequency in the absence of the mirrors: acceptor absorption (solid line), emission to the right (long-dashed line), and emission to the left (short-dashed line) in μeV (a) for $\tilde{\Delta}_{res} = 3.0$, (b) for $\tilde{\Delta}_{res} = 0.1$.

the losses as a function of the donor frequency clearly distinguishes the situations with and without the cavity.

### 5.3. STRONG COUPLING REGIME

An interesting case, which we have not considered yet, is that of the strong coupling between the acceptor and the cavity mode. This situation occurs when the acceptor absorption band width is sufficiently small and the acceptor–light coupling (in our model governed by the parameter $\tilde{\Delta}_{\text{res}}$) is sufficiently strong. In Figure 9 we plot the imaginary part of the response function Im $\chi_{TT}$ (in nm$^{-1}$ and meV, analogously to Figure 4) at $k = 0$ as a function of frequency for $\tilde{\Delta}_{\text{res}} = 3.0$, $\gamma_A = 20$ meV. It exhibits two distinct peaks, corresponding to the upper and lower polariton branches, formed by the cavity mode and the spatially coherent wave of the acceptor excitations. Integrating over **k**, we make a plot, analogous to those in Figure 7, which shows the point donor decay rate as a function of the donor frequency (Figure 10). As we see from the main plot, the transfer rate to the acceptors does not have the two-peaked shape, since the strong coupling is realized only in the narrow region of wave vectors around $k = 0$, and the splitting is washed out by the contributions from larger $k$. However, the maximum is shifted to higher frequencies due to the contribution of the upper polariton.

An interesting feature seen in Figure 10 is the suppression of the losses due to the mirrors around the acceptor resonance, especially pronounced for the metallic mirror. As it was already mentioned, below the mode cutoff frequency the dominant contribution comes from the surface wave at the metallic interface. It

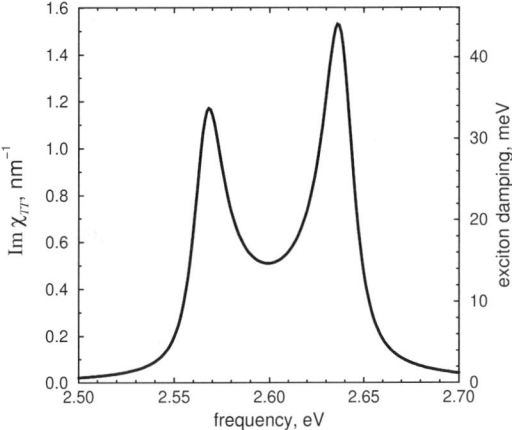

Fig. 9. The imaginary part of the transverse–transverse component of the response function Im $\chi_{TT}$ at **k** $= 0$ in nm$^{-1}$ (left axis) and rescaled to the exciton broadening (right axis), for $\tilde{\Delta}_{\text{res}} = 3.0$, $\gamma_A = 20$ meV (the strong-coupling regime).

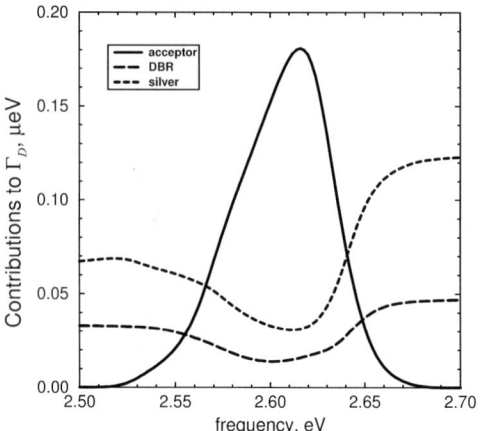

Fig. 10. Different contributions to the donor excitation damping $2\gamma_D$ (µeV) as a function of the donor frequency: from the acceptor (solid line), the DBR (long-dashed line), and the metallic mirror (short-dashed line) in nm$^{-1}$ µeV for $\tilde{\Delta}_{\text{res}} = 3.0$, $\gamma_A = 20$ meV. The inset shows the spectral function of the electromagnetic field Im $\chi_{ii}/3$ in nm$^{-1}$ at $k = 0$, as a function of $\omega$ (eV).

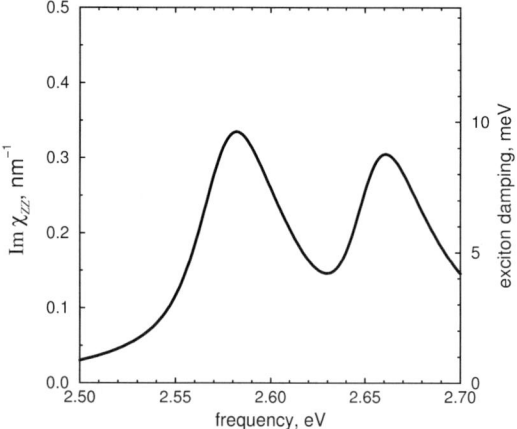

Fig. 11. The imaginary part of the $zz$-component of the response function Im $\chi_{ZZ}$ at $\hbar ck = 5.45$ eV in nm$^{-1}$ (left axis) and rescaled to the exciton broadening (right axis), for $\tilde{\Delta}_{\text{res}} = 3.0$, $\gamma_A = 20$ meV. The two peaks correspond to the coupled surface wave and acceptor excitations.

turns out that when the frequency of the surface wave is close to the narrow acceptor resonance, they become strongly coupled and the anticrossing-like behaviour is observed. This is demonstrated by Figure 11, where we plot Im $\chi_{ZZ}$ (the component, on which the projection of the surface excitation is the largest) at

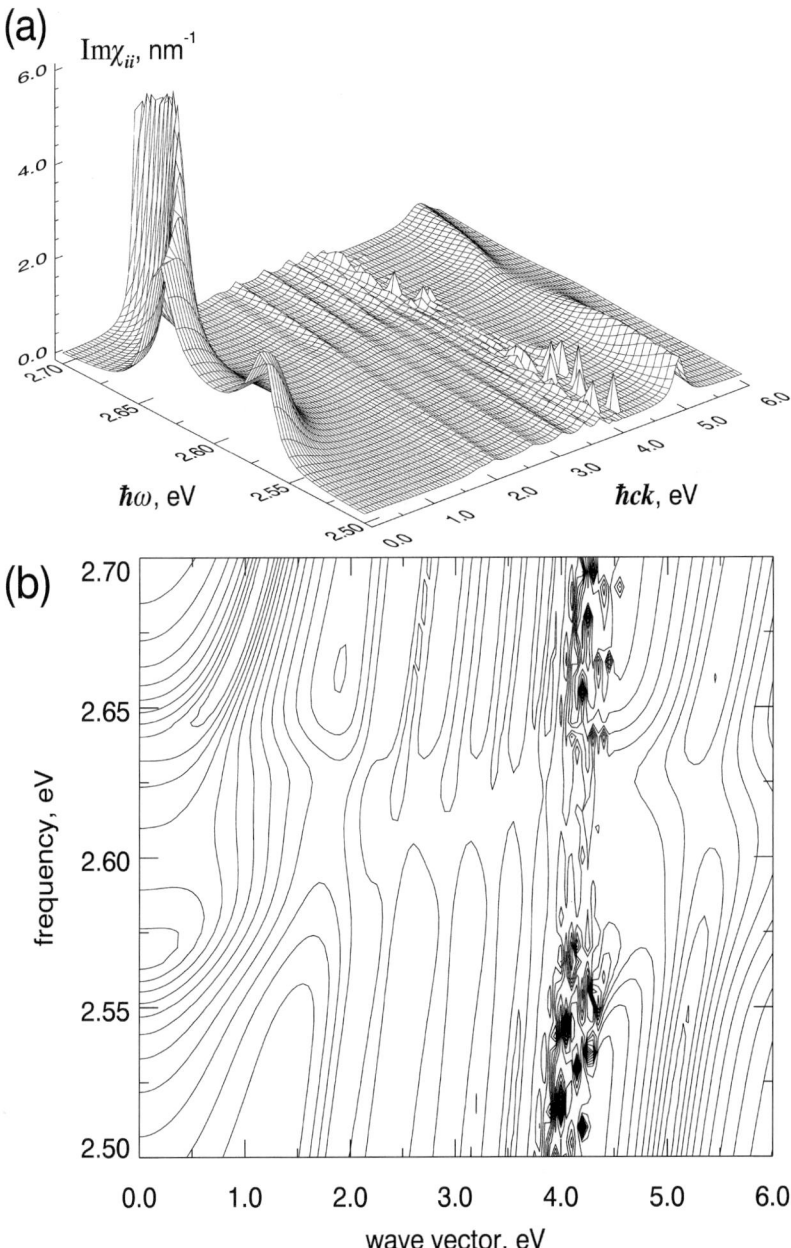

Fig. 12. The imaginary part of the sum of the diagonal components of the response function $\mathrm{Im}\,\chi_{ii}$ at $\hbar ck = 5.45$ eV in nm$^{-1}$ (a), and the contour plot of $\ln \mathrm{Im}\,\chi_{ii}$ (b). $\tilde{\Delta}_{\mathrm{res}} = 3.0$, $\gamma_A = 20$ meV.

$\hbar ck = 5.45$ eV as a function of frequency. The dip in the short-dashed curve in Figure 10 is nothing else than the gap between the two excitations, seen in the Figure 11 for a specific value of $k$. Note that the gap is shifted by about 20 meV above the acceptor energy $\hbar\omega_A = 2.6$ eV. We explain this feature by the fact that the surface excitation and acceptor excitation are coupled via an evanescent wave, whose spatial decrement decreases at larger $\omega$, thus increasing the coupling strength. An analogous suppression takes place for the losses via the leaky modes of the DBR: even though these modes are strongly damped, nonetheless some anticrossing-like decrease in the spectral function is observed at the energies slightly higher than the bare acceptor level $\hbar\omega_A$. This shift is due to the interaction of the acceptor with the electromagnetic field (i.e., the leaky modes are coupled to the bulk polaritons in the acceptor slab, which arise due to the poor reflection of the DBR). All these features may be seen in Figure 12, where we plot the sum of the diagonal components Im $\chi_{ii}$ as a function of $k$ and $\omega$ [a surface plot in Figure 12(a) and a contour plot for the logarithm of Im $\chi_{ii}$ in Figure 12(b)]. The sharp spikes around $\hbar ck = 4$ eV correspond to the guided modes of the DBR (see Figure 5 and the related discussion), which are not resolved by the grid in the $(k, \omega)$-plane, used for plotting.

### 5.4. SUMMARY

Summarizing the results of this lengthy section, we have considered the problem of energy transfer via the transverse electromagnetic field in a planar microcavity for a plane-wave-like donor (which may be a Wannier–Mott exciton in a semiconductor quantum well) and a point-like donor (a single molecule), treating acceptors as a continuous absorbing medium. Depending on the strength and shape of the acceptor absorption, one may identify three regimes: the regime of strong and broad absorption, the regime of weak absorption, and the regime of strong coupling between the field and the acceptors, when the acceptor absorption has the shape of a strong and narrow peak. In the first case, the cavity mode is practically destroyed, and instead of a sharp peak one obtains a broad asymmetric maximum in the spectral function of the electromagnetic field for a given in-plane wave vector **k**. In the case of the weak absorption, the cavity mode is still well-defined and the absorption introduces just some additional broadening of the peak in the spectral function. In the strong coupling regime, the spectral function has two sharp peaks, corresponding to the upper and lower polaritons, formed by the cavity mode and the excitations in the acceptor layer.

We have shown that taking into account the realistic structure of the mirrors is important. In particular, if the mirror is a DBR, then much of the energy is emitted outside the cavity through the leaky modes, in agreement with results previously known for semiconductor microcavities. If the mirror is metallic, an important process turns out to be the excitation of a surface wave at the metallic interface (which is subsequently absorbed in the metal). The decay rate of a point-like donor (i.e., integrated over **k**) has been studied as a function of the donor frequency. If the acceptor absorption band is broad, then all the decay rates for all channels exhibit a step-like feature around the cavity mode cutoff frequency, corresponding to "switching on" the mode contribution, with the characteristic width of the step determined by the width of the cavity mode. The step is practically washed out if the absorption is strong, and is more pronounced in the case of weak absorption, which may be detected by studying the total decay rate of the donor, or the luminescence from the acceptor. For the case of the acceptor-field strong coupling, the leakage through the mirrors turns out to be suppressed at the acceptor frequency due to the anticrossing of the acceptor excitations with leaky and surface modes, which may be observed if one looks at the light emitted from the cavity at the donor frequency.

## 6. Discussion

This section is dedicated to the discussion of some physical aspects of the results presented above and of some relevant experiments.

### 6.1. Decay: Not the End of the Story

The spectral function of the electromagnetic field for a given wave vector **k** (or integrated over **k**) determines the probability per unit time $\Gamma$ of the donor excitation decay into the states of the system "electromagnetic field + acceptors" for a plane-wave-like (or a localized point-like) donor, as prescribed by Fermi Golden Rule. If some other nonradiative decay processes are also present (with the corresponding rate $\Gamma'$), then the final states are those with energies close to $\hbar\omega_D$ within an interval of the characteristic width $\Gamma + \Gamma'$. If these nonradiative processes are not strongly dominating, then the changes in the total donor lifetime, calculated here, may be observed, for example, measuring the decay kinetics, or the saturation behaviour (if the donor is pumped optically). The actual state of excitation after the donor decay may be different, depending on the specific conditions: it may be localized on an acceptor molecule, may be a cavity photon for the weak absorption case, it may also be a cavity polariton if the acceptor-light coupling is strong. Now we ask ourselves a question: what will happen to this excitation afterwards?

We have separated different contributions to the spectral function, namely, from the continuum of states leaking through the mirrors, and the contribution from the states in the acceptor, interacting with its own reservoir (intramolecular vibrations, phonons in the matrix, etc.). These two contributions correspond to the probabilities of different dissipation processes the donor may be involved in. The rate $\Gamma^{(\mathrm{mirr})}$ of escape through the mirrors speaks for itself – it corresponds to the outgoing light for the DBR, and to the absorption in the metal for the thick metallic mirror. Thus the outgoing light at the donor luminescence frequency carries the information about the donor population (provided that some additional condition holds – see below the discussion of the role of dephasing in the acceptor).

The situation with the decay into the acceptors is more complicated. To understand what happens to the excitation afterwards, we should specify the dissipative mechanisms, contributing to the *homogeneous* broadening of each acceptor molecule. The nonradiative decay is the simplest possibility, however, of little interest. Another possibility is the vibronic relaxation, important for materials with large Stokes shift. From the formal point of view it is fully analogous to the case of the nonradiative decay, since the excitation effectively "disappears" from the frequency region of interest. The important point is that it "reappears" in a different frequency region, and the information about the transfer to the acceptors may be obtained by measuring the acceptor luminescence, especially if the acceptor luminescence spectrum is below the DBR stop band (as, e.g., in Ref. [4]). But even if it is inside the stop band, some light may still come out at oblique angles (the leaky modes).

A nontrivial case is when the dominant dissipative process in the acceptor is dephasing, not followed by any significant energy relaxation, however, contributing to the homogeneous broadening of the acceptor level. Then, after the dephasing has occurred, the excitation is still there, at an energy close to the donor frequency $\hbar\omega_D$, "waiting" for its chance to decay. If the material has strong Stokes shift, then usually the dominant mechanism is the vibronic relaxation, and we return to the previous case. However, if the Stokes shift is negligible (as it was in the work of Ref. [33]) and the nonradiative recombination does not play any significant role, the excitation in the acceptor may decay radiatively through the mirrors. We stress that this radiative decay *is not* included into the $\Gamma^{(\mathrm{mirr})}$, calculated above. Clearly, if one considers an ideal limiting case when all the dissipation in the acceptor is pure dephasing, then at the end all 100% of the donor energy will leave the cavity through the mirrors at the same frequency $\omega_D$. Separating this light from that originating directly from the donor, requires a proper experimental arrangement (possible signatures may be a change in the polarization or a different kinetics). Strictly speaking, in such a situation the excited acceptor should be treated as a new donor, and we find ourselves at the beginning of the same story. However, in this situation one cannot "get rid" of the Coulomb transfer, separating spatially donors and acceptors, since the new donor is situated inside the acceptor medium itself. The situation becomes even more complicated when the resulting state is a substantially mixed state of the system "acceptors + field", like a cavity polariton. Generally, we should conclude that the calculations performed above, are directly related to observable quantities only if the dominant broadening process in the acceptors is decay-like (including the vibronic relaxation).

## 6.2. Review of Relevant Experiments

At the time when this chapter was being written the author was aware of three experimental works where the microcavity effect on the energy transfer was studied.

The first one is the paper by Hopmeier et al. [4]. A planar cavity with one aluminum mirror and a DBR as the second mirror, was filled with polycarbonate homogeneously doped with poly(phenyl-$p$-phenylene vinylene) (PPPV) which served as the donor, and 4-dicyanmethilene-2-methyl-6-($p$-dimethylaminostyryl)-4$H$-pyran (DCM) as the acceptor. The donor was excited by a laser pulse through the DBR at a frequency above the DBR stop band. The subsequent luminescence was observed in two spectral regions: a sharp peak corresponding to the donor emission into the cavity mode (the acceptor did not emit at this frequency), and broad luminescence at frequencies below the DBR stop band, to which both donor and acceptor could contribute. The intensity of the total luminescence in this second spectral region was measured for different cavity mode frequencies $\omega_{k=0}$ (several cavities with different lengths were made).

What the authors observed was the enhancement of the detected luminescence with respect to that of a reference sample without mirrors. On increasing the cavity mode frequency the enhancement factor changed from 1 to 4, which was attributed to the enhancement of the transfer due to the cavity. However, this result seems strange if one recalls our considerations about the transfer in the weak absorption case (clearly identifiable from the presence of the narrow cavity mode) in Section 4. Indeed, as the mode cutoff frequency $\omega_{k=0}$ sweeps the region of the donor–acceptor spectral overlap from lower to higher frequencies, the enhancement factor should decrease, since the cavity mode can contribute to the transfer at $\omega > \omega_{k=0}$ (the transfer is mediated by the photons with $k > 0$), but not to the transfer at $\omega < \omega_{k=0}$. Another objection to the interpretation of Hopmeier et al. is that in their structure, when donors and acceptors are mixed together in a random manner, the short-range Förster transfer due to the Coulomb interaction should dominate over the cavity mode contribution, the latter not being competitive enough to produce such a large enhancement factor as 4.

Of course it does not make much sense to speculate about the reasons of such disagreement (both quantitative and qualitative) between the experiment and our simple theory without knowing the details of the experiment. However, one of possible effects that could affect the measured luminescence, is the change in the efficiency of the excitation as the cavity thickness is varied. Recalling that the excitation frequency was only slightly above the DBR stop band, and that at this frequency the DBR reflectance still is not small and behaves in a complicated way, the penetration of the exciting light into the cavity may depend on the cavity thickness. Then this might affect the acceptor luminescence since the acceptor absorption at the excitation frequency was not small, as mentioned in Ref. [4], and the direct excitation of the acceptor was not avoided.

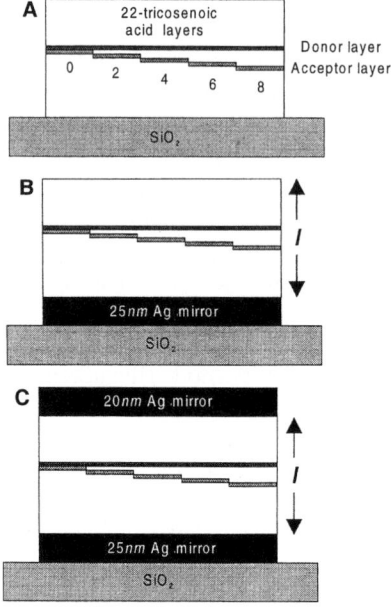

Fig. 13. The cavity samples used in the work of Ref. [5]. (Reprinted with permission from Ref. [5]. Copyright (2000) American Association for the Advancement of Science.)

A neat experimental work was reported by Andrew and Barnes [5]. They prepared several planar microcavity structures where the donors [$N$-hexadecyl pyridinium tetrakis (1,3-diphenyl-1,3-propandionato) europium (III) complexes with narrow luminescence spectrum and high quantum yield] and acceptors (1,1′-dioctadecyl-3,3,3′,3′-tetramethylindodicarbocyanine) were confined to 2D monolayers, whose separation $l'$ was controlled at nm scale (Figure 13). The quality of the cavities varied strongly: from open structures without any mirrors at all to complete cavities with two silver mirrors. Combined with different cavity lengths $L$, it provided a wide range of photonic mode densities at the positions of donor and acceptor layers. The main object of measurement was the donor luminescence decay rate $\Gamma(l')$ in different cavities for $l' = 0$–$25$ nm.

At such distances the energy transfer is dominated by the Coulomb contribution, and in the absence of the cavity the decay rate behaves as [17]:

$$\Gamma(l') = \Gamma^{(r)} + \Gamma^{(nr)} = \Gamma^{(r)}\left[1 + \left(\frac{l_0}{l'}\right)^4\right], \quad (6.1)$$

where $\Gamma^{(r)}$ is the radiative rate of the donor, $\Gamma^{(nr)}$ is the nonradiative decay rate (associated with the Förster transfer), and $l_0$ is called the critical distance. In the experiment each cavity was characterized by the donor pure radiative decay

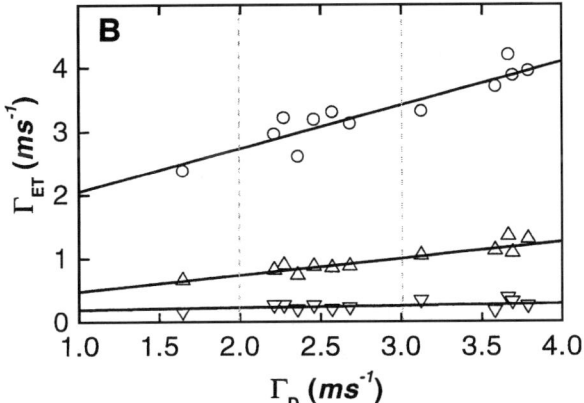

Fig. 14. Dependence of the donor nonradiative decay rate $\Gamma^{(nr)}$ on the radiative decay rate, as measured by the authors of Ref. [5] (our notations are different from the original ones). The three sets of data with the corresponding linear fits correspond to different donor–acceptor separations – 10.4 nm (circles), 15.6 nm (up triangles), 20.8 nm (down triangles). (Reprinted with permission from Ref. [5]. Copyright (2000) American Association for the Advancement of Science.)

rate $\Gamma^{(r)}$, identified with $\Gamma(l' \to \infty)$, i.e., measured at large donor–acceptor separation. This is correct if the presence of the acceptor layer does not affect the mode structure strongly, which is a reasonable assumption. Thus $\Gamma^{(r)}$ is a measure of the photonic mode density at the donor position for a given cavity. Then, subtracting the measured radiative rate from the total one, the authors obtained the transfer rate $\Gamma^{(nr)}$ for different cavities. The key result, obtained by Andrew and Barnes was that at fixed $l'$ the transfer rate $\Gamma^{(nr)}$ depended on $\Gamma^{(r)}$, the dependence being well described by a linear on (Figure 14). This, according to the authors, was an evidence that the Förster transfer rate depends on the photonic mode density, and was interpreted in terms of Eq. (6.1).

This conclusion is hard to accept, since we know that the short-range transfer via the longitudinal Coulomb field cannot depend on the density of the transverse photonic modes, and $\Gamma^{(nr)}$ should not depend on $\Gamma^{(r)}$ since these quantities are determined by contributions from different photons: large $k$ scalar photons in the first case, and the small $k \sim \omega/c$ transverse photons in the second case. The question, however, remains: if not what they claim, then what effect have Andrew and Barnes actually observed? Again, here we can only speculate about possible things that may have occurred. First, the dependence $\Gamma^{(nr)}(\Gamma^{(r)})$ obtained experimentally, is much weaker than if one assumes Eq. (6.1) to hold in a cavity: extrapolating the experimental straight lines $\Gamma^{(nr)}(\Gamma^{(r)})$ to $\Gamma^{(r)} \to 0$, we see that $\Gamma^{(nr)}$ tends to some constant, strongly dependent on $l'$. Second, the slope of these straight lines decreases with increasing $l'$. This may happen if the proximity of the acceptor layer induces some changes in the emitting properties of the donor, thus leading to some dependence $\Gamma^{(r)}(l')$. In this case the difference $\Gamma(l') - \Gamma(\infty)$ does not give the purely nonradiative transfer rate, as assumed by Andrew and Barnes, but contains some part of the radiative rate $\Gamma^{(r)}(l')$ as well, thus leading to some dependence on the photonic mode density. The nature of this proximity effect remains yet to be understood.

In the third experimental work [6] a 200 nm layer of donors (CdSe nanocrystals in polystyrene matrix) and a 40 nm layer of acceptors (1-ethyl-2-[(1-ethyl-2(1H)-quinolinylidene)methyl] quinolinium bromide) were placed into a high-quality cavity with one silver mirror and one DBR. The cavity had many modes due to the complicated behaviour of the DBR reflectivity. The authors have seen (i) enhancement of the acceptor luminescence with respect to the reference sample with the mirrors removed, and (ii) suppression of the acceptor luminescence with respect to the bilayer cavity when an additional spacer (20 nm SiO) layer is inserted between the donor and the acceptor layers to eliminate the Förster transfer. Again, the authors claim to have observed the enhancement of the Förster transfer due to the cavity effects.

The effect of the cavity in this case, however, should be reduced to the emission and reabsorption of the cavity photons, as the acceptor absorption is weak, and this experiment also falls in the category, described in Section 4. The suppression of the transfer after the insertion of the spacer may be due to the change in the cavity length, since this leads to the change in the frequencies of the cavity modes. The position of the cavity modes with respect to the donor and acceptor spectra is especially important in this experiment since the acceptor absorption spectrum was not smooth, having pronounced vibronic structure. In this case a small change in the cavity modes' frequencies may change the transfer strongly. This is a possible explanation of the results of Ref. [6].

## 7. Concluding Remarks

In this chapter we have discussed the problem of the energy transfer via electromagnetic field in a microcavity, focusing mainly on the theoretical aspects of the problem. We have studied different cases that may be realized depending on the optical properties of acceptor molecules and analyzed the distribution of energy between different possible decay channels for the donor excitation.

Having reviewed the relevant experimental results reported so far, we have seen that their interpretation is not so straightforward as it was presented in the corresponding works. Thus, this problem continues to be interesting and important, and more studies are needed to improve our understanding of the phenomenon.

## Acknowledgements

The author wishes to thank profs. V.M. Agranovich, G.C. La Rocca, and F. Bassani for guidance and collaboration during his work at Scuola Normale Superiore (Pisa, Italy).

## References

1. C. Weisbuch, H. Benisty, R. Houdré, *J. Lumin.* 85 (2000) 271.
2. T. Kobayashi, Q. Zheng, T. Sekiguchi, *Phys. Rev. A* 52 (1995) 2835.
3. G.S. Agarwal, S. Dutta Gupta, *Phys. Rev. A* 57 (1998) 667.
4. M. Hopmeier, W. Guss, M. Deussen, E.O. Göbel, R.F. Mahrt, *Phys. Rev. Lett.* 82 (1999) 4118.
5. P. Andrew, W.L. Barnes, *Science* 290 (2000) 785.
6. C.E. Finlayson, D.S. Ginger, N.C. Greenham, *Chem. Phys. Lett.* 338 (2001) 83.
7. V.M. Agranovich, H. Benisty, C. Weisbuch, *Solid State Commun.* 102 (1997) 631.
8. D.G. Lidzey, D.D.C. Bradley, A. Armitage, S. Walker, M.A. Skolnick, *Science* 288 (2000) 1620.
9. J. Wainstain, D. Gendt, C. Delalande, M. Voos, V. Thierry-Mieg, J. Bloch, R. Planel, *Physica E* 2 (1998) 925.
10. V.M. Agranovich, M.D. Galanin, Electronic Excitation Energy Transfer in Condensed Matter, North-Holland, Amsterdam, 1982.
11. Th. Förster, *Ann. Phys.* 2 (1948) 55.
12. D.M. Basko, F. Bassani, G.C. La Rocca, V.M. Agranovich, *Phys. Rev. B* 62 (2000) 15962.
13. L.D. Landau, E.M. Lifshitz, The Classical Theory of Fields, Pergamon Press, Oxford, 1962.
14. D.L. Dexter, *J. Chem. Phys.* 21 (1953) 836.
15. M.D. Galanin, I.M. Frank, *Zh. Eksp. Teor. Fiz.* 21.
16. A.D. McLachlan, *Mol. Phys.* 8 (1964) 409.
17. H. Kuhn, *J. Chem. Phys.* 53 (1970) 101.
18. L.D. Landau, E.M. Lifshitz, Statistical Physics, Pergamon Press, Oxford, 1980.
19. L.D. Landau, E.M. Lifshitz, Electrodynamics of Continuous Media, Pergamon Press, London, 1965.
20. G.S. Agarwal, *Phys. Rev. A* 11 (1975) 230; *Phys. Rev. A* 11 (1975) 243; *Phys. Rev. A* 11 (1975) 253; *Phys. Rev. A* 12 (1975) 1475; *Phys. Rev. A* 12 (1975) 1974; *Phys. Rev. A* 12 (1975) 1987.
21. A.A. Abrikosov, L.P. Gor'kov, I.Ye. Dzyaloshinskii, Quantum Field Theoretical Methods in Statistical Physics, Pergamon Press, Oxford, 1965.
22. L.C. Andreani, *Phys. Lett. A* 192 (1994) 99.
23. V. Savona, in: H. Benisty, et al. (Eds.), Confined Photon Systems: Fundamentals and Applications, Springer, Berlin, 1998.
24. M.S. Tomaš, Z. Lenac, *Phys. Rev. A* 60 (1999) 2431.
25. V.M. Agranovich, V.L. Ginzburg, Crystal Optics with Spatial Dispersion and Excitons, Springer, 1984.
26. L.C. Andreani, F. Tassone, F. Bassani, *Solid State Commun.* 77 (1991) 641.
27. C. Piermarocchi, F. Tassone, V. Savona, A. Quattropani, P. Schwendimann, *Il Nuovo Cimento* 17D (1995) 1663.
28. F. Tassone, C. Piermarocchi, V. Savona, A. Quattropani, P. Schwendimann, *Phys. Rev. B* 53 (1996) R7642.
29. M.S. Skolnik, T.A. Fisher, D.M. Whittaker, *Semicond. Sci. Technol.* 13 (1998) 645.
30. V. Savona, F. Tassone, C. Piermarocchi, A. Quattropani, P. Schwendimann, *Phys. Rev. B* 53 (1996) 13051.

31. D.G. Lidzey, D.D.C. Bradley, M.S. Skolnik, T. Virgili, S. Walker, D.M. Whittaker, *Nature* 395 (1998) 53.
32. E.D. Palik (Ed.), Handbook of Optical Constants of Solids, Academic Press, Orlando, 1985.
33. D.G. Lidzey, D.D.C. Bradley, T. Virgili, A. Armitage, M.S. Skolnik, S. Walker, *Phys. Rev. Lett.* 82 (1999) 3316.

Chapter 10

# Energy Transfer from a Semiconductor Nanostructure to an Organic Material and a New Concept for Light-Emitting Devices

D.M. BASKO

*ICTP Trieste, International Centre for Theoretic Physics, Strada Costiera 11, Trieste, I-34000, Italy*

1. Introduction . . . . . . . . . . . . . . . . . . . . . . . . . . . . . . . . . . . . . . . 447
2. The General Calculation Scheme for the Energy Transfer . . . . . . . . . . . . . . . . . . 449
3. Transfer in the Planar Geometry: Quantum Wells . . . . . . . . . . . . . . . . . . . . . 456
4. Transfer in the Spherical Geometry: Quantum Dots . . . . . . . . . . . . . . . . . . . . 469
5. Application to Light-Emitting Devices . . . . . . . . . . . . . . . . . . . . . . . . . 480
6. Summary . . . . . . . . . . . . . . . . . . . . . . . . . . . . . . . . . . . . . . . . 483
   References . . . . . . . . . . . . . . . . . . . . . . . . . . . . . . . . . . . . . . . 485

## 1. Introduction

In the chapter by V.M. Agranovich, V.I. Yudson and P. Reineker of the present book the authors studied hybridization of Wannier–Mott excitons in an inorganic semiconductor and Frenkel excitons in an organic material. Namely, they considered a high-quality hybrid nanostructure consisting of a semiconductor part and an organic part. If the two excitonic states belonging to the corresponding parts of the nanostructure, which are spatially separated, have close energies and are coupled (e.g., via Coulomb interaction), the eigenstates of the system may no longer be purely Frenkel or Wannier excitons, but rather coherent superpositions of the two. These new excitations would then have a number of peculiar properties, as discussed in the chapter by V.M. Agranovich, V.I. Yudson and P. Reineker.

For this effect to occur, the magnitude of the Coulomb coupling (typically, several meV) should be large compared to the broadenings of the excitonic states in both materials. However, for many organic materials the width of the excitonic resonances can be larger than the coupling. In such a situation the coherent superposition of excitonic states is destroyed by dissipative processes; instead, incoherent Förster-like energy transfer has to be considered. As the energy relaxation in the organics is assumed to be fast, it makes sense to think about energy transfer from the semiconductor part of the structure (donor, which is initially excited) to the organic material (acceptor).

The physical picture is analogous to that considered in the chapter on energy transfer in a microcavity. Here we will consider distances which are small compared to the resonant light wavelength, correspondingly retardation effects can be neglected, the electromagnetic interaction between donor and acceptor is dominated by the Coulomb part, and the transfer is purely nonradiative. Our purpose will be to give predictions for the Coulomb transfer in different structures, to look at the effect of the geometry on the transfer, to consider its dependence on the character of excited states, etc.

The overall scenario can be described as follows. Suppose, the semiconductor is excited. Then excitation transfer to the organics immediately starts with the characteristic time $\tau$. As a result, an excitation is created in the organics with the energy equal to the Wannier–Mott exciton energy in the semiconductor (as prescribed by Fermi Golden Rule). Then the competition starts between the back transfer to the semiconductor (with the same time constant $\tau$) and the vibronic relaxation in the organics (the characteristic time $\tau_{vib}$). If the excitation has relaxed in the organics, then it will never return to the semiconductor, since its energy is already low, lying deep in the semiconductor gap. Thus, if $\tau_{vib} \ll \tau$, there will be no back transfer. In the opposite limit the populations will equilibrate. Another relevant time scale is, of course, the excitation lifetime in the semiconductor $\tau_{sem}$ (without the transfer to organics). If $\tau_{sem} \ll \tau$, then the efficiency of the transfer is low.

The typical values are $\tau_{sem} \sim 1$ ns, $\tau_{vib} \sim 1$ ps. The estimation of $\tau$ for different configurations, as it was done in Refs. [1–6], is the focus of the present chapter. To anticipate the main result, in several configurations with realistic material and geometry parameters the transfer time from the Wannier exciton is $\tau \sim 10$–$100$ ps, which allows to say that (i) the transfer to organic is efficient, compared to the semiconductor intrinsic relaxation channels, and (ii) the back transfer to the semiconductor is suppressed. A possible application of this phenomenon, as discussed in Section 5, is to use the semiconductor part of the nanostructure for the electrical pumping of Wannier excitons, which, in turn, would pump organics via the transfer, described above. The organics, pumped in such indirect way (which, however, may be quite efficient), may be used as the active medium in a light-emitting device. Such pumping eliminates the problem of providing a good electric contact to the organic medium, which is one of the main problems in the technology of organic LEDs. This indirect way of pumping eliminates also the singlet–triplet problem in the organics, since only singlet states may participate in the Förster transfer, and all the energy, transferred to the organics, is transferred to singlets. Such a hybrid device might be advantageous also with respect to semiconductor LEDs, since the Förster transfer may compete with other nonradiative relaxation processes in the semiconductor, and the quantum efficiency of such a device would be larger than that of a single semiconductor, if the luminescence quantum yield in the organics is high.

The rest of the chapter is organized as follows. In Section 2 we describe the general scheme of calculation of the transfer rate, and in particular the input of the theory: models for the semiconductor and the organic parts of the structure. Sections 3, 4 are dedicated to the calculations of the transfer rate in two specific configurations: planar (semiconductor quantum well covered by an organic material overlayer) and spherical (semiconductor quantum dot surrounded by the organic material). Finally, in Section 5 we discuss possible application of the described phenomenon for light-emitting devices.

## 2. The General Calculation Scheme for the Energy Transfer

This section is dedicated to the description of the general calculation scheme of the rate of energy transfer from an inorganic semiconductor to an organic material due to Coulomb interaction. The appropriate point to start from is classical electrostatics. Consider two spatially separated systems of charges $a$ and $b$ with the charge densities $\rho^a(\mathbf{r})$ and $\rho^b(\mathbf{r})$ respectively. Then the interaction energy is given by

$$V = \int \rho^a(\mathbf{r}) G(\mathbf{r},\mathbf{r}') \rho^b(\mathbf{r}') d^3r\, d^3r', \tag{2.1}$$

where $G(\mathbf{r},\mathbf{r}')$ is the Green's function of the Poisson equation. In vacuum it is the usual $1/|\mathbf{r}-\mathbf{r}'|$, in an inhomogeneous anisotropic medium with the dielectric constant $\varepsilon_{ij}(\mathbf{r})$ it satisfies the equation

$$\frac{\partial}{\partial x_i} \varepsilon_{ij}(\mathbf{r}) \frac{\partial}{\partial x_j} G(\mathbf{r},\mathbf{r}') = -4\pi \delta(\mathbf{r}-\mathbf{r}'). \tag{2.2}$$

If the total charges of the systems $a$ and $b$ are zero, one may introduce the polarizations, vanishing outside the regions of space, occupied by the systems [7], such that

$$\rho^a(\mathbf{r}) = -\operatorname{div}\mathbf{P}^a(\mathbf{r}), \qquad \rho^b(\mathbf{r}) = -\operatorname{div}\mathbf{P}^b(\mathbf{r}). \tag{2.3}$$

Then the energy may be rewritten as

$$V = \int P_i^a(\mathbf{r}) P_j^b(\mathbf{r}') \frac{\partial^2 G(\mathbf{r},\mathbf{r}')}{\partial x_i \partial x'_j} d^3r\, d^3r' = -\int \mathbf{P}^a(\mathbf{r}) \cdot \mathcal{E}^b(\mathbf{r}) d^3r, \tag{2.4}$$

where $\mathcal{E}^b(\mathbf{r})$ is the electric field, produced by the system $b$:

$$\mathcal{E}_i^b(\mathbf{r}) = \int \bar{\chi}_{ij}(\mathbf{r},\mathbf{r}') P_j^b(\mathbf{r}') d^3r', \qquad \bar{\chi}_{ij}(\mathbf{r},\mathbf{r}') \equiv -\frac{\partial}{\partial x_i} \frac{\partial}{\partial x'_j} G(\mathbf{r},\mathbf{r}'). \tag{2.5}$$

The Coulomb response function $\bar{\chi}_{ij}(\mathbf{r},\mathbf{r}')$ gives the $i$th component of the electric field at the point $\mathbf{r}$, produced by the $j$th component of a point dipole, situated at the point $\mathbf{r}'$. It is very simply related to the electromagnetic field response function

$\chi_{ij}(\mathbf{r}, \mathbf{r}', \omega)$ considered in Chapter 8 of the present book where energy transfer in a microcavity was studied. $\bar{\chi}_{ij}(\mathbf{r}, \mathbf{r}')$ is nothing else but the limit of $\chi_{ij}(\mathbf{r}, \mathbf{r}', \omega)$ at $\omega \to 0$ (electrostatics) or, equivalently, $c \to \infty$ (no retardation). Working in this limit is justified if the characteristic distances in the problem are much shorter than the resonant light wave length, which is equivalent to $k \gg \omega/c$.

In quantum mechanics the expression (2.1) acquires the sense of the interaction Hamiltonian $\hat{H}_{\text{int}}$, the charge densities and the polarizations become operators $\hat{\rho}^{a,b}(\mathbf{r})$ and $\hat{\mathbf{P}}^{a,b}(\mathbf{r})$. The explicit form of these operators is determined by the microscopic structure of the systems considered and the approximations made (so far all the relations of this section have been exact). In principle, the polarizations contain the contributions both from electrons and nuclei, but usually the nuclear dipole moment may be neglected since the characteristic displacements of nuclei are much smaller than those of electrons due to large mass difference. One should keep in mind that speaking of two different electron densities for the systems $a$ and $b$ already completely disregards the exchange effects, which is justified if the wave function overlap between the systems $a$ and $b$ is negligible.

Generally, the input parameters of the theory for calculation of the energy transfer rate should be (i) the model for the energy levels, the dissipation, and the excitonic polarization in the semiconductor (donor), (ii) the same for the organics (acceptor), and (iii) the Coulomb response function $\bar{\chi}_{ij}(\mathbf{r}, \mathbf{r}')$ for the structure under consideration. In what follows we are going to describe the microscopic models for the semiconductor and the organic parts of the structure and the procedure of calculation of the transfer rate.

## 2.1. THE MODEL FOR SEMICONDUCTOR POLARIZATION

In this section we discuss in detail the explicit form of the polarization $\hat{\mathbf{P}}(\mathbf{r})$ in semiconductor nanostructures. In the subsequent considerations just the final expression (2.13) is used, and if the reader does not want to go into details, he may actually skip this section.

In inorganic semiconductors the electron–phonon interaction is weak and one may completely neglect vibronic transitions, focusing on the electronic states. The most interesting for us will be the zinc-blende semiconductors formed by elements of the groups III and V or groups II and VI of the periodic table (like GaAs or ZnSe). Taking into account the spin degeneracy, in these materials the lowest conduction band is twice degenerate, and the highest valence band is fourfold degenerate at $\mathbf{k} = 0$ due to the symmetry of the crystal, both bands having extrema at $\mathbf{k} = 0$ (the $\Gamma$ point). At wave vectors $\mathbf{k} \neq 0$ the four-dimensional manifold splits into two bands (each of them being twice degenerate), having different effective masses. They are called light hole and heavy hole bands [8–10]. For our purposes it will be sufficient to consider first one nondegenerate conduction band and one nondegenerate valence band, summing the contributions of different bands at the end, when needed.

A system of many identical particles is most conveniently described by the $\hat{\psi}$ field operators [11]. The generic form of the electronic charge density operator in a semiconductor is

$$\hat{\rho}^{\text{sem}}(\mathbf{r}) = -e\hat{\psi}^{\dagger}(\mathbf{r})\hat{\psi}(\mathbf{r}). \tag{2.6}$$

Taking into account only two nondegenerate bands, one can split the electronic $\psi$-operator into two parts, corresponding to the conduction and valence bands:

$$\hat{\psi}(\mathbf{r}) = \hat{\psi}_c(\mathbf{r}) + \hat{\psi}_v(\mathbf{r}) \equiv \hat{\psi}_e(\mathbf{r}) + \hat{\psi}_h^{\dagger}(\mathbf{r}), \tag{2.7}$$

where we introduced a hole creation operator as a valence electron destruction operator. In the operator of electron density

$$\hat{\psi}^{\dagger}(\mathbf{r})\hat{\psi}(\mathbf{r}) = \hat{\psi}_v^{\dagger}(\mathbf{r})\hat{\psi}_c(\mathbf{r}) + \hat{\psi}_c^{\dagger}(\mathbf{r})\hat{\psi}_v(\mathbf{r}) + \hat{\psi}_c^{\dagger}(\mathbf{r})\hat{\psi}_c(\mathbf{r})$$
$$+ \hat{\psi}_v^{\dagger}(\mathbf{r})\hat{\psi}_v(\mathbf{r}), \tag{2.8}$$

the first two terms correspond to the interband transitions, the last two – to the intraband transitions.

Consider the first term (the second one being just its hermitian conjugate). The matrix element of the operator $\hat{\psi}_v^{\dagger}(\mathbf{r}')\hat{\psi}_c(\mathbf{r})$ between the ground state $|0\rangle$ and some eigenstate $|s\rangle$ of a single electron–hole pair is

$$\langle 0|\hat{\psi}_v^{\dagger}(\mathbf{r}')\hat{\psi}_c(\mathbf{r})|s\rangle = v_0 \int \frac{d^3k}{(2\pi)^3} \frac{d^3k'}{(2\pi)^3} \Psi_s(\mathbf{k},\mathbf{k}') u_{c,\mathbf{k}}(\mathbf{r}) u_{v,-\mathbf{k}'}^*(\mathbf{r}') e^{i\mathbf{k}\mathbf{r}+i\mathbf{k}'\mathbf{r}'}$$
$$\approx v_0 u_c(\mathbf{r}) u_v^*(\mathbf{r}') \Psi_s(\mathbf{r},\mathbf{r}'). \tag{2.9}$$

Here $v_0$ is the unit cell volume, $u_c(\mathbf{r})$ and $u_v(\mathbf{r})$ are the conduction and valence band extrema Bloch functions,[1] $\Psi_s(\mathbf{r}_e,\mathbf{r}_h)$ is the envelope wave function of the electron–hole pair state $|s\rangle$, normalized to the unit integral

$$\int |\Psi_s(\mathbf{r}_e,\mathbf{r}_h)|^2 d^3\mathbf{r}_e d^3\mathbf{r}_h = 1.$$

Now consider the interband matrix element of the charge density, corresponding to the first term in the expression (2.8). One is interested in the long-wavelength Fourier components of the density, given by

$$\rho_{0s}(\mathbf{q}) \equiv \langle 0|\hat{\rho}^{\text{sem}}(\mathbf{q})|s\rangle$$
$$= -e \int \frac{d^3k}{(2\pi)^3} \Psi_s(\mathbf{k},\mathbf{q}-\mathbf{k}) \int_{u.c.} u_{c,\mathbf{k}}(\mathbf{r}) u_{v,\mathbf{k}-\mathbf{q}}^*(\mathbf{r}) d^3\mathbf{r}, \tag{2.10}$$

---

[1] We adopt the term "Bloch function" for the cell-periodic part of the full electron wave function in the crystal.

which is obtained from the first line of (2.9) using the periodicity of the Bloch functions. More specifically, due to the periodicity the product $u_{c,\mathbf{k}}(\mathbf{r})\,u^*_{v,\mathbf{k-q}}(\mathbf{r})$ contains plane waves with the wave vectors equal to either zero or a reciprocal lattice vector. Being interested only in the long-wavelength part of $\rho_{0s}$, one should pick up only the zero wave vector contribution, which corresponds to the integration over the unit cell, as done in (2.10).

If one simply approximates the Bloch functions in (2.10) by those for the band extrema, the result will be zero due to the orthogonality

$$\int_{u.c.} u_c(\mathbf{r})\,u_v^*(\mathbf{r})\,d^3\mathbf{r} = 0.$$

Hence one has to expand the Bloch functions using the $\mathbf{k}\cdot\mathbf{p}$ perturbation theory [8–10], to find the admixture to the function $u_{c,\mathbf{k}}(\mathbf{r})$ of the Bloch functions $u_b(\mathbf{r})$ of all the other bands $b$:

$$u_{c,\mathbf{k}}(\mathbf{r}) \approx u_c(\mathbf{r}) - ik_j\frac{\hbar^2}{m_0}\sum_b \frac{\langle b|\partial/\partial x_j|c\rangle}{E_c(0) - E_b(0)} u_b(\mathbf{r})$$

$$= u_c(\mathbf{r}) - ik_j\sum_b \langle b|x_j|c\rangle\,u_b(\mathbf{r}), \qquad (2.11)$$

where the symbol $\langle b|\mathcal{O}|c\rangle$ denotes

$$\int_{u.c.} u_b^*(\mathbf{r})\,\mathcal{O}\,u_c(\mathbf{r})\,d^3\mathbf{r}.$$

In the transition from the first to the second line of (2.11) the quantum-mechanical relation $\dot{\mathbf{r}} = -i(\hbar/m_0)\nabla$ for the bare electron in the crystal was used, $\dot{\mathbf{r}}$ being related to the commutator of $\mathbf{r}$ with the crystal Hamiltonian. Expanding $u^*_{v,\mathbf{k-q}}(\mathbf{r})$ analogously, and substituting them into (2.10), one obtains

$$\rho_{0s}(\mathbf{q}) = -i\mathbf{q}\cdot\mathbf{d}^{vc}\int\frac{d^3\mathbf{k}}{(2\pi)^3}\,\Psi_s(\mathbf{k},\mathbf{q}-\mathbf{k}),$$

$$\rho_{0s}(\mathbf{r}) = -\operatorname{div}\bigl(\mathbf{d}^{vc}\,\Psi_s(\mathbf{r},\mathbf{r})\bigr), \qquad (2.12)$$

where $\mathbf{d}^{vc} = \langle v|(-e\mathbf{r})|c\rangle$ is the transition dipole moment of the unit cell. The contributions of all the bands different from the conduction and valence bands vanish due to the orthogonality. According to (2.3), the obtained charge density corresponds to the interband polarization

$$\langle 0|\hat{\mathbf{P}}(\mathbf{r})|s\rangle = \mathbf{d}^{vc}\,\Psi_s(\mathbf{r},\mathbf{r}). \qquad (2.13)$$

The expression (2.13) is the basic one to be used in the present work. The Cartesian components of the dipole moments $d_i^{vc}$ ($i=x,y,z$) may be expressed in terms of the Kane's energy $E_0$ [9] as

$$|d_i^{vc}|^2 = \frac{e^2\hbar^2 E_0 c_i^2}{2m_0 E_g^2} = c_i^2(ea_0)^2 \frac{E_0 \text{Ry}_0}{E_g^2} \equiv c_i^2 |d^{vc}|^2, \qquad (2.14)$$

where $a_0$ and $\text{Ry}_0$ are the hydrogen atom Bohr radius and Rydberg constant, and $c_i$ are the appropriate symmetry coefficients. In semiconductors with the zinc-blend structure $c_x^{hh} = c_y^{hh} = 1$, $c_z^{hh} = 0$ (heavy holes), $c_x^{lh} = c_y^{lh} = 1/\sqrt{3}$, $c_z^{hh} = \sqrt{4/3}$ (light holes).

For an intraband transition between the states $|s\rangle$ and $|s'\rangle$ of the electron–hole pair one may simply average $\hat{\psi}_c^\dagger(\mathbf{r})\hat{\psi}_c(\mathbf{r})$ and $\hat{\psi}_v^\dagger(\mathbf{r})\hat{\psi}_v(\mathbf{r})$ over the unit cell using the Bloch functions at the band extrema, since the principal term does not vanish already. As a result, the corresponding matrix element of the charge density is given by the sum of electron and hole contributions:

$$\langle s|\hat{\rho}^{\text{sem}}(\mathbf{r})|s'\rangle = -e \int \Psi_s^*(\mathbf{r}, \mathbf{r}_h) \Psi_{s'}(\mathbf{r}, \mathbf{r}_h) d^3\mathbf{r}_h$$

$$+ e \int \Psi_s^*(\mathbf{r}_e, \mathbf{r}) \Psi_{s'}(\mathbf{r}_e, \mathbf{r}) d^3\mathbf{r}_e. \qquad (2.15)$$

## 2.2. THE MODEL FOR ORGANICS

The way of describing the organic material should be chosen according to two principal points. First, one expects it to have some general properties:

- The material has broad absorption band around the frequency $\omega_D$ of the Wannier–Mott exciton.
- The wave vector is not expected to be a good quantum number, e.g., because the plane waves are completely destroyed by scattering on phonons or disorder, or the material may consist simply of the acceptor molecules randomly distributed in a matrix in which case introducing the wave vector makes no sense at all. In brief terms, the spatial dispersion is neglected.
- The microscopic and the macroscopic length scales are assumed to be well-separated, which means that one can average the properties of the material over a volume, small enough compared to "long" scales of the problem (the characteristic structure size, the characteristic length of the electric field variation, etc.) on one hand, but on the other hand this volume should be large enough for this average not to vary strongly from point to point. If it is not so, the results would strongly depend on the specific realization. The slow variation of the parameters is allowed (e.g., such as the concentration of the impurity acceptor molecules). In other words, the material is assumed to be "statistically smooth", which allows to describe it as a continuous medium.

The second point is that once a rigorous microscopic description of the organics is strongly dependent on the specific material, and may be necessarily statistical because of the disorder, a justified thing to do is to introduce the simplest model, reproducing the general properties of the material and relevant averaged quantitative characteristics. As will become clear from the following discussion, these relevant characteristics are the optical ones, i.e., the dielectric function, which is directly available from experiments.

We assume that the excitations in the organic medium are localized, corresponding to the excited states of a molecule or a group of strongly coupled molecules. Thus, the organic subsystem may be described by the ground state $|g_A\rangle$, and the excited states $|\mathbf{r}, \nu\rangle$, where $\mathbf{r}$ is the continuous position of the excited state and $\nu$ is a continuous quantum number, labelling the excited states at the point $\mathbf{r}$. As we restrict ourselves to the linear regime, only "one-particle" excited states are considered, which means that two excitations $|\mathbf{r}, \nu\rangle$ and $|\mathbf{r}', \nu'\rangle$ are not allowed to exist simultaneously. The particular dissipation mechanism, determining the structure of these states need not be specified here, according to the above-said. We use the following normalization of the states

$$\langle g_A|g_A\rangle = 1, \tag{2.16}$$

$$\langle g_A|\mathbf{r}, \nu\rangle = 0, \tag{2.17}$$

$$\langle \mathbf{r}, \nu|\mathbf{r}', \nu'\rangle = \delta(\nu - \nu')\delta(\mathbf{r} - \mathbf{r}'), \tag{2.18}$$

$$\hat{1}_A = |g_A\rangle\langle g_A| + \int d^3\mathbf{r} \int d\nu\, |\mathbf{r}, \nu\rangle\langle \mathbf{r}, \nu|, \tag{2.19}$$

where $\hat{1}_A$ is the unit operator for the organic subsystem. The Hamiltonian and the polarization of the organic medium are written as

$$\hat{H}_A = |g_A\rangle 0 \langle g_A| + \int d^3\mathbf{r} \int d\nu |\mathbf{r}, \nu\rangle E_\nu(\mathbf{r})\langle \mathbf{r}, \nu|, \tag{2.20}$$

$$\hat{\mathbf{P}}^A(\mathbf{r}) = \int d\nu |\mathbf{r}, \nu\rangle \mathbf{d}^\nu(\mathbf{r})\langle g_A| + \text{h.c.}, \tag{2.21}$$

where the $E_\nu(\mathbf{r})$ is the energy of the corresponding state and $\mathbf{d}^\nu(\mathbf{r})$ is the matrix element of the dipole moment between the excited and the ground state:

$$\mathbf{d}^\nu(\mathbf{r}) = \langle \mathbf{r}, \nu| \left( \sum_i e_i \hat{\mathbf{r}}_i \right) |g_A\rangle, \tag{2.22}$$

where $e_i$ and $\hat{\mathbf{r}}_i$ are the charge and the position operator of the $i$th charge in the medium and the sum is taken over all charges, constituting the medium. Both $E_\nu(\mathbf{r})$ and $\mathbf{d}^\nu(\mathbf{r})$ are assumed to be slow varying in space.

## 2.3. CALCULATION OF THE TRANSFER RATE

Typically the broadening of the Wannier–Mott exciton resonance (several meV) is negligible compared to the width of the absorption spectrum of organics (several hundred meV). Thus, one may consider the excited state in the semiconductor to be discrete with the energy $\hbar\omega_D$. The excited state $|\Psi\rangle$ of the semiconductor is an electron–hole pair described by the two-particle envelope function $\Psi(\mathbf{r}_e, \mathbf{r}_h)$, assumed to be normalized to unity:

$$\int |\Psi_s(\mathbf{r}_e, \mathbf{r}_h)|^2 d^3\mathbf{r}_e \, d^3\mathbf{r}_h = 1. \qquad (2.23)$$

The effective Hamiltonian and the polarization for the semiconductor are simply given by

$$\hat{H}_D = |\Psi\rangle \hbar\omega_D \langle\Psi| + |g_D\rangle 0 \langle g_D|, \qquad (2.24)$$

$$\hat{\mathbf{P}}^D(\mathbf{r}) = \mathbf{d}^{vc} \Psi(\mathbf{r}, \mathbf{r}) |g_D\rangle \langle\Psi| + \text{h.c.}, \qquad (2.25)$$

where $|g_D\rangle$ is the semiconductor ground state, $\mathbf{d}^{vc}$ was defined in Section 2.1.

The Hamiltonian of the Coulomb interaction between the semiconductor and the organics is given by Eqs. (2.4), (2.5):

$$\hat{H}_{\text{int}} = -\int \hat{\mathbf{P}}^A(\mathbf{r}) \cdot \hat{E}^D(\mathbf{r}) d^3\mathbf{r}, \qquad \hat{\mathcal{E}}_i^D(\mathbf{r}) = \int \bar{\chi}_{ij}(\mathbf{r}, \mathbf{r}') \hat{P}_j^D(\mathbf{r}') d^3\mathbf{r}', \qquad (2.26)$$

where $\hat{E}^D(\mathbf{r})$ is the operator of the electric field, produced by the semiconductor polarization $\hat{\mathbf{P}}^D(\mathbf{r})$ in the organics. The Förster transfer rate $\Gamma$ (determining the partial electron–hole pair lifetime $\tau$ due to the transfer) is given by the Fermi Golden Rule:

$$\Gamma \equiv \frac{1}{\tau} = \frac{2\pi}{\hbar} \int d^3\mathbf{r} \int dv \left| \mathbf{d}^v(\mathbf{r}) \cdot E^{vc}(\mathbf{r}) \right|^2 \delta(E_v(\mathbf{r}) - \hbar\omega_D), \qquad (2.27)$$

where $E^{vc}(\mathbf{r}) = \langle g_D | \hat{E}^D(\mathbf{r}) | \Psi \rangle$. Considering the general expression for the dielectric function [12], which in our normalization of states may be written as

$$\varepsilon_{ij}(\mathbf{r}, \omega) = \delta_{ij} - 8\pi \int dv \frac{E_v(\mathbf{r}) d_i^v(\mathbf{r}) (d_j^v(\mathbf{r}))^*}{(\hbar\omega)^2 - (E_v(\mathbf{r}))^2 + i\eta\omega}, \qquad (2.28)$$

where the infinitesimal $\eta \to +0$ prescribes the position of the poles in the $\omega$ complex plane, the expression (2.27) may be identically rewritten as

$$\Gamma = \frac{1}{2\pi\hbar} \int \text{Im}\,\varepsilon_{ij}(\mathbf{r}, \omega_D) \mathcal{E}_i^{vc}(\mathbf{r}) (\mathcal{E}_j^{vc}(\mathbf{r}))^* d^3\mathbf{r}, \qquad (2.29)$$

where $\varepsilon_{ij}(\mathbf{r}, \omega_D)$ is the dielectric function of the organic medium at the frequency of the Wannier exciton frequency.

It is interesting to compare the expression (2.29) to the original Förster formula [13,14]. First, in Eq. (2.29) there is no integral over frequencies present in the Förster formula, since we assumed the luminescence spectrum of the semiconductor to be a sharp line (actually, a $\delta$-peak). Instead of the acceptor absorption coefficient $\mu(\omega)$ in the Förster formula, in Eq. (2.29) one has the imaginary part of the dielectric function, which is physically equivalent, since for the isotropic case $\varepsilon_{ij} = \varepsilon \delta_{ij}$ the two are connected by a simple relation [7]:

$$\mu(\omega) = \frac{2\omega}{c}\sqrt{(|\varepsilon(\omega)| - \mathrm{Re}\,\varepsilon(\omega))/2} \approx \frac{\omega}{c}\frac{\mathrm{Im}\,\tilde{\varepsilon}(\omega)}{\sqrt{\mathrm{Re}\,\tilde{\varepsilon}(\omega)}}.$$

Analogously to the discussion of the Förster formula [14], the phenomenological quantity $\varepsilon_{ij}(\mathbf{r}, \omega)$ already takes into account the local field effects. Finally, the geometrical factor $1/R^6$ appearing in the Förster formula (where $R$ is the donor-acceptor distance) is no longer present in (2.29), since the geometry of the structure is completely different.

An interesting feature is that Eq. (2.29), multiplied by the energy of the excitation $\hbar\omega_D$, coincides with the expression for the power dissipated in the organic medium in the presence of classical external electric field of the frequency $\omega_D$ and the amplitude, equal to $\mathcal{E}(\mathbf{r})$ [7]. This is related to the general fact that a quantum mechanical system in the linear regime is equivalent to a classical oscillator, producing a real electric field (instead of the off-diagonal matrix element for the quantum description) in the organics. This equivalence was used in Chapter 8 to calculate the rates of different dissipative processes in a microcavity.

Summarizing, we can give the following recipe for calculating the energy transfer rate. Suppose that inside the semiconductor one has the classical macroscopic polarization, oscillating with the frequency $\omega_D$:

$$\mathbf{P}(\mathbf{r}, t) = \mathbf{d}^{vc}\,\Psi(\mathbf{r}, \mathbf{r})\,e^{-i\omega_D t} + \mathrm{c.c.} \tag{2.30}$$

Then, solve the electrostatic problem (i.e., neglecting retardation) and find the corresponding electric field

$$E(\mathbf{r}, t) = E(\mathbf{r})\,e^{-i\omega_D t} + \mathrm{c.c.}$$

The latter, substituted into (2.29), will give the correct quantum mechanical decay rate if the complex dielectric function of the organics is known (independently of its microscopic structure). The expression (2.29) will be the basic one to be applied to specific configurations in the following sections.

## 3. Transfer in the Planar Geometry: Quantum Wells

In this section the Förster transfer in the planar geometry is studied, the donor being a semiconductor quantum well (QW), while the surrounding organic material

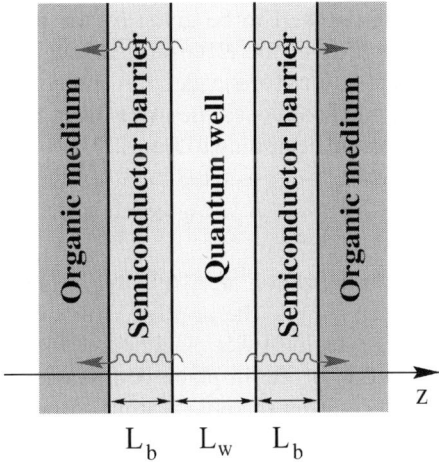

Fig. 1. A sketch of the planar structure under study.

plays the role of acceptor [1,2]. Three different possibilities are considered, corresponding to the initial excited state in the semiconductor being a free Wannier–Mott exciton, a localized Wannier–Mott exciton, or a dissociated electron–hole pair.

The geometry of the problem is the following (Figure 1). We consider a symmetric structure, consisting of a semiconductor QW of thickness $L_w$ between two barriers of thickness $L_b$ each, the whole semiconductor structure being surrounded by thick slabs of an organic material (actually, we assume each slab to be semi-infinite). We assume that in the frequency region here considered the semiconductor background dielectric constant $\varepsilon$ is real (including only the contribution of higher resonances with respect to the exciton resonance under consideration) and the same for the well and the barrier, while that of the organic material $\tilde{\varepsilon}$ is complex. For simplicity we assume the organic material to be isotropic (generalization to the anisotropic case is straightforward). So, the dielectric constant to be used in Eq. (2.29) as well as in the Poisson equation below, is

$$\varepsilon_{ij}(\mathbf{r}) = \begin{cases} \varepsilon\,\delta_{ij}, & |z| < L_w/2 + L_b, \\ \tilde{\varepsilon}\,\delta_{ij}, & |z| > L_w/2 + L_b, \end{cases} \quad (3.1)$$

where the $z$-axis is chosen to be along the growth direction, $z = 0$ corresponding to the center of the QW.

### 3.1. Free Excitons

We adopt a simplified microscopic quantum mechanical model of a 2D Wannier–Mott exciton, in which the polarization (Eq. (2.25)) can be taken to vanish for $|z| > L_w/2$ and inside the well to be given by the product of the 1s-wave function of the relative motion of the electron and hole at the origin, with the lowest subband envelope functions for the electron and hole in the approximation of the infinitely deep well and finally with the wave function of the center-of-mass motion, all of them normalized according to Eq. (2.23). Thus, we have

$$\mathbf{P}(\mathbf{r}) = \mathbf{d}^{vc} \sqrt{\frac{2}{\pi a_{B2}^2} \frac{2}{L_w}} \cos^2\left(\frac{\pi z}{L_w}\right) \frac{e^{i\mathbf{k}\mathbf{r}_\|}}{\sqrt{S}}, \qquad (3.2)$$

where $S$ is the in-plane normalization area, $\mathbf{k}$ is the in-plane wave vector of the center-of-mass motion, $\mathbf{r}_\| \equiv (x, y)$ – the in-plane component of $\mathbf{r}$ and $a_{B2}$ is the 2D 1s-exciton Bohr radius, which is twice smaller than the bulk Bohr radius [15]. We choose as $x$ the direction of the in-plane component of the exciton dipole moment $\mathbf{d}^{vc}$, preferring to consider the polarization not with respect to the wave vector, but to some fixed frame. This little complication is justified since next to the free exciton we intend to study the case of the localized exciton, i.e., a system with broken 2D translational symmetry. Evidently, we need to consider two cases: $\mathbf{d}^{vc}$ being parallel and perpendicular to the QW plane. We will refer to them as X and Z polarizations respectively. When dealing only with free excitons in a single well, three modes of different symmetry would be identified: longitudinal (L), transverse (T) and perpendicular (Z). The L and Z modes correspond to the X and Z polarizations above (their energies are split by the depolarization shift, but this is immaterial for the following). For the T mode the dipole–dipole interaction here considered vanishes [16].

The corresponding electric field $E(\mathbf{r}) \equiv -\nabla\phi(\mathbf{r})$ can be obtained from the solution of the Poisson equation for the potential $\phi(\mathbf{r})$ (the charge density being $\rho(\mathbf{r}) \equiv -\nabla \cdot \mathbf{P}(\mathbf{r})$)

$$\varepsilon(z)\nabla^2\phi(\mathbf{r}) = 4\pi \nabla \cdot \mathbf{P}(\mathbf{r}), \qquad (3.3)$$

with the appropriate boundary conditions at $z = \pm L_w/2$ and at $z = \pm(L_w/2 + L_b)$, i.e., continuity of the tangential component of the electric field $E(\mathbf{r})$ and of the normal component of the electric displacement $\mathbf{D}(\mathbf{r}) = \varepsilon(z)E(\mathbf{r})$. Writing $\phi(\mathbf{r}) = \phi(z) e^{i\mathbf{k}\mathbf{r}_\|}$, we have the equation for $\phi(z)$:

$$\left[\frac{d^2}{dz^2} - k^2\right]\phi(z) = \begin{cases} 4\pi\rho(z)/\varepsilon, & |z| < L_w/2, \\ 0, & |z| > L_w/2, \end{cases} \qquad (3.4)$$

where

$$\rho^{(X)}(z) = ik_x L_w \rho_0 (1 + \cos qz), \qquad (3.5)$$

$$\rho^{(Z)}(z) = -q L_w \rho_0 \sin qz, \qquad (3.6)$$

$$\rho_0 = \sqrt{\frac{2}{\pi a_{B2}^2}} \frac{d^{vc}}{\sqrt{S}L_w^2}, \quad q \equiv 2\pi/L_w \qquad (3.7)$$

with the boundary conditions that $\phi(z)$ and $\varepsilon(z)\, d\phi(z)/dz$ should be continuous at the four interfaces. The corresponding solution in the organic material (for $z > L_w/2 + L_b$) is given by

$$\phi(z) = \rho_0\, C_{\mathbf{k}}\, e^{-k(z - L_b - L_w/2)}, \tag{3.8}$$

$$C_{\mathbf{k}}^{(X)} = -\frac{ik_x}{k} \frac{8\pi^2 q}{k(k^2 + q^2)} \frac{\sinh(kL_w/2)}{\varepsilon \sinh(kL_b + kL_w/2) + \tilde{\varepsilon} \cosh(kL_b + kL_w/2)}, \tag{3.9}$$

$$C_{\mathbf{k}}^{(Z)} = \frac{8\pi^2 q}{k(k^2 + q^2)} \frac{\sinh(kL_w/2)}{\varepsilon \cosh(kL_b + kL_w/2) + \tilde{\varepsilon} \sinh(kL_b + kL_w/2)}. \tag{3.10}$$

Thus, the electric field penetrating the organic material is given by

$$E(\mathbf{r}) = [-i\mathbf{k} + k\mathbf{e}_z]\phi(z)\, e^{i\mathbf{k}\mathbf{r}_\parallel}. \tag{3.11}$$

Now we simply substitute this electric field into (2.29) and get the decay rate:

$$\Gamma \equiv \frac{1}{\tau} = \frac{S}{2\pi\hbar} \operatorname{Im} \tilde{\varepsilon} \int_{L_b + L_w/2}^{+\infty} 2k^2 |\phi(z)|^2\, dz$$

$$= \frac{\operatorname{Im} \tilde{\varepsilon}}{\pi^2 \hbar} \frac{|d^{vc}|^2}{a_{B2}^2} \frac{k|C_{\mathbf{k}}|^2}{L_w^4}, \tag{3.12}$$

where we have considered the absorption only at $z > L_w/2 + L_b$ (considering also the organic material in $z < -L_w/2 - L_b$, $\tau$ would be twice shorter).

For the numerical estimations we use the parameters, typical for II–VI semiconductor (e.g., ZnSe/ZnCdSe) quantum wells [17]: $\varepsilon \approx 6$, $d^{vc} \approx 0.1\, ea_{B2}$ (about 12 Debye, the Bohr radius is taken to be 25 Å). For the organic part one needs to know only the dielectric constant, taken to be $\tilde{\varepsilon} = 4 + 3i$. This value is not even the most optimistic one: for PTCDA one has $\tilde{\varepsilon} \approx 3.6 + 4.5i$ [18]. Two cases should be considered: $\mathbf{d}^{vc}$ lying in the QW plane, $\mathbf{k} \| \mathbf{d}^{vc}$ (L-exciton) and $\mathbf{d}^{vc}$ perpendicular to the QW plane (Z-exciton). Taking $L_w = 60$ Å, $L_b = 40$ Å, we plot $\tau_L$ and $\tau_Z$ as functions of $K$ for $\varepsilon = 6$, $\tilde{\varepsilon} = 4 + 3i$ and $\varepsilon = 4$, $\tilde{\varepsilon} = 6 + 3i$ on Figures 2(a), (b). It is seen from the plot, that the lifetime does not depend drastically on the polarization and the real parts of dielectric constants. Figure 3(a) shows that the dependence on $L_w$ is also weak (though at large $k$, $kL_w \gtrsim 2\pi$, $\tau \propto L_w^6$ from (3.10)), while $L_b$ (Figure 3(b)), when grows, gives an obvious exponential factor (clearly seen from the hyperbolic functions in the denominators of (3.9), (3.10)). The most interesting dependence is that on $k$. We see, that $\tau$ exhibits a minimum at $k_{\min} \sim 1/L_b$. This dependence may be easily understood if one recalls that the dipole–dipole interaction between two planes behaves like

$$V(\mathbf{k}, z) \propto k\, e^{-kz}, \tag{3.13}$$

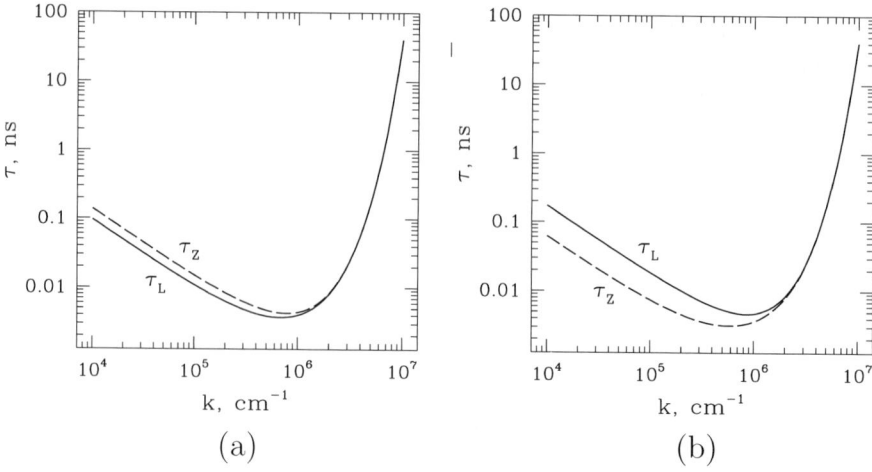

Fig. 2. Free L-exciton (solid line) and Z-exciton (dashed line) lifetime $\tau$ (ns) versus the in-plane wave vector $k$ (cm$^{-1}$). $d^{vc} = 0.1\,ea_{B2}$, $L_w = 60$ Å, $L_b = 40$ Å, $\varepsilon_b = 6$, $\tilde{\varepsilon} = 4 + 3i$ (a); the same, but $\varepsilon_b = 4$, $\tilde{\varepsilon} = 6 + 3i$ (b).

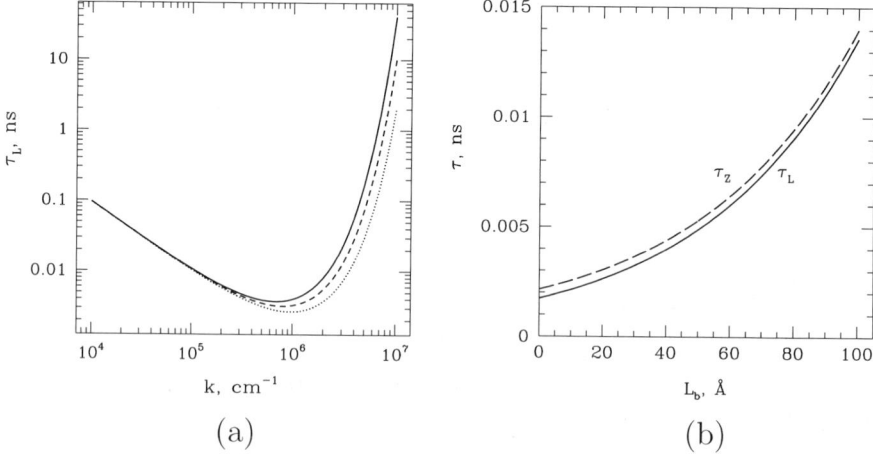

Fig. 3. (a) Free L-exciton lifetime $\tau$ (ns) versus the in-plane wave vector $k$ (cm$^{-1}$) for three well widths: $L_w = 20$ Å (dotted line), $L_w = 40$ Å (dashed line), $L_w = 60$ Å (solid line), other parameters being $L_b = 40$ Å, $\varepsilon_b = 6$, $\tilde{\varepsilon} = 4 + 3i$. (b) Free L-exciton (solid line) and Z-exciton (dashed line) lifetime $\tau$ (ns) versus the barrier width $L_b$ (Å). $k = 10^6$ cm$^{-1}$, $L_w = 60$ Å, $\varepsilon_b = 6$, $\tilde{\varepsilon} = 4 + 3i$.

which, when substituted into the Fermi Golden Rule, gives the correct asymptotics $\tau \sim 1/k$ at $k \to 0$ and exponential growth at $k \to \infty$.

Considering at first a quasi-thermalized exciton distribution, typical values of $k$ at a temperature $\sim 100$ K are $\sim 3 \cdot 10^6$ cm$^{-1}$. The corresponding energy transfer lifetime (tens of picoseconds) is much less than the exciton recombination lifetime which is about 200 ps in II–VI semiconductor QWs, as reported by different authors (Ref. [17] and references therein, Ref. [19]). We remark that for the case of free excitons in a quantum well, the effective radiative lifetime (which, assuming a thermal distribution, increases linearly with temperature) is determined by the population transfer from nonradiative excitons with large $k$ to small $k$ excitons undergoing a fast radiative decay [20,21]. Thus, the dipole–dipole energy transfer mechanism considered here proves to be efficient enough to quench a large fraction of the semiconductor excitons, thereby activating the organic medium luminescence. Moreover, the intraband relaxation of excitons due to the acoustic phonon scattering occurs at time scales of the order of 20–30 ps at 10 K [19], which is larger than the minimal transfer lifetime, obtained here (less then 10 ps for $k_{min} \sim 10^6$ cm$^{-1}$). This makes it possible to excite the QW in a way to produce the initial nonequilibrium distribution of excitons with $k = k_{min}$, tuning the frequency of the excitation pulse to exceed the energy $\hbar\omega_{exc}(k_{min})$ of the exciton with $k = k_{min}$ by one LO-phonon frequency $\Omega_{LO}$ (since in II–VI semiconductors the free-carrier-to-exciton relaxation is governed mainly by LO-phonon scattering and happens at times of about 1 ps [19–24]), or an integer multiple of $\Omega_{LO}$, if the exciton binding energy is larger than $\hbar\Omega_{LO}$. A numerical estimate for ZnSe gives $\hbar\omega_{exc}(k_{min}) - \hbar\omega_{exc}(k = 0) \sim 1$ meV, while $\hbar\Omega_{LO} \approx 31$ meV [19], so that the following kinetics of excitons at $k \sim k_{min}$ is governed mainly by the acoustic phonons. Finally, another possibility would be to resonantly pump excitons with the appropriate $k$ by using a coupling grating configuration [25].

Analogous calculations may be performed for the case of III–V semiconductor materials. We take $\varepsilon \approx 11$, $d^{vc} \approx 0.05\, ea_{B2}$ and plot the L-exciton lifetime versus the wave vector $k$ for several values of $L_w$ (Figure 4, analogous to Figure 3(a) for II–VI materials). All other parameters are the same as on Figure 3(a). We see that the lifetime is longer compared to that on Figure 3(a) by about an order of magnitude, which is due to the larger values of $a_{B2}$ and $\varepsilon$. However, the energy transfer discussed here is still efficient enough because the effective exciton recombination time in III–V materials is also larger (about 1 ns [26]).

## 3.2. Localized Excitons

Now we turn to the situation when the QW width fluctuations, alloy disorder or impurities localize the 2D exciton (such a situation is more frequent for II–VI semiconductor quantum wells than for III–V ones). Then, the wave function of the center-of-mass exciton motion $\Phi(\mathbf{r}_\parallel)$ is no longer just a plane wave,

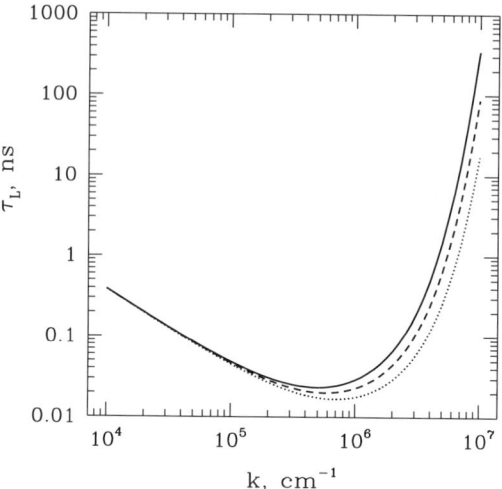

Fig. 4. The same as on Figure 3, but for the III–V semiconductor compounds ($\varepsilon_b = 11$, $d^{vc} = 0.05\, ea_{B2}$), all other parameters being the same as on Figure 3(a).

and the corresponding polarization is given by

$$\mathbf{P}(\mathbf{r}) = \mathbf{d}^{vc} \sqrt{\frac{2}{\pi a_{B2}^2}} \frac{2}{L_w} \cos^2\left(\frac{\pi z}{L_w}\right) \Phi(\mathbf{r}_\parallel), \qquad (3.14)$$

which implies that $\Phi(\mathbf{r}_\parallel)$ is normalized according to

$$\int d^2 \mathbf{r}_\parallel \left|\Phi(\mathbf{r}_\parallel)\right|^2 = 1. \qquad (3.15)$$

The solution of the Schrödinger equation for a particle in the random potential, caused by the QW width fluctuations and the alloy disorder is beyond the scope of the present paper (much work has been done in this field, e.g., see [27] and references therein). We can mention only some general properties that $\Phi(\mathbf{r}_\parallel)$ should have: (i) it should be localized within some distance $L \gtrsim L_w$, (ii) it should be smooth and without nodes. As a consequence, its spatial Fourier expansion should contain mainly the components with wave vectors $k \lesssim 1/L$.

It is convenient to expand the wave function $\Phi(\mathbf{r}_\parallel)$ into plane waves:

$$\Phi(\mathbf{r}_\parallel) = \int \frac{d^2\mathbf{k}}{(2\pi)^2} \Phi_\mathbf{k} e^{i\mathbf{k}\mathbf{r}_\parallel}, \qquad (3.16)$$

and analogously the charge density $\rho(\mathbf{r})$ and the potential $\phi(\mathbf{r})$. Then one again obtains Eq. (3.4), but the charge density is now given by

$$\rho_\mathbf{k}^{(X)}(z) = i k_x L_w \, \tilde{\rho}_0 \, L \Phi_\mathbf{k} (1 + \cos qz), \qquad (3.17)$$

$$\rho_{\mathbf{k}}^{(Z)}(z) = -qL_w\,\tilde{\rho}_0\,L\Phi_{\mathbf{k}}\sin qz, \tag{3.18}$$

$$\tilde{\rho}_0 = \sqrt{\frac{2}{\pi a_{B2}^2}}\frac{d^{vc}}{LL_w^2}. \tag{3.19}$$

The solution is

$$\phi_{\mathbf{k}}(z) = \tilde{\rho}_0\,L\Phi_{\mathbf{k}}C_{\mathbf{k}}\,e^{-k(z-L_b-L_w/2)}, \tag{3.20}$$

with the same $C_{\mathbf{k}}$, given by (3.9), (3.10). For the decay rate we obtain:

$$\Gamma = \frac{1}{\tau} = \frac{\operatorname{Im}\tilde{\varepsilon}}{2\pi\hbar}\int_{L_b+L_w/2}^{+\infty} dz \int \frac{d^2\mathbf{k}}{(2\pi)^2} 2k^2 |\phi_{\mathbf{k}}(z)|^2 \tag{3.21}$$

$$= \frac{\operatorname{Im}\tilde{\varepsilon}}{\pi^2\hbar}\frac{|d^{vc}|^2}{a_{B2}^2}\frac{1}{L_w^4}\int\frac{d^2\mathbf{k}}{(2\pi)^2} k|\Phi_{\mathbf{k}}|^2|C_{\mathbf{k}}|^2. \tag{3.22}$$

It is possible to get some information about the decay rate (3.22) based only on general properties of the wave function, mentioned above. We have three length scales in our problem: $L_w$, $L_b$ and $L$. First, we have the condition $L_w \lesssim L$. Since wave vectors with $kL \gtrsim 1$, being cut off by $|\Phi_{\mathbf{k}}|^2$, do not contribute to the integral, we may set $kL_w \to 0$. The subsequent analysis depends on the relation between $L_b$ and $L$.

If $L_b \ll L$, we may put $kL_b \to 0$ as well. Then we have

$$\frac{C_{\mathbf{k}}^{(X)}}{L_w^2} \to -\frac{2\pi i}{\tilde{\varepsilon}}\frac{k_x}{k}, \qquad \frac{C_{\mathbf{k}}^{(Z)}}{L_w^2} \to \frac{2\pi}{\varepsilon} \tag{3.23}$$

and the integral may be estimated as

$$\frac{1}{\tau_X} = A\,\frac{2}{\hbar}\frac{\operatorname{Im}\tilde{\varepsilon}}{|\tilde{\varepsilon}|^2}\frac{|d^{vc}|^2}{a_{B2}^2}\frac{1}{L}, \qquad \frac{1}{\tau_Z} = A\,\frac{4}{\hbar}\frac{\operatorname{Im}\tilde{\varepsilon}}{\varepsilon^2}\frac{|d^{vc}|^2}{a_{B2}^2}\frac{1}{L} \tag{3.24}$$

up to a numerical factor $A \sim 1$, determined by the detailed shape of $\Phi_{\mathbf{k}}$. We have set the average value of $k$ over the wave function $\Phi_{\mathbf{k}}$ to be $A/L$ and, if for the X polarization we assume $\Phi_{\mathbf{k}}$ to be cylindrically symmetric (which may be considered as the average over the realizations of disorder), then the numerical factor $A$ is the same for both cases.

In the opposite limit, $L_b \gg L$ (which also implies $L_b \gg L_w$), we may set $\Phi_{\mathbf{k}} = \Phi_{\mathbf{k}=0}$ since the values of $\mathbf{k}$, contributing to the integral, are determined by $C_{\mathbf{k}}$, (namely, $k \lesssim 1/L_b$) which in this limit takes the form

$$\frac{C_{\mathbf{k}}^{(X)}}{L_w^2} \to -\frac{ik_x}{k}\frac{4\pi}{(\tilde{\varepsilon}+\varepsilon)e^{kL_b}+(\tilde{\varepsilon}-\varepsilon)e^{-kL_b}}, \tag{3.25}$$

$$\frac{C_{\mathbf{k}}^{(Z)}}{L_w^2} \to \frac{4\pi}{(\tilde{\varepsilon}+\varepsilon)e^{kL_b}-(\tilde{\varepsilon}-\varepsilon)e^{-kL_b}}. \tag{3.26}$$

Estimating the integral, we have

$$\frac{1}{\tau_X} = A \frac{1}{\pi\hbar} \frac{\operatorname{Im}\tilde{\varepsilon}}{|\tilde{\varepsilon}+\varepsilon|^2} \frac{|d^{vc}|^2}{a_{B2}^2} \frac{|\Phi_{\mathbf{k}=0}|^2}{L_b^3}, \qquad (3.27)$$

where $|\Phi_{\mathbf{k}=0}|^2 \sim L^2$, which follows from the normalization condition. The expression for $1/\tau_Z$ differs from this by an additional factor of 2 and the factor $A$ may be different in the two cases. It is determined by the values of $\tilde{\varepsilon}$, $\varepsilon$:

$$A = \int_0^\infty \frac{4\xi^2\, d\xi}{\left|\frac{\tilde{\varepsilon}+\varepsilon}{|\tilde{\varepsilon}+\varepsilon|}e^\xi \pm \frac{\tilde{\varepsilon}-\varepsilon}{|\tilde{\varepsilon}+\varepsilon|}e^{-\xi}\right|^2}, \qquad (3.28)$$

and is bounded by

$$\frac{\pi^2}{12} = \int_0^\infty \frac{\xi^2\, d\xi}{\cosh^2\xi} < A < \int_0^\infty \frac{\xi^2\, d\xi}{\sinh^2\xi} = \frac{\pi^2}{6}. \qquad (3.29)$$

So we see that at $L \ll L_b$ the decay rate is proportional to $L^2$, at $L \gg L_b$ – to $L^{-1}$, therefore it has a maximum at some $L \sim L_b$. This is in agreement with the results of the previous section, since the plane waves, giving the largest contribution to the wave function $\Phi(\mathbf{r}_\parallel)$ and thus determining the decay rate, have the values of wave vector of the order of $k \sim 1/L$ and we have seen that wave vectors, corresponding to the shortest lifetimes were $k_{\min} \sim 1/L_b$.

To illustrate these considerations, we choose a specific example of the localized wave function – that of the ground state in the isotropic parabolic potential:

$$\Phi_{\mathbf{k}} = \sqrt{4\pi}\, L\, e^{-k^2 L^2/2}, \qquad (3.30)$$

which obviously has all the necessary features mentioned in the beginning of this section. For this wave function the integral in (3.22) may be evaluated numerically for arbitrary parameters $L$, $L_w$, $L_b$ (we remind that only physically relevant are $L \gtrsim L_w$). The results of the calculation ($\tau$ versus $L$) are plotted in Figure 5 along with the asymptotic dependencies for $L_b = 40$ Å, $\varepsilon = 6$, $\tilde{\varepsilon} = 4 + 3i$ (we have set $L_w \to 0$ for the plots on Figure 5(a), but a more specific value $L_w = 10$ Å was chosen for Figure 5(b)). In the limit $L \gg L_b$ the coefficient $A = \sqrt{\pi}/2$, the coefficient for $L \ll L_b$, given by (3.28) was calculated numerically.

We also plot the dependence, analogous to that on Figure 5(b), for parameters typical of III–V semiconductors: $d^{vc}/ea_{B2} = 0.05$ and $\varepsilon = 11$ (Figure 6); analogously to the previous section, we obtain larger lifetimes, than those for II–VI semiconductors.

## 3.3. FREE CARRIERS

Finally, we consider the situation, when the carriers are not bound into excitons, thus forming a 2D plasma. We assume electrons and holes to be quasi-thermalized

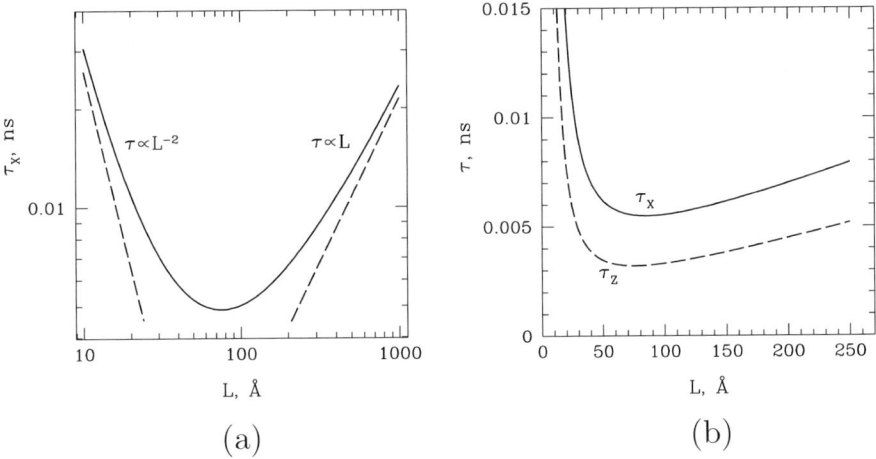

Fig. 5. (a) Localized X-exciton lifetime $\tau$ (ns) versus the localization length $L$ (Å) (solid line) along with the limiting cases $L \ll L_b$ and $L \gg L_b$ (dashed lines), $L_w \ll L$, $L_b = 40$ Å, $\varepsilon_b = 6$, $\tilde{\varepsilon} = 4 + 3i$. (b) Localized X-exciton (solid line) and Z-exciton (dashed line) lifetime $\tau$ (ns) versus the localization length $L$ (Å), $L_w = 10$ Å, $L_b = 40$ Å, $\varepsilon_b = 6$, $\tilde{\varepsilon} = 4 + 3i$.

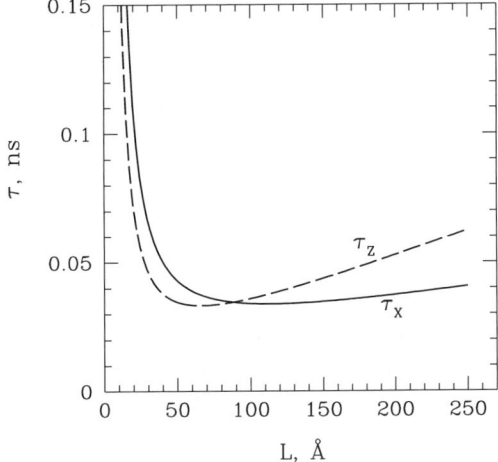

Fig. 6. The same as on Figure 5(b), but for the III–V semiconductor compounds, other parameters being the same as on Figure 5(b).

within corresponding bands (considering only one subband for each band), distributed according to the ideal Fermi gas law with the temperature $T$ and the chemical potentials $\mu^e$ and $\mu^h$ for electrons and holes respectively (we set the

Boltzmann constant equal to unity everywhere in this section, measuring the temperature in the units of energy). For an ideal 2D Fermi gas the chemical potential $\mu$ is related to the concentration $n$ via

$$e^{\mu(n,T)/T} = e^{\mu_0(n)/T} - 1, \quad \mu_0(n) = \frac{2\pi\hbar^2 n}{m}, \qquad (3.31)$$

where $\mu_0$ is the chemical potential at the zero temperature, $m$ is the corresponding (electron or hole) mass, spin degeneracy is not taken into account (i.e., all concentrations are those of particles with a given spin).

If $N_e$ ($N_h$) is the total number of electrons (holes) in the system, then the total recombination rate due to the energy transfer here considered can be written as

$$-\frac{dN_e}{dt} = -\frac{dN_h}{dt} = \sum_{\mathbf{k},\mathbf{k'}} \frac{f^e_{\mathbf{k}_e} f^h_{\mathbf{k}_h}}{\tau(\mathbf{k}_e + \mathbf{k}_h)}, \qquad (3.32)$$

where $f^e_{\mathbf{k}_e}$, $f^h_{\mathbf{k}_h}$ are the Fermi occupation numbers and $\tau(\mathbf{k}_e + \mathbf{k}_h)$ is the corresponding recombination time for a single electron–hole pair with 2D wave vectors $\mathbf{k}_e$ and $\mathbf{k}_h$. It depends only on the total momentum $\mathbf{k} = \mathbf{k}_e + \mathbf{k}_h$ since the wave function of the pair is simply

$$\Psi(\mathbf{r},\mathbf{r}) = \frac{2}{L_w}\cos^2\left(\frac{\pi z}{L_w}\right)\frac{e^{i(\mathbf{k}_e+\mathbf{k}_h)\mathbf{r}_\parallel}}{S} = \frac{2}{L_w}\cos^2\left(\frac{\pi z}{L_w}\right)\frac{e^{i\mathbf{k}\mathbf{r}_\parallel}}{S}, \qquad (3.33)$$

the in-plane motion being described by a direct product of two plane waves. The corresponding polarization is given by (3.2), where the factor $\sqrt{2/(\pi a_{B2}^2)}$ must be replaced by $1/\sqrt{S}$. For the pair recombination rate we obtain the expression (3.12), multiplied by the factor $\pi a_{B2}^2/2S$:

$$\frac{1}{\tau(\mathbf{k})} = \frac{1}{2\pi\hbar} \frac{|d^{vc}|^2}{SL_w^4} \operatorname{Im}\tilde{\varepsilon} k|C_\mathbf{k}|^2, \qquad (3.34)$$

which depends on the normalization area. Such a dependence is however immaterial: rewriting (3.32) for the 2D densities $n_{e,h} = N_{e,h}/S$ and transforming the sum into an integral, the total recombination rate becomes:

$$-\frac{dn_e}{dt} = -\frac{dn_h}{dt} = \int \frac{d^2\mathbf{k}_e\, d^2\mathbf{k}_h}{(2\pi)^4} \frac{S f^e_{\mathbf{k}_e} f^h_{\mathbf{k}_h}}{\tau(\mathbf{k}_e + \mathbf{k}_h)}. \qquad (3.35)$$

In the general case this integral cannot be calculated analytically, but the situation simplifies significantly in the classical limit ($T \gg \mu_0^e, \mu_0^h$). The occupation numbers then reduce to the Boltzmann distribution:

$$f^{e,h}_{\mathbf{k}_{e,h}} = \exp\left(-\frac{\hbar^2 k_{e,h}^2}{2m_{e,h}T} + \frac{\mu^{e,h}(n_{e,h},T)}{T}\right) \qquad (3.36)$$

and, if we perform the change of the integration variables with the unit Jacobian:

$$\mathbf{k}_e, \quad \mathbf{k}_h \rightarrow \mathbf{k} = \mathbf{k}_e + \mathbf{k}_h, \quad \tilde{\mathbf{k}} = \frac{m_h \mathbf{k}_e - m_e \mathbf{k}_h}{m_e + m_h} \quad (3.37)$$

($\tilde{\mathbf{k}}$ being the momentum of the relative motion) and use the fact that $\tau$ does not depend on $\tilde{\mathbf{k}}$, the integral is factorized. The Gaussian integral over $\tilde{\mathbf{k}}$ gives a normalization factor and finally we obtain

$$-\frac{dn_{e,h}}{dt} = \frac{\hbar n_e n_h}{(m_e + m_h)T} \frac{\operatorname{Im}\tilde{\varepsilon} \; |d^{vc}|^2}{L_w^4} \int \frac{d^2 k}{(2\pi)^2} k |C_{\mathbf{k}}|^2 \exp\left(-\frac{\hbar^2 k^2}{2(m_e + m_h)T}\right). \quad (3.38)$$

If we introduce the "thermal length" $L_T$ defined by

$$L_T^2 \equiv \frac{\hbar^2}{2(m_e + m_h)T}, \quad (3.39)$$

the integral in (3.38) formally coincides with that in (3.22) for the Gaussian wave function. Then the recombination rate may be written as

$$-\frac{dn_{e,h}}{dt} = n_e n_h S \int \frac{d^2 k}{(2\pi)^2} \frac{4\pi L_T^2}{\tau(\mathbf{k})} e^{-k^2 L_T^2} = \frac{\pi a_{B2}^2}{2} \frac{n_e n_h}{\tau_{\text{loc}}}, \quad (3.40)$$

where $\tau_{\text{loc}}$ is the lifetime of the exciton from the previous section with the Gaussian wave function (3.30), localized at the length $L_T$. Of course, the recombination rate of free carriers cannot depend on the exciton Bohr radius and $a_{B2}^2$ in the numerator (3.40) just cancels the analogous factor in $\tau_{\text{loc}}$.

The process under consideration is a bimolecular decay rather then a monomolecular one. To determine the appropriate characteristic time we write down the kinetic equations (3.35) in the form:

$$\frac{dn_e}{dt} = \frac{dn_h}{dt} = -n_e n_h \, g(n_e, n_h, T). \quad (3.41)$$

In the classical limit $g = g(T)$ does not depend on the concentrations at all. Suppose for a while that this situation takes place. If the concentrations are not equal initially, their difference will preserve in time and the solution of the kinetic equations is

$$n_e(t) = \frac{n_h(0) - n_e(0)}{(n_h(0)/n_e(0))\exp[(n_h(0) - n_e(0))gt] - 1}, \quad n_h(t) = (e \leftrightarrow h). \quad (3.42)$$

If the concentrations are equal, $n_e = n_h = n$ (e.g., this is the case for an optically pumped undoped well), then the solution for $n(t)$ has a different character and is given by

$$n(t) = \frac{n(0)}{1 + gn(0)t}. \quad (3.43)$$

This latter case is more relevant for the present problem since the carriers in excess do not participate in the transfer. So, the transfer rate may be defined as $ng$ in the classical case and in the situation when the degeneracy of the plasma becomes significant, $ng(n)$ has the meaning of the instantaneous transfer rate. In the following we analyze the behaviour of the inverse recombination rate constant $g(n, T)^{-1}$ whose dependence on $n$ is then a measure of degeneracy of the electron–hole gas, and which, divided by $n$, gives the effective lifetime.

The results of the numerical integration of (3.35) for the II–VI parameters ($\varepsilon = 6$, $m_e = 0.16 m_0$, heavy holes with $m_h = 0.6 m_0$, X-polarization), 60 Å well, 40 Å barrier, $\tilde{\varepsilon} = 4 + 3i$, are shown in Figure 7. We plot $g(n, T)^{-1}$ as a function of temperature in the range $3\,\mathrm{K} \leqslant T \leqslant 300\,\mathrm{K}$ for several values of the concentrations $10^{11}$ cm$^{-2}$, $3 \cdot 10^{11}$ cm$^{-2}$, $10^{12}$ cm$^{-2}$ along with the limiting case (3.40), which is reached even at low temperatures for concentrations lower than $n_0 = 10^{10}$ cm$^{-2}$ (actually, the curve for $n = n_0$ is indistiguishable from the classical curve in all the temperature range). This is in agreement with the fact that the corresponding "Fermi energies" for electrons and holes are $\mu_0^e(n_0) = 3.5\,\mathrm{K}$, $\mu_0^h(n_0) = 0.93\,\mathrm{K}$. The "thermal length" at $T = 3\,\mathrm{K}$ is $L_T = 140$ Å (correspondingly, for $T = 300\,\mathrm{K}$, $L_T = 14$ Å), which explains the monotony of the classical

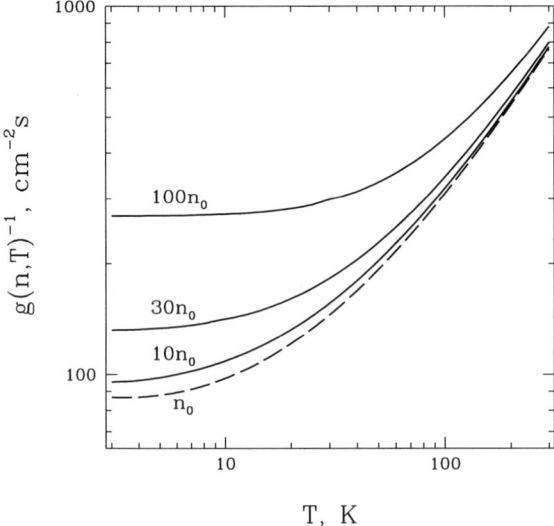

Fig. 7. The inverse recombination rate constant $g(n, T)^{-1}$ (cm$^{-2}$ s) versus temperature $T$ (K) for different carrier concentrations $n_e = n_h = n = 10 n_0$, $30 n_0$, $100 n_0$ (solid lines, $n_0 \equiv 10^{10}$ cm$^{-2}$). The dashed line ($n = n_0$) also represents the classical limit, which is concentration independent. The parameters are $\varepsilon = 6$, $m_e = 0.16 m_0$, $m_h = 0.6 m_0$, $L_w = 60$ Å, $L_b = 40$ Å, $\tilde{\varepsilon} = 4 + 3i$.

dependence on $T$: it is nothing different, but the *left* part of the plot on Figure 5(a), properly renormalized to the case of free carriers.

We see that the corresponding effective lifetime ($\tau = 1/(ng)$, according to the above considerations) is of the order of 1 ns. The relaxation to excitons by means of LO-phonon emission, as already has been mentioned, takes about 1 ps in II–VI materials [24], if the density is not too high ($n \ll 1/a_{B2}^2$). This means that the carriers actually do not even reach the thermal quasi-equilibrium, quickly binding into excitons; the results of the two previous sections are thus much more relevant. However, at higher densities, $n \gtrsim 1/a_{B2}^2$, when the plasma phase is favoured with respect to the excitonic phase [28,29] and, on the other hand, the effective energy transfer rate itself is increased ($1/\tau \propto n$, for arbitrary momentum distribution, not necessarily equilibrium), the rate of the dipole energy transfer from the free carriers may be comparable with rates of other processes, giving some information on the carrier kinetics as well.

Calculations were also performed for the case of III–V-type materials. We do not present them on a separate figure since they have no significantly different features with respect to what has already been mentioned in the previous sections.

## 4. Transfer in the Spherical Geometry: Quantum Dots

This section is dedicated to the estimation of the Förster transfer rate in the spherical geometry [4,5]. The donor is a semiconductor quantum dot (QD), the acceptor is the organic material, surrounding it. The explicit analytical expression for a wave function of an electron–hole pair in a QD is not known for an arbitrary relation between the dot radius $R$ and the Bohr radius $a_B$. To calculate the wave function numerically would not be a justified effort, since the structure itself has yet not been realized experimentally, and what one needs is just a simple estimate aimed at understanding the general tendencies. Thus in this section, after the derivation of the general expression for the transfer rate in the spherical geometry, the two limiting cases of strong and weak confinement (when $R \ll a_B$ and $R \gg a_B$ respectively) are considered, and then a simple variational interpolation between these limiting cases is made.

The issue of the electron/hole relaxation in quantum dots deserves some attention, since it happens in a way qualitatively different from that in higher-dimensionality nanostructures due to the discreteness of states in a QD. In particular, under certain circumstances this relaxation may be suppressed as predicted [30] and observed [31]. Thus the problem of the energy transfer not only from the lowest excited state, but also from higher levels of the dot is relevant. We also show that the carrier relaxation itself may be affected by Förster transfer.

### 4.1. THE ELECTRIC FIELD IN THE ORGANICS IN THE SPHERICAL GEOMETRY

The nanostructure considered here (Figure 8) consists of a spherical QD of radius $R$ with dielectric constant $\varepsilon$, surrounded by a concentric semiconductor barrier with the thickness $L_b$ and the same dielectric constant. The barrier is assumed to be infinitely high, thus carriers cannot penetrate it. The space outside the barrier is filled with the organic substance with dielectric constant $\tilde{\varepsilon}$. As in Section 3, $\varepsilon$ is real, $\tilde{\varepsilon}$ is complex. The radius $R$ is considered to be large enough (as compared to the semiconductor lattice constant) for the envelope function approximation to be valid.

In the spherical geometry the states of the electron–hole pair may be classified by the total angular momentum quantum numbers $l, m$, which determine the angular dependence of the electron–hole pair wave function

$$\Psi(\mathbf{r}, \mathbf{r}) \propto Y_{lm}(\theta, \varphi),$$

where $Y_{lm}(\theta, \varphi)$ are the spherical harmonic functions. Putting the $z$-axis along the vector $\mathbf{d}^{vc}$, for the polarization we may write

$$P_z(\mathbf{r}) = P^{(r)}(r) Y_{lm}(\theta, \varphi), \tag{4.1}$$

which corresponds to the charge density:

$$\rho(\mathbf{r}) = -\operatorname{div}\mathbf{P}(r) = -\frac{\partial P}{\partial z} = \rho^{(r)}_{l-1}(r) Y_{l-1,m} + \rho^{(r)}_{l+1}(r) Y_{l+1,m}. \tag{4.2}$$

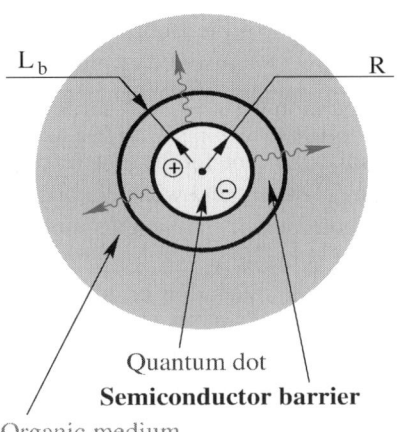

Fig. 8. A sketch of the spherical structure under study.

After some algebra the functions $\rho_{l\mp1}^{(r)}(r)$ may be related to $P^{(r)}(r)$:

$$i\rho_{l-1}^{(r)}(r) = \sqrt{\frac{l^2 - m^2}{4l^2 - 1}} \left[ \frac{dP^{(r)}}{dr} + (l+1)\frac{P^{(r)}}{r} \right], \quad (4.3)$$

$$i\rho_{l+1}^{(r)}(r) = \sqrt{\frac{(l+1)^2 - m^2}{4(l+1)^2 - 1}} \left[ -\frac{dP^{(r)}}{dr} + l\frac{P^{(r)}}{r} \right]. \quad (4.4)$$

The electrostatic potential $\phi(\mathbf{r})$, satisfying the Poisson equation, may be also decomposed into the spherical components analogously to Eq. (4.2). The terms with $Y_{l-1,m}$ and $Y_{l+1,m}$ separate in the Poisson equation and one has two equations for the radial parts of the potential $\phi_\lambda^{(r)}(r)$ with $\lambda = l-1, l+1$:

$$\frac{1}{r^2}\frac{d}{dr}\left(r^2\frac{d\phi_\lambda^{(r)}}{dr}\right) - \frac{\lambda(\lambda+1)}{r^2}\phi_\lambda^{(r)} = -4\pi\rho_\lambda^{(r)}(r), \quad (4.5)$$

where $\rho_\lambda^{(r)}(r)$ vanishes at $r > R$ (as $P(r)$ does). A general solution in a region, free of charges, corresponds to the $2^\lambda$-pole potential [32]:

$$\phi_\lambda^{(r)}(R < r < R + L_b) = \sqrt{\frac{4\pi}{2\lambda+1}} \left[ \frac{Q_\lambda^{\text{exc}}}{\varepsilon r^{\lambda+1}} + Q_\lambda^{\text{surf}} r^\lambda \right], \quad (4.6)$$

$$\phi_\lambda^{(r)}(r > R + L_b) = \sqrt{\frac{4\pi}{2\lambda+1}} \frac{Q_\lambda^{\text{eff}}}{r^{\lambda+1}}, \quad (4.7)$$

with some constants $Q_\lambda$. The coefficient $Q_\lambda^{\text{exc}}$ is just the *bare* $2^\lambda$-pole moment of the exciton polarization:

$$Q_{l+1}^{\text{exc}} = -i\sqrt{(l+1)^2 - m^2} \sqrt{\frac{4\pi}{2l+1}} \int_0^R r^l P^{(r)}(r) r^2 \, dr, \quad (4.8)$$

while the contribution of the moment $\lambda = l - 1$ turns out to be identically zero when integrated over the dot volume, due to the fact that $P^{(r)}(r)$ vanishes at $r = R$. The background dielectric screening by the polarization charges inside the dot leads to the correction $Q_{l+1}^{\text{exc}} \to Q_{l+1}^{\text{exc}}/\varepsilon$, taken into account in the expression (4.6). The coefficient $Q_{l+1}^{\text{surf}}$ is determined by the surface polarization charges at the interface $r = R + L_b$ between the media with different dielectric constants $\varepsilon$ and $\tilde{\varepsilon}$. The coefficient $Q_{l+1}^{\text{eff}}$ is the effective multipole, which determines the field outside the dot. One may relate $Q_{l+1}^{\text{surf}}$ and $Q_{l+1}^{\text{eff}}$ to $Q_{l+1}^{\text{exc}}$ requiring the continuity of $\phi_{l+1}^{(r)}(r)$ and $\varepsilon(r) d\phi_{l+1}^{(r)}(r)/dr$ at $r = R + L_b$, analogously to the procedure of Ref. [14], which gives

$$Q_{l+1}^{\text{eff}} = \frac{2l+3}{(l+1)\varepsilon + (l+2)\tilde{\varepsilon}} Q_{l+1}^{\text{exc}}. \quad (4.9)$$

Having calculated the potential, one obtains the decay rate[2]:

$$\Gamma = \frac{1}{\tau} = \frac{\operatorname{Im}\tilde{\varepsilon}}{2\pi\hbar}\int_{|\mathbf{r}|>R+L_b}\left(\nabla\phi^*\cdot\nabla\phi\right)d^3\mathbf{r}$$

$$= \frac{2l+4}{2l+3}\frac{\operatorname{Im}\tilde{\varepsilon}}{\hbar}\frac{|Q^{\text{eff}}_{l+1}|^2}{(R+L_b)^{2l+3}}. \tag{4.10}$$

It is interesting to compare this expression with that for the *radiative* lifetime of a multipole. In vacuum it is given by [33]:

$$\Gamma^{(\text{rad})} = \frac{2l+4}{(l+1)(2l+3)[(2l+1)!!]^2}\left(\frac{\omega_D}{c}\right)^{2l+3}\frac{|Q_{l+1}|^2}{\hbar},$$

where $Q_{l+1}$ is the transition multipole matrix element. The same expression for the multipole in a medium with the real dielectric constant $\bar{\varepsilon}$ may be obtained by rescaling $c \to c/\sqrt{\bar{\varepsilon}}$, $|Q_{l+1}|^2 \to |Q_{l+1}|^2/\bar{\varepsilon}$, and reads as[3]:

$$\Gamma^{(\text{rad})} = \frac{2l+4}{(l+1)(2l+3)[(2l+1)!!]^2}\left(\frac{\sqrt{\bar{\varepsilon}}\omega_D}{c}\right)^{2l+3}\frac{|Q_{l+1}|^2}{\hbar\bar{\varepsilon}},$$

where $\bar{\varepsilon}$ is real (otherwise the radiative decay cannot be separated from the Förster transfer). In our configuration one should substitute there $Q_{l+1} = \bar{\varepsilon}\bar{Q}^{\text{eff}}_{l+1}$, where $\bar{Q}^{\text{eff}}_{l+1}$ is obtained from the same expression (4.9) using $\bar{\varepsilon}$ instead of $\tilde{\varepsilon}$ (as long as $\sqrt{\bar{\varepsilon}}\omega_D R/c \ll 1$), which follows from comparison of the Coulomb fields. Thus, we obtain

$$\Gamma = \Gamma^{(\text{rad})}(l+1)[(2l+1)!!]^2\frac{\operatorname{Im}\tilde{\varepsilon}}{\bar{\varepsilon}}\left[\frac{c}{\sqrt{\bar{\varepsilon}}\omega_D(R+L_b)}\right]^{2l+3}\frac{|Q^{\text{eff}}_{l+1}|^2}{|\bar{Q}^{\text{eff}}_{l+1}|^2}. \tag{4.11}$$

---

[2] The easy way of calculation is to reduce the bulk integral in (4.10) to a surface one employing the formula

$$\int_{\mathcal{O}}\nabla f(\mathbf{r})\cdot\nabla g(\mathbf{r})\,dV + \int_{\mathcal{O}}f(\mathbf{r})\nabla^2 g(\mathbf{r})\,dV = \int_{\partial\mathcal{O}}f(\mathbf{r})\nabla g(\mathbf{r})\cdot d\mathbf{s}$$

for two arbitrary functions $f(\mathbf{r})$ and $g(\mathbf{r})$, where the integrals on the l.h.s. are performed over some region $\mathcal{O}$ in space (the volume element $dV$), and the integral in the r.h.s. is over the boundary $\partial\mathcal{O}$ with the element of the surface $d\mathbf{s}$, oriented outside. This formula is just the Gauss–Ostrogradsky theorem for the vector field $f\nabla g$. In our case the region $\mathcal{O}$ is defined by $|\mathbf{r}| > R + L_b$, consequently $d\mathbf{s}$ is directed to the center $\mathbf{r} = 0$.

[3] The latter substitution may be understood if one thinks of the radiative decay as a result of the interaction of the multipole with the electric field produced by itself: in the medium the amplitude of the electric field, produced by a distribution of charges, is $\bar{\varepsilon}$ times less than in vacuum.

For $l = 0$ (the dipole transition) it gives simply

$$\Gamma = \Gamma^{(\text{rad})} \frac{\text{Im}\,\tilde{\varepsilon}}{\bar{\varepsilon}} \left[ \frac{c}{\sqrt{\bar{\varepsilon}}\,\omega_D(R + L_b)} \right]^3 \frac{|Q_1^{\text{eff}}|^2}{|\tilde{Q}_1^{\text{eff}}|^2}. \tag{4.12}$$

If one takes the light wavelength corresponding to the frequency $\omega_D$ in vacuum to be 314 nm, $R + L_b = 5$ nm, $\bar{\varepsilon} = 4$, then the coefficient in the square brackets is equal to 5, which gives two orders of magnitude for the dipole transition. Thus, if the radiative relaxation is the dominant electron–hole pair decay mechanism in the absence of the Förster transfer, it will be completely suppressed by the transfer. However, it is not necessarily the case, so one has to perform numerical estimates.

One may want to average (4.10) over $m$, which makes sense since the energy does not depend on the magnetic quantum number $m$. This is certainly relevant for the case when the dot is pumped electrically, while for the case of excitation by polarized light one should choose the state, corresponding to the given polarization. Such averaging corresponds just to the substitution in (4.8):

$$\sqrt{(l+1)^2 - m^2} \to \sqrt{(l+1)(2l/3 + 1)}, \tag{4.13}$$

since the radial part of the polarization cannot depend on the quantum number $m$.

### 4.2. The Wave Function: Limiting Cases

The exciton wave function $\Psi(\mathbf{r}_e, \mathbf{r}_h)$ is determined by two interactions: (i) the confinement potential, which we consider to be infinite for $r > R$ and zero at $r < R$, and (ii) the Coulomb attraction of the electron and the hole. The characteristic length scales corresponding to these interactions are $R$—the quantum dot radius, and $a_B$—the exciton bulk Bohr radius. Solution of the Schrödinger equation for arbitrary $R$ and $a_B$ is quite a complicated problem (see, e.g., [34,35]), but the situation becomes much simpler in two limiting cases.

If $R \ll a_B$ (strong confinement), Coulomb interaction may be completely neglected and the electron–hole pair wave function will be simply the product of two one-particle wave functions. Each one-particle state is labeled by three quantum numbers: the orbital quantum number $l = 0, 1, \ldots$, the magnetic quantum number $m = -l, \ldots, l$, and the principal quantum number $n = 1, 2, \ldots$ [11]. Denoting this set by a single symbol $\nu$, one may write

$$\Psi_{\nu_e, \nu_h}(\mathbf{r}_e, \mathbf{r}_h) = \chi_{\nu_e}(\mathbf{r}_e) \chi_{\nu_h}(\mathbf{r}_h), \tag{4.14}$$

where the single-particle wave function is given by [11]

$$\chi_{nlm}(\mathbf{r}) \equiv \chi_\nu(\mathbf{r}) = \sqrt{\frac{2}{R^3}} \frac{j_l(\alpha_{ln} r/R)}{j_{l+1}(\alpha_{ln})} Y_{lm}(\theta, \varphi), \tag{4.15}$$

where $j_l(x)$ is the $l$th spherical Bessel function, $\alpha_{ln}$ is its $n$th zero ($n = 1, 2, 3, \ldots$).

However, to apply the results of the previous section, we have to form the linear combinations corresponding to states with definite total angular momentum. The new set of quantum numbers is $\{l_e, n_e, l_h, n_h, l, m\}$ and the radial part of the corresponding wave function is expressed in terms of Wigner $3j$-symbols as

$$\psi_{lm}^{(r)}(r) = \frac{2 \cdot (-1)^{(l_e+l_h+l)/2}}{R^3}$$

$$\times \begin{pmatrix} l_e & l_h & l \\ 0 & 0 & 0 \end{pmatrix} \sqrt{\frac{(2l_e+1)(2l_h+1)}{4\pi}}$$

$$\times \frac{j_{l_e}(\alpha_{l_e n_e} r/R) \, j_{l_h}(\alpha_{l_h n_h} r/R)}{j_{l_e+1}(\alpha_{l_e n_e}) \, j_{l_h+1}(\alpha_{l_h n_h})}. \quad (4.16)$$

For $l_e = l_h = l = 0$ this expression gives the dipole moment and the transfer rate

$$Q_1^{\text{eff}} = \frac{3d^{vc}}{\varepsilon + 2\tilde{\varepsilon}}, \quad \frac{1}{\tau} = \frac{12 \, \text{Im}\, \tilde{\varepsilon}}{|\varepsilon + 2\tilde{\varepsilon}|^2} \frac{1}{\hbar} \frac{|d^{vc}|^2}{(R+L_b)^3}. \quad (4.17)$$

If $R \gg a_B$ (weak confinement), then the distance between the single-particle levels in the spherical potential well (even for the lighter particle) is much less than the bulk exciton binding energy. In this case the exciton may be considered a rigid particle moving in a spherical well. The relative motion of the electron and the hole and the center-of-mass motion are effectively separated and the wave function is factorized. For $1s$-exciton state

$$\psi_{v,1s}(\mathbf{r}, \mathbf{r}) = \frac{1}{\sqrt{\pi a_B^3}} \chi_v(\mathbf{r}). \quad (4.18)$$

For $l = 0$ one obtains

$$Q_1^{\text{eff}} = \frac{1}{\pi} \left(\frac{2R}{a_B}\right)^{3/2} \frac{3d^{vc}}{\varepsilon + 2\tilde{\varepsilon}},$$

$$\frac{1}{\tau} = \frac{96 \, \text{Im}\, \tilde{\varepsilon}}{\pi^2 |\varepsilon + 2\tilde{\varepsilon}|^2} \frac{1}{\hbar} \frac{|d^{vc}|^2}{a_B^3} \left(\frac{R}{R+L_b}\right)^3. \quad (4.19)$$

One may note the similarity of the expressions (4.17), (4.19) to those for localized excitons in a quantum well (3.24), (3.27), taking into account that in Section 3.2 the confinement was strong in one dimension and weak in two dimensions (there it was assumed that the well width was $L_w \ll a_B$, while the 2D localization length was $L \gg a_B$).

As in Section 3, for the numerical estimations we use the parameters, typical for II–VI semiconductors: $a_B \simeq 50$ Å, $d^{vc} \simeq 12$ D, $\varepsilon \simeq 6$, $\tilde{\varepsilon} \simeq 4 + 3i$ for the organics. Having fixed $L_b = 30$ Å, in Figure 9(a) we plot the transfer times for several strongly confined states, averaged over the magnetic quantum number, as

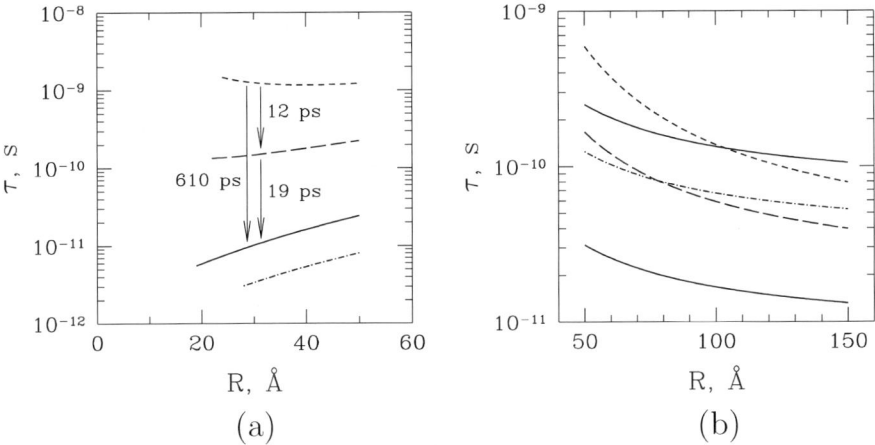

Fig. 9. The Förster transfer time corresponding to the interband transition for II–VI semiconductors: (a) in the strong confinement limit from the states $\{0, 1, 0, 1\}$ (solid line), $\{0, 1, 1, 1\}$ (long-dashed line), $\{0, 1, 2, 1\}$ (short-dashed line), $\{1, 1, 1, 1\}$ ($l = 0$, dash-dotted line) versus the dot radius $R$, $L_b = 30$ Å, the vertical arrows show schematically the intraband transition times (see the text for the details); (b) in the weak confinement limit from the states $\{0, 1\}$ ($1s$—the lower solid line, $2s$—the upper solid line), and $1s$ states $\{1, 1\}$ (long-dashed line), $\{2, 1\}$ (short-dashed line), $\{0, 2\}$ (dash-dotted line) versus the dot radius $R$, $L_b = 30$ Å.

mentioned in the end of Section 4.1, as a function of the dot radius $R$. These states are $\{l_e = 0, n_e = 1, l_h = 0, n_h = 1\}$ (dipole transition, solid line), $\{0, 1, 1, 1\}$ (quadrupole transition, long-dashed line), $\{0, 1, 2, 1\}$ (octupole transition, short-dashed line)—the lowest ones, and we take also the $l = 0$ component of $\{1, 1, 1, 1\}$ (the seventh level), the next one after the lowest state, possessing a nonzero dipole moment (dash-dotted line). The upper limit of $R$ is taken to be $a_B$, since for $R > a_B$ the strong-confinement approximation definitely breaks down. The lower limit of $R$ is different for each state and is taken to be the radius, at which the confinement energy of the state reaches 1 eV, since higher confinement energies are not realistic. We see that for the lowest state the transfer time is quite short – of the order of 10 ps. The states with larger $l$ have longer times, one may roughly say that every unit of $l$ "costs" about an order of magnitude of $\tau$. This fact, however, might even not spoil the efficiency of the energy transfer since the carrier radiative recombination time will also increase for higher multipole transitions, provided that other relaxation channels are quenched.

We have already mentioned in the beginning of this section that the carrier relaxation in quantum dots may be strongly inhibited due to the discreteness of states. In this situation, any material, not necessarily an organic one, with nonzero absorption at the frequency of an *intraband* transition may affect the intraband relaxation of carriers. The estimation of the relaxation time due to the Förster transfer may be performed along the same lines as for the interband transition, starting from the point that the matrix element of the charge density for the electron (hole) transition from the state $\{l_1, m_1, n_1\}$ to $\{l_2, m_2, n_2\}$ is given by Eq. (2.15):

$$\rho^{exc}(\mathbf{r}) = \mp e \, \chi^*_{l_2 m_2 n_2}(\mathbf{r}) \, \chi_{l_1 m_1 n_1}(\mathbf{r}). \tag{4.20}$$

This expression should be expanded into spherical harmonics $Y_{lm}$ with $l = |l_1 - l_2|, \ldots, l_1 + l_2$, $m = m_1 - m_2$. The intraband transfer rate is given by the same formulas (4.9), (4.10) with $l + 1$ substituted by $l$ (since there is no differentiation $\partial/\partial z$), but the bare exciton multipole moment, corresponding to the transition rate, averaged over $m_1$ and summed over $m_2$ (see the end of the previous section), is given by

$$\bar{Q}_l^{exc} = 2eR^l \sqrt{\frac{2l_2 + 1}{2l + 1}} \begin{pmatrix} l_1 & l_2 & l \\ 0 & 0 & 0 \end{pmatrix}$$
$$\times \int_0^1 dx \, x^{l+2} \, \frac{j_{l_1}(\alpha_{l_1 n_1} x)}{j_{l_1+1}(\alpha_{l_1 n_1})} \frac{j_{l_2}(\alpha_{l_2 n_2} x)}{j_{l_2+1}(\alpha_{l_2 n_2})}. \tag{4.21}$$

The intraband transitions correspond to the infrared spectral range, where the absorption is usually much weaker than in the visible range. Hence, we set $\tilde{\varepsilon} = 4 + 0.3i$ and evaluate the intraband transition times for the transitions $\{l_1 = 1, n_1 = 1\} \to \{l_2 = 0, n_2 = 1\}$ (dipole transition—19 ps), $\{2, 1\} \to \{1, 1\}$ (dipole transition—12 ps), $\{2, 1\} \to \{0, 1\}$ (quadrupole transition—610 ps) for $R = 30$ Å, $L_b = 30$ Å, shown by the vertical arrows in Figure 9(a) (in our model the times are the same for electrons and holes). As we see, dipole transitions are much more intensive than higher multipole ones and the times may be relatively short (a few tens of ps). Thus, if other relaxation processes are quenched, the Förster transfer may play some role.

In this connection one may also consider a single-carrier relaxation process, related both to the above considered Förster transfer and to the Auger relaxation, dealt in Ref. [36]. If one considers an array of quantum dots of characteristic size $R$, separated by distance $r$ from each other on the wetting layer of thickness $L_w$ (no surrounding organics is assumed), then the following process may occur. Two carriers in the neighbouring dots interact via Coulomb interaction (the first nonvanishing multipole component is the dipole–dipole one), then as a result of this interaction, one of the carrier relaxes to a lower state in the dot, while the other is promoted to the wetting layer continuum. This process is adequately described by the Fermi Golden Rule and a simple estimate gives the corresponding time of the order of

$$\tau \simeq \frac{\hbar^3 \varepsilon^2}{me^4} \frac{r^6}{R^5 L_w}. \tag{4.22}$$

Estimating the first fraction as 10–100 fs, one may obtain numbers of the order of 10–100 ps in a favorable geometry. However, a detailed treatment is beyond the scope of this chapter.

Now we return to the energy transfer when the electron–hole pair is annihilated (interband transition), considering the case $R \gg a_B$. In Figure 9(b) we plot the transfer times for several weakly confined $1s$-exciton states, the lowest ones— $\{l = 0, n = 1\}$ (the lower solid line), $\{1, 1\}$ (long-dashed line), $\{2, 1\}$ (short-dashed line), $\{0, 2\}$ (dash-dotted line), and the $2s$-state with $\{l = 0, n = 1\}$ (the upper solid line) as a function of $R$ at fixed $L_b = 30$ Å. The lower limit of $R$ is taken to be $a_B$, as the upper limit we choose 150 Å, since at larger dot radii the times do not change significantly. From the plots we see that (i) the transfer from the $2s$-state is less rapid than that from the $1s$ (by a factor of 8), (ii) the difference between dipole, quadrupole and octupole transitions is not very large (due to the fact that the ratio $R/(R + L_b)$ is not much smaller than 1) and (iii) the dipole transfer from higher excited states with $l = 0$ is slower due to a partial cancellation in the radial integrals.

We have also performed similar transfer rate calculations for parameters typical for III–V materials (e.g., InGaAs): $a_B \simeq 160$ Å, $d^{vc} \simeq 50$ D, $\varepsilon \simeq 9$. Figure 10(a) shows the transfer times for the same states as in Figure 9(a) as a function of $R$ for $L_b = 30$ Å for the strong-confinement limit. The times are shorter than those for II–VI semiconductors since the latter have smaller dipole moments $d^{vc}$. We also plot the transfer time as a function of the barrier thickness $L_b$ for $R = 100$ Å in Figure 10(b). As one may expect, the higher multipoles, whose electric field

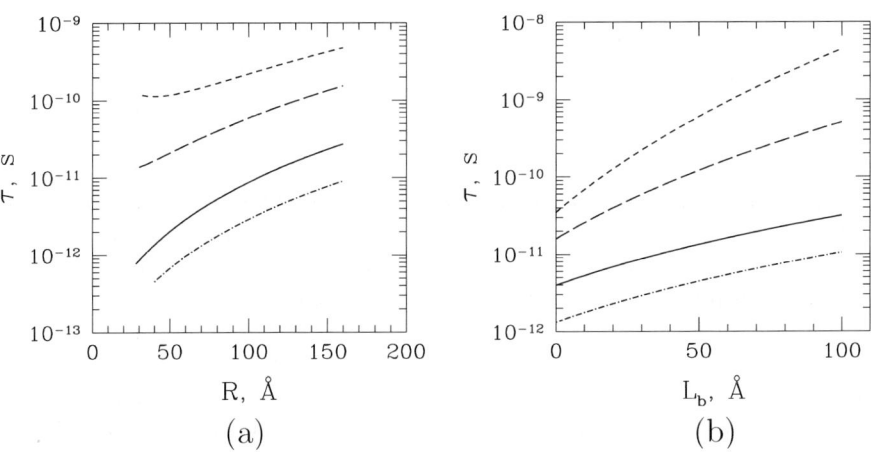

Fig. 10. The Förster transfer time corresponding to the interband transition for III–V semiconductors in the strong confinement limit from the states $\{0, 1, 0, 1\}$ (solid line), $\{0, 1, 1, 1\}$ (long-dashed line), $\{0, 1, 2, 1\}$ (short-dashed line), $\{1, 1, 1, 1\}$ ($l = 0$, dash-dotted line): (a) versus the dot radius $R$, $L_b = 30$ Å (analogously to Figure 9(a)); (b) versus the barrier thickness $L_b$, for $R = 100$ Å.

decreases more rapidly in space, are more sensitive to the barrier thickness than the lower ones.

At this stage it is worth mentioning an experiment where the evidence was found for the Förster transfer from porous silicon to an organic dye that was put into the pores [37]. The silicon was pumped optically, then its luminescence was quenched on the time scale of several ten nanoseconds. One can roughly estimate the transfer rate using the expression (4.12). Setting $R \sim 20$ Å, $\hbar\omega_D \sim 2$ eV, $\tau_{rad} \sim 10^{-4}$ s$^{-1}$ [38], Im $\tilde{\varepsilon} \sim 1$, one obtains $\tau \sim 1$ ns. It is shorter then that observed, which may be attributed to the simple fact that the pores occupied only 65% of the material, thus the expressions for the decay rate for a single dot immersed into an infinite organic matrix certainly overestimate it.

### 4.3. THE WAVE FUNCTION: VARIATIONAL CALCULATION

To investigate the crossover region between the two limiting cases, that is $R \sim a_B$, we take a simple variational wave function for the lowest excited state:

$$\psi_{a,b}(\mathbf{r}_e, \mathbf{r}_h) = A(a, b_e, b_h) \frac{\chi_0(r_e)}{\sqrt{\chi_0(r_e) + b_e}} \frac{\chi_0(r_h)}{\sqrt{\chi_0(r_h) + b_h}}$$
$$\times \exp\left[-\frac{|\mathbf{r}_e - \mathbf{r}_h|}{a}\right], \quad (4.23)$$

where $a$, $b_e$, and $b_h$ are positive variational parameters, $A(a, b_e, b_h)$ is the normalization coefficient. The wave function $\chi_0(r)$ is the one given by the equation (4.15) for $l = m = 0, n = 1$:

$$\chi_0(r) = \frac{1}{\sqrt{2\pi R}} \frac{\sin(\pi r/R)}{r}. \quad (4.24)$$

The wave function (4.23) is slightly more general than one proposed long ago by Kayanuma [39]. The latter function did not contain the denominators with the square roots as in the function (4.23) and depended on the single parameter $a$. We have chosen the wave function (4.23) since it reproduces correctly the shape of the *polarization* $\psi(\mathbf{r}, \mathbf{r})$ in both limiting cases: for $R \ll a_B$ both $a$ and $b$'s become large and one obtains the wave function (4.14), while for $R \gg a_B$ we have $a \simeq a_B$ (the Wannier exciton "is formed"), $b_e, b_h \sim a_B/R^{5/2}$ (the surface corrections to the exciton wave function) and one gets the same polarization as for the wave function (4.18). The exponential factor is very important since it enhances the probability of finding the electron and hole at the same point, which affects the polarization. The variable "exciton radius" $a$ measures the role of Coulomb correlations between the electron and the hole. The Hamiltonian we consider is given by

$$\hat{H} = \frac{\hat{\mathbf{p}}_e^2}{2m_e} + \frac{\hat{\mathbf{p}}_h^2}{2m_h} + V(r_e) + V(r_h) - \frac{e^2}{\varepsilon_0|\mathbf{r}_e - \mathbf{r}_h|}, \quad (4.25)$$

where $V(r)$ is zero for $r < R$ and infinity for $r > R$, $\varepsilon_0$ is the semiconductor background dielectric constant at low frequencies.

To test the wave function (4.23), we compare the energies obtained from it for different values of $R/a_B$ with those obtained by the exact diagonalization (Ref. [34]) for $m_e = m_h$. The discrepancy is quite small (less than 10% of the excitonic Rydberg). We were not able to repeat the calculation for the value $m_e/m_h = 0.01$ to compare it with the results of Ref. [34], due to computational problems (for strongly different masses the energy to minimize becomes a very inconvenient function to treat by standard methods). However, having performed the calculations for $m_e/m_h = 0.29$ (the case of CdSe, heavy holes), we see that though the energy is sensitive to the mass ratio, the dipole moment is not (the two corresponding curves in Figure 9 would merge). Thus, we may hope that the wave function (4.23) reproduces the dipole moment of the dot reasonably well.

Figure 11 shows the transfer times from the lowest state of e–h pair (discontinuous solid line) and the second excited state (discontinuous dashed line) in both limiting cases versus the dot radius for the same parameters as in Figure 9. The lowest state is the $\{0, 1, 0, 1\}$ state for the strong confinement and the $\{0, 1\}$ state of $1s$-exciton for the weak confinement, while the second excited state is $\{0, 1, 1, 1\}$ and $\{1, 1\}$ respectively. The continuous solid line represents the result of the variational calculation for the lowest state. For $R \sim a_B$ it gives even a more optimistic result. The behaviour of the variational curve at $R > 2a_B$ is somewhat unexpected, however, the detailed inspection of the numbers shows that this curve

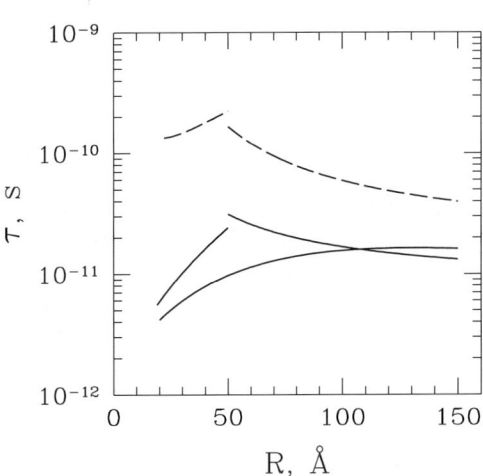

Fig. 11. The transfer times from the lowest state of e–h pair (discontinuous solid line) and the second excited state (discontinuous dashed line) in both limiting cases versus the dot radius for the same parameters as in Figure 9. The continuous solid line represents the result of the variational calculation for the lowest state.

has a maximum at $R \simeq 3a_B$ and then converges to the limiting curve. We cannot be sure whether this feature is indeed present in the exact solution, or is just due to the poorness of the chosen function. In any case, the discrepancy is not very large (the factor is about 1.3). So, one can say that the estimations using the limiting expressions for the strong and weak confinement describe the situation reasonably well (within a factor of 3).

## 5. Application to Light-Emitting Devices

In this section we discuss potential application of the considered phenomenon of energy transfer to light-emitting devices (LEDs) [6]. The progress in the development of organic light-emitting devices was considerable in the past few years. In all such devices the fundamental role is played by electroluminescence, the generation of light by electric excitations. Already in early studies [40] it was established that the process responsible for the electroluminescence requires injection of electrons from one electrode and of holes from the other, transport of one or both charges, capture of oppositely charged carriers on the same molecule (or recombination center), and radiative decay of the resulting excited electron–hole state. For inorganic semiconductors, where the recombining electron–hole state may be a Wannier–Mott exciton, all the above mentioned processes are investigated and well documented [41]. In organics, in contrast to inorganic semiconductors, scientists have met many problems in the use of electroluminescence for the creation of devices conceptually similar to what was done with the use of semiconductor materials. The main reasons are the small mobility of carriers and strong chemical interaction between organics and metals which prevents the injection of charges into organics. Due to the small mobility of charge carriers in organic materials the current flow is bulk-limited, principally through the built-up of a space charge.

In principle, one may think of circumventing these problems by combining comparatively good transport properties of semiconductors (e.g., pumped electrically) and good light-emitting properties of organic substances in a single hybrid nanostructure. In a semiconductor, as a result of electrical pumping, electron–hole pairs appear which transform very rapidly into free and localized 2D Wannier–Mott excitons. As discussed in the previous sections, the resonant Coulomb interaction between the semiconductor and the organics in combination with the fast dephasing, common for many organic substances (e.g., due to the scattering by phonons) will result in the nonradiative transfer of the semiconductor excitation energy to the organics. The key parameter, determining the efficiency of this transfer, is its characteristic time, as compared to other decay channels of the semiconductor excitation. As the results of the previous sections show, the energy transfer mechanism we consider will be fast enough to efficiently quench

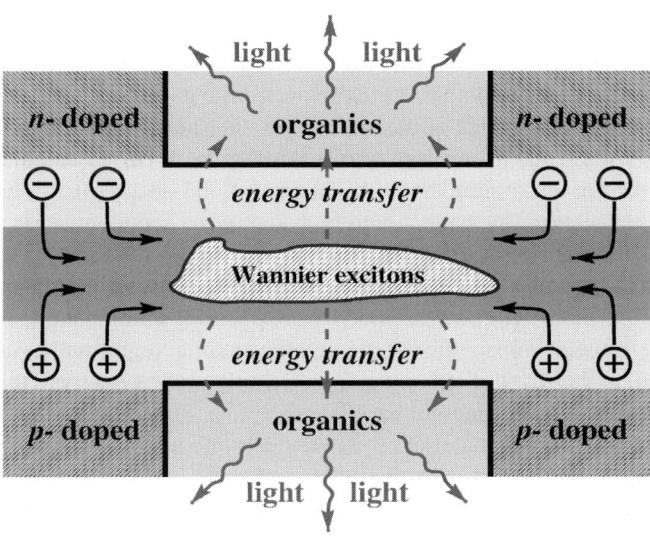

Fig. 12. The schematic view of the hybrid light-emitting device proposed. The carriers are injected from $p$- and $n$-doped regions into the semiconductor part of the structure, where they form excitons. The latter are transferred nonradiatively to the organic light-emitting medium, where they eventually recombine radiatively.

the Wannier–Mott exciton luminescence and to turn on the organic molecule light emission, assuming that the luminescence quantum yield of the organic is high. Usually, the relaxation of the electronic excited states in the organics is much faster then the back transfer rate.

A possible configuration for such a device based on a semiconductor quantum well is sketched in Figure 12. Electrons and holes are injected into the QW from $n$- and $p$-doped regions respectively. In the QW they quickly form excitons (the transfer rate from uncorrelated electrons and holes is low, as shown in Section 3). The energy transfer from excitons is efficient in a rather wide region of wave vectors, determined by the geometry of the structure. In particular, for a thermalized exciton population with the temperature of about 100 K the typical wave vector would be several $10^{-6}$ cm$^{-1}$, e.g., for II–VI semiconductors the corresponding transfer time is several tens of picoseconds. Another important point is that the localization of excitons does not have any dramatic effect on the energy transfer time. Since the typical lifetime of Wannier–Mott excitons in the absence of the nonradiative (Förster) transfer is of the order of several hundreds picoseconds in II–VI materials, the most part of the excitations in the semiconductor may be transferred to the organics. The energy transfer rate decreases with increasing thickness of the barrier. It means that in order to optimize energy transfer the barrier should be taken as thin as possible, limited only by the requirement of providing the carrier confinement and good structural quality.

The model used throughout this chapter was quite general, giving the possibility to estimate the magnitude of the effect for a wide class of materials. However, it was simplified, and, if one wants to make a more accurate calculation for a specific structure, the model should be improved and adapted for the particular materials used. Especially, care should be taken of the semiconductor part of the structure, as the transfer is sensitive to the nature of the excited states (as one can see from the comparison of the results for excitons and free electrons and holes). The organic part may still be described by the phenomenological dielectric function, however, specific properties of the organic material (e.g., such as anisotropy) may also affect the result. Such developments only make sense, of course, when related to a specific experimental work.

One of the promising materials for optoelectronic applications is gallium nitride and the alloy $Ga_{1-x}Al_xN$. The energy gap of GaN is about $E_g \approx 3.5$ eV, corresponding to the near ultraviolet, for AlN it is about 6.2 eV. Combined with the organics having the Stokes shift of a fraction of eV, GaN may be used to produce violet or blue light. A peculiar feature of nitrides is large spontaneous polarization (permitted by the symmetry for the wurtzite crystal structure) and large piezoelectric coupling constants [42]. For quantum wells it results in the constant electric field inside the well, appearing due to the spontaneous polarization gradients at the interfaces and to the strain of the well material. Consequently, the electric field tends to separate the electron and the hole spatially (in the growth direction), which decreases the oscillator strength of the excitonic transition. The electric field $\mathcal{E}_0 \sim 400$ kV/cm [43] is too strong to treat the problem perturbatively (for a QW of the width $L_w = 25$ Å it gives $\mathcal{E}_0 L_w = 100$ mV). Alternative approaches are to use a variational function [9] or to impose a fictitious periodicity as proposed by Singh [44], which works for quantum wells thicker than 15 Å [43].

Wide band-gap II–VI semiconductors are also frequently used [17]. If one is interested in the nearest ultraviolet range, ZnS ($E_g \simeq 3.7$ eV) and its alloys, such as $Zn_{1-x}Cd_xS$, $ZnS_{1-x}Se_x$, $ZnS_{1-x}Te_x$ may be useful (all having smaller gaps). Each of the materials has its individual features. For example, ZnS has two modifications – cubic (the usual one) and hexagonal, while CdS has hexagonal lattice. This fact, as well as the properties of the substrate, should be taken into account when considering the properties of the alloy. The characteristic feature of the $ZnS_{1-x}Se_x$ is the small conduction band offset (5% of the gap difference [17]). This makes the electron confinement weak and the Coulomb attraction between the electron and the hole is very important in determining the exciton properties also in the growth direction (which was totally neglected in the considerations of Section 3). A peculiar property of $ZnS_{1-x}Te_x$ alloy is strong intrinsic self-trapping of excitons due to the lattice distortion they induce around Te atoms. This results in large Stokes shift between the energy gap, seen in the absorption, and the luminescence energy [45] (in the latter work Stokes shifts > 100 meV for $ZnSe_{0.99}Te_{0.01}$ were observed). An analogous phenomenon was observed in $ZnS_{1-x}Se_x$ [46], and even at low Te concentrations the luminescence energy was not higher than 2.5 eV (i.e., the Stokes shift exceeded 1 eV). In principle, this process may affect the relative motion of the electron and the hole, thus the effect of the self-trapping on the oscillator strength cannot be neglected *a priori* either. The self-trapping introduces also an extra time scale—that of the lattice relaxation in the semiconductor, which should be compared with the Förster transfer time.

Silicon is the second most widespread chemical element on Earth (after oxygen) and the most important material for high technology. Silicon technology is quite cheap and well-developed. However, the crystalline silicon is an extremely bad light-emitter due to the indirect gap, which prohibits the radiative recombination of carriers, relaxed to the band edges, by the momentum conservation (if the process is not accompanied by the emission/absorption of a phonon). A remedy for this obstacle is to "spoil" the crystalline structure of the material, e.g., by electro-chemical etching or by other techniques [47], which results in the porous structure with the characteristic size of the pores being several nanometers. In such a structure the momentum conservation restriction is relaxed and light-emitting diodes based on the porous silicon microcavities have already been constructed [48,49].

The experiment of Ref. [37] was already mentioned in Section 4. In this work organic dye molecules were introduced into the pores, and the evidence for the Förster transfer has been found. Thus, developing the theory for such a system is a relevant problem. This task is, however, complicated since the theoretical description of luminescence from porous silicon is not yet fully satisfactory [47]. Thus, apart from the obvious technological interest, studying this phenomenon from both theoretical and experimental point of view may help to the basic understanding of the processes occurring in such systems. The technological advantages with respect to III–V- or II–VI-semiconductor-based structures would be small spatial separation of the semiconductor and organic components (which is important for the transfer efficiency) as well as much easier production technology.

## 6. Summary

In the present chapter we have done a theoretical study of effects arising in hybrid organic-inorganic systems due to the resonant Coulomb interaction in the weak coupling regime. The main objective was to describe incoherent energy transfer from excited inorganic semiconductor nanostructure (the donor) to an organic material (acceptor), whose electronic excitation energy is close to that of the donor. This work may be viewed as the extension of Förster's picture of the transfer between two point molecules to systems, whose wave functions are essentially delocalized in space. The main physical ingredients, however, remain the same: the interaction is assumed to be weak in comparison to the characteristic broadening of the acceptor excited level (which ensures the validity of the Fermi Golden Rule, as in the Förster theory), and the result may be expressed via the spectral overlap of the donor luminescence and the acceptor absorption spectra – the quantities that represent the basic optical properties of the constituting materials, and may be measured independently. The acceptors were treated in the continuous medium approximation, which is valid as long as the electric field, produced by the donor, varies slowly on the molecular scale. Once the properties of the materials are known, the problem is reduced to the problem of classical electrostatics: finding the electric field, produced by a given distribution of charges in an inhomogeneous dielectric medium.

This theoretical scheme was applied to two possible system configurations: a plane semiconductor quantum well, sandwiched between two organic acceptor layers [2,3], and a spherical quantum dot surrounded by the acceptor medium [4, 5]. The energy transfer rate depends on the actual initial state of the donor. If it is a freely moving Wannier exciton with the 2D wave vector $\mathbf{k}$ in a quantum well, the transfer rate vanishes for small, as well as for large $k$, reaching the maximum for the $k$, corresponding to the characteristic size of the structure (the donor-acceptor separation). For realistic values of several tens of angstroms and parameters typical for II–VI semiconductors and organic luminescent materials, the transfer rate obtained was as fast as several tens of picoseconds, which is much shorter than the intrinsic lifetime of the Wannier exciton in quantum wells. The corresponding values of the wave vectors are of the same order as those typical for a thermal distribution at temperatures of the order of 100 K, which eliminates the problem of the relaxation bottleneck, well known in the technology of semiconductor light-emitting devices.

For a quantum well Wannier–Mott exciton, localized due to the interface roughness or alloy disorder, the transfer rate depends on the characteristic localization length $L$. It is determined by the plane-wave Fourier components, corresponding to the given localization length, and may be also short. We have also considered the case of the unbound electron–hole pairs in the quantum well. The transfer rate depends strongly on the carrier density and degree of degeneracy, but even for high enough densities the transfer from unbound carriers is much less efficient than that from excitons.

The calculations for quantum dots show that the Förster energy transfer from the lowest state of the electron–hole pair in the dot is also fast enough compared to the carrier recombination time (with the characteristic numbers similar to those for the quantum well). The transfer times from higher excited states may be longer if the corresponding transition is dipole-forbidden. The presence of the acceptor substance may also affect the carrier intraband relaxation in the case when the organics has a nonzero absorption coefficient at the frequency, corresponding to the given intraband transition. Unlike intradot Auger processes, such relaxation mechanism does not require more than one carrier inside each dot. Of course, such a process may occur in the presence of any absorbing substance, not necessarily organic.

A possible application of this phenomenon would be a light-emitting device combining comparatively good transport properties of semiconductors (e.g., pumped electrically) and good light-emitting properties of organic substances in a single hybrid nanostructure. The main goal would be to inject carriers into the inorganic part of the structure (which is quite efficient due to the good transport properties), where they are quickly bound into excitons. Then the excitation would be transferred to the organic part. The excitation in the organics would then recombine, emitting light. Such an "indirect" pumping of organics might prove to be more efficient than the direct electrical pumping, on one hand, and than the direct use of the semiconductor for light emission, on the other. In particular, one of the important problems in the technology of organic LEDs, is to provide a good electric contact between the electrodes and the active medium, which would be circumvented in the hybrid device, since the direct contact between the semiconductor and the organic parts of the structure is not necessary for the energy transfer, considered here, to occur. Also if such a transfer is competitive with the nonradiative decay channels in the semiconductor, the efficiency of the hybrid device may be higher than that of a purely inorganic LED.

## References

1. V.M. Agranovich, G.C. La Rocca, F. Bassani, *JETP Lett.* 66 (1997) 748.
2. D.M. Basko, G.C. La Rocca, F. Bassani, V.M. Agranovich, *Eur. Phys. J. B* 8 (1999) 353.
3. V.M. Agranovich, D.M. Basko, G.C. La Rocca, F. Bassani, *J. Phys.: Condens. Matter* 10 (1998) 9369.
4. V.M. Agranovich, D.M. Basko, *JETP Lett.* 69 (1999) 250.
5. D.M. Basko, V.M. Agranovich, F. Bassani, G.C. La Rocca, *Eur. Phys. J. B* 13 (2000) 653–659.
6. V.M. Agranovich, D.M. Basko, G.C. La Rocca, F. Bassani, *Synth. Met.* 116 (2001) 349.
7. L.D. Landau, E.M. Lifshitz, Electrodynamics of Continuous Media, Pergamon Press, London, 1965.
8. P.Y. Yu, M. Cardona, Fundamentals of Semiconductors: Physics and Materials Properties, Springer, Berlin, 1996.
9. G. Bastard, Wave Mechanics Applied to Semiconductor Heterostructures, Les Editions de Physique, Les Ulis, 1988.
10. F. Bassani, G. Pastori Parravicini, Electronics States and Optical Transitions in Solids, Pergamon, New York, 1975.
11. L.D. Landau, E.M. Lifshitz, Quantum Mechanics, Non-Relativistic Theory, Pergamon Press, 1965.
12. V.M. Agranovich, V.L. Ginzburg, Crystal Optics with Spatial Dispersion and Excitons, Springer, 1984.
13. Th. Förster, *Ann. Phys.* 2 (1948) 55.

14. V.M. Agranovich, M.D. Galanin, Electronic Excitation Energy Transfer in Condensed Matter, North-Holland, Amsterdam, 1982.
15. M. Shinada, S. Sugano, *J. Phys. Soc. Jpn.* 21 (1966) 1936.
16. V.M. Agranovich, R. Atanasov, F. Bassani, *Solid State Commun.* 92 (1994) 295.
17. R. Cingolani, in: Semiconductors and Semimetals, Vol. 44, Academic Press, 1997, pp. 163–226.
18. M. Hoffmann, private communication.
19. M. Umlauff, J. Hoffmann, H. Kalt, W. Langbein, J.M. Hvam, M. Scholl, J. Söllner, M. Heuken, B. Jobst, D. Hommel, *Phys. Rev. B* 57 (1998) 1390.
20. L.C. Andreani, F. Tassone, F. Bassani, *Solid State Commun.* 77 (1991) 641.
21. B. Deveaud, F. Clerot, N. Roy, K. Satzke, B. Sermage, D.S. Katzer, *Phys. Rev. Lett.* 67 (1991) 2355.
22. S. Permogorov, *Phys. Stat. Sol. (b)* 68 (1975) 9.
23. S. Permogorov, V. Travnikov, *Solid State Commun.* 29 (1979) 615.
24. N. Pelekanos, J. Ding, Q. Fu, A.V. Nurmikko, S.M. Durbin, M. Kobayashi, R.L. Gunshor, *Phys. Rev. B* 43 (1991) 9354.
25. M. Kohl, D. Heitmann, P. Grambow, K. Ploog, *Phys. Rev. B* 42 (1990) 2941.
26. J. Shah, Ultrafast Spectroscopy of Semiconductors and Semiconductor Nanostructures, Springer-Verlag, 1996.
27. R. Zimmerman, *Pure and Applied Chemistry* 69 (1997) 1179.
28. R. Zimmerman, Many-Particle Theory of Highly Excited Semiconductors, 1987.
29. R. Cingolani, L. Calcagnile, G. Golì, D. Greco, R. Rinaldi, M. Lomascolo, M. Di Dio, A. Franciosi, L. Sorba, G.C. La Rocca, D. Campi, *J. Opt. Soc. Am. B* 13 (1996) 1268.
30. U. Bockelmann, G. Bastard, *Phys. Rev. B* 42 (1990) 8947.
31. J. Urayama, T.B. Norris, J. Singh, P. Bhattacharya, *Phys. Rev. Lett.* 86 (2001) 4930.
32. L.D. Landau, E.M. Lifshitz, The Classical Theory of Fields, Pergamon Press, Oxford, 1962.
33. V.B. Berestetski, E.M. Lifshitz, L.P. Pitaevskii, Relativistic Quantum Theory, Pergamon Press, Oxford, 1971.
34. Y.Z. Hu, M. Lindberg, S.W. Koch, *Phys. Rev. B* 42 (1990) 1713.
35. F. Bassani, R. Buczko, G. Czajkowski, *Nuovo Cimento* 19D (1997) 1565.
36. R. Ferreira, G. Bastard, *Appl. Phys. Lett.* 74 (1999) 2818.
37. S. Létant, J.C. Vial, *J. Appl. Phys.* 82 (1997) 397.
38. C. Delerue, G. Allan, E. Martin, M. Lannoo, in: J.C. Vial, J. Derrien (Eds.), Porous Silicon Science and Technology, Les Editions de Physique/Springer, Les Ulis, 1995.
39. Y. Kayanuma, *Solid State Commun.* 59 (1986) 405.
40. M. Pope, H. Kalmann, P. Magnante, *J. Chem. Phys.* 38 (1963) 2042.
41. C. Weisbuch, B. Vinter, Quantum Semiconductor Structures: Fundamentals and Applications, Academic Press, Boston, 1991.
42. V. Fiorentini, F. Bernardini, F. Della Sala, A. Di Carlo, P. Lugli, *Phys. Rev. B* 60 (1999) 8849.
43. M. Leroux, N. Grandjean, M. Laügt, J. Massies, B. Gil, P. Lefebvre, P. Bigenwald, *Phys. Rev. B* 58 (1998) R13371.
44. J. Singh, *Appl. Phys. Lett.* 64 (1994) 2694.
45. D. Lee, A. Mysyrowicz, A.V. Nurmikko, B.J. Fitzpatrick, *Phys. Rev. Lett.* 58 (1987) 1475.
46. I.K. Sou, K.S. Wong, Z.Y. Yang, H. Wang, G.K.L. Wong, *Appl. Phys. Lett.* 66 (1995) 1915.
47. A.G. Cullis, L.T. Canham, P.D.J. Calcott, *J. Appl. Phys.* 82 (1997) 909.
48. M. Araki, H. Koyama, N. Koshida, *Appl. Phys. Lett.* 69 (1996) 2956.
49. S. Chan, P.M. Fauchet, *Appl. Phys. Lett.* 75 (1999) 274.

# Subject Index

0-dimensional system 344
4TBPPZn 374, 382, 399
– dye 380
– exciton modes 376
– molecules 369, 374

absorption coefficient 99, 278
absorption edge 100, 105, 109
absorption probability 108, 111
additional boundary conditions (ABC) 146
additional propagating wave 134
additional wave in CdS 141
aggregates 4
AlN 482
aluminum 365, 441
amplifiers 398
angular momentum 470
anisotropy (of crystals) 53
anti-crossing 367, 376, 394, 395
anti-Stokes 391
aromatic crystals 20
Asymmetric Quantum Wells 161
attenuated total reflection (ATR) 146
Auger relaxation 476

band gap 98
band structure 98, 99
band width 14
barrier penetration 112
Bessel function 473
biexciton 32, 123, 124
binding energy 3, 109, 112, 356
bistability effect 178
bistable energy transmission 216
Bloch function 417, 451–453
Bloch states 98
blue shift 122
Bogoliubov–Tyablikov transformation 19, 22

Bohr radius 97, 101, 103, 318, 417, 453, 458, 469, 473
Boltzmann distribution 418, 466
Boltzmann equation 45, 49
Born–Oppenheimer approximation 246
Bose–Einstein condensation 98, 125, 179, 180
bosonization 15, 17
boundary condition 107, 414, 417, 458, 459
bulk cavity polaritons 174, 178
bulk crystal 20

cavity 377
– mode 168
– polariton 365, 375, 398
– – dispersion 172
– resonant modes 167
cavity-photon 356
cavity-polariton 355, 383
– states 396
charge transfer excitons 3, 63
– nonlinear optical response 311–314
– phase transition to conducting state 293–304
– – analytical approach 293–297
– photogenerated static electric field 313
– resonant optical nonlinearity 311–313
charge transfer state 230, 270
charge-transfer crystal 230
charge-transfer integral 278, 281
chlorosomes 89
circular dichroism 89
coherence 115, 116, 366, 379
– length 45
coherent 355, 365, 396
– excitons 49, 59, 87
– potential approximation 78
cold photoconductivity 293
collisional broadening 121

487

# Subject Index

commutation rules  116
compound oscillator  378
conduction band  98, 99, 450
– extrema  417
confinement  473, 474
– energy  107
correlation effects  118
Coulomb attraction  105
Coulomb excitons  35
Coulomb interaction  117, 120, 332, 405, 409, 413, 427, 447, 449, 455, 473, 476, 480, 483
coupled oscillator  378
cumulative photovoltage in D-A superlattices  304–310
current densities  140
cyanine dye  67, 346, 370, 372
cyanine monomer  391

Davydov bands  21
Davydov component  276
Davydov splitting  9, 23, 24, 42
DBR  359–361, 364, 394, 399, 400
DBR mirrors  174
delocalization length  75
density matrix  139
– approach  139
detuning  327
dielectric constant  97, 103, 105, 111, 276
– tensor  25, 35, 186
dielectric response  24
dielectric theory of Frenkel excitons  34
diffusion coefficient  50, 55
diffusion constant  52
diffusion length  47
diffusion of excitons  45
diffusion of polaritons  56
diffusion tensor  47
dimer  6
dipole–dipole interaction  104, 319, 332, 394
direct gap  99
disorder  349, 453, 461
disorder (static)  45, 74
dispersion relation  326
distributed Bragg reflector (DBR)  166, 358, 414, 426
donor–acceptor complex  63
Double Quantum Wells  161
driving electric field  333

dynamic exciton–exciton interaction  12
Dyson equation  412

effective mass  102
– approximation  103
– equation  110, 119
effective nonlinear susceptibility  196
effective Rydberg  103
electro-absorption  64
– spectroscopy  66
electromagnetic interaction  448
electron–hole
– Coulomb attraction  103
– exchange contribution  151
– pair  451, 453, 455, 457, 466, 469, 473, 477, 480, 484
electron–phonon
– coupling  270
– interaction  103
– scattering  118
emission spectrum  282
energy transport  1
energy-transfer  393
envelope function  97, 103, 109, 110
exact transformation from paulion to boson  17
exchange interaction  104
exchange narrowing  68, 74
excitation induced dephasing  118, 121, 122
exciton  129, 318
– band  2
– – structure  235, 281
– binding energies  97, 101, 113, 391
– blue-shift  121
– fusion and fission  32
– oscillator strength  114, 156
– radius  330
– state  102
– superradiance  68
– transfer integral  228, 229, 251, 278, 283
exciton–exciton
– annihilation  32, 86
– interaction  11, 14, 31, 80, 84
– – at D-A interface  295, 297, 311
– interaction (dynamic)  16
exciton–phonon
– coupling  245, 286
– – constant  248
– interaction  33, 59, 331

exciton-polariton 27
– in microcavities 166
excitonic absorption 102
excitonic polaritons 129

Fabry–Perot frequency 168
Fermi Golden Rule 405, 410, 412, 428, 432, 439, 448, 476, 484
Fermi resonance 207
– modes 215
figure of merit 319, 320
first-order perturbation theory 336
fluorescence 363, 381
– emission 381
Förster dipole–dipole description 396
Förster effect 178
Förster energy transfer 45, 54, 394
Förster formula 405, 409
Förster transport 87
Franck–Condon energy 248
Franck–Condon factor 249
Frenkel exciton 2, 10, 97, 130, 318, 356, 365, 398, 447
– character 235, 256, 272, 278

$Ga_{1-x}Al_xN$ 482
GaAlAs 331
GaAs/InGaAs 367
GaN 482
gas-condensed matter shift 7, 12, 37, 202
giant oscillator strength 72
Green's function 323, 412, 449
gyration tensor 192

H-aggregate 72
hard-core boson approach 80
hard-core repulsion 120
Hartree–Fock 118
– approximation 333
– decoupling 336
heavy hole 112, 450
Heavy Hole (H.H.) and Light Hole (L.H.) excitons 150
Heitler–London approximation 7, 18, 22
Holstein Hamiltonian 251, 271
HOMO 241
homogeneous and inhomogeneous broadening 378
homogeneous broadening 5, 377

homogeneous energy linewidths 366
hopping integral 228
hybrid Frenkel and Wannier–Mott excitons
– at interface 321–329
– – linear optical response 329–331
– in quantum dots 344, 345
– in quantum wires 339–344
hybrid semiconductor microcavities 393
hybrid states 343
hybrid structure 318
hybridization in microcavity 345–349
hybridization parameter 340

II–VI semiconductor 459, 482
III–V quantum well 380, 382, 400
III–V semiconductor 355, 361, 367, 382
impurity 39
incident electromagnetic wave 329
incoherent energy transfer 54
incoherent excitons 53, 59
indirect exciton 124
indium tin-oxide 364
inhomogeneity 6
inhomogeneous broadening 111
inhomogeneous linewidth 366
inorganic quantum well (IQW) 321
interaction Hamiltonian 323
interaction matrix element 324
interaction parameter 325
interband polarization 118
interband transition 100, 101, 451, 477
interwire coupling 343
intraband transition 451, 453, 475, 476

J band 67
J-aggregate 67, 72, 370–372, 381–385, 389, 394, 396
J-aggregated dyes 395
joint density of states 101, 108
Jordan–Wigner transformation 14, 19, 81

Kane's energy 325, 453
kinematic interaction 11, 14, 16, 80
Kramers–Krönig relation 408, 415
Kravets integral 40

Langmuir–Blodgett technique 86
large binding energy 398
lateral modulation 340

lattice matching 320
lattice sum 228
layer Lorentz-factor tensor 199
leaky modes 363, 426, 432
LED 364, 365
lifetimes of quantum wire polaritons 165
light emitting diode or LED 364
light harvesting complex 396
light hole 112, 450
light-emitting device 480, 448
light-harvesting complex 2, 69, 84, 87, 393, 396
linear susceptibility 119, 336
linewidth 105, 111, 118, 122
local field 34, 417, 456
– effects 100
local-field
– approximation 17
– corrections for impurities 39
– method 31
– tensor 37
localization 53, 60, 69, 74
– length 75, 474, 484
longitudinal excitons 130
Lorentz oscillator model 132
Lorentz–Lorenz formula 36
low-dimensional exciton systems 67
lower polariton 132
LUMO 241
Luttinger matrix 106

material parameter 347
matrix 368
matrix polymer 368
Maxwell's equation 376, 407, 408, 411, 423
mean field approximation 118
mean free path 45
mechanical excitons 43
memory kernels 54
MePTCDI 222, 244, 276, 278
Merrifield model 231
metal phtalocyanine 331
metallic mirror 394
metallic thin films 358, 360
microcavity 166, 355–358, 361, 362, 364, 369, 374, 376, 382, 384, 388, 389, 391, 392, 395, 396, 399
– configurations 345
– polaritons 346

– with disordered organics 349
– – coexistence of coherent and incoherent states 351
mixing of molecular configurations 40
mobility of excitons 45
mobility of ST excitons 61
molecular aggregates 67
molecular beam epitaxy (MBE) 317
molecular vibron model 261, 267
momentum space 227
motional narrowing 68
multi-exciton bands 13
multi-exciton states 80
multi-level molecules 23
multipole 472

N-N'-dimethylperylene-3,4,9,10-dicarboximide 223
new vacuum of the polaritons 137
non-local dielectric response 111
nonlinear absorption 30
nonlinear effects 98, 115
nonlinear exciton equations 33
nonlinear optical experiments 14
nonlinear optical response 7, 79
nonlinear optics 331
nonlinear processes 392, 398
nonlinear spectroscopy 30
normal-incidence reflectivity 159
NTCDA 331
numerical simulation 78

on-site energy 226, 228, 283
one-dimensional system 339
one-electron approximation 98
one-exciton states 14, 71
optical
– bistability 177
– constant 101
– parametric oscillator 179
– properties of superlattices 194
– resonators 356
– switches 398
organic molecular beam deposition (OMBD) 320
organic quantum well (OQW) 321, 149
organic-inorganic heterostructures 346

oscillator strength   5, 98, 101, 104, 109, 110, 112, 120, 239, 248, 257, 318, 330, 366, 367, 377, 380, 395, 399, 400
– reduction   122
oscillator strength (transfer)   24

Pauli blocking   119, 120
Pauli exclusion   11
– gap   83
Pauli operator   10, 15
paulions   11–17
Pekar's ABC   174
pentacene   331
perovskites   372
perylenetetracarboxylic dianhydride   223
phase space filling   117, 120, 319
phonon   33, 48, 453, 469, 480
– cloud   253, 258, 286
– scattering   112
photoconductivity   63
photoluminescence (PL)   177, 178, 370, 374, 381, 383–385, 388
pillar microcavities   177
plasma enhanced chemical vapour deposition   359
Pointing vector   409, 411, 419, 429
Poisson equation   449, 457, 458, 471
polariton   28, 33, 35, 73, 378, 379, 381, 386, 388–391, 394, 395, 397, 414, 435, 438
– in CuCl   140
– in the quantum well   155
– lasers   179
– MC lasers   177
– state   398
– stopgap   29
– vacuum   137
polarization functions   333
polarization intensity relations   42
polaron   58
polystyrene   374, 427
population   118
– transfer   385
porous silicon   478, 483
porphyrin   368
– dyes   346
PTCDA   223, 276, 278, 331, 459
pump-probe technique   82
PVA film   372

Q-band   368
$Q$-factor of the cavity   361
quantization of the center of mass motion of the exciton   175
quantum confinement   97, 106, 107, 109
– effect   286
quantum dot   113, 114, 320, 345, 449, 469, 484
quantum theory of polaritons   135
– in MC   176
quantum well
– exciton   98
– polaritons   147
quantum well (QW)   106, 107, 148, 317, 320, 372, 386, 417, 422, 429, 438, 449, 456, 481, 484
quantum wire   113, 114, 320, 339
– polaritons   161
quantum yield   448, 481
QW exciton   355, 367, 381, 393, 399
QW polariton   153
– lifetime   155

Rabi splitting   170, 177, 178, 346, 365–368, 376, 378, 381, 383, 387, 391, 393, 396, 398, 400
radiation coupling   178
radiative decay   408, 440, 472
radiative lifetime   472
Raman scattering   385, 391, 392
random walk   54
real photon   405
real-space density matrix approach   141
reduced mass   103
reduction of the oscillator strength   121
reflection coefficient   414
reflectivity   361, 373, 374
– spectra   361, 373, 375
relaxation time   49
resonant excitation   356, 389
resonant interaction   7
resonant two-photon absorption   140
response function   411, 412, 419, 449, 450
retardation effects   153
retarded interaction   35
rotating wave approximation   208
Rydberg constant   453

saturation density   98, 115, 117
Schrödinger equation   462, 473

screening 103, 118
second harmonic generation (SHG) 30, 317
second quantization 9, 17
second-order nonlinear polarizability 63
second-order susceptibility 338
self-trapped exciton 58
self-trapping 58, 482
– of charge-transfer excitons 60
– of Frenkel excitons 58
semiconductor Bloch equations 116, 118, 332
silicon dioxide 426
silicon nitride 426
silver 426, 429
silver-halide microcrystals 86
singlet excitons 47, 370
size-enhanced, nonlinear susceptibilities 79
Sommerfeld factor 106, 110
space charge 480
spatial dependence 176
spatial dispersion 27, 131, 145
spectral function 412, 414, 423, 438
spherical harmonics 470, 476
spin degeneracy 332
spin orbit coupling 105
spin singlet 104
spin triplet 104
spontaneous emission 363, 366
Stark effect 64
statistics 117
stimulated scattering 389, 398
Stokes & anti-Stokes Raman 391
Stokes shift 367, 406, 416, 440, 483
stop band 168, 359
strong confinement 469, 473, 477
strong coupling regime 170, 320, 346, 365, 368, 369, 371
strong optical coupling 355, 356, 367, 369, 372–375, 381
subband 107, 108
superlattice 186
superlattice dielectric tensor 191
superradiance (of excitons) 73
superradiant decay 151
superradiant radiative decay 202
surface polaritons 141, 144
surface wave 189, 429, 431
susceptibility 25, 335

susceptibility ($n$th order) 30
symmetry classification of excitons in QW (groups $D_{2d}$) 150

tetracene 331
thermal evaporation 370
third-order nonlinearity 334
total reflectivity 174
transfer integral 226
transfer matrix approach 168
transfer matrix model 378
transformation (exact) from paulion to boson 17
transient grating 56
transition
– density 243
– dipole 238, 242, 248, 257, 276, 407
– matrix element 101
– probability 99
translational symmetry 98
transport of excitation energy 45
transverse excitons 130
triplet excitons 47, 59
two-dimensional exciton 109
– radius 110
two-dimensional system 321
two-exciton states 14, 80
two-level molecule 5
two-photon absorption 79

upper polariton 132

valence 417
– band 98, 99, 450
– – degeneracy 112
Van der Waals interaction 20
vibrations 33, 45
vibronic relaxation 406, 440, 448
vibronic states 5
virtual excitations 115, 119
virtual photons 405, 423

Wannier–Mott exciton 2, 97, 101, 130, 222, 233, 318, 356, 365, 398, 418, 438, 447, 448, 453, 455, 457, 478, 480, 481, 484
weak and strong-coupling regime 355, 356
weak confinement 469, 474
weak coupling 170, 320, 346, 363, 367

## Subject Index

weighting coefficients  322
width of the cavity mode  170
Wigner $3j$-symbols  474

zinc-blende semiconductors  450
$Zn_{1-x}Cd_xS$  482

ZnCdSe  331
ZnS  482
$ZnS_{1-x}Se_x$  482
$ZnS_{1-x}Te_x$  482
ZnSe  331
ZnSe/ZnCdSe  459

## Recent Volumes In This Series

Maurice H. Francombe and John L. Vossen, *Physics of Thin Films*, Volume 16, 1992.

Maurice H. Francombe and John L. Vossen, *Physics of Thin Films*, Volume 17, 1993.

Maurice H. Francombe and John L. Vossen, *Physics of Thin Films, Advances in Research and Development, Plasma Sources for Thin Film Deposition and Etching*, Volume 18, 1994.

K. Vedam (guest editor), *Physics of Thin Films, Advances in Research and Development, Optical Characterization of Real Surfaces and Films*, Volume 19, 1994.

Abraham Ulman, *Thin Films, Organic Thin Films and Surfaces: Directions for the Nineties*, Volume 20, 1995.

Maurice H. Francombe and John L. Vossen, *Homojunction and Quantum-Well Infrared Detectors*, Volume 21, 1995.

Stephen Rossnagel and Abraham Ulman, *Modeling of Film Deposition for Microelectronic Applications*, Volume 22, 1996.

Maurice H. Francombe and John L. Vossen, *Advances in Research and Development*, Volume 23, 1998.

Abraham Ulman, *Self-Assembled Monolayers of Thiols*, Volume 24, 1998.

Subject and Author Cumulative Index, Volumes 1–24, 1998.

Ronald A. Powell and Stephen Rossnagel, *PVD for Microelectronics: Sputter Deposition Applied to Semiconductor Manufacturing*, Volume 26, 1998.

Jeffrey A. Hopwood, *Ionized Physical Vapor Deposition*, Volume 27, 2000.

Maurice H. Francombe, *Frontiers of Thin Film Technology*, Volume 28, 2001.

Maurice H. Francombe, *Non-Crystalline Films for Device Structures*, Volume 29, 2002.

Maurice H. Francombe, *Advances in Plasma-Grown Hydrogenated Films*, Volume 30, 2002.

ISBN 0-12-533031-6